SOCIETÀ ITALIANA DI FISICA

RENDICONTI
DELLA
SCUOLA INTERNAZIONALE DI FISICA
«ENRICO FERMI»

CXXII Corso
a cura di J. SILK e N. VITTORIO
Direttori del Corso
VARENNA SUL LAGO DI COMO
VILLA MONASTERO
21-31 Luglio 1992

Formazione delle galassie

1994

SOCIETÀ ITALIANA DI FISICA
BOLOGNA-ITALY

ITALIAN PHYSICAL SOCIETY

PROCEEDINGS
OF THE
INTERNATIONAL SCHOOL OF PHYSICS
«ENRICO FERMI»

COURSE CXXII
edited by J. SILK and N. VITTORIO
Directors of the Course
VARENNA ON LAKE COMO
VILLA MONASTERO
21-31 July 1992

Galaxy Formation

1994

NORTH-HOLLAND
AMSTERDAM · OXFORD · NEW YORK · TOKYO

Copyright ©, 1994, by Società Italiana di Fisica

All rights reserved. No part of this publication may be reproduced, stored in a retrieval system, or transmitted, in any form or by any means, electronic, mechanical, photocopying, recording or otherwise, without the prior permission of the copyright owner.

PUBLISHED BY
North-Holland
Elsevier Science Publishers B.V.
P.O. Box 211
1000 AE Amsterdam
The Netherlands

SOLE DISTRIBUTORS FOR THE USA AND CANADA:
Elsevier Science Publishing Company, Inc.
655 Avenue of the Americas
New York, N.Y. 10010
U.S.A.

Technical Editor
P. PAPALI

Library of Congress Cataloging-in-Publication Data

International School of Physics "Enrico Fermi" (1991 : Varenna, Italy)
  Galaxy formation : Varenna on Lake Como, Villa Monastero, 21-31 July 1992 / edited by J. Silk and N. Vittorio.
      p.   cm. -- (Proceedings of the International School of Physics "Enrico Fermi" ; course 122)
   At head of title: Italian Physical Society.
   ISBN 0-444-82160-0
   1. Galaxies--Formation--Congresses. 2. Stars--Formation--Congresses. 3. Galaxies--Clusters--Congresses.   I. Silk, Joseph, 1942-   . II. Vittorio, N. III. Società italiana di fisica.
IV. Title. V. Series: International School of Physics "Enrico Fermi." Proceedings of the International School of Physics "Enrico Fermi" ; course 122.
QB857.5.E96I58 1991
523.1'12--dc20   94-43402
                                                                    CIP

Proprietà Letteraria Riservata
Printed in Italy

INDICE

J. SILK and N. VITTORIO – Introduction .... pag. XI

Gruppo fotografico dei partecipanti al Corso   fuori testo

M. S. TURNER – Inflation after COBE. Lectures on inflationary cosmology.

1. Hot big bang: successes and challenges .... pag. 1
   1`1. Successes .... » 1
       1`1.1. The expansion .... » 1
       1`1.2. The cosmic background radiation .... » 2
       1`1.3. Primordial nucleosynthesis .... » 4
       1`1.4. Et cetera—and the age crisis? .... » 6
   1`2. Basics of the big-bang model .... » 7
       1`2.1. Friedmann equation and the first law .... » 8
       1`2.2. A short diversion concerning the present mass density ...... » 9
       1`2.3. The early, radiation-dominated Universe .... » 11
       1`2.4. The earliest history .... » 13
       1`2.5. The matter- and curvature-dominated epochs .... » 13
       1`2.6. One last thing: horizons .... » 14
   1`3. The challenge: development of structure .... » 15
       1`3.1. The general picture: gravitational instability .... » 15
       1`3.2. CBR temperature fluctuations .... » 16
       1`3.3. An initial-data problem .... » 18
2. Inflation: an overview .... » 19
   2`1. Shortcomings of the standard cosmology .... » 19
   2`2. Generic aspects of inflation .... » 21
   2`3. Current status of inflationary models .... » 24
       2`3.1. Types of inflation .... » 24
       2`3.2. Viable models .... » 26
   2`4. Initial conditions: no-hair theorems .... » 29

3.	Inflation: the fundamentals	pag.	30
	3˙1. Metric perturbations and CBR anisotropy	»	39
	3˙2. Worked examples	»	44
	3˙2.1. Exponential potentials	»	45
	3˙2.2. Chaotic inflation	»	45
	3˙2.3. New inflation	»	46
	3˙2.4. Natural inflation	»	47
	3˙2.5. Lessons	»	49
4.	Structure formation after COBE	»	50
	4˙1. The Universe observed	»	51
	4˙1.1. Normalization: the great leap forward!	»	53
	4˙2. CDM models	»	55
	4˙2.1. MCDM	»	56
	4˙2.2. BCDM	»	56
	4˙2.3. Tilt	»	57
	4˙2.4. Best-fit models	»	57
	4˙3. The scorecard and future	»	59
5.	Concluding remarks	»	60

P. L. RICHARDS – Cosmic microwave background.

1. Introduction	»	69
2. Receivers for direct measurements	»	70
3. Sources of interference	»	74
4. Microwave measurements of the CMB spectrum	»	76
5. Direct measurements at frequencies beyond the peak	»	78
6. Rocket measurements of the spectrum of the CMB	»	80
7. CMB spectral measurements from molecular lines	»	81
8. The Far Infrared Absolute Spectrophotometer experiment	»	81
9. Anisotropy of the CMB	»	82
10. Anisotropy measurements	»	83
11. COBE Differential Microwave Radiometer experiment	»	83
12. Power spectrum of fluctuations	»	84
13. Present and future measurements of CMB anisotropy	»	87
14. The MAX experiment	»	88
15. Summary	»	90

B. MELCHIORRI, L. BRANCA and F. MELCHIORRI – Cosmic background radiation: present knowledge and future programs.

1. Cosmic background radiation: how Planckian is it?	»	95
2. Considerations about the dipole anisotropy	»	99
3. Considerations about the CBR anisotropy	»	103
3˙1. Problems in detecting the CBA: one-channel experiments	»	103
3˙2. Problems in detecting the CBA: multichannel experiments	»	121
3˙3. Data analysis and discussion	»	127
3˙4. Considerations on CBR anisotropies and primordial molecules	»	128

F. LUCCHIN, L. MOSCARDINI, S. MATARRESE and A. MESSINA – The large-scale structure of the Universe in non-Gaussian CDM models.

1. Introduction.. pag.	141
2. Skewed CDM models... »	143
2`1. Chi-squared models.. »	145
2`2. Lognormal models .... »	146
3. Results... »	148
3`1. Probability distributions... »	149
3`2. Correlation function... »	150
3`3. Topology of isodensity contours... »	151
3`4. Void probability function .... »	154
4. Discussion... »	155

S. MATARRESE, O. PANTANO, F. LUCCHIN, L. MOSCARDINI and D. SAEZ – Nonlinear evolution of self-gravitating collisionless matter.

1. Introduction.. »	159
2. Newtonian dynamics of self-gravitating collisionless matter................. »	160
3. Adding artificial viscosity... »	168
4. Relativistic dymamics of a self-gravitating collisionless fluid............... »	171
5. Conclusions.. »	174

A. BLANCHARD – The cosmological history of the baryons.

1. Introduction.. »	179
1`1. The quantity of seen baryons... »	179
1`2. Estimates of the mean density of the Universe............................. »	180
1`3. A sensible extrapolated value for the observed amount of baryons.. »	181
1`4. Early history .... »	181
2. Basic elements... »	184
2`1. Time scale considerations.. »	184
2`2. The neutral period.. »	185
2`3. A short comparison with observation.. »	186
3. Cooling and heating mechanisms.. »	187
3`1. Time evolution.. »	188
3`2. Inverse Compton cooling.. »	188
3`3. Bremsstrahlung cooling.. »	189
3`4. Recombinative cooling.. »	189
3`5. Collisional-excitation cooling... »	190
3`6. The cooling function.. »	190
3`7. Ionization equilibrium... »	191
4. The history of the cosmological gas and galaxy formation.................. »	192

2. The age of galaxies .... pag. 316
   2˙1. The age of galactic globular clusters. .... » 316
   2˙2. Dating galaxies by evolutionary population synthesis...................... » 319
   2˙3. Forming elliptical galaxies .... » 324
   Exercises .... » 326
3. The iron bound on elliptical-galaxy formation .... » 326
   3˙1. The iron-mass-to-light ratio in clusters of galaxies .... » 327
        3˙1.1. The ICM iron-mass-to-light ratio .... » 328
        3˙1.2. The stellar iron-mass-to-light ratio .... » 328
   3˙2. Supernova type Ia *vs.* type II iron production .... » 329
        3˙2.1. Type-Ia supernovae .... » 329
        3˙2.2. Type-II supernovae.. » 331
        3˙2.3. The solar supernova proportion .... » 333
        3˙2.4. Producing the observed IMLR... » 333
   3˙3. Supernova heating *vs.* binding energy of galaxies .... » 334
   3˙4. The iron-light connection .... » 337
   3˙5. The galaxy-ICM chemical asymmetry .... » 338
   Exercises .... » 342
   Answers to all the exercises.. » 342

J. SILK – The role of star formation in galaxy formation.

1. Introduction... » 345
2. Gravitational instability of cold disks .... » 346
3. Star formation rate in disks .... » 346
4. Starbursts... » 350
5. Implications for galaxies and galaxy evolution .... » 352
   5˙1. Tully-Fisher relation .... » 352
   5˙2. Galaxy clusters .... » 353
   5˙3. Galaxy peculiar velocities .... » 354
   5˙4. Large-scale structure... » 354
   5˙5. Low-redshift galaxy evolution .... » 355
6. Galaxy formation .... » 356
   6˙1. Star formation rate in protogalaxies .... » 356
   6˙2. Protoelliptical galaxies .... » 356
7. Conclusions... » 357

Introduction.

Eighty students from sixteen countries assembled on the shores of sunny Lake Como to hear a remarkable series of lectures on galaxy formation. The lecturers were chosen to be the pioneers and pacesetters in this rapidly developing field. The formation of galaxies is one of the major unresolved problems in cosmology, and poses one of the greatest challenges. New generations of telescopes and detectors are producing data that probe the Universe to unsurpassed depths. Understanding these data requires an amalgam of astronomical specialties that comprise, among many topics, star formation, interstellar matter, radio galaxies, X-ray clusters, quasars, inflationary cosmology and the cosmic microwave background.

These were the subjects covered by our Varenna lecturers. The broad range of material captivated the attention of the students, who intentionally were chosen from a variety of backgrounds, over the 10 day duration of the course. At the end of the course, the students completed a questionnaire. As a guide to future schools, the response to our survey on student satisfaction is shown as fig. 1. Clearly, their satisfaction level was high! The only occasional complaint was that students would have benefitted by more free time between lectures. The students gave high ratings to the choice of topics, the stimulation of new ideas and the utility and level of the lecturers. The moral that one may draw is that the students experienced a wonderful summer school, with expert lecturers, and with excellent support provided by the Italian Physical Society.

Our lecturers were chosen to represent diverse areas of astrophysics and cosmology. Hans BÖHRINGER and Gunther HASINGER provided an in-depth overview of the revolution in extragalactic X-ray astronomy triggered by the ROSAT satellite. Paul RICHARDS and Francesco MELCHIORRI lectured on the current status of the cosmic microwave background. Star formation in galaxies and galaxy evolution were covered in lectures by Colin NORMAN and Alvio RENZINI, as well as by one of the directors (J. S.). Michael TURNER laid a sound basis for the course by reviewing inflation and the early universe, the connection between intergalactic clouds seen in absorption towards quasars and forming galaxies was developed in the lectures by Don YORK, while Alain BLANCHARD gave a review of the thermal history of the cosmological gas and galaxy formation. Only three lecturers (Richard ELLIS, Alex SZALAY and Simon WHITE) defaulted on their committments to provide written lecture notes.

However, all of the lecturers are immortalized in two ways. Figure 2 displays the response to our questionnaire on the lecturer's competence and ability. Figure 3 displays the speaker baud rate, displayed as the number of trans-

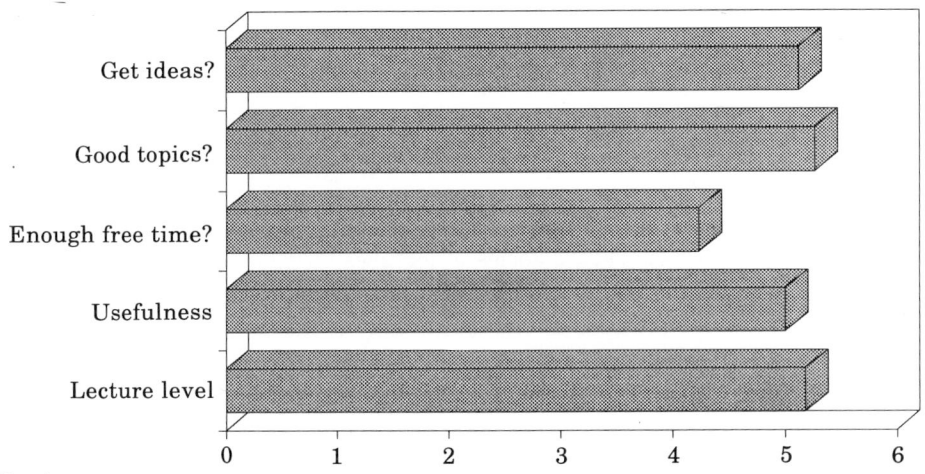

Fig. 1.

parencies and slides projected each 5 minutes for 6 randomly chosen lectures by different lecturers. We are indebted to Chris METZLER and Max TEGMARK for taking the data and conducting a first cut analysis. Note that the spectral range is remarkably broad, ranging from a culminating frenzied burst to monotonic quasi-white noise. A key to the speaker code is available on request.

The organization of a school could not be accomplished without the generous help and support of many individuals and agencies. This school could not have been started without the help of Liú CATENA of the Physics Department of the University of Rome «Tor Vergata», and could not have run properly without

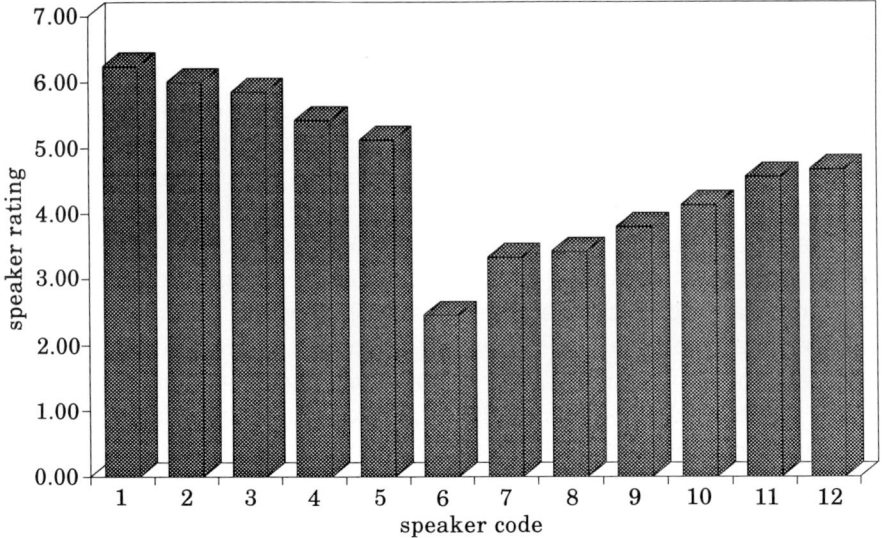

Fig. 2.

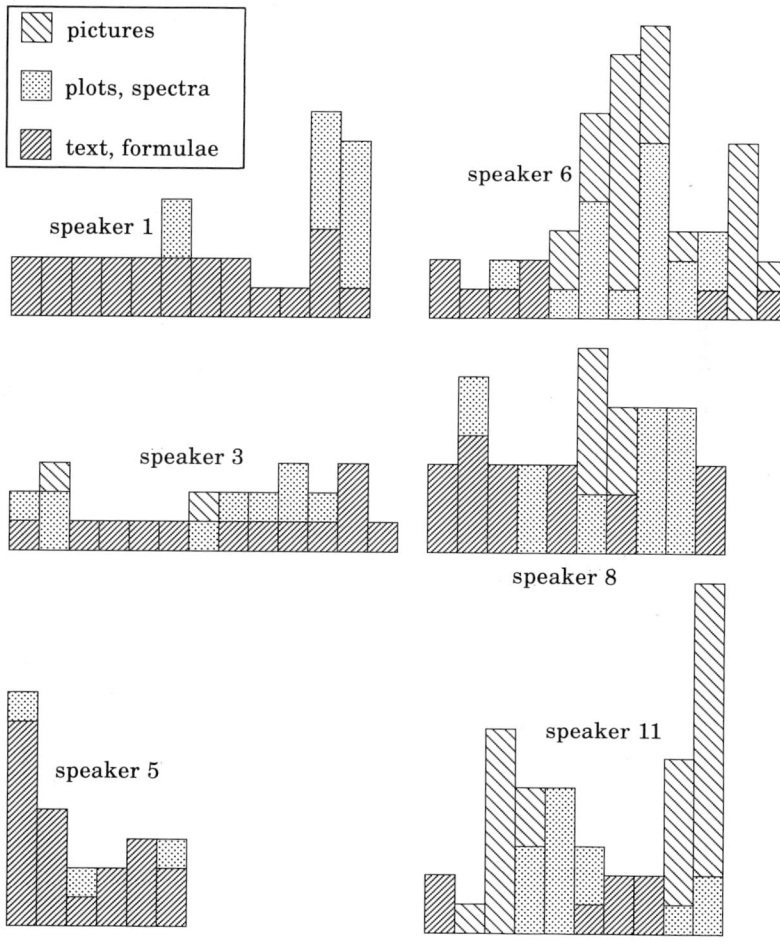

Fig. 3.

the continuous efforts and the professionality of Enrica MAZZI, Secretary of the Italian Physical Society. We are also indebted with Carmen VASINI for her patience in collecting the manuscripts on «cosmological» time scales. The school received financial support from NATO, as NATO Advanced Institute, and from the Italian Physical Society.

J. SILK
N. VITTORIO

INTERNATIONAL SCHOOL OF PHYSICS «ENRICO FERMI»
COURSE «Galaxy formation»
Varenna, 21-31 July 1992

STUDENT QUESTIONNAIRE

To help us plan future Schools, please answer the following questions on a scale of 1 to 7, such that 1 = **poor** and 7 = **excellent**:

A. The level of the lectures was generally: 1 2 3 4 5 6 7

B. How useful were the seminars?   1 2 3 4 5 6 7
    1 = I went swimming
    7 = very useful

C. Was there sufficient time for social interactions? 1 2 3 4 5 6 7
    1 = too little time
    7 = too much time

D. Topics covered   1 2 3 4 5 6 7
    1 = important topics left out
    7 = excellent distribution

E. Did you come away with ideas for research problems? 1 2 3 4 5 6 7
    1 = no ideas
    7 = many

F. Please rank the lecturers according to how much you learnt:
    Blanchard
    Böhringer
    Ellis
    Hasinger
    Melchiorri
    Norman
    Renzini
    Richards
    Szalay
    Turner
    White
    York

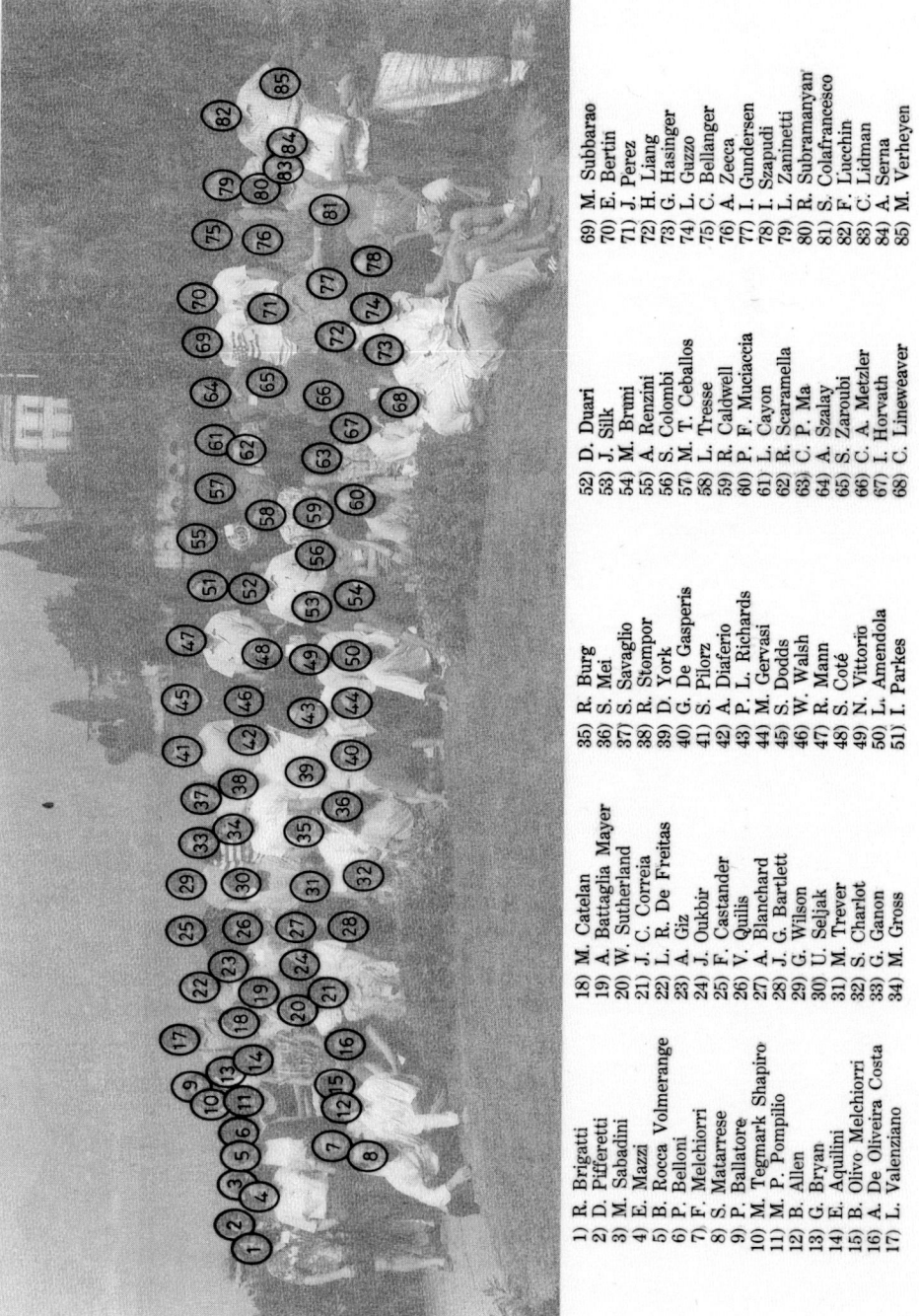

SOCIETÀ ITALIANA DI FISICA

SCUOLA INTERNAZIONALE DI FISICA «E. FERMI»

CXXII CORSO - VARENNA SUL LAGO DI COMO - VILLA MONASTERO - 21-31 Luglio 1992

1) R. Brigatti
2) D. Pifferetti
3) M. Sabadini
4) E. Mazzi
5) B. Rocca Volmerange
6) P. Belloni
7) F. Melchiorri
8) S. Matarrese
9) P. Ballatore
10) M. Tegmark Shapiro
11) M. P. Pompilio
12) B. Allen
13) G. Bryan
14) E. Aquilini
15) B. Olivo Melchiorri
16) A. De Oliveira Costa
17) L. Valenziano
18) M. Catelan
19) A. Battaglia Mayer
20) W. Sutherland
21) J. C. Correia
22) L. R. De Freitas
23) A. Giz
24) J. Oukbir
25) F. Castander
26) V. Quilis
27) A. Blanchard
28) J. G. Bartlett
29) G. Wilson
30) U. Seljak
31) M. Trever
32) B. Charlot
33) G. Ganon
34) M. Gross
35) R. Burg
36) S. Mei
37) S. Savaglio
38) R. Stompor
39) D. York
40) G. De Gasperis
41) S. Pilorz
42) A. Diaferio
43) P. L. Richards
44) M. Gervasi
45) S. Dodds
46) W. Walsh
47) R. Mann
48) S. Coté
49) N. Vittorio
50) L. Amendola
51) I. Parkes
52) D. Duari
53) J. Silk
54) M. Bruni
55) A. Renzini
56) S. Colombi
57) M. T. Ceballos
58) L. Tresse
59) R. Caldwell
60) P. F. Muciaccia
61) L. Cayon
62) R. Scaramella
63) C. P. Ma
64) A. Szalay
65) S. Zaroubi
66) C. A. Metzler
67) I. Horvath
68) C. Lineweaver
69) M. Subbarao
70) E. Bertin
71) J. Perez
72) H. Liang
73) G. Hasinger
74) L. Guzzo
75) C. Bellanger
76) A. Zecca
77) I. Gundersen
78) I. Szapudi
79) L. Zaninetti
80) R. Subramanyan
81) S. Colafrancesco
82) F. Lucchin
83) C. Lidman
84) A. Serna
85) M. Verheyen

Inflation after COBE.
Lectures on Inflationary Cosmology (*).

M. S. TURNER

Departments of Physics and Astronomy and Astrophysics, Enrico Fermi Institute
The University of Chicago - Chicago, IL 60637-1433
NASA/Fermilab Astrophysics Center, Fermi National Accelerator Laboratory
Batavia, IL 60510-0500

1. – Hot big bang: successes and challenges.

1˙1. *Successes.* – The hot big-bang model, more properly the Friedmann-Robertson-Walker (FRW) cosmology or standard cosmology, is spectacularly successful: In short, it provides a reliable and tested accounting of the history of the Universe from about 0.01 s after the bang until today, some 15 billion years later. The primary pieces of evidence that support the model are: 1) the expansion of the Universe, 2) the cosmic background radiation and 3) the primordial abundances of the light elements D, ^{3}He, ^{4}He and ^{7}Li [1].

1˙1.1. The expansion. Although the precise value of the Hubble constant is not known to better than a factor of two, $H_0 = 100\,h$ km s^{-1} Mpc^{-1} with $h = 0.4$–1, there is little doubt that the expansion obeys the «Hubble law» out to red shifts approaching unity [2, 3] (see fig. 1). As is well appreciated, the fundamental difficulty in determining the Hubble constant is the calibration of the cosmic-distance scale as «standard candles» are required [4, 5].

The Hubble law allows one to infer the distance to an object from its red shift z: $d = zH_0^{-1} \simeq 3000\,zh^{-1}$ Mpc (for $z \ll 1$, the galaxy's recessional velocity $v \simeq zc$), and hence «maps of the Universe» constructed from galaxy positions and red shifts are referred to as red-shift surveys. Ordinary galaxies and clusters of galaxies are seen out to red shifts of order unity; more unusual and rarer objects, such as radiogalaxies and quasars, are seen out to red shifts of almost

---

(*) Supported in part by the DOE (at Chicago and Fermilab) and by the NASA through grant NAGW-2381 (at Fermilab).

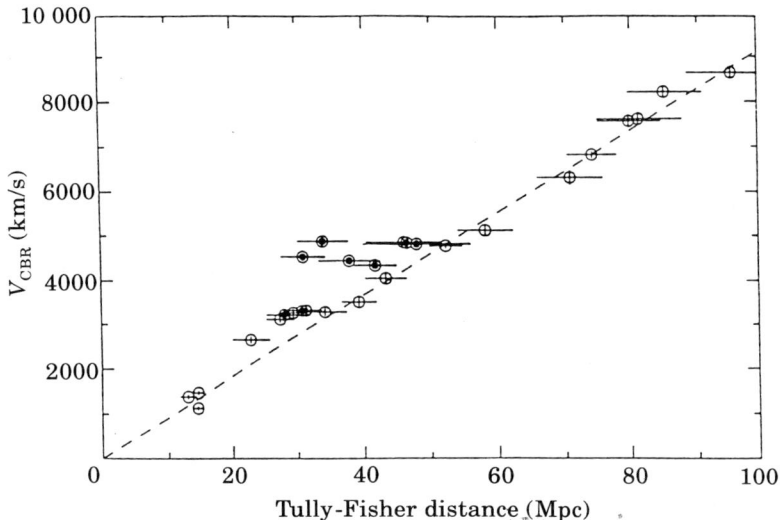

Fig. 1. – Hubble diagram (from [3]). The deviation from a linear relationship around 40 Mpc is due to peculiar velocities.

five (the current record holder is a quasar with red shift 4.9). Thus we can probe the Universe with visible light to within a few billion years of the big bang.

1'1.2. The cosmic background radiation. – The spectrum of the cosmic background radiation (CBR) is consistent with that of a black body at temperature 2.73 K over more than three decades in wavelength ($\lambda \sim (0.03-100)$ cm) (see fig. 2). The most accurate measurement of the temperature and spectrum is that by the FIRAS instrument on the COBE satellite which determined its

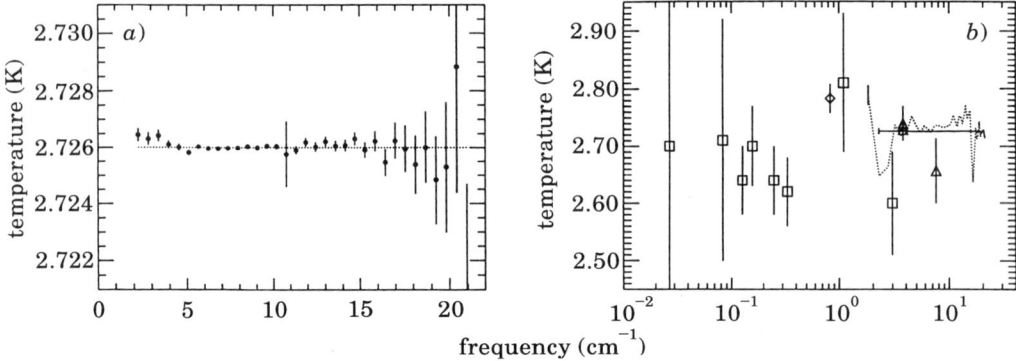

Fig. 2. – a) COBE FIRAS measurements of the CBR temperature, b) summary of the other CBR temperature measurements (from [6]); the dotted curve indicates the data from the other high-precision measurement, by the UBC rocket-borne COBRA instrument [7].

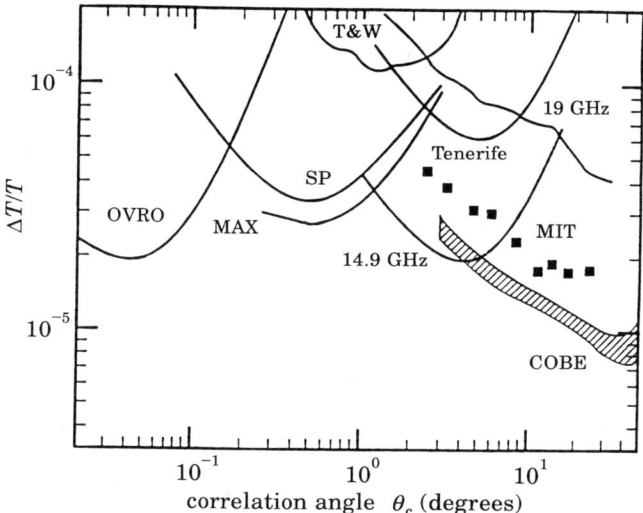

Fig. 3. – Summary of recent high-sensitivity CBR anisotropy measurements; with the exception of COBE all results are upper limits (from [8,9]). The solid boxes (MIT balloon experiment) have recently been reanalyzed and shown to be a detection which is consistent with the COBE DMR result[10].

temperature to be (2.726 ± 0.01) K [6]. It is difficult to come up with a process other than an early hot and dense phase in the history of the Universe that would lead to such a precise black body [11]. According to the standard cosmology, the surface of last scattering for the CBR is the Universe at a red shift of about 1100 and an age of about $180\,000 \, (\Omega_0 h^2)^{-1/2}$ y. It is possible that the Universe became ionized again after this epoch, or due to energy injection never recombined; in this case the last-scattering surface is even «closer», $z_{\text{LSS}} \simeq$ $\simeq 10 \, [\Omega_B h / \sqrt{\Omega_0}]^{-2/3}$.

The temperature of the CBR is very uniform across the sky, to better than a part in 10^4 on angular scales from tens of arcseconds to 90 degrees (see fig. 3). Three forms of temperature anisotropy—two spatial and one temporal—have now been detected: 1) A dipole anisotropy of about a part in 10^3, generally believed to be due to the motion of galaxy relative to the cosmic rest frame, at a speed of about 620 km s^{-1} [12]; 2) a yearly modulation in the temperature in a given direction on the sky of about a part in 10^4, due to our orbital motion around the Sun at 30 km s^{-1} (see fig. 4)[13]; and 3) the temperature anisotropies detected by the Differential Microwave Radiometer (DMR) on the Cosmic Background Explorer (COBE) satellite, $\langle (\Delta T/T)^2 \rangle_{10°}^{1/2} = (1.1 \pm 0.2) \cdot 10^{-5}$ and $(\Delta T/T)_Q = (6 \pm 2) \cdot 10^{-6}$, where the first measurement refers to the r.m.s. temperature fluctuation averaged over the entire sky as measured by a beam of width 10°, and the second is the magnitude of the quadrupole temperature anisotropy[14]. The 10° and quadrupole anisotropies provide strong evidence

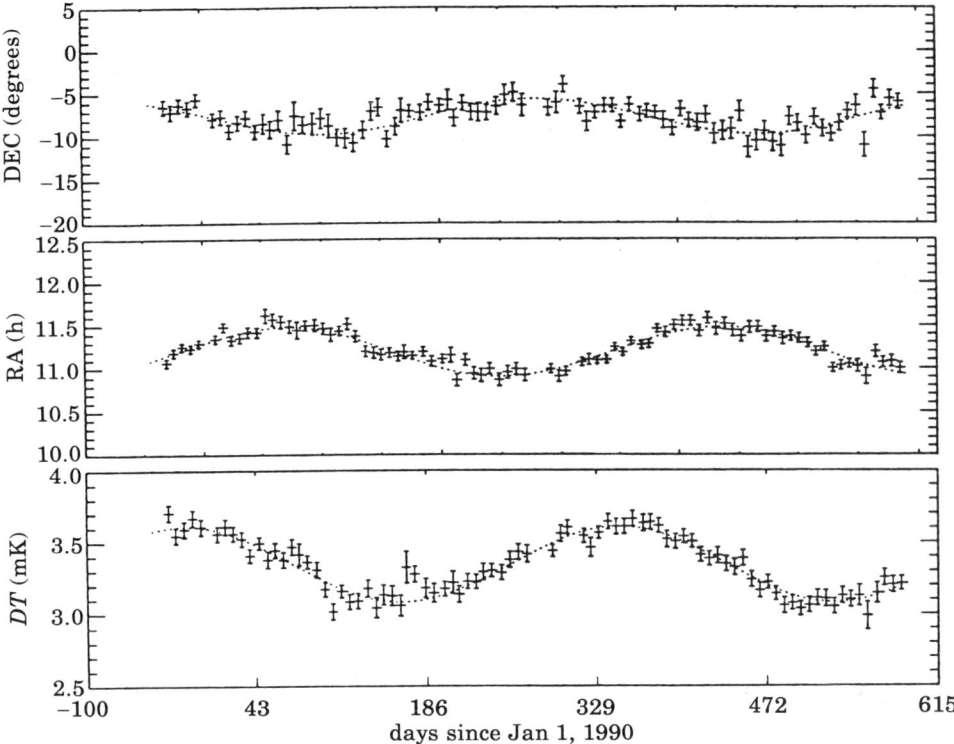

Fig. 4. – Yearly modulation of the CBR temperature—the Earth really orbits the Sun (!) (from [14]).

for primeval density inhomogeneities of the same magnitude, which, amplified by gravity, grew into the structures that we see today: galaxies, clusters of galaxies, superclusters, voids, walls, etc.

1˙1.3. Primordial nucleosynthesis. Last, but certainly not least, there are the abundances of the light elements. According to the standard cosmology, when the age of the Universe was measured in seconds, the temperatures were of order MeV, and the conditions were right for nuclear reactions which ultimately led to the synthesis of significant amounts of D, ^{3}He, ^{4}He and ^{7}Li. The yields of primordial nucleosynthesis depend upon the baryon density, quantified as the baryon-to-photon ratio η, and the number of very light ($\lesssim$ MeV) particle species, often quantified as the equivalent number of light neutrino species, N_ν. The predictions for the primordial abundances of all four light elements agree with their measured abundances provided that $3 \cdot 10^{-10} \lesssim \eta \lesssim 5 \cdot 10^{-10}$ and $N_\nu \lesssim 3.4$ (see fig. 5) [15]

Accepting the success of the standard model of nucleosynthesis, our precise knowledge of the present temperature of the Universe allows us to convert η

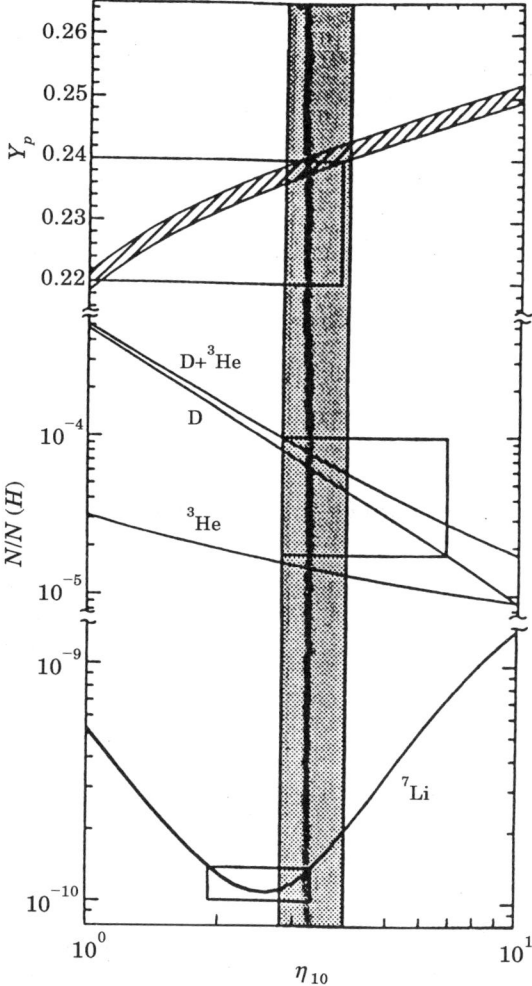

Fig. 5. – Predicted light-element abundances and inferred abundances (from [15]); $\tau_n = (889 \pm 7)$ s, (2σ), $N_\nu = 3.0$. The measured primordial abundances are indicated and the concordance region is shaded.

to a mass density, and, by dividing by the critical density, $\rho_{\text{crit}} \simeq 1.88 h^2 \times 10^{-29}$ g cm^{-3}, to the fraction of critical density contributed by ordinary matter:

(1) $$0.011 \lesssim \Omega_B h^2 \lesssim 0.019 \Rightarrow 0.011 \lesssim \Omega_B \lesssim 0.12;$$

this is the most accurate determination of the baryon density. Note, the uncertainty in the value of the Hubble constant leads to most of the uncertainty in Ω_B.

The nucleosynthesis bound to N_ν, and more generally to the number of light degrees of freedom in thermal equilibrium at the epoch of nucleosynthesis, is consistent with precision measurements of the properties of the Z^0 boson, which give $N_\nu = 3.0 \pm 0.05$; further, the cosmological bound predates these accelerator measurements! The nucleosynthesis bound provides a stringent limit to the existence of new, light particles (even beyond neutrinos), and even provides a bound to the mass of the tau neutrino, excluding a tau-neutrino mass between 0.5 MeV and 25 MeV [16]. Primordial nucleosynthesis provides a beautiful illustration of the powers of the Heavenly Laboratory, though it is outside the focus of this lecture.

The remarkable success of primordial nucleosynthesis gives us confidence that the standard cosmology provides an accurate accounting of the Universe at least as early as 0.01 s after the bang, when the temperature was about 10 MeV.

1`1.4. Et cetera—and the age crisis? There are additional lines of reasoning and evidence that support the standard cosmology [11]. I mention two: age of the Universe and structure formation. I will discuss the basics of structure formation a bit later; for now it suffices to say that the standard cosmology provides a basic framework for understanding the formation of structure—amplification of small primeval density inhomogeneities through gravitational instability—which has recently been confirmed by COBE [14]. Here I focus on the age of the Universe.

The expansion age of the Universe—time back to zero size—depends upon the present expansion rate, energy content and equation of state: $t_{\exp} = f(\rho, p) H_0^{-1} \simeq 9.8 h^{-1} f(\rho, p)$ Gy. For a matter-dominated Universe, f is between 1 and 2/3 (for Ω_0 between 0 and 1), so that the expansion age is somewhere between 7 Gy and 20 Gy. There are other independent measures of the age of the Universe, *e.g.*, based upon long-lived radioisotopes, the oldest stars and the cooling of white dwarfs. These «ages», ranging from 13 to 18 Gy, span the same interval (!). This was not always the case; as late as the early 1950's it was believed that the Hubble constant was 500 km s^{-1} Mpc^{-1}, implying an expansion age of at most 2 Gy—less than the age of the Earth. This discrepancy was an important motivation for the steady-state cosmology.

While there is *general* agreement between the expansion age and other determinations of the age of the Universe, some cosmologists are worried that cosmology is on the verge of another age crisis [5]. Let me explain, while SANDAGE and a few others continue to obtain values for the Hubble constant around 50 km s^{-1} Mpc^{-1} [2], a variety of different techniques seem to be converging on a value around (80 ± 10) km s^{-1} Mpc^{-1} [5]. If $H_0 = 80$ km s^{-1} Mpc^{-1}, then $t_{\exp} = 12 f(\rho, p)$ Gy and, for $\Omega_0 = 1$, $t_{\exp} = 8$ Gy, which is clearly inconsistent with other measures of the age. *If* $H_0 = 80$ km s^{-1} Mpc^{-1}, one is almost forced to consider the radical alternative of a cosmological constant. For example, even with $\Omega_0 = 0.2$, $f \simeq 0.85$, corresponding to $t_{\exp} \simeq 10$ Gy; on the other hand, for a flat Universe with $\Omega_\Lambda = 0.8$, $f \simeq 1.1$ and the expansion age $t_{\exp} \simeq 13.5$ Gy. As I shall discuss later, structure formation provides another motivation for a cosmological constant. My own gut-level feeling is that, when the dust settles, we will find that $H_0 = 50$ km s^{-1} Mpc^{-1} or less; then again, maybe not.

1`2. *Basics of the big-bang model.* – The standard cosmology is based upon the maximally spatially symmetric Robertson-Walker line element

$$(2) \qquad ds^2 = dt^2 - R(t)^2 \left[\frac{dr^2}{1 - kr^2} + r^2(d\theta^2 + \sin^2\theta\, d\dot\phi^2) \right],$$

where $R(t)$ is the cosmic-scale factor, $R_{\text{curv}} \equiv R(t)|k|^{-1/2}$ is the curvature radius, and $k/|k| = -1, 0, 1$ is the curvature signature. All three models are without boundary: the positively curved model is finite and «curves» back on itself; the negatively curved and flat models are infinite in extent (though finite versions of both can be constructed by imposing a periodic structure: identifying all points in space with a fundamental cube). The Robertson-Walker metric embodies the observed isotropy and homogeneity of the Universe. It is interesting to note that this form of the line element was originally introduced for the sake of mathematical simplicity; we now know that it is well justified at early times or today on large scales ($\gg 10$ Mpc), at least within our Hubble volume.

The coordinates, r, θ and ϕ, are referred to as comoving coordinates: A particle at rest in these coordinates remains at rest, *i.e.* constant r, θ and ϕ. A freely moving particle eventually comes to rest in these coordinates, as its momentum is red-shifted by the expansion, $p \propto R^{-1}$. Motion with respect to the comoving coordinates (or cosmic rest frame) is referred to as peculiar velocity; unless «supported» by the inhomogeneous distribution of matter peculiar velocities decay away as R^{-1}. Thus the measurement of peculiar velocities, which is not easy as it requires independent measures of both the distance and velocity of an object, can be used to probe the distribution of mass in the Universe.

Physical separations (*i.e.* measured by meter sticks) between freely moving particles scale as $R(t)$; or, said another way, the physical separation between two points is simply $R(t)$ times the coordinate separation. The momenta of freely propagating particles decrease, or «red-shift», as $R(t)^{-1}$, and thus the wavelength of a photon stretches as $R(t)$, which is the origin of the cosmological red shift. The red shift suffered by a photon emitted from a distant galaxy $1 + z = R_0/R(t)$; that is, a galaxy whose light is red-shifted by $1 + z$ emitted that light when the Universe was a factor of $(1 + z)^{-1}$ smaller. Thus, when the light from the most distant quasar yet seen ($z = 4.9$) was emitted, the Universe was a factor of almost six smaller; when CBR photons last scattered, the Universe was about 1100 times smaller.

1´2.1. Friedmann equation and the first law. The evolution of the cosmic-scale factor is governed by the Friedmann equation

$$H^2 \equiv \left(\frac{\dot{R}}{R}\right)^2 = \frac{8\pi G \rho_{tot}}{3} - \frac{k}{R^2}, \tag{3}$$

where ρ_{tot} is the total energy density of the Universe, matter, radiation, vacuum energy, and so on. A cosmological constant is often written as an additional term ($= \Lambda/3$) on the r.h.s.; I will choose to treat it as a constant energy density («vacuum energy density»), where $\rho_{vac} = \Lambda/8\pi G$. (My convention in this regard is not universal.) The evolution of the energy density of the Universe is governed by

$$d(\rho R^3) = -p \, dR^3, \tag{4}$$

which is the first law of thermodynamics for a fluid in the expanding Universe. (In the case that the stress energy of the Universe is comprised of several, non-interacting components, this relation applies to each separately; *e.g.*, to the matter and radiation separately today.) For $p = \rho/3$, ultra-relativistic matter, $\rho \propto R^{-4}$; for $p = 0$, very nonrelativistic matter, $\rho \propto R^{-3}$; and for $p = -\rho$, vacuum energy, $\rho = $ const. If the r.h.s. of the Friedmann equation is dominated by a fluid with equation of state $p = \gamma\rho$, it follows that $\rho \propto R^{-3(1+\gamma)}$ and $R \propto t^{2/3(1+\gamma)}$.

We can use the Friedmann equation to relate the curvature of the Universe to the energy density and expansion rate:

$$\frac{k/R^2}{H^2} = \Omega - 1, \quad \Omega = \frac{\rho_{tot}}{\rho_{crit}}; \tag{5}$$

and the critical density today $\rho_{crit} = 3H^2/8\pi G = 1.88 \, h^2 \, \mathrm{g \, cm^{-3}} \simeq 1.05 \cdot 10^4 \, \mathrm{eV \, cm^{-3}}$. There is a one-to-one correspondence between Ω and the spatial curvature of the Universe: positively curved, $\Omega_0 > 1$; negatively curved, $\Omega_0 < 1$; and flat, $\Omega_0 = 1$. Further, the «fate of the Universe» is determined by the curvature: model universes with $k \leq 0$ expand forever, while those with $k > 0$ necessarily recollapse. The curvature radius of the Universe is related to the Hubble radius and Ω by

$$R_{curv} = \frac{H^{-1}}{|\Omega - 1|^{1/2}}. \tag{6}$$

In physical terms, the curvature radius sets the scale for the size of spatial separations where the effects of curved space become «pronounced». And in the case of the positively curved model it is just the radius of the 3-sphere.

The energy content of the Universe consists of matter and radiation (today, photons and neutrinos). Since the photon temperature is accurately known, $T_0 = (2.73 \pm 0.01)$ K, the fraction of critical density contributed by radiation is also accurately known: $\Omega_{rad} h^2 = 4.18 \cdot 10^{-5}$. The matter content is another matter.

1˙2.2. A short diversion concerning the present mass density. The matter density today, *i.e.* the value of Ω_0, is not nearly so well known [17]. Stars contribute less than 1% of critical density; based upon nucleosynthesis, we can infer that baryons contribute between 1% and 10% of critical. The dynamics of various systems allow astronomers to infer their gravitational mass. With their telescopes they measure the amount of light, and form a mass-to-light ratio. Multiplying this by the measured luminosity density of the Universe gives a determination of the mass density. (The critical mass-to-light ratio is $1200 h\ M_\odot / \mathscr{L}_\odot$.)

The motions of stars and gas clouds in spiral galaxies indicate that most of the mass of spiral galaxies exists in the form of dark (*i.e.* no detectable radiation), extended halos, whose full extent is still not known. Many cite the flat rotation curves of spiral galaxies, which indicate that the halo density decreases as r^{-2}, as the best evidence that most of the matter in the Universe is dark. Taking the mass-to-light ratio inferred for spiral galaxies to be typical of the Universe as a whole and remembering that the full extent of the dark-matter halos is not known, one infers $\Omega_{halo} \gtrsim 0.03$–$0.1$.

The masses of clusters of galaxies can be estimated using the virial theorem, and these mass estimates too indicate the presence of large amounts of dark matter. Taking cluster mass-to-light ratios to be typical of the Universe as a whole, in spite of the fact that only about 1 in 10 galaxies resides in a cluster, one infers $\Omega_{cluster} \sim 0.1$–$0.3$.

Most galaxies are found in associations of a few galaxies known as small groups. Estimating the masses of these systems using dynamics is tricky because of the problem of «interlopers», galaxies that happen to be in the same part of the sky, but are not associated with the group [18]. This Fall, however, ROSAT detected the weak X-ray emission from the hot gas in the small group NGC 2300 [19]; from their measurements they were able to infer the shape of the gravitational potential—and hence total mass of the group—as well as the mass of the X-ray-emitting gas and the visible mass in galaxies. They found that the total mass of the group was about 20 times that in ordinary matter(!). If one takes this to be a universal ratio of the total amount of matter to that in baryons and $\Omega_B \sim 0.05$, one concludes that $\Omega_0 \sim 1$.

Not one of these methods is wholly satisfactory: Rotation curves of spiral galaxies are still «flat» at the last measured points, indicating that the mass is still increasing; likewise, cluster virial mass estimates are insensitive to material that lies beyond the region occupied by the visible galaxies—and, moreover, only about one galaxy in ten resides in a cluster. What one would like is a measurement of the mass of a very big sample of the Universe, say a cube of $100 h^{-1}$ Mpc on a side, which contains tens of thousands of galaxies.

Over the past five years or so progress has been made toward such a measurement. It involves the peculiar motion of our own Galaxy, at a speed of about 620 km s^{-1} in the general direction of Hydra-Centaurus. This motion is due to the lumpy distribution of matter in our vicinity. By using gravitational-perturbation theory (actually, not much more than Newtonian physics) and the distribution of galaxies in our vicinity (as determined by the IRAS catalogue of infrared selected galaxies), one can infer the average mass density in a very large volume and thereby Ω_0.

The basic physics behind the method is simple: the net gravitational pull on our Galaxy depends both upon how inhomogeneous the distribution of galaxies is and how much mass is associated with each galaxy; by measuring the distribution of galaxies and our peculiar velocity one can infer the «mass per galaxy» and Ω_0.

The value that has been inferred is big(!): $\Omega_{IRAS} \sim 1 \pm 0.2$ [20]. Moreover, the measured peculiar velocities of other galaxies in this volume, more than thousand, have been used in a similar manner and indicate a similarly large value for Ω_0 [21]. While this technique is very powerful, it does have its drawbacks: One has to make simple assumptions about how accurately mass is traced by light (the observed galaxies); one has to worry whether or not a significant portion of our Galaxy's velocity is due to galaxies outside the IRAS sample—if so, this would lead to an overestimate of Ω_0, and so on. This technique is not only very promising—but also provides the «correct» answer (in my opinion!).

The so-called classical kinematic tests—Hubble diagram, angle-red shift relation, galaxy count-red shift relation—can, in principle, provide a determination of Ω_0 [22]. However, all these methods require standard candles, rulers, or galaxies, and for this reason have proved inconclusive. However, that has not stopped efforts to use these tests, particularly the galaxy number-count test [23], and one or more of these classical tests may one day provide a definitive measurement.

To summarize this aside on the mass density of the Universe:

1) Most of the matter is dark.

2) Baryons provide between about 1% and 10% of the mass density.

3) Ω_0 could conceivably be as small as 0.1—in which case all the dark matter could be baryons (*e.g.*, neutron stars, «jupiters», and so on).

4) If asked for the value of Ω_0, a typical astronomer would respond with a number in the interval 0.2 ± 0.1.

5) The evidence continues to mount for a gap between Ω_B and Ω_0—in which case nonbaryonic dark matter is required.

The current prejudice—and certainly that of this author—is a flat Universe

($\Omega_0 = 1$) with nonbaryonic dark matter, $\Omega_X \sim 1 \gg \Omega_B$. However, I shall continue to display the Ω_0 dependence of important quantities.

1'2.3. *The early, radiation-dominated Universe.* – In any case, at present, matter outweighs radiation by a wide margin. However, since the energy density in matter decreases as R^{-3}, and that in radiation as R^{-4} (the extra factor due to the red-shifting of the energy of relativistic particles), at early times the Universe was radiation dominated—indeed the calculations of primordial nucleosynthesis provide excellent evidence for this. Denoting the epoch of matter-radiation equality by subscript «EQ», and using $T_0 = 2.73$ K, it follows that

$$R_{EQ} = 4.18 \cdot 10^{-5} (\Omega_0 h^2)^{-1}, \qquad T_{EQ} = 5.62 (\Omega_0 h^2) \,\text{eV}, \tag{7}$$

$$t_{EQ} = 4.17 \cdot 10^{10} (\Omega_0 h^2)^{-2} \,\text{s}. \tag{8}$$

At early times the expansion rate and age of the Universe were determined by the temperature of the Universe and the number of relativistic degrees of freedom:

$$\rho_{\text{rad}} = g_*(T) \frac{\pi^2 T^4}{30}, \qquad H \simeq 1.67 g_*^{1/2} T^2 / m_{\text{Pl}}, \tag{9}$$

$$\Rightarrow R \propto t^{1/2}, \qquad t \simeq 2.42 \cdot 10^{-6} g_*^{-1/2} (T/\text{GeV})^{-2} \,\text{s}, \tag{10}$$

where $g_*(T)$ counts the number of ultra-relativistic degrees of freedom ($\approx$ the sum of the internal degrees of freedom of particle species much less massive than the temperature) and $m_{\text{Pl}} \equiv G^{-1/2} = 1.22 \cdot 10^{19}$ GeV is the Planck mass. For example, at the epoch of nucleosynthesis, $g_* = 10.75$ assuming three, light ($\ll$ MeV) neutrino species; taking into account all the species in the standard model, $g_* = 106.75$ at temperatures much greater than 300 GeV (see fig. 6).

A quantity of importance related to g_* is the entropy density in relativistic particles,

$$s = \frac{\rho + p}{T} = \frac{2\pi^2}{45} g_* T^3,$$

and the entropy per comoving volume,

$$S \propto R^3 s \propto g_* R^3 T^3.$$

By a wide margin most of the entropy in the Universe exists in the radiation bath. The entropy density is proportional to the number density of relativistic particles. At present, the relativistic-particle species are the photons and neutrinos, and the entropy density is a factor of 7.04 times the photon number density: $n_\gamma = 413 \,\text{cm}^{-3}$ and $s = 2905 \,\text{cm}^{-3}$.

In thermal equilibrium—which provides a good description of most of the

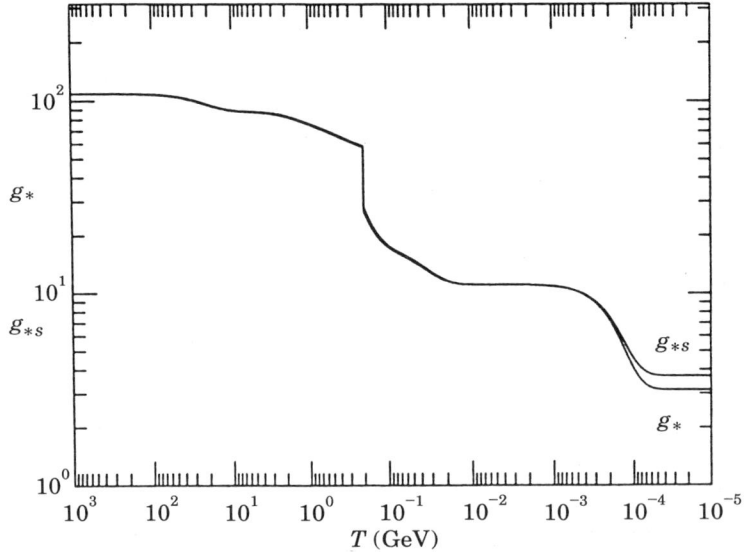

Fig. 6. – The total effective number of relativistic degrees of freedom $g_*(T)$ in the standard model of particle physics as a function of temperature.

history of the Universe—the entropy per comoving volume S remains constant. This fact is very useful. First, it implies that the temperature and scale factor are related by

$$T \propto g_*^{-1/3} R^{-1}, \tag{11}$$

which for $g_* = $ const leads to the familiar $T \propto R^{-1}$.

Second, it provides a way of quantifying the net baryon number (or any other particle number) per comoving volume:

$$N_B \equiv R^3 n_B = \frac{n_B}{s} \simeq (4\text{--}7) \cdot 10^{-11}. \tag{12}$$

The baryon number of the Universe tells us two things: 1) The entropy per particle in the Universe is extremely high, about 10^{10} or so compared to about 10^{-2} in the Sun and a few in the core of a newly formed neutron star. 2) The asymmetry between matter and antimatter is very small, about 10^{-10}, since at early times quarks and antiquarks were roughly as abundant as photons. One of the great successes of particle cosmology is baryogenesis, the idea that B, C and CP violating interactions occurring out of equilibrium early on allow the Universe to develop a net baryon number of this magnitude [24].

Finally, the constancy of the entropy per comoving volume allows us to characterize the size of comoving volume corresponding to our present Hubble volume in a very physical way: by the entropy it contains,

$$S_\text{U} = \frac{4\pi}{3} H_0^{-3} s \simeq 10^{90}. \tag{13}$$

1˙2.4. The earliest history. The standard cosmology is tested back to times as early as about 0.01 s; it is only natural to ask how far back one can sensibly extrapolate. Since the fundamental particles of Nature are pointlike quarks and leptons whose interactions are perturbatively weak at energies much greater than 1 GeV, one can imagine extrapolating as far back as the epoch where general relativity becomes suspect, *i.e.* where quantum gravitational effects are likely to be important: the Planck epoch, $t \sim 10^{-43}$ s and $T \sim 10^{19}$ GeV. Of course, at present, our firm understanding of the elementary particles and their interactions only extends to energies of the order of 100 GeV, which corresponds to a time of the order of 10^{-11} s or so. We can be relatively certain that at a temperature of (100–200) MeV ($t \sim 10^{-5}$ s) there was a transition (likely a second-order phase transition) from quark/gluon plasma to very hot hadronic matter, and that some kind of phase transition associated with the symmetry breakdown of the electroweak theory took place at a temperature of the order of 300 GeV ($t \sim 10^{-11}$ s).

It is interesting to look at the progress that has taken place since Weinberg's classic text on cosmology was published in 1972 [25]; at that time many believed that the Universe had a limiting temperature of the order of several hundred MeV, due to the exponentially rising number of particle states, and that one could not speculate about earlier times. Today, based upon our present knowledge of physics and powerful mathematical tools (*e.g.*, gauge theories, grand unified theories and superstring theory) we are able to make quantitative speculations back to the Planck epoch—and even earlier. Of course, these speculations could be totally wrong, based upon a false sense of confidence (arrogance?). As I shall discuss, inflation is one of these well-defined—and well-motivated—speculations about the history of the Universe well after the Planck epoch, but well before primordial nucleosynthesis.

1˙2.5. The matter- and curvature-dominated epochs. After the equivalence epoch, the matter density exceeds that of radiation. During the matter-dominated epoch the scale factor grows as $t^{2/3}$ and the age of the Universe is related to red shift by

$$t = 2.06 \cdot 10^{17} (\Omega_0 h^2)^{-1/2} (1+z)^{-3/2} \text{ s}. \tag{14}$$

If $\Omega_0 < 1$, the matter-dominated epoch is followed by a «curvature-dominated» epoch where the r.h.s. of the Friedmann equation is dominated by the $|k|/R^2$ term. When the Universe is curvature dominated, it is said to expand freely, no longer decelerating since the gravitational effect of matter has become negligible: $\ddot{R} \approx 0$ and $R \propto t$. The epoch of curvature dominance begins when the matter and curvature terms are equal:

$$(15) \qquad R_{\mathrm{CD}} = \frac{\Omega_0}{1 - \Omega_0} \to \Omega_0, \qquad z_{\mathrm{CD}} = \Omega_0^{-1} - 2 \to \Omega_0^{-1},$$

where the limits shown are for $\Omega_0 \to 0$. By way of comparison, in a flat Universe with a cosmological constant, the Universe becomes «vacuum dominated» when $R = R_{\mathrm{vac}}$:

$$(16) \qquad R_{\mathrm{vac}} = \left(\frac{\Omega_0}{1 - \Omega_0}\right)^{1/3} \to \Omega_0^{1/3}, \qquad z_{\mathrm{vac}} = \left(\frac{1 - \Omega_0}{\Omega_0}\right)^{1/3} - 1 \to \Omega_0^{-1/3}.$$

For a given value of Ω_0, the transition occurs much more recently, which has important implications for structure formation since small density perturbations only grow during the matter-dominated era.

1˙2.6. One last thing: horizons. In spite of the fact that the Universe was vanishingly small at early times, the rapid expansion precluded causal contact from being established throughout. Photons travel on null paths characterized by $dr = dt/R(t)$; the physical distance that a photon could have traveled since the bang until time t, the distance to the horizon, is

$$(17) \qquad d_{\mathrm{H}}(t) = R(t) \int_0^t \frac{dt'}{R(t')} =$$

$$= t/(1-n) = nH^{-1}/(1-n) \qquad \text{for } R(t) \propto t^n, \qquad n < 1.$$

Note, in the standard cosmology the distance to the horizon is finite, and, up to numerical factors, equal to the age of the Universe or the Hubble radius, H^{-1}. For this reason, I will use horizon and Hubble radius interchangeably (*).

An important quantity is the entropy within a horizon volume: $S_{\mathrm{HOR}} \sim H^{-3} T^3$; during the radiation-dominated epoch $H \sim T^2/m_{\mathrm{Pl}}$, so that

$$(18) \qquad S_{\mathrm{HOR}} \sim \left(\frac{m_{\mathrm{Pl}}}{T}\right)^3;$$

from this we conclude that at early times the comoving volume that encompasses all that we can see today (characterized by an entropy of 10^{90}) was comprised of a very large number of causally disconnected regions.

(*) In inflationary models the horizon and Hubble radius are not roughly equal as the horizon distance grows exponentially relative to the Hubble radius; in fact, at the end of inflation they differ by e^N, where N is the number of e-folds of inflation. However, I will slip and use «horizon» and «Hubble radius» interchangeably, though I will always mean Hubble radius.

1˙3. *The challenge: development of structure.* – This brings us to what I believe is the major challenge of the standard cosmology at present: a detailed understanding of the formation of structure in the Universe. We have every indication that the Universe at early times, say $t \ll 300\,000\,\text{y}$, was very homogeneous; however, today inhomogeneity (or structure) is ubiquitous: stars ($\delta\rho/\rho \sim 10^{30}$), galaxies ($\delta\rho/\rho \sim 10^{5}$), clusters of galaxies ($\delta\rho/\rho \sim 10\text{--}10^{3}$), superclusters, or «clusters of clusters» ($\delta\rho/\rho \sim 1$), voids ($\delta\rho/\rho \sim -1$), great walls, and so on.

For some 25 years the standard cosmology has provided a general framework for understanding this: Once the Universe becomes matter dominated (around $1000\,\text{y}$ after the bang) primeval density inhomogeneities ($\delta\rho/\rho \sim 10^{-5}$) are amplified by gravity and grow into the structure we see today [26]. The fact that a fluid of self-gravitating particles is unstable to the growth of small inhomogeneities was first pointed out by JEANS and is known as the Jeans instability. The existence of these inhomogeneities was confirmed in spectacular fashion by the COBE DMR discovery of CBR anisotropy this past Spring: The temperature anisotropies detected almost certainly owe their existence to primeval density inhomogeneities, as causality precludes microphysical processes from producing anisotropies on angular scales larger than about 1°, the angular size of the horizon at last scattering.

At last, the basic picture has been put on firm ground (whew!). Now the challenge is to fill in the details—origin of the density perturbations, precise evolution of the structure, and so on. As I shall emphasize, such an understanding may well be within reach, and offers a window on the early Universe.

1˙3.1. *The general picture: gravitational instability.* Let us begin by expanding the perturbation to the matter density in plane waves

$$(19) \qquad \frac{\delta\rho_M(\boldsymbol{x},t)}{\rho_M} = \frac{1}{(2\pi)^3} \int d^3k\, \delta_k(t) \exp[-i\boldsymbol{k}\cdot\boldsymbol{x}],$$

where $\lambda = 2\pi/k$ is the comoving wavelength of the perturbation and $\lambda_{\text{phys}} = R\lambda$ is the physical wavelength. The comoving wavelengths of perturbations corresponding to bright galaxies, clusters and the present horizon scale are, respectively, about 1 Mpc, 10 Mpc and $3000\,h^{-1}\,\text{Mpc}$, where $1\,\text{Mpc} \simeq 3.09 \cdot 10^{24}\,\text{cm} \simeq$
$\simeq 1.56 \cdot 10^{38}\,\text{GeV}^{-1}$.

The growth of small matter inhomogeneities of wavelength smaller than the Hubble scale ($\lambda_{\text{phys}} \lesssim H^{-1}$) is governed by a Newtonian equation:

$$(20) \qquad \ddot{\delta}_k + 2H\dot{\delta}_k + v_s^2 k^2 \delta_k/R^2 = 4\pi G \rho_M \delta_k,$$

where $v_s^2 = dp/d\rho_M$ is the square of the sound speed. Competition between the pressure term and the gravity term on the r.h.s. determines whether or not pressure can counteract gravity: Perturbations with wave number larger than the Jeans wave number, $k_J^2 = 4\pi G R^2 \rho_M / v_s^2$, are Jeans stable and just oscillate; perturbations with smaller wave number are Jeans unstable and can grow. For cold dark matter $v_s \simeq 0$ and all scales are Jeans unstable; even for baryonic matter, after decoupling k_J corresponds to a baryon mass of only about $10^5 M_\odot$. All the scales of interest here are Jeans unstable and we will ignore the pressure term.

Let us discuss solutions to this equation under different circumstances. First, consider the Jeans problem, evolution of perturbations in a static fluid, i.e. $H = 0$. In this case Jeans-unstable perturbations grow exponentially, $\delta_k \propto \exp[t/\tau]$, where $\tau = 1/\sqrt{4\pi G \rho_M}$. Next, consider the growth of Jeans-unstable perturbations in a matter-dominated Universe, i.e. $H^2 = 8\pi G \rho_M / 3$ and $R \propto t^{2/3}$. Because the expansion tends to «pull particles away from one another», the growth is only power law, $\delta_k \propto t^{2/3}$; i.e. at the same rate as the scale factor. Finally, consider a radiation- or curvature-dominated Universe, i.e. $8\pi G \rho_{\text{rad}}/3$ or $|k|/R^2$ much greater than $8\pi G \rho_M / 3$. In this case, the expansion is so rapid that matter perturbations grow very slowly, as $\ln R$ in radiation-dominated epoch, or not at all, $\delta_k = \text{const}$ in the curvature-dominated epoch.

The growth of nonlinear perturbations is another matter; once a perturbation reaches an overdensity of order unity or larger it «separates» from the expansion—i.e. becomes its own self-gravitating system and ceases to expand any further. In the process of virial relaxation, its size decreases by a factor of two—density increases by a factor of 8; thereafter, its density contrast grows as R^3 since the average matter density is decreasing as R^{-3}, though smaller scales can become Jeans unstable and collapse further to form smaller objects of higher density, stars, etc.

From this we learn that structure formation begins when the Universe becomes matter dominated and ends when it becomes curvature dominated (at least the growth of linear perturbations). The total growth available for linear perturbations is $R_{\text{CD}}/R_{\text{EQ}} \simeq 2.4 \cdot 10^4 \Omega_0^2 h^2$; since nonlinear structures have evolved by the present epoch, we can infer that primeval perturbations of the order $(\delta \rho_M / \rho_M)_{\text{EQ}} \sim 4 \cdot 10^{-5} (\Omega_0 h)^{-2}$ are required. Note that in a low-density Universe larger initial perturbations are necessary as there is less time for growth («the low-Ω_0 squeeze»). Further, in a baryon-dominated Universe things are even more difficult as perturbations in the baryons cannot begin to grow until after decoupling since matter is tightly coupled to the radiation. (In a flat, low-Ω_0 model with a cosmological constant the growth of linear fluctuations continues until almost today since $z_\Lambda \sim \Omega_0^{-1/3}$, and so the total growth factor is about $2.4 \cdot 10^4 (\Omega_0 h^2)$. We will return to this model later.)

1˙3.2. CBR temperature fluctuations. – The existence of density inhomogeneities has another important consequence: fluctuations in the temperature of the CBR of a similar amplitude [27]. The temperature difference measured between two points separated by a large angle ($\geqslant 1°$) arises due to a very simple physical effect(*): The difference in the gravitational potential between the two points on the last-scattering surface, which in turn is related to the density perturbation, determines the temperature anisotropy on the angular scale subtended by that length scale,

$$\left(\frac{\delta T}{T}\right)_\theta = -\left(\frac{\delta\phi}{3}\right)_\lambda \approx \frac{1}{2}\left(\frac{\delta\rho}{\rho}\right)_{\text{HOR},\lambda}, \qquad (21)$$

where the scale $\lambda \sim 100\, h^{-1}\,\text{Mpc}\,(\theta/\text{degree})$ subtends an angle θ on the last-scattering surface. This is known as the Sachs-Wolfe effect [28].

The quantity $(\delta\rho/\rho)_{\text{HOR},\lambda}$ is the amplitude with which a density perturbation crosses inside the horizon, i.e. when $R\lambda \sim H^{-1}$. Since the fluctuation in the gravitational potential $\delta\phi \sim (R\lambda/H^{-1})^2 (\delta\rho/\rho)$, the horizon-crossing amplitude is equal to the gravitational potential (or curvature) fluctuation. The horizon-crossing amplitude $(\delta\rho/\rho)_{\text{HOR}}$ has several nice features: i) during the matter-dominated era the potential fluctuation on a given scale remains constant, and thus the potential fluctuations at decoupling on scales that crossed inside the horizon after matter-radiation equality, corresponding to angular scales $\gtrsim 0.1°$, are just given by their horizon-crossing amplitude; ii) because of its relationship to $\delta\phi$ it provides a dimensionless, geometrical measure of the size of the density perturbation on a given scale, and its effect on the CBR; iii) by specifying perturbation amplitudes at horizon crossing one can effectively avoid discussing the evolution of density perturbations on scales larger than the horizon, where a Newtonian analysis does not suffice and where gauge subtleties (associated with general relativity) come into play; and finally iv) the density perturbations generated in inflationary models are characterized by $(\delta\rho/\rho)_{\text{HOR}} \simeq \text{const}$.

On angular scales smaller than about 1° two other physical effects lead to CBR temperature fluctuations: the motion of the last-scattering surface (Doppler) and the intrinsic fluctuations in the local photon temperature. These fluctuations are much more difficult to compute, and depend on microphysics—the ionization history of the Universe and the damping of perturbations in the photon-baryon fluid due to photon streaming. Not only are the Sachs-Wolfe fluctuations simpler to compute, but they also accurately mirror the primeval fluctuations since at the epoch of decoupling microphysics is restricted to angular scales less than about a degree.

In sum, on large angular scales the Sachs-Wolfe effect dominates; on the scale of about 1° the total CBR fluctuation is about twice that due to the Sachs-Wolfe effect; on smaller scales the Doppler and intrinsic fluctuations dominate. CBR temperature fluctuations on scales smaller than

---

(*) Large angles mean those larger than the angle subtended by the horizon scale at decoupling, $\theta \sim H_{\text{DEC}}^{-1}/H_0^{-1} \sim z_{\text{DEC}}^{-1/2} \sim 1°$.

about 0.1° are severely reduced by the smearing effect of the finite thickness of last-scattering surface.

Details aside, in the context of the gravitational-instability scenario density perturbations of sufficient amplitude to explain the observed structure lead to temperature fluctuations in the CBR of characteristic size,

$$\frac{\delta T}{T} \approx 10^{-5} (\Omega_0 h)^{-2} \,. \tag{22}$$

To be sure I have brushed over important details, but this equation conveys a great deal. First, the overall amplitude is set by the inverse of the growth factor, which is just the ratio of the radiation energy density to matter density at present. Next, it explains why theoretical cosmologists were so relieved when the COBE DMR detected temperature fluctuations of this amplitude, and conversely why one heard offhanded remarks before the COBE DMR detection that the standard cosmology was in trouble because the CBR temperature was too uniform to allow for the observed structure to develop. Finally, it illustrates one of the reasons why cosmologists who study structure formation have embraced the flat-Universe model with such enthusiasm: If we accept the Universe that meets the eye, $\Omega_0 \sim 0.1$ and baryons only, then the simplest models of structure formation predict temperature fluctuations of the order of 10^{-3}, far too large to be consistent with observation. Later, I will mention Peebles' what-you-see-is-what-you-get model[29], also known as PIB for primeval baryon isocurvature fluctuation, which is still viable because the spectrum of perturbations decreases rapidly with scale, so that the perturbations that give rise to CBR fluctuations are small (which is no mean feat). Historically, it was fortunate that one started with a low-Ω_0, baryon-dominated Universe: the theoretical predictions for the CBR fluctuations were sufficiently favorable that experimentalists were stirred to try to measure them—and then, slowly, theorists lowered their predictions. Had the theoretical expectations begun at 10^{-5}, experimentalists might have been too discouraged to even try!

1˙3.3. *An initial-data problem.* With the COBE DMR detection in hand we can praise the success of the gravitational-instability scenario; however, the details now remain to be filled in. The structure formation problem is now one of initial data, namely:

1) The quantity and composition of matter in the Universe, Ω_0, Ω_B and Ω_{other}.

2) The spectrum of initial density perturbations: for the purist, $(\delta\varrho/\varrho)_{\text{EQ}}$, or, for the simulator, the Fourier amplitudes at the epoch of matter-radiation equality.

In a statistical sense, these initial data provide the «blueprint» for the formation of structure.

The initial data are the challenge and the opportunity. Although the gravitational-instability picture has been around since the discovery of the CBR itself, the lack of specificity in initial data has impeded progress. With the advent of the serious study of the earliest history of the Universe a new door was opened. We now have several well-motivated early-Universe blueprints: Inflation-produced density perturbations and nonbaryonic dark matter, cosmic-string-produced perturbations and nonbaryonic dark matter[30], texture-produced density perturbations and nonbaryonic dark matter[31], a baryon-dominated Universe with isocurvature fluctuations(*)[29]. Structure formation also provides the opportunity to probe the earliest history of the Universe, by testing these interesting, if not bold, blueprints. I will be focusing on the blueprints motivated by inflation.

2. – Inflation: an overview.

2˙1. *Shortcomings of the standard cosmology.* – By now the shortcomings of the standard cosmology are well appreciated: the horizon or large-scale smoothness problem, the small-scale inhomogeneity problem (origin of density perturbations), the flatness or oldness problem and the monopole problem. I will only briefly review them here. They do not indicate any logical inconsistencies of the standard cosmology; rather, that very special initial data seem to be required for evolution to a universe that is qualitatively similar to ours today. Nor is inflation the first attempt to address these shortcomings: Over the past two decades cosmologists have pondered this question and proposed other solutions[32]. Inflation is a solution based upon well-defined, albeit speculative, early-Universe microphysics describing the post-Planck epoch.

The uniformity of the CBR temperature, to better than a part in 10^4, implies that the Universe on the largest scales (say $\geqslant 100 h^{-1}$ Mpc) is very smooth as density inhomogeneities induce temperature fluctuations of a similar magnitude. The existence of particle horizons in the standard cosmology precludes explaining the smoothness as a result of microphysical events: The horizon at decoupling, the last time one could imagine temperature fluctuations being smoothed by particle interactions, corresponds to an angular scale on the sky of about 1°, which precludes temperature variations on larger scales from being erased. In terms of entropy, the presently observed Universe corresponds to a comoving volume containing an entropy of order 10^{90}; during the early radiation-dominated epoch the horizon volume contained an entropy of order

---

(*) Isocurvature baryon number fluctuations correspond at early times to fluctuations in the local baryon number but not in the energy density. At late times, when the Universe is matter dominated, they become fluctuations in the mass density of a comparable amplitude.

$(m_{Pl}/T)^3$, implying that at early times our current Hubble volume consisted of countless, causally distinct regions.

To account for the small-scale lumpiness of the Universe today, density perturbations with horizon-crossing amplitudes of 10^{-5} on scales of 1 Mpc to 10^4 Mpc or so are required. As can be seen in fig. 7, in the standard cosmology the physical size of a perturbation, which grows as the scale factor, begins larger than the horizon and relatively late in the history of the Universe crosses inside the horizon,

$$(23) \qquad t_{HOR} \simeq \begin{cases} 3 \cdot 10^8 (\lambda/\text{Mpc})^2 \text{ s}, & \lambda \lesssim 13 \, h^{-2} \text{ Mpc}; \\ 3 \cdot 10^7 (\lambda/\text{Mpc})^3 \text{ s}, & \lambda \gtrsim 13 \, h^{-2} \text{ Mpc}. \end{cases}$$

This precludes a causal microphysical explanation for the origin of the required density perturbations(*).

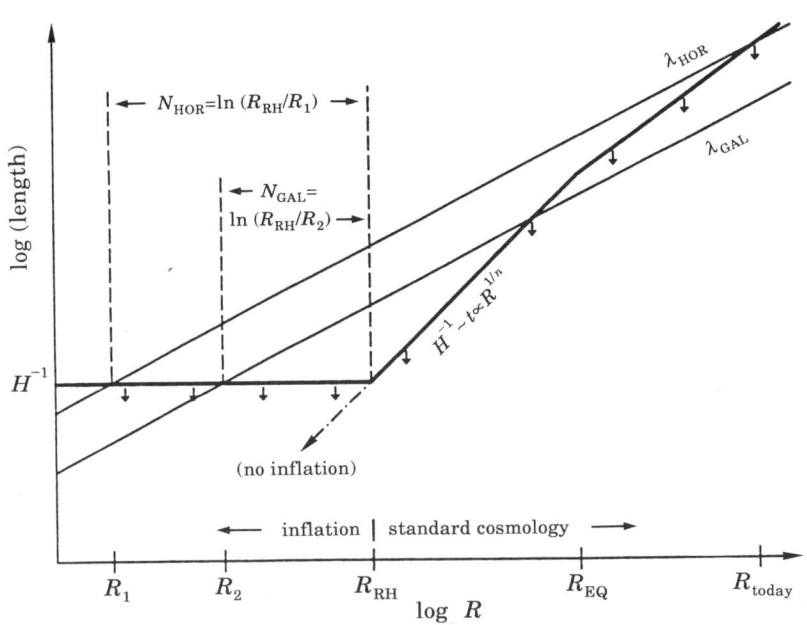

Fig. 7. – The physical wavelength of a density perturbation and the horizon size H^{-1} as a function of scale factor; λ_{GAL} indicates a galactic-sized perturbation ($\lambda \sim 1$ Mpc) and λ_{HOR} corresponds to the present Hubble radius (horizon). Microphysics operates on scales $\lesssim H^{-1}$; without inflation scales cross the Hubble radius but once.

(*) Of course, it is possible to produce the perturbations at very late times, when the relevant scale has already crossed inside the horizon[33]; the motivation for the nonstandard microphysics required to do so is lacking at present. It is also possible for microphysics to produce isocurvature perturbations by producing a pressure wave that eventually propagates to large scales; this is the type of perturbation that is generated by cosmic strings or textures.

The fact that Ω_0 is of order unity means that the curvature radius is comparable to the Hubble radius. Had that been the case at the initial epoch, the Universe would be a very different place today: Since the curvature term in the Friedmann equation decreases only as R^{-2}, while the matter and radiation densities decrease as R^{-3} and R^{-4}, respectively, a curvature radius comparable to the Hubble radius early on would have led to a Universe that quickly became curvature dominated. For positive curvature, recollapse would follow, and, for negative curvature, a coasting phase that would lead to a Universe that cools too quickly (for $t_{\text{initial}} \sim 10^{-43}$ s, the temperature reaches 3 K at an age of 10^{-11} s). Put another way, Ω is an unstable fixed point:

(24)
$$\begin{cases} \Omega(t) = \dfrac{1}{1 - x(t)}, \\ x(t) = \dfrac{k/R^2}{8\pi G\rho/3}; \end{cases}$$

the deviation of $\Omega(t)$ from unity increases as $x(t) \propto R^n$, $n = 2$ (radiation-dominated epoch), $n = 1$ (matter-dominated epoch). In order that Ω still be close to unity today, it must have been extremely close to unity early on; for $t_{\text{initial}} \sim 10^{-43}$ s, $|\Omega(t_{\text{initial}}) - 1| \lesssim 10^{-60}$ is necessary. Thus for most of its history the Universe must have been extremely flat, i.e. $R_{\text{curv}} \gg H^{-1}$; if Ω_0 is not equal to unity, then the Universe just today is beginning to exhibit its curvature. Why now?

Last, I mention the monopole problem: The simplest grand unified theories and the standard cosmology lead to a disastrous prediction, the extreme overproduction of magnetic monopoles [34]. This overproduction traces to the smallness of the horizon at very early times: magnetic monopoles are produced as defects of the GUT phase transition at an abundance of about 1 per horizon volume which corresponds to a present monopole-to-photon ratio of order $(T_{\text{GUT}}/m_{\text{Pl}})^3$.

The first three problems do not involve logical inconsistencies: The initial data for a perturbed FRW model that is extremely flat exist. Rather, it is the fact that such initial data are «very special» which is disturbing. COLLINS and HAWKING quantified it: The set of initial data that evolve to a state qualitatively similar to our Universe is of measure zero [35]. Maybe the Creator had a lucky day! Or better yet, perhaps the present state of the Universe traces to events that took place early on. Inflation provides an interesting example of the latter.

2'2. *Generic aspects of inflation.* – Inflationary cosmology has become a very mature subject in the decade since GUTH wrote his influential paper [36] that launched the inflationary-cosmology boom. While there are a multitude

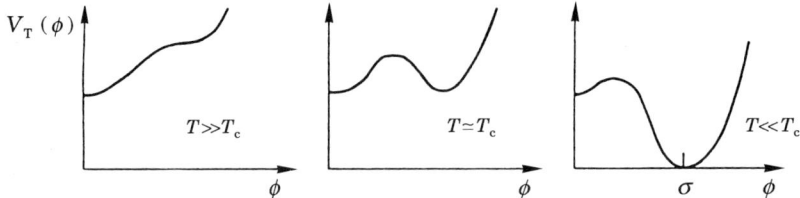

Fig. 8. – The free-energy density as a function of temperature for a first-order phase transition.

of different kinds of inflation (see below), two features are common to all models of inflation [37]:

superluminal expansion,

massive entropy production.

Superluminal expansion refers to accelerated growth of the scale factor ($\ddot{R} > 0$ which implies $R \propto t^n$ with $n > 1$), and its necessity is easy to understand. In order that the physical size of a comoving scale, $d_{\text{phys}} \propto R(t)$, begin sub-Hubble size and become super-Hubble size, $R(t)$ must increase faster than t since $H^{-1} \propto R(t)^{1/n}$. Thus «superluminal» expansion is a necessary kinematic requirement if one is to both solve the horizon and create density perturbations (see fig. 7).

The reason for the second requirement is equally simple: In the absence of entropy production the entropy per comoving volume $S \propto (RT)^3$ remains constant; rapid expansion can create a «very large» smooth patch, but the entropy within that patch remains constant. As discussed above, at early times the entropy within a horizon-sized patch is very small, too small to account for the entropy within our present Hubble volume. Only massive entropy production can change this [37].

To illustrate, consider Guth's original model of inflation based upon a first-order phase transition [36]. The basic idea is that the Higgs field responsible for the spontaneous breakdown of the GUT symmetry gets «hung up» in a local, high-energy, minimum of its potential (more precisely, free-energy density). At high temperatures the state of minimum free energy is characterized by $\phi = 0$, indicating that the full GUT symmetry is manifest; as the temperature drops below the critical temperature, the state of minimum free energy is characterized by $\phi \neq 0$, the state that exhibits broken symmetry. In a first-order transition $\phi = 0$ can remain a local minimum of the free energy, separated from the global minimum by a potential barrier (see fig. 8). During the time that ϕ is hung up the large vacuum energy density, $\rho = V(\phi = 0) \equiv \mathcal{M}^4$, drives very rapid expansion ($\mathcal{M}$ is the energy scale that characterizes the symmetry breaking).

For definiteness, take $\mathcal{M} = 10^{14}$ GeV, a typical scale for inflation; the Hubble time associated with the false-vacuum energy $H^{-1} \sim 10^{-34}$ s. The size of a region that one might expect to be smooth is of order $ct \sim 10^{-23}$ cm; the entropy within such a patch is of order 10^{14}. While the Higgs field is trapped in the false vacuum, the temperature of the Universe continues to decrease as R^{-1}; very soon the thermal-energy density becomes insignificant compared to the constant false-vacuum energy density. At this point, the Universe enters a de Sitter phase of exponential expansion since $\rho \sim \mathcal{M}^4 = \text{const}$; this is the superluminal expansion. As the Universe expands, it cools exponentially with the entropy per comoving volume remaining constant; the smooth horizon-sized patch continues to contain an entropy of only 10^{14} as it grows exponentially in size.

During inflation the scale factor undergoes many e-folds; the precise number is determined by how long the Higgs field is hung up: $N = H \Delta t$. Again, for definiteness, suppose that the Universe gets hung up for a mere 10^{-32} s; then, during inflation the patch grows in linear size by $e^{100} \sim 10^{43}$ and its temperature drops by the same factor. Thus far, inflation has done little. When the Higgs field does make its way to the true vacuum, the enormous false-vacuum energy is released and ultimately thermalized, reheating the patch to a temperature of order $\mathcal{M} \sim 10^{14}$ GeV, thereby increasing the entropy of the patch by a factor of $e^{3N} \sim 10^{129}$. This is the massive entropy production. After «reheating» the patch contains an entropy of order 10^{143}, and can easily contain the comoving volume that corresponds to our present Hubble radius, which is characterized by an entropy of «only» 10^{90}.

It is clear that the smoothness problem has been solved. The kinematic requirement for producing density perturbations on astrophysically interesting scales has been satisfied; the mechanism that produces density perturbations, quantum fluctuations in the ϕ field, will be discussed later. What about the flatness problem? Suppose for definiteness that the curvature radius at the beginning of inflation is of order the Hubble radius (which corresponds to Ω just beginning to deviate from unity); at the end of inflation the curvature radius has grown by a factor of e^N, while the energy density has remained constant. This means that $\Omega_{\text{end}} = 1/[1 - (k/R^2)/(8\pi G\rho/3)] \simeq 1 \pm e^{-2N}$ has been reset to a value exponentially close to unity. Using our fiducial numbers, at the end of inflation the curvature radius is order 10^{20} cm; from then until today it grows by a factor of $\mathcal{M}/3\text{K} \sim 10^{27}$, reaching a present size of order 10^{47} cm. This is enormous compared to the present Hubble radius and implies that Ω is still very close to unity today. The flatness problem has clearly been solved and a flat Universe predicted.

Consider the fate of monopoles—or any other «cosmic pollutant»—in the pre-inflationary Universe. The number of monopoles within the patch ($= N_M$) remains constant; however, the number per comoving volume, $n_M/s = N_M/S$, decreases by a factor of $e^{3N} \sim 10^{129}$ due to the massive entropy production. Undesirables are diluted away! Of course, this also implies that the baryon number of the Universe, $n_B/s \sim 10^{-10}$, must be produced after inflation.

Finally, a simple exercise; what is the minimum amount of inflation needed to solve the smoothness problem? Start with a Hubble-sized patch at the beginning of inflation; it contains an entropy of $S_{\text{initial}} \sim H^{-3} T^3 \sim (m_{\text{Pl}}/\mathcal{M})^3$. Assuming perfect conversion of vacuum energy to radiation, after inflation the entropy contained within the patch is $e^{3N} S_{\text{initial}} \sim e^{3N} m_{\text{Pl}}^3/\mathcal{M}^3$. To solve the smoothness problems this must be greater than 10^{90}, which implies

(25) $$N \gtrsim N_{\min} = 56 + \ln(\mathcal{M}/10^{14}\text{ GeV}).$$

Equivalently, one can express the size of the patch today relative to the present Hubble radius,

(26) $$d_{\text{patch}} = \exp[N - N_{\min}] H_0^{-1}.$$

What about the flatness problem? It is simple to show the present value of Ω is related to that at the beginning of inflation and the size of the patch today:

(27) $$|\Omega_0 - 1| = \left(\frac{H_0^{-1}}{d_{\text{patch}}}\right)^2 |\Omega_{\text{preinflation}} - 1|.$$

Remarkably enough, the amount of inflation required to solve the flatness and smoothness problems is the same. Put another way, if one comfortably solves the smoothness problem, Ω_0 is necessarily very, very close to unity. This means that a flat Universe is an unequivocal prediction of inflation.

2˙3. *Current status of inflationary models.*

2˙3.1. Types of inflation. In this very brief overview I divide models of inflation into three broad classes: old, slow rollover and first order (or extended). By old inflation I mean Guth's original model, which I forgot to mention was a nonstarter! Let me explain; once trapped in the false vacuum, the Higgs field must quantum-mechanically tunnel to the true vacuum; in order to ensure a sufficient amount of inflation, this transition must not occur until 60 or so Hubble times after inflation has begun. As we shall see, this is essentially impossible to arrange.

The decay of the false vacuum is well understood [38]: It proceeds via the nucleation of bubbles of true vacuum that expand outward at the speed of light. For a given potential the bubble nucleation rate (per unit volume) Γ is straightforward to calculate [38]. Roughly speaking, bubbles convert all of space into the true vacuum when Γ/H^4, the number of bubbles nucleated in a Hubble volume in a Hubble time, exceeds order unity; since each bubble nucleated during a Hubble time liberates about a Hubble volume, $\Gamma/H^4 \sim 1$ ensures that all of space is converted to true vacuum in a Hubble time (before the expansion «creates» more false vacuum). The false-vacuum energy is converted into «heat» by the collision of vacuum bubbles [39].

The recipe for successful old inflation is for Γ/H^4 to remain less than unity for 60 or so Hubble times and then increase to greater than unity. Unfortunately, shortly after inflation begins Γ, like the expansion rate, becomes constant, as the temperature of the Universe rapidly approaches zero and becomes irrelevant. This is the fundamental problem with old inflation; Γ/H^4 is constant. The Universe can either get hung up in the false vacuum and inflate, or make the transition to the true vacuum, not both!

Slow-rollover inflation solved this problem, but at a price. The fix, suggested independently by LINDE [40] and ALBRECHT and STEINHARDT [41], is for inflation to occur as the scalar field slowly rolls down the potential. They proposed using very flat potentials with small or nonexistent barriers between the false- and true-vacuum states; the vacuum-driven expansion takes place as the scalar field slowly (time scale $\geqslant 60 H^{-1}$), but inevitably rolls toward the true-vacuum state. When the scalar field responsible for inflation (often called the inflaton) reaches the true minimum of its potential, it oscillates about it, the large vacuum energy having been converted into coherent inflaton oscillations. These oscillations ultimately decay into light-particle states reheating the Universe. From the quantum view, these coherent field oscillations correspond to zero-momentum inflaton particles; the decay of the scalar-field oscillations corresponds to the decay of massive inflaton particles [42].

Slow rollover led to the first viable models of inflation. There was, however, a price: In all models of slow rollover the inflaton field must be very weakly coupled (dimensionless self-coupling of order 10^{-14} or so); as we shall see, this is dictated by achieving density perturbations of size 10^{-5} or so. Because of this fact, the inflaton cannot be directly responsible for GUT symmetry breaking as loop corrections from the inflaton-gauge interaction would spoil the flatness of the potential. The decoupled nature of the scalar field responsible for inflation gave birth to its name.

In the broadest sense, slow-rollover inflation refers to any model of inflation where a scalar field is displaced from the minimum of its potential and slowly rolls toward it, with the Universe inflating as it does. The slowly rolling scalar field responsible for inflation can begin near the origin at a high-energy local minimum of the potential (often referred to as «new inflation»), as it might with potentials used for spontaneous symmetry breaking, or it can begin far from the origin, away from a local minimum of the potential (often referred to as «chaotic inflation»), as it would with a potential such as $V(\phi) = \lambda \phi^4$ or $V(\phi) = m^2 \phi^2 /2$.

The latest and perhaps most interesting development in inflationary models is first-order (or extended) inflation [43]. In many ways it combines the best features of old inflation—intimate connection to particle physics phenomenology—and slow-rollover inflation—it works! As the name suggests, these models are associated with a first-order phase transition; how then do these models solve the Guth dilemma—the constancy of Γ/H^4? The first model of this type was due to LA and STEINHARDT [44]; their new twist was to use the Brans-Dicke theory of gravity rather than general relativity. In Brans-Dicke the gravitational constant $G_{\text{eff}} = \Phi^{-2}$ evolves as the Brans-Dicke field Φ evolves. Because of this, for constant energy density *the scale factor only increases as a power of time*, $R(t) \propto t^{\omega+1/2}$, *and H decreases with time*; here ω is the coefficient of the kinetic-energy term for Φ. Thus the efficiency of bubble nucleation $\Gamma/H^4 \propto t^4$ increases during inflation; at early times it can be much less than unity (so that the Universe remains trapped in the false vacuum) and then exceeds unity triggering the end of inflation via the nucleation and percolation of bubbles of true vacuum.

Models based on variations of this idea have been proposed. For example, if the Higgs field couples to other fields which are evolving during inflation, then Γ will vary during inflation, leading to the variation of Γ/H^4 [45]. In first-order inflation models the Higgs field plays a relatively passive role, remaining trapped in the false vacuum during inflation; further, it need not be weakly coupled, nor is the shape of its potential particularly relevant.

By means of a conformal transformation extended inflation can be recast as slow-rollover inflation with an exponential potential with ln Φ field playing the role of the inflaton [46]. In first-order inflation models there is another problem one has to worry about: If Γ/H^4 does not change rapidly enough, then too many bubbles will be nucleated long before the end of inflation; these bubbles eventually grow to astrophysical size and can have disastrous consequences (large anisotropies in the CBR, interference with primordial nucleosynthesis, and so on) [47]. To avoid «the big-bubble problem» in extended inflation ω must be less than about 20; that there be an upper limit to ω is not surprising since in the limit $\omega \to \infty$ Brans-Dicke goes to general relativity.

2˙3.2. Viable models. There is no standard model of inflation; nor is there a model of inflation without some flaw. There are a number of «proof of existence» models, models that successfully implement inflation, but are only beautiful in the eyes of their creators. Of course, this situation should be viewed in light of our general ignorance about physics at energy scales $\gg 10^3$ GeV (most inflation models involve an energy scale of order 10^{14} GeV). Moreover, the same criticism—lack of a standard model—applies to baryogenesis, and applied to primordial nucleosynthesis until the early 1970's!

Slow rollover. There are numerous viable models; I will mention but a representative few. There is an almost decade old model based upon an ordinary GUT due to SHAFI and VILENKIN [48] and PI [49]. This model has the virtue that the inflaton field does more than cause inflation; it also breaks Peccei-Quinn symmetry and induces GUT symmetry breaking (by producing a negative mass squared for the GUT Higgs field). After inflation the Universe reheats to a temperature of order 10^7 GeV, and a scenario for baryogenesis is included. In short, it is a complete model.

In passing, let me mention a similar model just proposed by KNOX and myself[50]. The new twist is that the scale of inflation can be as small as the electroweak scale(!), and the inflaton field can be used to induce electroweak-symmetry breaking and other low-energy phenomena (*e.g.*, right-handed-neutrino masses). In principle, this model can be tested in laboratory experiments. Of course, this model is only viable provided one believes that the baryon asymmetry of the Universe can be produced at the weak scale or below.

There are many supersymmetric implementations of slow-rollover inflation[51]; a particularly elegant one is that of Holman, Ramond and Ross[52]. The superpotential for their inflaton is very simple, $W(\phi) = (\Delta^2/M)(\phi - M)^2$; here $M = m_{Pl}/\sqrt{8\pi}$ and Δ is the GUT scale. In this model, the self-coupling of the inflaton in its scalar potential is given by the fourth power of the ratio of the GUT to Planck scales, $(\Delta/M)^4$, and the canonical small number arises because of the discrepancy between the GUT and Planck scales. The reheat temperature in this model is order 10^6 GeV, and the details of baryogenesis are spelled out.

There is a model called (by the authors) «natural inflation»[53]. The primary purpose of this model is to address the small self-coupling of the inflaton. To wit, the inflaton is a pseudo Nambu-Goldstone boson akin to the axion; a Nambu-Goldstone boson has an absolutely flat potential, *i.e.* is massless, and becomes a pseudo Nambu-Goldstone boson due to explicit symmetry-breaking terms. The potential, $V(\phi) = \Lambda^4[1 + \cos(\phi/f)]$, has two energy scales: $f \sim m_{Pl}$, the scale of the spontaneous symmetry breaking, and $\Lambda \sim 10^{-5} f$, the scale of explicit symmetry breaking (GUT scale?). (In the axion analogy, $\Lambda = \Lambda_{QCD} \simeq \simeq 200$ MeV and f is the PQ symmetry-breaking scale.) Some superstring adherents have taken interest in this model as superstring theories often have pseudo Nambu-Goldstone bosons with Planck-scale symmetry breaking.

There is a broad class of slow-rollover models referred to as chaotic inflation; they illustrate the simplicity of inflation and were pioneered by LINDE[54]. The simplest such models are based upon potentials that have nothing to do with spontaneous symmetry breaking, $V(\phi) = \lambda \phi^4$ or $V(\phi) = m^2 \phi^2/2$, with the inflaton initially displaced far from the origin, $\phi_{initial} \gtrsim 5 m_{Pl}$. (Chaotic inflation can also be implemented with more complicated potentials whose minima are not at the origin[55].) As with all slow-rollover models, there is a small, dimensionless number: $\lambda \sim 10^{-14}$ or $m^2 \simeq 10^{-12} m_{Pl}^2$. While the simplest models of chaotic inflation are not tied to specific particle physics theories, some have been[56].

There are models where the inflaton field is not actually a scalar field; *e.g.*, where it is related to the size of the compactified dimensions in models with extra dimensions[57], or is related to the scalar curvature $\mathcal{R}$ in higher-derivative theories of gravity[58].

The common undesirable feature of all slow-rollover models is a small, dimensionless number of order 10^{-14}, typically the self-coupling of the inflaton; as we shall discuss, this small number is necessary to guarantee density perturbations of the appropriate size. To ensure the stability of the flatness of the potential against quantum (radiative) corrections the inflaton must be weakly coupled to the «rest of the world», and in this sense *all* the models mentioned are natural. However, weak coupling works at cross purposes with reheating and baryogenesis. Slow-rollover models liberate only a tiny fraction of the false-vacuum energy to radiation and have a relatively low reheat temperature, which is problematic for baryogenesis as it must proceed after inflation. The second problem lies in the name «inflaton»; because the field responsible for inflation is so weakly coupled, without heroic efforts it is difficult to make it an integral part of a more encompassing particle physics theory.

First order. These models have the potential (no pun intended) to incorporate the best aspects of both slow-rollover and old inflation. Inflation is again intimately connected to a cosmological phase transition at a scale of order the GUT scale and no special flatness is required of the Higgs potential. Moreover, reheating proceeds via vacuum-bubble collisions which guarantees good reheating and a unique signature of first-order inflation, a background of gravitational waves preceded by bubble collisions, $\Omega_{GW} \sim 10^{-8}$ at a frequency determined by the scale of inflation, $f_{GW} \sim 10^6$ Hz ($\mathcal{M}/10^{12}$ GeV)[59].

The simplest first-order inflation model is extended inflation. First the good news: Brans-Dicke gravity exhibits conformal (scale) invariance (the Planck scale is replaced by a field). Conformal invariance is «the Hallmark» of superstring theory, which has stimulated new interest in Brans-Dicke-like theories. Now the bad news: in order to avoid «the big-bubble problem», the Brans-Dicke parameter ω must be less than about 20, while solar-system tests set a *lower limit* of about 500[60]. In its simplest form, extended inflation is not viable. Several variants have been put forth[43]; the simplest fix is to give the Brans-Dicke field a mass[46]. (A mass for the Brans-Dicke field anchors at the right value and makes the theory immune to solar-system tests.) Any mass less than about 10^9 GeV and greater than a tiny fraction of an eV will do. Moreover, this simple fix involves something that string theorists must do anyway: break conformal invariance (the world is not conformally invariant, it has a multitude of energy scales).

In sum, inflation provides a very attractive early-Universe paradigm. Models of inflation are based upon well-defined, albeit very speculative, physics at energy scales well below the Planck scale. At present there is no standard model, or even a particularly compelling model; there are, however, a variety of models that work. Given our general ignorance about physics at energy scales $\gg 10^3$ GeV, perhaps that should be enough for the time being. In any case, while elegance, simplicity and mathematical beauty often provide guidance to the theorist, in the end, experiment and observation are the final arbiters. As I will discuss toward the end, observations involving structure formation are starting to do just that.

2˙4. *Initial conditions: no-hair theorems.* – Inflation is cosmologically attractive because it promises to account for our present nearly FRW space-time starting from very general initial conditions. Somewhat paradoxically, inflation is usually analyzed in the context of the isotropic and homogeneous FRW cosmology. I will now explain the *apparent* paradox and discuss to what extent inflation lessens the dependence of the present state of the Universe upon its initial state.

To begin consider the anisotropic but homogeneous (Bianchi) models; the mean expansion rate of the Universe can be written as

$$(28) \qquad H^2 \equiv (\dot{\overline{R}}/\overline{R})^2 = \frac{8\pi G \rho}{3} + F(\dot{\overline{R}}, \overline{R}),$$

where $\overline{R}$ is the mean scale factor and ρ is the usual energy density and the function F accounts for the additional terms that arise due to anisotropy. In general, the function F decreases at least as rapidly as $1/\overline{R}^2$, that is, as rapidly as the spatial curvature term in the FRW cosmology or faster. The false-vacuum energy density appears in the energy density term and is, of course, constant. Provided that F is positive, the Universe will eventually become vacuum energy dominated; once it does, the $F(\dot{\overline{R}}, \overline{R})$ term will quickly decrease and become insignificant and the space-time becomes isotropic(*). This justifies the usual FRW analysis of inflation.

Not all anisotropic space-times will inflate; if F is sufficiently large and negative it will prevent inflation; the simplest noninflating model is a very positively curved FRW model that recollapses before it can inflate. The strongly positively curved models preclude a true cosmological no-hair theorem; however, it has been shown that all spatially homogeneous, but anisotropic models eventually inflate, except for the very positively curved models[62]. And further, it has been shown that «smooth regions» of inhomogeneous models of sufficient size and that are negatively curved will inflate[63, 64]. While not all space-times will inflate, the class of space-times that do is not special, but very generic[64]. Thus inflation does indeed lessen the dependence of the present state of the Universe on its initial state.

Does inflation render a generic space-time isotropic and homogeneous forever? The answer is clearly no; the most one can expect in an inhomogeneous

---

(*) There is one worry: namely that the inflaton field will evolve to the minimum of its potential before the vacuum-dominated phase begins. In general, this does not occur as anisotropy increases the expansion rate, and thus the friction term in the equation of motion for the inflaton (see [61]).

space-time is that negatively curved regions inflate. Further, once inflation is over, inhomogeneity and anisotropy will «grow back». Consider spatial curvature; if the Universe was not flat before inflation, it will not be flat after inflation. However, inflation exponentially postpones the epoch when spatial curvature becomes important because the value of Ω after inflation becomes exponentially close to unity. Likewise, in the exponential distant future our Hubble volume will become larger than the generic smooth patch created by inflation and we will in principle be able to see the inhomogeneity beyond our inflationary patch [65].

Finally, there are the initial data for the scalar field responsible for inflation itself. In first-order inflation, as in old inflation, this is a dynamical issue: the initial value of the scalar field is determined by thermal considerations. However, in slow-rollover inflation the story is very different; the initial value of the inflaton field (and its spatial and temporal derivatives) are not so determined, and at the classical level must be considered to be initial data. While this has become a subject unto itself, some very general statements can be made. First, the inflaton field must be smooth on a scale comparable to the Hubble radius, otherwise the energy density associated with spatial gradients will dominate over the vacuum energy preventing inflation. Second, the value of the scalar field must be small enough in models of «new inflation» or large enough in models of «chaotic inflation» so that it takes the field more than 60 Hubble times to roll to the bottom of the potential. Finally, the initial velocity of the inflaton (*i.e.* $\dot{\phi}$) must be small enough so that it does not rapidly speed to the bottom of the potential. For a given inflationary model, all of these considerations can be studied and quantitative statements made about the necessary initial data for the inflaton field [66]; further, attempts have been made using the wave function of the Universe to quantify the quantum expectation for the initial state of the inflaton field [67].

In the final analysis it cannot be said that all initial space-times undergo inflation and become isotropic and homogeneous for all time; further, the initial data for the inflaton itself must now be considered. The strongest statement that one can make is to say that inflation greatly lessens the dependence of the present state of the Universe upon its initial state. In my mind, that is no mean feat and inflation should be considered a great success.

3. – Inflation: the fundamentals.

In this section I discuss how to analyze an inflationary model, given the scalar potential. In two sections hence I will work through a number of examples. The focus will be on the metric perturbations—density fluctuations [68] and gravity waves [69]—that arise due to quantum fluctuations, and the CBR temperature anisotropies that result from them (*). Perturbations on all astrophysically interesting scales, say 1 Mpc to 10^4 Mpc, are produced during an interval of about 8 e-folds around 50 e-folds before the end of inflation, when these scales crossed outside the horizon during inflation. I will show how the density perturbations and gravity waves can be related to three features of the inflationary potential: its value V_{50}, its steepness $x_{50} \equiv (m_{\text{Pl}} V'/V)_{50}$ and the change in its steepness x'_{50}, evaluated in the region of the potential where the scalar field was about 50 e-folds before the end of inflation. In principle, cosmological observations, most importantly CBR anisotropy, can be used to determine the characteristics of the density perturbations and gravitational waves and thereby V_{50}, x_{50} and x'_{50}.

All viable models of inflation are of the slow-rollover variety, or can be recast as such [71]. In slow-rollover inflation a scalar field that is initially displaced from the minimum of its potential rolls slowly to that minimum, and, as it does, the cosmic-scale factor grows very rapidly. Once the scalar field reaches the minimum of the potential, it oscillates about it, so that the large potential energy has been converted into coherent scalar-field oscillations, corresponding to a condensate of nonrelativistic scalar particles. The eventual decay of these particles into lighter-particle states and their subsequent thermalization lead to the reheating of the Universe to a temperature $T_{\text{RH}} \simeq \sqrt{\Gamma m_{\text{Pl}}}$, where Γ is the decay width of the scalar particle [42, 71]. Here, I will focus on the classical evolution of the inflaton field during the slow-roll phase and the small quantum fluctuations in the inflaton field which give rise to density perturbations and those in the metric which give rise to gravity waves.

To begin, let us assume that the scalar field driving inflation is minimally coupled so that its stress-energy tensor takes the canonical form

(29) $$T_{\mu\nu} = \partial_\mu \phi \partial_\nu \phi - \mathscr{L} g_{\mu\nu},$$

where the Lagrangian density of the scalar field $\mathscr{L} = (1/2) \partial_\mu \phi \partial^\mu \phi - V(\phi)$. If we make the usual assumption that the scalar field ϕ is spatially homogeneous, or at least so over a Hubble radius, the stress-energy tensor takes the perfect-fluid form with energy density, $\rho = (1/2) \dot\phi^2 + V(\phi)$, and isotropic pressure, $p = (1/2) \dot\phi^2 - V(\phi)$. The classical equations of motion for ϕ can be obtained from the first law of thermodynamics, $d(R^3 \rho) = -p \, dR^3$, or by taking the four-divergence of $T^{\mu\nu}$:

(30) $$\ddot\phi + 3H\dot\phi + V'(\phi) = 0;$$

the $\Gamma\dot\phi$ term responsible for reheating has been omitted since we shall only be interested in the slow-rollover phase. In addition, there is the Friedmann equa-

---

(*) Isocurvature perturbations can arise due to quantum fluctuations in other massless fields, e.g., the axion field, if it exists [70].

tion, which governs the expansion of the Universe,

$$(31) \qquad H^2 = \frac{8\pi}{3m_{\text{Pl}}^2}\left(V(\phi) + \frac{1}{2}\dot{\phi}^2\right) \simeq \frac{8\pi V(\phi)}{3m_{\text{Pl}}^2},$$

where we assume that the contribution of all other forms of energy density, e.g., radiation and kinetic energy of the scalar field, and the curvature term (k/R^2) are negligible. The justifications for discussing inflation in the context of a flat FRW model with a homogeneous scalar field driving inflation were discussed earlier (and at greater length in ref. [72]); including the $\dot{\phi}$ kinetic term increases the right-hand side of eq. (31) by a factor of $1 + x^2/48\pi$, a small correction for viable models.

In the next section I will be more precise about the amplitude of density perturbations and gravitational waves; for now, let me briefly discuss how these perturbations arise and give their characteristic amplitudes. The metric perturbations produced in inflationary models are very nearly «scale invariant», a particularly simple spectrum which was first discussed by HARRISON and ZEL'DOVICH [73], and arise due to quantum fluctuations. In de Sitter space all massless scalar fields experience quantum fluctuations of amplitude $H/2\pi$. The graviton is massless and can be described by two massless scalar fields, $h_{+,\times} = \sqrt{16\pi G}\,\phi_{+,\times}$ (+ and × are the two polarization states). The inflaton by virtue of the flatness of its potential for all practical purposes can be treated as massless.

Fluctuations in the inflaton field lead to density fluctuations because of its scalar potential, $\delta\rho \sim HV'$; as a given mode crosses outside the horizon, the density perturbation on that scale becomes a classical metric perturbation. While, outside the horizon, the description of the evolution of a density perturbation is beset with subtleties associated with the gauge freedom in general relativity, there is, however, a simple gauge-invariant quantity, $\zeta \simeq \delta\rho/(\rho + p)$, which remains constant outside the horizon. By equating the value of ζ at post-inflation horizon crossing with its value as the scale crosses outside the horizon it follows that $(\delta\rho/\rho)_{\text{HOR}} \sim HV'/\dot{\phi}^2$ (note: $\rho + p = \dot{\phi}^2$) (see fig. 7).

The evolution of a gravity wave perturbation is even simpler; it obeys the massless Klein-Gordon equation

$$(32) \qquad \ddot{h}_k^i + 3H\dot{h}_k^i + k^2 h_k^i/R^2 = 0,$$

where k is the wave number of the mode and $i = +, \times$. For superhorizon-sized modes, $k \lesssim RH$, the solution is simple: $h_k^i = \text{const}$. Like their density perturbation counterparts, gravity wave perturbations become classical metric perturbations as they cross outside the horizon; they are characterized by an amplitude $h_k^i \simeq \sqrt{16\pi G}(H/2\pi) \sim H/m_{\text{Pl}}$. At post-inflation horizon crossing their amplitude is unchanged.

Finally, let me write the horizon-crossing amplitudes of the scalar and tensor metric perturbations in terms of the inflationary potential,

$$(33) \qquad (\delta\rho/\rho)_{\text{HOR},\lambda} = c_S \left(\frac{V^{3/2}}{m_{\text{Pl}}^3 V'} \right)_1,$$

$$(34) \qquad h_{\text{HOR},\lambda} = c_T \left(\frac{V^{1/2}}{m_{\text{Pl}}^2} \right)_1,$$

where $(\delta\rho/\rho)_{\text{HOR},\lambda}$ is the amplitude of the density perturbation on the scale λ when it crosses the Hubble radius during the post-inflation epoch, $h_{\text{HOR},\lambda}$ is the dimensionless amplitude of the gravitational-wave perturbation on the scale λ when it crosses the Hubble radius, and c_S, c_T are numerical constants of order unity. Subscript 1 indicates that the quantity involving the scalar potential is to be evaluated when the scale in question crossed outside the horizon during the inflationary era. The metric perturbations produced by inflation are characterized by almost scale-invariant horizon-crossing amplitudes; the slight deviations from scale invariance result from the variation of V and V' during inflation which enter through the dependence upon t_1. (In eq. (33) I got ahead of myself and used the slow-roll approximation (see below) to rewrite the expression $(\delta\rho/\rho)_{\text{HOR},\lambda} \simeq HV'/\dot\phi$ in terms of the potential only.)

Equations (30)-(33) are the fundamental equations that govern inflation and the production of metric perturbations. It proves very useful to recast these equations using the scalar field as the independent variable; we then express the scalar and tensor perturbations in terms of the value of the potential, its steepness, and the rate of change of its steepness when the interesting scales crossed outside the Hubble radius during inflation, about 50 e-folds in scale factor before the end of inflation, defined by

$$V_{50} \equiv V(\phi_{50}), \qquad x_{50} \equiv \frac{m_{\text{Pl}} V'(\phi_{50})}{V(\phi_{50})}, \qquad x_{50}' = \frac{m_{\text{Pl}} V''(\phi_{50})}{V(\phi_{50})} - \frac{m_{\text{Pl}} [V'(\phi_{50})]^2}{V^2(\phi_{50})}.$$

To evaluate these three quantities 50 e-folds before the end of inflation we must find the value of the scalar field at this time. During the inflationary phase the $\ddot\phi$ term is negligible (the motion of ϕ is friction dominated), and eq. (30) becomes

$$(35) \qquad \dot\phi \simeq \frac{-V'(\phi)}{3H};$$

this is known as the slow-roll approximation [74]. While the slow-roll approximation is almost universally applicable, there are models where the slow-roll approximation cannot be used; e.g., a potential where during the crucial 8 e-folds the scalar field rolls uphill, «powered» by the velocity it had when it hit the incline.

The conditions that must be satisfied in order that $\ddot\phi$ be negligible are

(36) $$|V''| < 9H^2 \simeq 24\pi V/m_{Pl}^2\,,$$

(37) $$|x| \equiv |V' m_{Pl}/V| < \sqrt{48\pi}\,.$$

The end of the slow roll occurs when either or both of these inequalities are saturated, at a value of ϕ denoted by ϕ_{end}. Since $H \equiv \dot R/R$, or $H\,dt = d\ln R$, it follows that

(38) $$d\ln R = \frac{8\pi}{m_{Pl}^2} \frac{V(\phi)\,d\phi}{-V'(\phi)} = -\frac{8\pi\,d\phi}{m_{Pl}\,x}\,.$$

Now express the cosmic-scale factor in terms of its value at the end of inflation, R_{end}, and the number of e-foldings before the end of inflation, $N(\phi)$,

$$R = \exp[-N(\phi)]R_{end}\,.$$

The quantity $N(\phi)$ is a timelike variable whose value at the end of inflation is zero and whose evolution is governed by

(39) $$\frac{dN}{d\phi} = \frac{8\pi}{m_{Pl}\,x}\,.$$

Using eq. (39) we can compute the value of the scalar field 50 e-folds before the end of inflation ($\equiv \phi_{50}$); the values of V_{50}, x_{50} and x'_{50} follow directly.

As ϕ rolls down its potential during inflation, its energy density decreases, and so the growth in the scale factor is not exponential. By using the fact that the stress-energy of the scalar field takes the perfect-fluid form, we can solve for evolution of the cosmic-scale factor. Recall, for the equation of state $p = \gamma\rho$, the scale factor grows as $R \propto t^q$, where $q = 2/3(1+\gamma)$. Here,

(40) $$\gamma = \frac{(1/2)\dot\phi^2 - V}{(1/2)\dot\phi^2 + V} = \frac{x^2 - 48\pi}{x^2 + 48\pi}\,,$$

(41) $$q = \frac{1}{3} + \frac{16\pi}{x^2}\,.$$

Since the steepness of the potential can change during inflation, γ is not in general constant; the power law index q is more precisely the logarithmic rate of the change of the logarithm of the scale factor, $q = d\ln R/d\ln t$.

When the steepness parameter is small, corresponding to a very flat potential, γ is close to -1 and the scale factor grows as a very large power of time. To solve the horizon problem the expansion must be «superluminal» ($\ddot R > 0$), corresponding to $q > 1$, which requires that $x^2 < 24\pi$. Since $(1/2)\dot\phi^2/V = x^2/48\pi$, this implies that $(1/2)\dot\phi^2/V(\phi) < 1/2$, justifying neglect of the scalar-field kinetic energy in computing the expansion rate for all but the steepest potentials. (In fact there are much stronger constraints;

the COBE DMR data imply that $n \geq 0.5$, which restricts $x_{50}^2 \leq 4\pi$, $(1/2)\dot\phi^2/V \leq 1/12$, and $q \geq 4$.)

Next, let us relate the size of a given scale to when that scale crosses outside the Hubble radius during inflation, specified by $N_1(\lambda)$, the number of e-folds before the end of inflation. The physical size of a perturbation is related to its comoving size, $\lambda_{phys} = R\lambda$; with the usual convention, $R_{today} = 1$, the comoving size is the physical size today. When the scale λ crosses outside the Hubble radius $R_1 \lambda = H_1^{-1}$. We then assume that 1) at the end of inflation the energy density is $\mathcal{M}^4 \simeq V(\phi_{end})$, 2) inflation is followed by a period where the energy density of the Universe is dominated by coherent scalar-field oscillations which decrease as R^{-3}, and 3), when value of the scale factor is R_{RH}, the Universe reheats to a temperature $T_{RH} \simeq \sqrt{m_{Pl}\Gamma}$ and expands adiabatically thereafter. The «matching equation» that relates λ and $N_1(\lambda)$ is

$$(42) \qquad \lambda = \frac{R_{today}}{R_1} H_1^{-1} = \frac{R_{today}}{R_{RH}} \frac{R_{RH}}{R_{end}} \frac{R_{end}}{R_1} H_1^{-1}.$$

Adiabatic expansion since reheating implies $R_{today}/R_{RH} \simeq T_{RH}/2.73$ K; and the decay of the coherent scalar-field oscillations implies $(R_{RH}/R_{end})^3 = (\mathcal{M}/T_{RH})^4$. If we define $\bar q = \ln(R_{end}/R_1)/\ln(t_{end}/t_1)$, the mean power law index, it follows that $(R_{end}/R_1) H_1^{-1} = \exp[N_1(\bar q - 1)/\bar q] H_{end}^{-1}$, and eq. (42) becomes

$$(43) \quad N_1(\lambda) =$$

$$= \frac{\bar q}{\bar q - 1}\left[48 + \ln\lambda_{Mpc} + \frac{2}{3}\ln(\mathcal{M}/10^{14}\text{ GeV}) + \frac{1}{3}\ln(T_{RH}/10^{14}\text{ GeV})\right].$$

In the case of perfect reheating, which probably only applies to first-order inflation, $T_{RH} \simeq \mathcal{M}$.

The scales of astrophysical interest today range roughly from that of galaxy size, $\lambda \sim$ Mpc, to the present Hubble scale, $H_0^{-1} \sim 10^4$ Mpc; up to the logarithmic corrections these scales crossed outside the horizon between about $N_1(\lambda) \sim 48$ and $N_1(\lambda) \simeq 56$ e-folds before the end of inflation. *That is, the interval of inflation that determines its all observable consequences covers only about 8 e-folds.*

Except in the case of strict power law inflation q varies during inflation; this means that the $(R_{end}/R_1)H_1^{-1}$ factor in eq. (42) cannot be written in closed form. Taking account of this, the matching equation becomes a differential equation,

$$(44) \qquad \frac{d\ln\lambda_{Mpc}}{dN_1} = \frac{q(N_1) - 1}{q(N_1)},$$

subject to the «boundary condition»

$$\ln \lambda_{\text{Mpc}} = -48 - \frac{4}{3} \ln(\mathscr{M}/10^{14} \text{ GeV}) + \frac{1}{3} \ln(T_{\text{RH}}/10^{14} \text{ GeV})$$

for $N_1 = 0$, the matching relation for the mode that crossed outside the Hubble radius at the end of inflation. Equation (44) allows one to obtain the precise expression for when a given scale crossed outside the Hubble radius during inflation. To actually solve this equation, one would need to supplement it with the expressions $dN/d\phi = 8\pi/m_{\text{Pl}} x$ and $q = 16\pi/x^2$. For our purposes we need only know: 1) The scales of astrophysical interest corresponding to $N_1 \sim$ «50 ± 4», where for definiteness we will throughout take this to be an equality sign. 2) The expansion of eq. (44) about $N_1 = 50$,

$$(45) \qquad \Delta N_1(\lambda) = \frac{q_{50} - 1}{q_{50}} \Delta \ln \lambda_{\text{Mpc}},$$

which, with the aid of eq. (39), implies that

$$(46) \qquad \Delta \phi = \frac{q_{50} - 1}{q_{50}} \frac{x_{50}}{8\pi} \Delta \lambda_{\text{Mpc}}.$$

We are now ready to express the perturbations in terms of V_{50}, x_{50} and x'_{50}. First, we must solve for the value of ϕ, 50 e-folds before the end of inflation. To do so we use eq. (39),

$$(47) \qquad N(\phi_{50}) = 50 = \frac{8\pi}{m_{\text{Pl}}^2} \int_{\phi_{\text{end}}}^{\phi_{50}} \frac{V \, d\phi}{V'}.$$

Next, with the help of eq. (46) we expand the potential V and its steepness x about ϕ_{50}:

$$(48) \qquad V \simeq V_{50} + V'_{50}(\phi - \phi_{50}) = V_{50}\left[1 + \frac{x_{50}^2}{8\pi} \frac{q_{50}}{q_{50} - 1} \Delta \ln \lambda_{\text{Mpc}}\right],$$

$$(49) \qquad x \simeq x_{50} + x'_{50}(\phi - \phi_{50}) = x_{50}\left[1 + \frac{m_{\text{Pl}} x'_{50}}{8\pi} \frac{q_{50}}{q_{50} - 1} \Delta \ln \lambda_{\text{Mpc}}\right];$$

of course, these expansions only make sense for potentials that are smooth. We note that additional terms in either expansion are $\mathcal{O}(\alpha_i^2)$ and beyond the accuracy we are seeking.

Now recall the equations for the amplitude of the scalar and tensor perturbations,

$$(50) \qquad (\delta \rho/\rho)_{\text{HOR}, \lambda} = c_{\text{S}} \left(\frac{V^{1/2}}{m_{\text{Pl}}^2 x}\right)_1,$$

$$(51) \qquad h_{\text{HOR}, \lambda} = c_{\text{T}} \left(\frac{V^{1/2}}{m_{\text{Pl}}^2}\right)_1,$$

where subscript 1 means that the quantities are to be evaluated where the scale λ crossed outside the Hubble radius, $N_1(\lambda)$ e-folds before the end of inflation. The origin of any deviation from scale invariance is clear: For tensor perturbations it arises due to the variation of the potential, and for scalar perturbations it arises due to the variation of both the potential and its steepness.

Using eqs. (45)-(50) it is now simple to calculate the power law exponents α_S and α_T that quantify the deviations from scale invariance,

$$\alpha_T = \frac{x_{50}^2}{16\pi} \frac{q_{50}}{q_{50}-1} \simeq \frac{x_{50}^2}{16\pi}, \tag{52}$$

$$\alpha_S = \alpha_T - \frac{m_{Pl} x_{50}'}{8\pi} \frac{q_{50}}{q_{50}-1} \simeq \frac{x_{50}^2}{16\pi} - \frac{m_{Pl} x_{50}'}{8\pi}, \tag{53}$$

where

$$q_{50} = \frac{1}{3} + \frac{16\pi}{x_{50}^2} \simeq \frac{16\pi}{x_{50}^2}, \tag{54}$$

$$h_{\text{HOR},\lambda} = c_T \left(\frac{V_{50}^{1/2}}{m_{Pl}^2}\right) \lambda_{\text{Mpc}}^{\alpha_T}, \tag{55}$$

$$(\delta\rho/\rho)_{\text{HOR},\lambda} = c_S \left(\frac{V_{50}^{1/2}}{x_{50} m_{Pl}^2}\right) \lambda_{\text{Mpc}}^{\alpha_S}. \tag{56}$$

The spectral indices α_i are defined as $\alpha_S = [d \ln(\delta\rho/\rho)_{\text{HOR},\lambda}/d \ln \lambda_{\text{Mpc}}]_{50}$ and $\alpha_T = [d \ln h_{\text{HOR},\lambda}/d \ln \lambda_{\text{Mpc}}]_{50}$, and in general vary slowly with scale. Note too that the deviations from scale invariance, quantified by α_S and α_T, are of the order of x_{50}^2, $m_{Pl} x_{50}'$. In the expressions above we retained only lowest-order terms in $\mathcal{O}(\alpha_i)$. The next-order contributions to the spectral indices are $\mathcal{O}(\alpha_i^2)$; those to the amplitudes are $\mathcal{O}(\alpha_i)$ and are given two sections hence. The justification for truncating the expansion at lowest order is that the deviations from scale invariance are expected to be small—and are required by astrophysical data to be small.

As I discuss in more detail two sections hence, the more intuitive power law indices α_S, α_T are related to the indices that are usually used to describe the power spectra of scalar and tensor perturbations, $P_S(k) = |\delta_k|^2 = Ak^n$ and $P_T(k) = |h_k|^2 = A_T k^{n_T - 3}$,

$$n = 1 - 2\alpha_S = 1 - \frac{x_{50}^2}{8\pi} + \frac{m_{Pl} x_{50}'}{4\pi}, \tag{57}$$

$$n_T = -2\alpha_T = -\frac{x_{50}^2}{8\pi}. \tag{58}$$

CBR temperature fluctuations on large angular scales ($\theta \gtrsim 1°$) due to metric perturbations arise through the Sachs-Wolfe effect; very roughly, the temperature fluctuation on a given angular scale θ is related to the metric fluctuation on the length scale that subtends that angle at last scattering, $\lambda \sim 100\,h^{-1}\,\mathrm{Mpc} \cdot (\theta/\mathrm{degree})$,

$$\left(\frac{\delta T}{T}\right)_\theta \sim \left(\frac{\delta \rho}{\rho}\right)_{\mathrm{HOR},\lambda}, \tag{59}$$

$$\left(\frac{\delta T}{T}\right)_\theta \sim h_{\mathrm{HOR},\lambda}, \tag{60}$$

where the scalar and tensor contributions to the CBR temperature anisotropy on a given scale add in quadrature. Let me be more specific about the amplitude of the quadrupole CBR anisotropy. For small α_S, α_T the contributions of each to the quadrupole CBR temperature anisotropy:

$$\left(\frac{\Delta T}{T_0}\right)^2_{\mathrm{Q\text{-}S}} \approx \frac{32\pi}{45} \frac{V_{50}}{m_{\mathrm{Pl}}^4 x_{50}^2}, \tag{61}$$

$$\left(\frac{\Delta T}{T_0}\right)^2_{\mathrm{Q\text{-}T}} \approx 0.61 \frac{V_{50}}{m_{\mathrm{Pl}}^4}, \tag{62}$$

$$\frac{T}{S} \equiv \frac{(\Delta T/T_0)^2_{\mathrm{Q\text{-}T}}}{(\Delta T/T_0)^2_{\mathrm{Q\text{-}S}}} \approx 0.28\, x_{50}^2, \tag{63}$$

where expressions have been evaluated to lowest order in x_{50}^2 and $m_{\mathrm{Pl}} x'_{50}$ [75]. These quantities represent the ensemble averages of the scalar and tensor contributions to the quadrupole temperature anisotropy, which in terms of the spherical-harmonic expansion of the CBR temperature anisotropy on the sky are given by $5\langle |a_{2m}|^2\rangle/4\pi$. Further, the scalar and tensor contributions to the *measured* quadrupole anisotropy add in quadrature, and are subject to «cosmic variance». (Cosmic variance refers to the dispersion in the values measured by different observers in the Universe.)

Before going on, some general remarks [76]. The steepness parameter x_{50}^2 must be less than about 24π to ensure superluminal expansion. For «steep» potentials, the expansion rate is «slow», *i.e.* q_{50} closer to unity, the gravity wave contribution to the quadrupole CBR temperature anisotropy becomes comparable to, or greater than, that of density perturbations, and both scalar and tensor perturbations exhibit significant deviations from scale invariance. For «flat» potentials, *i.e.* small x_{50}, the expansion rate is «fast», *i.e.* $q_{50} \gg 1$, the gravity wave contribution to the quadrupole CBR temperature anisotropy is much smaller than that of density perturbations, and the tensor perturbations are scale invariant. Unless the steepness of the potential changes rapidly, *i.e.* large x'_{50}, the scalar perturbations are also scale invariant.

3`1. Metric perturbations and CBR anisotropy. – I was purposefully vague when discussing the amplitudes of the scalar and tensor modes, except when specifying their contributions to the quadrupole CBR temperature anisotropy; in fact, the spectral indices α_S and α_T, together with the scalar and tensor contributions to the CBR quadrupole, serve to provide all the information necessary. Here I will fill in more details about the metric perturbations.

The scalar and tensor metric perturbations are expanded in harmonic functions, in the flat Universe predicted by inflation, plane waves,

$$(64) \qquad h_{\mu\nu}(\mathbf{x}, t) = \frac{1}{(2\pi)^3} \int d^3k \, h_k^i(t) \, \varepsilon_{\mu\nu}^i \exp[-i\mathbf{k} \cdot \mathbf{x}],$$

$$(65) \qquad \frac{\delta\rho(\mathbf{x}, t)}{\rho} = \frac{1}{(2\pi)^3} \int d^3k \, \delta_k(t) \exp[-i\mathbf{k} \cdot \mathbf{x}],$$

where $h_{\mu\nu} = R^{-2} g_{\mu\nu} - \eta_{\mu\nu}$, $\varepsilon_{\mu\nu}^i$ is the polarization tensor for the gravity wave modes, and $i = +, \times$ are the two polarization states. Everything of interest can be computed in terms of h_k^i and δ_k. For example, the r.m.s. mass fluctuation in a sphere of radius r is obtained in terms of the window function for a sphere and the power spectrum $P_S(k) \equiv \langle |\delta_k|^2 \rangle$ (see below),

$$(66) \qquad \langle (\delta M/M)^2 \rangle_r = \frac{9}{2\pi^2 r^2} \int_0^\infty [j_1(kr)]^2 \, P_S(k) \, dk,$$

where $j_1(x)$ is the spherical Bessel function of first order. If $P_S(k)$ is a power law, it follows roughly that $(\delta M/M)^2 \sim k^3 |\delta_k|^2$, evaluated on the scale $k = r^{-1}$. This is what I meant by $(\delta\rho/\rho)_{\text{HOR}, \lambda}$: the r.m.s. mass fluctuation on the scale λ when it crossed inside the horizon. Likewise, by $h_{\text{HOR}, \lambda}$ I meant the r.m.s. strain on the scale λ as it crossed inside the Hubble radius, $(h_{\text{HOR}, \lambda})^2 \sim$ $\sim k^3 |h_k^i|^2$.

In the previous discussions I have chosen to specify the metric perturbations for the different Fourier modes when they crossed inside the horizon, rather than at a common time. I did so because scale invariance is made manifest, as the scale independence of the metric perturbations at post-inflation horizon crossing. Recall, in the case of scalar perturbations $(\delta\rho/\rho)_{\text{HOR}}$ is up to a numerical factor the fluctuation in the Newtonian potential, and, by specifying the scalar perturbations at horizon crossing, we avoid the discussion of scalar perturbations on superhorizon scales, which is beset by the subtleties associated with the gauge noninvariance of δ_k.

It is, however, necessary to specify the perturbations at a common time to carry out most calculations; e.g., an N-body simulation of structure formation or the calculation of CBR anisotropy. To do so, one has to take account of the evolution of the perturbations after they enter the horizon. After entering the horizon tensor perturbations behave like gravitons, with h_k decreasing as R^{-1} and the energy density associated with a given mode, $\rho_k \sim m_{\rm Pl}^2 k^5 |h_k|^2/R^2$, decreasing as R^{-4}. The evolution of scalar perturbations is slightly more complicated; modes that enter the horizon while the Universe is still radiation dominated remain essentially constant until the Universe becomes matter dominated (growing only logarithmically); modes that enter the horizon after the Universe becomes matter dominated grow as the scale factor. (The gauge noninvariance of δ_k is not an important issue for subhorizon-size modes; here a Newtonian analysis suffices, and there is only one growing mode, corresponding to a density perturbation.)

The method for characterizing the scalar perturbations is by now standard: The spectrum of perturbations is specified at the present epoch (assuming linear growth for all scales); the spectrum at earlier epochs can be obtained by multiplying δ_k by $R(t)/R_{\rm today}$. The inflationary metric perturbations are Gaussian; thus δ_k is a Gaussian, random variable. Its statistical expectation value is

(67) $$\langle \delta_k \delta_q \rangle = P_{\rm S}(k)(2\pi)^3 \delta^{(3)}(\boldsymbol{k} - \boldsymbol{q}),$$

where the power spectrum today is written as

(68) $$P_{\rm S}(k) \equiv A k^n T(k)^2 \,;$$

$n = 1 - 2\alpha_{\rm S}$ ($= 1$ for scale-invariant perturbations), and $T(k)$ is the «transfer function» which encodes the information about the post-horizon-crossing evolution of each mode and depends upon the matter content of the Universe, e.g., baryons plus cold dark matter, baryons plus hot dark matter, baryons plus hot and cold dark matter, and so on. The transfer function is defined so that $T(k) \to 1$ for $k \to 0$ (long-wavelength perturbations); an analytic approximation to the cold-dark-matter transfer function is given by [77]

(69) $$T(k) = \frac{\ln(1 + 2.34\,q)/2.34\,q}{[1 + (3.89\,q) + (16.1\,q)^2 + (5.46\,q)^3 + (6.71\,q)^4]^{1/4}},$$

where $q = k/(\Omega_0 h^2 \,{\rm Mpc}^{-1})$. Inflationary power spectra for different dark-matter possibilities are shown in fig. 9.

The overall normalization factor

(70) $$A = \frac{1024\,\pi^3}{75\,H_0^4} \frac{V_{50}}{m_{\rm Pl}^4 x_{50}^2} \times$$

$$\times \frac{[1 + (7/6)\,n_{\rm T} - (1/3)(n-1)]\{\Gamma[3/2 - (1/2)(n-1)]\}^2}{2^{n-1}[\Gamma(3/2)]^2} k_{50}^{1-n},$$

where the $\mathcal{O}(\alpha_i)$ correction to A has been included [78]. The quantity $n_{\rm T} = -2\alpha_{\rm T} = -x_{50}^2/8\pi$, $n - 1 = -2\alpha_{\rm S} = n_{\rm T} + x'_{50}/4\pi$, k_{50} is the comoving wave number of the scale that crossed outside the horizon 50 e-folds before the end of inflation. All the formulae below simplify if this scale corresponds to the

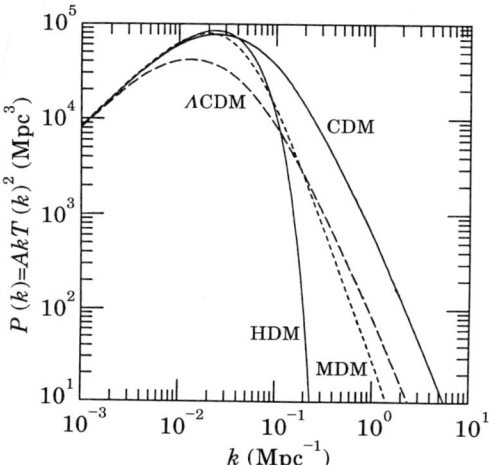

Fig. 9. – Power spectra for cold dark matter (CDM), hot dark matter (HDM), mixed dark matter (MDM = 30% hot + 70% cold) and cold dark matter with a cosmological constant (ΛCDM = 20% CDM + 80% Λ). All spectra are normalized to the COBE DMR quadrupole temperature anisotropy; $h = 0.5$ for all models except ΛCDM ($h = 0.8$).

present horizon scale, specifically, $k_{50} = H_0/2$. (Equation (70) can be simplified by expanding $\Gamma(3/2 + x) = \Gamma(3/2)[1 + x(2 - 2 \ln 2 - \gamma)]$, valid for $|x| \ll 1$; $\gamma \simeq$ $\simeq 0.577$ is Euler's constant.)

From this expression it is simple to compute the Sachs-Wolfe contribution of scalar perturbations to the CBR temperature anisotropy; on angular scales much greater than about 1° (corresponding to multipoles $l \ll 100$) it is the dominant contribution. If we expand the CBR temperature on the sky in spherical harmonics,

$$(71) \qquad \frac{\delta T(\theta, \phi)}{T_0} = \sum_{l \geq 2, m = -l}^{l = \infty, m = l} a_{lm} Y_{lm}(\theta, \phi),$$

where $T_0 = 2.73$ K is the CBR temperature today, then the ensemble expectation for the multipole coefficients is given by

$$(72) \qquad \langle |a_{lm}|^2 \rangle \simeq \frac{H_0^4}{2\pi} \int_0^\infty k^{-2} P_S(k) |j_l(kr_0)|^2 \, dk,$$

$$(73) \qquad \langle |a_{lm}|^2 \rangle \simeq \frac{A 2^{n-1} H_0^4 r_0^{1-n}}{16} \frac{\Gamma(l + (1/2)n - 1/2) \Gamma(3 - n)}{\Gamma(l - (1/2)n + 5/2)[\Gamma(2 - (1/2)n)]^2},$$

where $r_0 \approx 2H_0^{-1}$ is the comoving distance to the last-scattering surface, and this expression is for the Sachs-Wolfe contribution from scalar perturbations only. For n not too different from one, the ensemble expectation for the quadrupole CBR temperature anisotropy is

$$(74) \qquad \left(\frac{\Delta T}{T_0}\right)^2_{\text{Q-S}} \equiv \frac{5\,|a_{2m}|^2}{4\pi} \approx \frac{32\pi}{45}\,\frac{V_{50}}{m_{\text{Pl}}^4 x_{50}^2}\,(k_{50}r_0)^{1-n}.$$

(By choosing $k_{50} = r_0^{-1} = (1/2)H_0$, the last factor becomes unity.)

The ensemble expectation values for the multipole amplitudes are often referred to as the angular power spectrum. Further, the r.m.s. temperature fluctuation on a given angular scale is related to the multipole amplitudes

$$(75) \qquad \left(\frac{\Delta T}{T}\right)^2_\theta \sim l^2 \langle |a_{lm}|^2 \rangle \qquad \text{for } l \simeq 200°/\theta.$$

The procedure for specifying the tensor modes is similar, cf. ref.[79, 80]. For the modes that enter the horizon after the Universe becomes matter dominated, $k \lesssim 0.1\,h^2\,\text{Mpc}$, which are the only modes that contribute significantly to CBR anisotropy on angular scales greater than a degree,

$$(76) \qquad h_k^i(\tau) = a^i(\mathbf{k})\left(\frac{3j_1(k\tau)}{k\tau}\right),$$

where $\tau = r_0(t/t_0)^{1/3}$ is conformal time. (For the modes that enter the horizon during the radiation-dominated era, $k \gtrsim 0.1\,h^2\,\text{Mpc}^{-1}$, the factor $3j_1(k\tau)/k\tau$ is replaced by $j_0(k\tau)$ for the remainder of the radiation era. In either case, the factor involving the spherical Bessel function quantifies the fact that tensor perturbations remain constant while outside the horizon, and after horizon crossing decrease as R^{-1}.)

The tensor perturbations too are characterized by a Gaussian, random variable, here written as $a^i(\mathbf{k})$; the statistical expectation

$$(77) \qquad \langle h_k^i h_q^j \rangle = P_T(k)(2\pi)^6\,\delta^{(3)}(\mathbf{k} - \mathbf{q})\,\delta_{ij},$$

where the power spectrum

$$(78) \qquad P_T(k) = A_T k^{n_T - 3}\left[\frac{3j_1(k\tau)}{k\tau}\right]^2,$$

$$(79) \qquad A_T = \frac{8}{3\pi}\,\frac{V_{50}}{m_{\text{Pl}}^4}\,\frac{(1 + (5/6)n_T)[\Gamma(3/2 - (1/2)n_T)]^2}{2^{n_T}[\Gamma(3/2)]^2}\,k_{50}^{-n_T},$$

where the $\mathcal{O}(\alpha_i)$ correction to A_T has been included. Note that $n_T = -2\alpha_T$ is zero for scale-invariant perturbations.

Finally, the contribution of tensor perturbations to the multipole amplitudes, which arise solely due to the Sachs-Wolfe effect [28, 79, 80], is given by

$$\langle |a_{lm}|^2 \rangle \simeq 36\pi^2 \frac{\Gamma(l+3)}{\Gamma(l-1)} \int_0^\infty k^{n_{\rm T}+1} A_{\rm T} |F_l(k)|^2 \, dk, \tag{80}$$

where

$$F_l(k) = -\int_{r_{\rm D}}^{r_0} dr \frac{j_2(kr)}{kr} \frac{j_l(kr_0 - kr)}{(kr_0 - kr)^2}, \tag{81}$$

and $r_{\rm D} = r_0/(1+z_{\rm D})^{1/2} \simeq r_0/35$ is the comoving distance to the horizon at decoupling (= conformal time at decoupling). Equation (80) is approximate in that very-short-wavelength modes, $kr_0 \gg 100$, that crossed inside the horizon before matter-radiation equality have not been properly taken into account; to take them into account, the integrand must be multiplied by a transfer function,

$$T(k) \simeq 1.0 + 1.44\,(k/k_{\rm EQ}) + 2.54\,(k/k_{\rm EQ})^2, \tag{82}$$

where $k_{\rm EQ} \equiv H_0/(2\sqrt{2}-2)\,R_{\rm EQ}^{1/2}$ is the scale that entered the horizon at matter-radiation equality [76]. In addition, for $l \gtrsim 1000$, the finite thickness of the last-scattering surface must be taken into account.

The tensor contribution to the quadrupole CBR temperature anisotropy for $n_{\rm T}$ not too different from zero is

$$\left(\frac{\Delta T}{T_0}\right)^2_{\rm Q-T} \equiv \frac{5|a_{2m}|^2}{4\pi} \simeq 0.61 \frac{V_{50}}{m_{\rm Pl}^4} (k_{50}\,r_0)^{-n_{\rm T}}, \tag{83}$$

where the integrals in the previous expressions have been evaluated numerically.

Both the scalar and tensor contributions to a given multipole are dominated by wave numbers $kr_0 \sim l$. For scale-invariant perturbations and small l, both the scalar and tensor contributions to $(l+1/2)^2 \langle |a_{lm}|^2 \rangle$ are approximately constant. The contribution of scalar perturbations to $(l+1/2)^2 \langle |a_{lm}|^2 \rangle$ begins to decrease for $l \sim 150$ because the scalar contribution to these multipoles is dominated by modes that entered the horizon before matter domination (and hence are suppressed by the transfer function). The contribution of tensor modes to $(l+1/2)^2 \langle |a_{lm}|^2 \rangle$ begins to decrease for $l \sim 30$ because the tensor contribution to these multipoles is dominated by modes that entered the horizon before decoupling (and hence decayed as R^{-1} until decoupling). Figure 10 shows the contribution of scalar and tensor perturbations to the CBR anisotropy multipole amplitudes (and includes both the tensor and scalar transfer functions); the expected variance in the CBR multipoles is given by the sum of the scalar and tensor contributions.

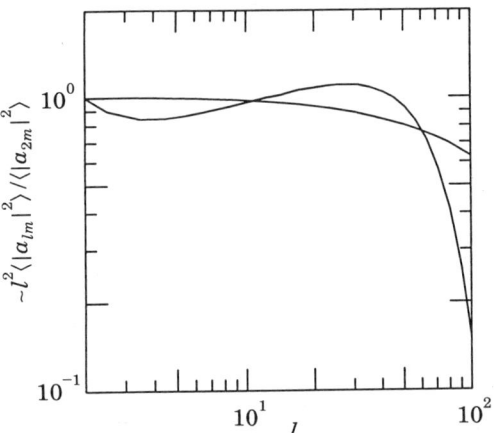

Fig. 10. – Scalar and tensor contributions to the CBR multipole moments: $l(l+1) \cdot \langle |a_{lm}|^2 \rangle / 6 \langle |a_{2m}|^2 \rangle$ for the scalar and $l(l+1/2) \langle |a_{lm}|^2 \rangle / 5 \langle |a_{2m}|^2 \rangle$ for the tensor. The tensor contribution begins to fall off for $l \sim 30$; here $n - 1 = n_T = 0$, $z_{\text{DEC}} = 1000$ and $h = 0.5$ (from [81]).

3˙2. *Worked examples.* – In this section I apply the formalism developed in the two previous sections to four specific models. So that I can, where appropriate, solve numerically for model parameters, I will 1) assume that the astrophysically interesting scales crossed outside the horizon 50 e-folds before the end of inflation; and 2) use the COBE DMR quadrupole measurement, $\langle (\Delta T)_Q^2 \rangle^{1/2} \approx (16 \pm 2)\,\mu\text{K}$ [14], to normalize the scalar perturbations; using eq. (61) this implies

(84) $$V_{50} \approx 1.6 \cdot 10^{-11} \, m_{\text{Pl}}^4 x_{50}^2 \,.$$

Of course, it is entirely possible that a significant portion of the quadrupole anisotropy is due to tensor-mode perturbations, in which case this normalization must be reduced by a factor of $(1 + T/S)^{-1}$. And it is straightforward to change «50» to the number appropriate to a specific model, or to normalize the perturbations another way.

Before going on let us use the COBE DMR quadrupole anisotropy to bound the tensor contribution to the quadrupole anisotropy and thereby the energy density that drives inflation:

(85) $$V_{50} \lesssim 6 \cdot 10^{-11} \, m_{\text{Pl}}^4 \,.$$

Thus the tensor contribution to the CBR quadrupole implies that the vacuum energy that drives inflation must be much less than the Planck energy density, strongly suggesting that inflation is not a quantum-gravitational phenomenon.

3˙2.1. Exponential potentials. There are a class of models that can be described in terms of an exponential potential,

(86) $$V(\phi) = V_0 \exp[-\beta\phi/m_{Pl}].$$

This type of potential was first invoked in the context of power law inflation [82], and has recently received renewed interest in the context of extended inflation [44]. In the simplest model of extended, or first-order, inflation, that based upon the Brans-Dicke-Jordan theory of gravity [44], β is related to the Brans-Dicke parameter: $\beta^2 = 64\pi/(2\omega + 3)$.

For such a potential the slow-roll conditions are satisfied provided that $\beta^2 \lesssim 24\pi$; thus inflation does not end until the potential changes shape, or, in the case of extended inflation, until the phase transition takes place. In either case we can relate ϕ_{50} to ϕ_{end},

(87) $$N(\phi_{50}) = 50 = \frac{8\pi}{m_{Pl}^2} \int_{\phi_{50}}^{\phi_{end}} \frac{V d\phi}{-V'} \Rightarrow \phi_{50} = \phi_{end} - 50\beta/8\pi.$$

Since ϕ_{end} is in effect arbitrary, the overall normalization of the potential is irrelevant. The two other parameters, x_{50} and x'_{50}, are easy to compute:

(88) $$x_{50} = -\beta, \qquad x'_{50} = 0.$$

Using the COBE DMR normalization, we can relate V_{50} and β:

(89) $$V_{50} = 1.6 \cdot 10^{-11} m_{Pl}^4 \beta^2.$$

Further, we can compute q, α_S, α_T and T/S:

(90) $$q = 16\pi/\beta^2, \qquad T/S = 0.28\beta^2, \qquad \alpha_T = \alpha_S = 1/(q-1) \simeq \beta^2/16\pi.$$

Note, for the exponential potential, q, $\alpha_T = \alpha_S$ are independent of epoch. In the case of extended inflation, $\alpha_S = \alpha_T = 4/(2\omega + 3)$; since ω must be less than about 20 [47], this implies significant tilt: $\alpha_S = \alpha_T \gtrsim 0.1$.

3˙2.2. Chaotic inflation. The simplest chaotic-inflation models are based upon potentials of the form

(91) $$V(\phi) = a\phi^b;$$

$b = 4$ corresponds to Linde's original model of chaotic inflation and a is dimensionless [54], and $b = 2$ is a model based upon a massive scalar field and $m^2 = 2a$ [83]. In these models ϕ is initially displaced from $\phi = 0$, and inflation occurs as ϕ slowly rolls to the origin. The value of ϕ_{end} is easily found: $\phi_{end}^2 = b(b - $

$-1) m_{\text{Pl}}^2/24\pi$, and

$$N(\phi_{50}) = 50 = \frac{8\pi}{m_{\text{Pl}}^2} \int_{\phi_{\text{end}}}^{\phi_{50}} \frac{V d\phi}{V'} ; \tag{92}$$

$$\Rightarrow \phi_{50}^2/m_{\text{Pl}}^2 = 50b/4\pi + b^2/48\pi \simeq 50b/4\pi ; \tag{93}$$

the value of ϕ_{50} is a few times the Planck mass.

For purposes of illustration consider $b = 4$; $\phi_{\text{end}} = m_{\text{Pl}}/\sqrt{2\pi} \simeq 0.4 m_{\text{Pl}}$, $\phi_{50} \simeq$ $\simeq 4 m_{\text{Pl}}$, $\phi_{46} \simeq 3.84 m_{\text{Pl}}$, and $\phi_{54} \simeq 4.16 m_{\text{Pl}}$. In order to have sufficient inflation, the initial value of ϕ must exceed about $4.2 m_{\text{Pl}}$; inflation ends when $\phi \approx 0.4 m_{\text{Pl}}$; and the scales of astrophysical interest cross outside the horizon over an interval $\Delta\phi \simeq 0.3 m_{\text{Pl}}$.

The values of the potential, its steepness and the change in steepness are easily found:

$$V_{50} = a m_{\text{Pl}}^b \left(\frac{50b}{4\pi}\right)^{b/2}, \quad x_{50} = \sqrt{\frac{4\pi b}{50}}, \quad m_{\text{Pl}} x_{50}' = \frac{-4\pi}{50}, \tag{94}$$

$$q_{50} = 200/b, \quad T/S = 0.07 b, \quad \alpha_T \simeq b/200, \quad \alpha_S = \alpha_T + 0.01. \tag{95}$$

Unless b is very large, scalar perturbations dominate tensor perturbations [84], α_T, α_S are very small, and q is very large. Further, when α_T, α_S become significant, they are equal. Using the COBE DMR normalization we find

$$a = 1.6 \cdot 10^{-11} b^{1-b/2} (4\pi/50)^{b/2+1} m_{\text{Pl}}^{4-b}. \tag{96}$$

For the two special cases of interest: $b = 4$, $a = 6.4 \cdot 10^{-14}$, and $b = 2$, $m^2 \equiv 2a = 2.0 \cdot 10^{-12} m_{\text{Pl}}^2$.

3'2.3. New inflation. These models entail a very flat potential where the scalar field rolls from $\phi \approx 0$ to the minimum of the potential at $\phi = \sigma$. The original models of slow-rollover inflation [85] were based upon potentials of the Coleman-Weinberg form:

$$V(\phi) = B\sigma^4/2 + B\phi^4 \left[\ln(\phi^2/\sigma^2) - \frac{1}{2}\right], \tag{97}$$

where B is a very small dimensionless coupling constant. Other very flat potentials also work (*e.g.*, $V = V_0 - \alpha\phi^4 + \beta\phi^6$ [74]). As before we first solve for ϕ_{50}:

$$N(\phi_{50}) = 50 = \frac{8\pi}{m_{\text{Pl}}^2} \int_{\phi_{\text{end}}}^{\phi_{50}} \frac{V d\phi}{V'} \Rightarrow \phi_{50}^2 = \frac{\pi\sigma^4}{100 |\ln(\phi_{50}^2/\sigma^2)| m_{\text{Pl}}^2}, \tag{98}$$

where the precise value of ϕ_{end} is not relevant, only the fact that it is much larger than ϕ_{50}. Provided that $\sigma \lesssim m_{Pl}$, both ϕ_{50} and ϕ_{end} are much less than σ; we then find

$$(99) \qquad V_{50} \simeq B\sigma^4/2, \qquad x_{50} \simeq -\frac{(\pi/25)^{3/2}}{\sqrt{|\ln(\phi_{50}^2/\sigma^2)|}} \left(\frac{\sigma}{m_{Pl}}\right)^2 \ll 1,$$

$$(100) \qquad m_{Pl} x'_{50} \simeq -24\pi/100, \qquad q_{50} \simeq \frac{2.5 \cdot 10^5 |\ln(\phi_{50}^2/\sigma^2)|}{\pi^2} \left(\frac{m_{Pl}}{\sigma}\right)^4 \gg 1,$$

$$(101) \qquad \alpha_T \simeq \frac{1}{q_{50}} \ll 1, \qquad \alpha_S = \alpha_T + 0.03, \qquad \frac{T}{S} \simeq \frac{6 \cdot 10^{-4}}{|\ln(\phi_{50}^2/\sigma^2)|} \left(\frac{\sigma}{m_{Pl}}\right)^4.$$

Provided that $\sigma \lesssim m_{Pl}$, x_{50} is very small; this means that q is very large, gravity waves and density perturbations are very nearly scale invariant, and T/S is small. Finally, using the COBE DMR normalization, we can determine the dimensionless coupling constant B:

$$(102) \qquad B \simeq 6 \cdot 10^{-14}/|\ln(\phi_{50}^2/\sigma^2)| \approx 3 \cdot 10^{-15}.$$

3˙2.4. *Natural inflation*. This model is based upon a potential of the form [53]

$$(103) \qquad V(\phi) = \Lambda^4 [1 + \cos(\phi/f)].$$

The flatness of the potential (and requisite small couplings) arise because the ϕ particle is a pseudo Nambu-Goldstone boson (f is the scale of spontaneous symmetry breaking and Λ is the scale of explicit symmetry breaking; in the limit that $\Lambda \to 0$ the ϕ particle is a massless Nambu-Goldstone boson). It is a simple matter to show that ϕ_{end} is of the order of πf.

This potential is difficult to analyze in general; however, there are two limiting regimes: i) $f \gg m_{Pl}$, and ii) $f \lesssim m_{Pl}$ [74]. In the first regime, the 50 or so relevant e-folds take place close to the minimum of the potential, $\sigma = \pi f$, and inflation can be analyzed by expanding the potential about $\phi = \sigma$,

$$(104) \qquad V(\psi) \simeq m^2 \psi^2/2,$$

$$(105) \qquad m^2 = \Lambda^4/f^2, \qquad \psi = \phi - \sigma.$$

In this regime natural inflation is equivalent to chaotic inflation with $m^2 = \Lambda^4/f^2 \simeq 2 \cdot 10^{-12} m_{Pl}^2$.

In the second regime, $f \lesssim m_{Pl}$, inflation takes place when $\phi \lesssim \pi f$, so that we can make the following approximations: $V \simeq 2\Lambda^4$ and $V' = -\Lambda^4 \phi/f^2$. Taking

$\phi_{\text{end}} \sim \pi f$, we can solve for $N(\phi)$:

$$N(\phi) = \frac{8\pi}{m_{\text{Pl}}^2} \int_\phi^{\pi f} \frac{V d\phi}{-V'} \simeq \frac{16\pi m_{\text{Pl}}^2}{f^2} \ln(\pi f/\phi), \tag{106}$$

from which it is clear that achieving 50 e-folds of inflation places a lower bound to f, very roughly $f \gtrsim m_{\text{Pl}}/3$ [53, 74].

Now we can solve for ϕ_{50}, V_{50}, x_{50} and x'_{50}:

$$\phi_{50}/\pi f \simeq \exp[-50\, m_{\text{Pl}}^2/16\pi f^2] \lesssim \mathcal{O}(0.1), \qquad V_{50} \simeq 2\Lambda^4, \tag{107}$$

$$x_{50} \simeq \frac{1}{2}\frac{m_{\text{Pl}}}{f}\frac{\phi_{50}}{f} \lesssim \mathcal{O}(0.1), \qquad x'_{50} \simeq -\frac{1}{2}\left(\frac{m_{\text{Pl}}}{f}\right)^2. \tag{108}$$

Using the COBE DMR normalization, we can relate Λ to f/m_{Pl}:

$$\Lambda/m_{\text{Pl}} = 6.7 \cdot 10^{-4} \sqrt{\frac{m_{\text{Pl}}}{f}} \exp[-25\, m_{\text{Pl}}^2/16\pi f^2]. \tag{109}$$

Further, we can solve for T/S, α_T and α_S:

$$\frac{T}{S} \simeq 0.07 \left(\frac{m_{\text{Pl}}}{f}\right)^2 \left(\frac{\phi_{50}}{f}\right)^2 \lesssim 0.1, \tag{110}$$

$$\alpha_T = \frac{1}{16\pi}\frac{q_{50}}{q_{50}-1}\left(\frac{1}{4}\frac{m_{\text{Pl}}^2}{f^2}\frac{\phi_{50}^2}{f^2}\right) \simeq \frac{1}{64\pi}\left(\frac{m_{\text{Pl}}}{f}\right)^2\left(\frac{\phi_{50}}{f}\right)^2 \ll 0.1, \tag{111}$$

$$\alpha_S = \frac{1}{16\pi}\frac{q_{50}}{q_{50}-1}\left(\frac{1}{4}\frac{m_{\text{Pl}}^2}{f^2}\frac{\phi_{50}^2}{f^2} + \frac{m_{\text{Pl}}^2}{f^2}\right) \simeq \frac{1}{16\pi}\left(\frac{m_{\text{Pl}}}{f}\right)^2, \tag{112}$$

$$q_{50} = 64\pi\left(\frac{f}{m_{\text{Pl}}}\right)^2\left(\frac{f}{\phi_{50}}\right)^2 \gg 1. \tag{113}$$

Regime ii) provides the exception to the rule that $\alpha_S \simeq \alpha_T$ and large α_S implies large T/S. For example, taking $f = m_{\text{Pl}}/2$, we find

$$\phi_{50}/f \sim 0.06, \qquad x_{50} \sim 0.06, \qquad x'_{50} = -2, \qquad q_{50} \sim 10^4, \tag{114}$$

$$\alpha_T \sim 10^{-4}, \qquad \alpha_S \sim 0.08, \qquad T/S \sim 10^{-3}. \tag{115}$$

The gravitational-wave perturbations are very nearly scale invariant, while the density perturbations deviate significantly from scale invariance. I note that regime ii), i.e. $f \lesssim m_{\text{Pl}}$, occupies only a tiny fraction of parameter space because f must be greater than about $m_{\text{Pl}}/3$ to achieve sufficient inflation; further, regime ii) is «fine tuned» and «unnatural» in the sense that the required value of Λ is exponentially sensitive to the value of f/m_{Pl}.

Finally, I note that the results for regime ii) apply to any inflationary model whose Taylor expansion in the inflationary region is similar; e.g., $V(\phi) = -m^2\phi^2 + \lambda\phi^4$, which was originally analyzed in ref. [74].

3˙2.5. Lessons.

To summarize the general features of our results. In all examples the deviations from scale invariance enhance perturbations on large scales. The only potentials that have significant deviations from scale invariance are either very steep or have rapidly changing steepness. In the former case, both the scalar and tensor perturbations are tilted by a similar amount; in the latter case, only the scalar perturbations are tilted.

For «steep» potentials, the expansion rate is «slow», i.e. q_{50} close to unity, the gravity wave contribution to the CBR quadrupole anisotropy becomes comparable to, or greater than, that of density perturbations, and both scalar and tensor perturbations are tilted significantly. For flat potentials, i.e. small x_{50}, the expansion rate is «fast», i.e. $q_{50} \gg 1$, the gravity wave contribution to the CBR quadrupole is much smaller than that of density perturbations, and, unless the steepness of the potential changes significantly, large x'_{50}, both spectra are very nearly scale invariant; if the steepness of the potential changes rapidly, the spectrum of scalar perturbations can be tilted significantly. The models that permit significant deviations from 'scale invariance involve exponential or low-order polynomial potentials; the former by virtue of their steepness, the latter by virtue of the rapid variation of their steepness. Exponential potentials are of interest because they arise in extended-inflation models; potentials with rapid steepness include $V(\phi) = -m^2\phi^2 + \lambda\phi^4$ or $\Lambda^4[1 + \cos(\phi/f)]$.

Finally, to illustrate how observational data could be used to determine the properties of the inflationary potential and test the consistency of the inflationary hypothesis, suppose observations determined the following:

(116) $\qquad (\Delta T)_Q \simeq 16 \,\mu\text{K}, \quad T/S = 0.24, \quad n = 0.9;$

that is, the COBE DMR quadrupole anisotropy, a four-to-one ratio of scalar-to-tensor contribution to the CBR quadrupole, and spectral index of 0.9 for the scalar perturbations. From T/S, we determine the steepness of the potential: $x_{50} \simeq 0.94$. From the steepness and the quadrupole anisotropy the value of the potential: $V_{50}^{1/4} \simeq 2.4 \cdot 10^{16}\,\text{GeV}$. From the spectral index the change in steepness: $x'_{50} \simeq -0.81/m_{\text{Pl}}$. These data can also be expressed in terms of the value of the potential and its first two derivatives:

(117) $\quad V_{50} = 1.4 \cdot 10^{-11} m_{\text{Pl}}^4, \quad V'_{50} = 1.5 \cdot 10^{-11} m_{\text{Pl}}^3, \quad V''_{50} = 1.0 \cdot 10^{-12} m_{\text{Pl}}^2.$

Further, they then lead to the prediction $n_T = -0.035$, which, when «measured», can be used as a consistency check for inflation.

4. – Structure formation after COBE.

Filling in the details of structure formation is one of the pressing challenges of the standard cosmology. In order to do so one must have the «initial data» for the structure formation problem: the spectrum of density perturbations and the quantity and composition of matter in the Universe. With initial data in hand one can hope to carry out detailed numerical simulations which can be compared to the observations(*). While neither the observational data nor the simulations are perfect, the situation in both regards is improving rapidly. In particular, the discovery of CBR anisotropy by the COBE DMR has provided the first direct evidence for the existence of density perturbations and thereby opened the door for their study.

Over the past decade or so many cosmologists have come to believe that required initial data trace to events that took place during the earliest history of the Universe ($t \ll 10^{-2}$ s). Thus the study of structure formation has the potential to test theories of the early Universe and the underlying particle physics. Inflation leads to two limiting scenarios: hot dark matter and cold dark matter, both with scale-invariant density perturbations.

In the hot-dark-matter scenario the streaming of neutrinos from regions of higher density to lower density erases perturbations on small scales ($\lesssim 13\, h^{-2}$ Mpc); therefore, structure forms from the top down: superclusters must form first and fragment into smaller objects. Therein lies the fundamental problem: Since we know that superclusters are just forming today, galaxies form too late to be consistent with the abundance of galaxies observed at red shifts of unity or so[86].

Cold dark matter looks much more promising; cold dark matter refers to dark-matter particles that move very slowly, either by virtue of their large mass (e.g., 10 GeV to 2 TeV neutralino) or the fact that they were born cold (10^{-5} eV axion). This means that perturbations on small scales are not erased and that structure forms from the bottom up. Cold dark matter has been subject to intense scrutiny over the past decade and has thus far survived, albeit with a number of scratches and bruises[87]. CDM models will be the focus of this section.

That is not to say that cold-dark-matter models are the only promising possibilities. There are scenarios where the density perturbations arise due to topological (and nontopological) defects such as strings[30], global monopoles and textures[31] with hot or cold dark matter. Scenarios have been discussed where the density perturbations arise in a rather recent phase transition (since decoupling!), due to new physics in the neutrino sector[33].

(*) Since the fluctuations predicted by inflation and other theories are only specified in a statistical sense, this comparison can only be done statistically; in the case of inflation, the fluctuations are Gaussian and so all predictions can be specified in terms of the power spectrum, $\langle |\delta_k|^2 \rangle$.

Finally, perhaps the most interesting alternative is Pleebles' PIB model or what-you-see-is-what-you-get model [29]. In PIB $\Omega_0 = \Omega_B \sim 0.2$, $h \sim 0.8$, and the density perturbations are isocurvature perturbations (variations in the local baryon-to-photon ratio and not in the energy density). PIB is not motivated by what early-Universe theorists would like, rather by «what we see» (though it violates the primordial nucleosynthesis bound by large factor since $\Omega_B h^2 \simeq$ $\simeq 0.13 \gg 0.02$). Remarkably, the scenario is still viable, though measurements of CBR anisotropy on scales of 1° to 90° are really putting it to the test: normalizing to the COBE 10° measurement, its predictions for the quadrupole are a factor of two small, while its predictions of scales of about five degrees exceed current upper limits [88].

4'1. *The Universe observed.* – By now we know a lot—and a little—about the structure that exists in the Universe today. A resurgence of interest in structure formation, brought about in part by the very intriguing early-Universe suggestions for initial data, has resulted in an explosion of observations that bear on the issue over the past decade. They include red-shift surveys (large-angle, pencil-beam and sparsely sampled surveys), the spectrum and spatial variation of the CBR temperature, peculiar-velocity measurements, QSO absorption line systems, studies of clusters and superclusters, determinations of the distribution and quantity of dark matter, studies of galactic evolution, catalogues of millions of galaxies on the sky, and on and on.

To place things in perspective, we know much about the distribution of light (bright galaxies)—as opposed to mass (which is what theorists like to discuss); the largest red-shift survey, the CfA$_2$ slices of the Universe, contains only about 20 000 galaxies with median red shift of about 0.02 [89]; and the total number of red shifts measured for all purposes is only about 50 000. We have no definitive evidence as to the epoch of galaxy formation, or how the neutral hydrogen left between galaxies became ionized (if it were not, we would not be able to see emission from distant QSOs shortward of Lyman-alpha, 1215 Å in the rest frame of the QSO). We probably only know the mean density of galaxies to within 20%; we have no fair sample of clusters; and so on.

Let me briefly try to summarize some of the data that can be used to test models of structure formation. Within the spirit of my broad-brush description, I will group the observations into three classes: small scale, observations that probe the Universe on scales less than order $30\,h^{-1}$ Mpc or so; intermediate scale, observations that probe the Universe on scales of $(30\,h^{-1}$–$300\,h^{-1})$ Mpc or so; and large scale, observations that probe the Universe on the very largest scales accessible (*).

(*) I warn the reader that my nomenclature is not universal; many refer to what I call intermediate scales as large scales.

Small-scale structure ($\lambda \lesssim 30\,h^{-1}\,\mathrm{Mpc}$): Our knowledge of these scales is the most extensive and well developed, though largely restricted to the distribution of bright galaxies like our own. These are also the scales on which astrophysical effects—star formation, blast waves, and so on—are potentially most important and poorly understood. The Universe on these scales is organized into galaxies and clusters, whose properties have been studied and quantified; for galaxies, number density and morphology—*i.e.* spiral, elliptical, etc.—, rotation curves, and so on; and for clusters, number density, velocity dispersions, richness class, and so on. Both galaxies and clusters, with measured two-point correlation functions, $\xi_{gg}(r) \simeq (r/5h^{-1}\,\mathrm{Mpc})^{-1.8}$ and $\xi_{cc}(r) \simeq (r/25h^{-1}\,\mathrm{Mpc})^{-1.8}$, though the cluster correlation function is less well known and depends upon cluster richness [90]. At some level we know the distribution of dark matter: spiral galaxies have large halos with unknown spatial extent and the bulk of the mass in clusters is dark [17]. We also know the pairwise galaxy velocity dispersion (line-of-sight velocity dispersion), $\langle (v_1 - v_2)^2 \rangle^{1/2} |_{10\,\mathrm{Mpc}} \simeq (300\text{–}400)\,\mathrm{km\,s^{-1}}$ [91]. (As discussed earlier, the peculiar motions of galaxies depend upon the amplitude of density perturbations and the amount of matter in the Universe, and thus are indicative of such.) On scales less than about $8h^{-1}\,\mathrm{Mpc}$ the Universe is nonlinear: specifically, the r.m.s. fluctuation in the number density of bright galaxies measured in a sphere of radius of $8h^{-1}\,\mathrm{Mpc}$ is unity.

Intermediate-scale structure (($30h^{-1}\text{–}300h^{-1}$) Mpc): These are the scales on which our knowledge is the most fragmentary and often more qualitative than quantitative (*). Observations include the voids and «Great Wall» seen in the CfA$_2$ red-shift survey; the reoccurring walls seen in the pencil-beam survey of Broadhurst *et al.* [92]; the angular-correlation function of galaxies $w(\theta)$, which is related to $\xi_{gg}(r)$, measured by EFSTATHIOU *et al.* [93] in the APM catalogue of 2 million galaxies on the sky (effective depth of $400\,h^{-1}\,\mathrm{Mpc}$); the peculiar velocities of galaxies measured by the Seven Samurai and others [94], about $400\,\mathrm{km\,s^{-1}}$ on the scale of $50\,h^{-1}\,\mathrm{Mpc}$; Great Attractors, and on and on. From red-shift surveys like the CfA$_2$ slices of the Universe, the IRAS 1.2 Jy survey of infrared-selected galaxies [95] and the APM-Stromlo 1 in 20 red-shift survey [96], the fluctuations in the galaxy number density have been measured on scales out to a few hundred Mpc (see fig. 11). By the year 2000 the Sloan Digital Sky Survey [97] will produce a «Map of the Universe», from the red shifts of a million galaxies (mean red shift of about 0.15 and survey depth of $500\,h^{-1}\,\mathrm{Mpc}$). With the exception of the peculiar-velocity measurements all these observations probe the distribution of light not mass. CBR anisotropy measurements on angular scales of a few degrees down to a few arcminutes also have the po-

---

(*) I often call these the *NY Times* scales, as new observations and their extravagant interpretation are reported there almost weekly!

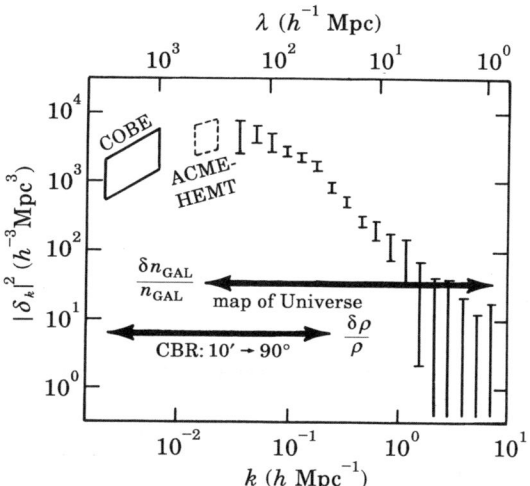

Fig. 11. – Summary of observational knowledge of the power spectrum $|\delta_k|^2$ based upon the IRAS 1.2 Jy red-shift survey and CBR anisotropy measurements (from [95]). ACME-HEMT indicates the South-Pole experiment that has detected anisotropy that may or may not be intrinsic to the CBR.

tential to probe the distribution of matter on these scales, as the CBR anisotropy on a given angular scale is related to the fluctuations in the mass density on a range of length scales around the characteristic length scale that subtends that angular size on the last-scattering surface: $\lambda \sim 100 h^{-1}$ Mpc · $\cdot (\theta/\text{degree})$. Very sensitive experiments are being done on these angular scales; with the important exception of the COBE DMR detection, there are now only upper limits, at the level of a few times 10^{-5} (see fig. 3). I believe that more detections are just around the corner!

Large-scale structure ($\gtrsim 300 h^{-1}$ Mpc): These scales are probed primarily by CBR anisotropy, though the Sloan Digital Sky Survey should provide some information about the distribution of galaxies on these scales. On angular scales much greater than about 1° the anisotropy arises due to the fluctuations in the gravitational potential on the last-scattering surface (Sachs-Wolfe effect), while on small angular scales the situation is more complicated as the velocity of the matter, temperature fluctuations intrinsic to the radiation and the ionization history of the Universe become important. On large angular scales it is very simple to relate the CBR anisotropy to the «virgin spectrum» of density fluctuations. It is these scales that were probed by the COBE DMR detection, proving the first direct information about the existence of the density inhomogeneity that seeded structure formation.

4˙1.1. Normalization: the great leap forward! Lacking a definite prediction for the overall normalization for inflationary density perturbations, those who study formation of structure have historically used data on small scales to normalize the spectrum of density perturbations, typically on the scale of $8h^{-1}$ Mpc. In so doing it is useful to define

$$\sigma_8 \equiv \langle (\delta M/M)^2 \rangle_{8h^{-1} \text{ Mpc}}^{1/2}, \tag{118}$$

which is the r.m.s. mass fluctuation in spheres of radius $8h^{-1}$ Mpc. The simplest (and most naive) procedure is to assume that light faithfully traces mass, i.e. $\delta\rho/\rho = \delta n_{\text{GAL}}/n_{\text{GAL}}$, and set $\sigma_8 = 1$ since the r.m.s. fluctuation in galaxy number in spheres of radius $8h^{-1}$ Mpc is unity; I will refer to this minimal cold dark matter (MCDM). I should remark that there is no *a priori* reason to expect light to trace mass, except on the very largest scales where only gravity is important.

Because this normalization leads to a galaxy pairwise velocity dispersion that is about a factor of two too large, the concept of «biasing» was introduced; namely, that light is a biased tracer of mass [98]. If light does not trace mass, the simplest *ansatz* is a linear factor between the two:

$$\delta n_{\text{GAL}}/n_{\text{GAL}} = b(\delta\rho/\rho). \tag{119}$$

Of course, there is every reason to expect that the real relationship is more complicated, $b = b(\lambda)$. In biased CDM models (BCDM), $\sigma_8 = b^{-1}$. In principle, the bias factor $b(\lambda)$ can be measured on scales where there is information about both the distribution of galaxies and of mass (cf. fig. 11).

(Another way of understanding the motivation for bias is the so-called Ω problem: Why do the dynamical measurements indicate $\Omega_0 = 0.2$, if Ω_0 is really unity? The biasing explanation is that most of the mass in the Universe is in low-surface-brightness galaxies that are too faint to see and that are less strongly clustered than the bright galaxies. Bright galaxies are more strongly clustered and account for only 20% of the mass density.)

Until the COBE DMR detection, a bias factor of 1.5 to 2 was in vogue to resolve the discrepancy in the galaxy pairwise velocity dispersion; $b \sim 1.5-2$ was known as the *standard* CDM model. Since the peculiar velocities of galaxies arise due to the lumpy distribution of matter, larger b implies a smoother mass distribution and thus smaller peculiar velocities. (Likewise, reducing the matter content, or Ω_0, can help.) Unfortunately, the predictions of BCDM on intermediate scales could not account for the level of inhomogeneity seen—voids, galaxy-galaxy angular-correlation function, peculiar velocities, and so on—since the mass distribution was smoother. Thus CDM was faulted for predicting too little power on «large scales» (in my nomenclature, intermediate scales).

The COBE DMR detection of CBR anisotropy changed the situation overnight by providing a new, more direct normalization of the density perturbations! Assuming the correctness of the result, we now have a measurement of the inhomogeneity in the mass distribution on large scales—and at last a «physics normalization». Remarkably, the COBE normalization (with scale-invariant perturbations) corresponds to the simplest CDM model: $\sigma_8 = 1.2 \pm \pm 0.2$ [99], *i.e.* no biasing (*).

The COBE DMR normalization has changed the way we view inflation and structure formation: Intermediate (and large) scales seem to be OK; the problem is with small scales. Addressing this problem is the focus of the brief discussion of CDM models that follows.

4˙2. *CDM models.* – The initial data for structure formation include i) spectrum of primeval density perturbations—amplitude on a given scale (normalization) and spectral index n; ii) composition of the Universe—Ω_i, i = baryons, cold dark matter, hot dark matter, vacuum energy, and so on; and iii) Hubble constant which sets the time/length scale for the Universe. With these in hand one can compute the spectrum of density perturbations at the equivalence epoch and let gravity run its course. Of course, astrophysics—cooling of baryons, star formation, etc.—is important too, but more difficult to model. Progress here too is being made with large N-body codes that include both gravity and hydrodynamics for the baryons [100]. The list of wanted cosmological parameters for a numerical simulation is: σ_8 (in the simple biasing prescription $\sigma_8 = b^{-1}$), n, Ω_B, Ω_{other} and h.

What predictions does inflation make for these parameters? The firmest is a flat Universe, in my notation $\Omega_0 = 1.0$, which implies nonbaryonic dark matter dominates. As mentioned earlier, hot dark matter (30 eV or so neutrinos) was ruled out early on; and so the cold-dark-matter scenario *appeared* to be the unique inflationary blueprint for structure formation [101]. Let me explain; to get the age of the Universe right we must have $h \sim 0.5$. This fact together with the primordial nucleosynthesis determination of $\Omega_B h^2$ implies $\Omega_B \simeq 0.04\text{--}0.10$. In *most* inflationary models the density perturbations are very nearly scale invariant, implying $n = 1$. Finally, the variance in galaxy counts on $8h^{-1}$ Mpc suggests $\sigma_8 = 1$. This is the minimal cold-dark-matter model (MCDM); it is certainly the simplest CDM scenario, though it is no longer the unique CDM model.

Partly due to problems with MCDM, partly due to the improvement in the observations that test models of structure formation, and partly due to the passage of time we now realize that there are other possibilities, some just as well motivated, some less well motivated. I will characterize the different models by

---

(*) For HDM the COBE DMR normalization implies $\sigma_8 = 0.7$. This drives another nail in the coffin, as it implies that only about 1% of the material in the Universe is in nonlinear structures.

their values for the key cosmological parameters for structure formation: σ_8, n, Ω_{other}, h and Ω_B.

4˙2.1. MCDM. This is the simplest and the original CDM model; it is characterized by $b = 1$, $n = 1$, $\Omega_{\text{other}} = \Omega_{\text{cold}} \simeq 0.9$, $\Omega_B \sim 0.1$ and $h = 0.5$. It is consistent with the COBE DMR data, which for $n = 1$ imply $\sigma_8 = 1.2 \pm 0.2$, and intermediate-scale structure. However, it has too much power on small scales, quantified by a galaxy pairwise velocity dispersion of about 1000 km s^{-1} compared to the observed 400 km s^{-1}. A comparison of MCDM power spectrum with the observations is shown in fig. 12.

Since MCDM is the simplest and most well-motivated model perhaps inflationists should sit tight and wait for the data (or their interpretations) to change. After all, the disagreement is on small scales where the Universe is highly nonlinear and astrophysics can play an important role.

From fig. 12 we see that concordance with the data can be achieved by three different «symmetry operations»: translation of the predictions of the model in the vertical direction, corresponding to biasing; rotation, corresponding to deviation from scale invariance; and horizontal translation, corresponding to a change in the transfer function.

4˙2.2. BCDM. This is the CDM model with biasing, imposed to solve the problem of too much power on small scales. The parameters of this model are: $b \sim 1.5\text{-}2$, $n = 1$, $\Omega_{\text{other}} = \Omega_{\text{cold}} \sim 0.9$, $\Omega_B \sim 0.1$ and $h = 0.5$. This model is disfavored for two reasons: 1) The COBE DMR results imply $b = 0.8 \pm 0.2$, and 2) (apparent) insufficient power on intermediate scales to account for peculiar velocities, the galaxy-galaxy angular correlation function, etc. However, one

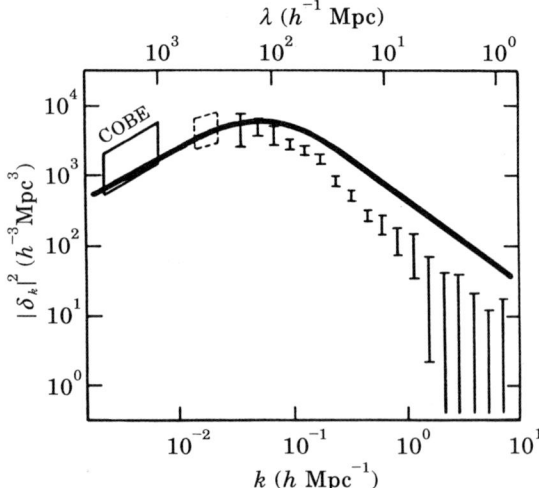

Fig. 12. – MCDM vs. observation.

should keep in mind that the COBE DMR results are new and may still change, and that our knowledge of intermediate scales is the least secure. Perhaps the truth is somewhere in between MCDM and BCDM; both are well motivated. Shifting the MCDM power spectrum in fig. 12 downward by a factor of 2–4 corresponds to $\sigma_8 \sim 0.5$-0.7.

4˙2.3. Tilt. Tilted CDM (TCDM) models are characterized by $\sigma_8 \sim 0.5$, $n \sim 0.8$, $\Omega_{\text{other}} = \Omega_{\text{cold}} \sim 0.9$, $\Omega_B \sim 0.1$ and $h = 0.5$ [102]. From the beginning it was realized that the inflationary perturbations were not precisely scale invariant, typically with more power on large scales ($n < 1$) [74], and so tilted models too are well motivated. Relative to scale-invariant perturbations ($n = 1$) the density perturbation in a tilted model is

$$(120) \qquad \left(\frac{\delta\rho}{\rho}\right) \propto \left(\frac{\delta\rho}{\rho}\right)_{n=1} \lambda^{(1-n)/2}.$$

The COBE DMR result provides a normalization on very large scales, $\lambda \sim 10^4$ Mpc; relative to MCDM, the density perturbations on intermediate scales, $\lambda \sim 300$ Mpc, are only a factor of about 1.4 smaller, while on small scales, $\lambda \sim 10$ Mpc, they are about a factor of 2 smaller (see fig. 13a)).

If tilt is the truth, two kinds of inflationary potentials are singled out: exponential and low-order polynomial potentials [103]. Further, for exponential potentials, the contribution of gravity waves to the CBR anisotropy on large angular scales is significant, which lowers the overall normalization of density perturbations further, by a factor of $(1 + T/S)^{1/2}$ [8].

4˙2.4. Best-fit models. These models address the problem of too much

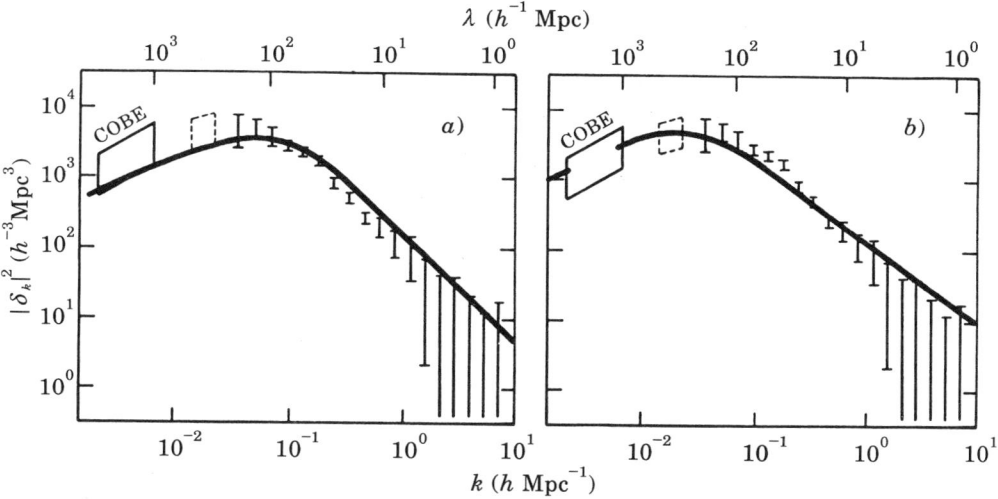

Fig. 13. – a) TCDM vs. observation, b) MDM/ΛCDM vs. observation.

small-scale power by changing the transfer function. The models considered thus far are: cold dark matter with a cosmological constant (ΛCDM), $n = 1$, $\Omega_B \sim 0.05$, $\Omega_{cold} \sim 0.15$, $\Omega_\Lambda \sim 0.8$ and $h \sim 0.8$ [104]; and mixed dark matter (MDM) —«the neutrino cocktail»—$n = 1$, $\Omega_B \sim 0.1$, $\Omega_{cold} \sim 0.6$, $h = 0.5$ and $\Omega_{hot} \sim 0.3$, corresponding to a 7 eV or so mass neutrino [105].

How is the transfer function changed? It is simplest to see in MDM; since part of the dark matter is in the form of neutrinos which freestream out of density fluctuations on small scales, perturbations on small scales are depressed (see fig. 9), just what the doctor ordered. In ΛCDM the story is a little more complicated; the bend in the transfer function is set by the scale that crosses inside the horizon at matter-radiation equality, $k_{EQ} \sim 0.5\,(\Omega_{matter}h^2)$ Mpc, where $\Omega_{matter} = \Omega_B + \Omega_{cold} \simeq 0.2$. Relative to MCDM, k_{EQ} is a factor of two smaller, shifting the spectrum to smaller k and decreasing power on small scales (see fig. 9).

(The ΛCDM model, which I once called «the best-fit Universe», has a number of other nice features. It automatically solves the Ω problem since 80% of the energy density is in vacuum energy which is uniformly distributed and thus does not «show up» in dynamical measurements of the mass density. It allows one to accommodate the higher values of the Hubble constant which are favored by many measurements. Likewise, MDM also addresses the Ω problem: there is not enough phase space in galaxies for neutrinos to account for halo masses; further, neutrinos probably move too fast to be captured even in clusters. This would explain why dynamical measurements of Ω_0 based on galactic rotation curves or cluster virial masses do not lead to values of Ω_0 close to unity.)

Because both ΛCDM and MDM have less power on small scales than MCDM, when normalized to the COBE DMR result, they are a much better fit to the small-scale data, and on large scales they are very similar to MCDM. They fit the present data very well (see fig. 13b)). One should, however, recall the words of Francis CRICK (of DNA fame); loosely quoted,

A theory that agrees with all of the data at any given time is necessarily wrong, since at any given time some of the data are incorrect.

The weak point of the «best-fit models» is motivation; however, let me try to make the best case for each. ΛCDM: A cosmological constant that contributes an energy density of about 10^{-46} GeV^4 would be very surprising. Since there is no physical mechanism known that explains why the present vacuum energy is not of order m_{Pl}^4 (perhaps with the help of supersymmetry only of order G_F^{-2}), one cannot rigorously say that $\rho_{vac} \simeq 10^{-46}$ GeV^4 is fine-tuned in the technical sense. MDM: Cold dark matter has so many good features that it must be *part* of the truth; neutrinos exist—in three varieties—and the see-saw mechanism suggests nondegenerate masses in «the eV range» (meaning 10^{-6} eV to tens of eV), it could well be that one of the three neutrinos has a mass of order 10 eV.

Certainly neither case for motivation is strong. Why is the cosmological constant just today becoming dynamically significant (recall, $\rho_{\text{vac}}/\rho_{\text{matter}} \propto R^3$)? And, if history is any guide, cosmologists beginning with EINSTEIN have too often invoked a cosmological constant to solve their problems. For MDM, one has to posit two kinds of nonbaryonic dark matter that each contribute comparably to the energy density of the Universe. If nonbaryonic dark matter exists it is already puzzling that baryons and dark matter each contribute similar amounts to the mass density [106].

4˙3. *The scorecard and future.* – Whereas cosmologists used to talk about «the CDM model» and its «uniqueness» (words that were also once used to describe the superstring), there is now a menu of CDM models. How do they stand, and which measurements can discriminate between them? In discussing their models, it is often said that theorists have many hands; on the one hand, on the other hand, on the other other hand, and so on. Let me try my hand at it.

Occam's razor points to the simplest model, MCDM. Moreover, it was vindicated by COBE and only differs from the observational data by a factor of two or so on small scales where complicated astrophysics can be very important. Perhaps theorists should sit tight and wait. On the other hand (here we go), biasing at some level is likely to be a fact of life, arguing for BCDM; BCDM resolves the small-scale problems of CDM, but COBE indicates that $b \sim 1$. Maybe the truth is somewhere in between MCDM and BCDM; the COBE normalization could come down a bit, making $b \sim 1.3$ or so viable.

On the other other hand, deviation from scale invariance was in the cards from the beginning, and so TCDM is well motivated too. Moreover, the tilt required points to a smaller class of inflationary models, exponential potentials and low-order polynomial potentials, which can be discriminated between by the size of their tensor perturbations.

On my final hands are the best-fit models, ΛCDM and MDM. They are not as well motivated, but agree better with the data at hand. Of the two, my first final hand has to go to MDM, and my second final hand to ΛCDM.

There will be a variety of observations that can be used to discriminate between the different CDM models; I will focus on CBR measurements on the 0.5°–2° scale, as there are several experiments with the sensitivity to probe CDM models which will be announcing results soon [107]. These experiments add roughly another order of magnitude to the range of scales probed by CBR anisotropy (recall, COBE probes 10° to 90°). For reference, the MCDM prediction for these angular scales is $\delta T/T \sim \textit{few} \times 10^{-5}$; the current upper limits are just above this level! Let me describe possible outcomes.

Scenario 1: The upper limits become detections. MCDM, ΛCDM and MDM are in; BCDM and TCDM are out. (Because the tilted models have less power on small scales than MCDM, their predictions for small-angle anisotropy are smaller.)

Scenario 2: Detections are announced below the 10^{-5} level. BCDM and TCDN are in; the rest are out. If the detections are much below the 10^{-5} level, exponential potentials are strongly favored as in these tilted models much of the COBE signal is due to tensor perturbations whose contribution to CBR anisotropy falls dramatically around a few degrees [12, 81, 108].

Finally, let me mention a very different test, the value of the Hubble constant. Suppose all parties agree on the currently popular value $h = 0.8$; all CDM models except ΛCDM fall by the way side, based on the age of the Universe. Conversely, suppose that evidence for $h = 0.5$ becomes overwhelming; ΛCDM is out.

5. – Concluding remarks.

Inflation is an extremely attractive cosmological paradigm; in spite—no, because—of its beauty it must be put to the ultimate test: confrontation with observation. In testing inflation one must focus on its robust predictions: In order of robustness, spatially flat Universe ($\Omega_0 = 1$), very nearly scale-invariant spectrum of density perturbations and nearly scale-invariant spectrum of gravitational waves. In addition, for first-order inflation there should be a «spike» in the stochastic background of gravitational waves of very significant energy density, $\Omega_{GW} \sim 10^{-9}$ or so, at a frequency $f \sim 10^4$ Hz ($\mathscr{M}/10^{10}$ GeV) [59]. It is also possible that fluctuations in other fields lead to primeval magnetic fields [109] or isocurvature perturbations (*e.g.*, in axions or baryons) [70].

There are a variety of means of testing these predictions. For example, there are kinematic and dynamic techniques for measuring Ω_0. The density perturbations lead to temperature fluctuations in the CBR. The tensor (gravity wave) perturbations also lead to CBR anisotropies, or may be detected directly by the next generation of gravity wave detectors, Laser Interferometer Gravitational-wave Observatories (LIGOs) [110].

From the «primary predictions», a series of secondary predictions follow. For example, since primordial nucleosynthesis restricts $\Omega_B \lesssim 0.1$, nonbaryonic dark matter is a necessity, and a host of experiments are under way to search for nonbaryonic dark matter [111]. In a flat, matter-dominated Universe $H_0 t_0 = 2/3$ or $t_0 \simeq 6.5 h^{-1}$ Gy, which implies that h must be greater than 0.65 to ensure that the Universe is older than 10 Gy (the absolute minimum age that is consistent with other measures of the age of the Universe). The spectrum of density perturbations, together with the matter content, provide the initial data for the structure formation problem, leading to another test.

In the near term I believe that structure formation will provide the most powerful test of inflation and probe of inflationary models. On balance, the inflation-inspired CDM models are doing quite well so far compared to the alternatives: Texture and cosmic-string models required a high level of biasing

($b \sim 4$) to be compatible with COBE DMR results [112]; and PIB not only strongly violates the primordial nucleosynthesis constraint to Ω_B but also seems to be inconsistent with CBR anisotropy bounds on the scale of five degrees [88]. Measurements of CBR anisotropy on the degree scale will further discriminate between CDM and the alternatives; relative to CDM, PIB predicts a very large anisotropy, and the texture/cosmic-string models predict a very small anisotropy whose properties are very non-Gaussian (most of the CBR is very quiet with a few hot and cold spots).

Great efforts have been made and are being made to further test the CDM scenarios; they involve many different techniques, CBR anisotropy, red-shift surveys, peculiar-velocity measurements, and so on. These observations not only have the power to falsify CDM, but could also reveal much about the inflationary potential: the value of the potential, its steepness and the change in steepness, which in turn can be used to learn about the underlying model [76]. For example, suppose that density perturbations do deviate significantly from scale invariance, then two classes of models are ruled out—chaotic and new inflation—and two types of models are ruled in—exponential potentials (as found in extended inflation) or low-order polynomial potentials (as found in natural inflation). The ratio of tensor to scalar perturbations can further narrow the field: large-tensor contribution to the CBR quadrupole points to exponential potentials and small-tensor contribution points to low-order polynomial potentials.

The moment of truth for inflation may be near!

* * *

I wish to thank E. MAZZI and her staff for taking care of all the important details that made the School both productive and enjoyable for the students and lecturers. I also thank N. VITTORIO and J. SILK for arranging such a fine scientific program.

REFERENCES

[1] For a textbook treatment of the standard cosmology see, *e.g.*, S. WEINBERG: *Gravitation and Cosmology* (Wiley, New York, N.Y., 1972); E. W. KOLB and M. S. TURNER: *The Early Universe* (Addison-Wesley, Redwood City, Cal., 1990).
[2] A. SANDAGE: *Phys. Scr. T*, **43**, 22 (1992).
[3] J. MOULD, L. STAVELEY-SMITH, R. A. SCHOMMER, G. D. BOTHUN, P. J. HALL, MING SHENG HAN, J. P. HUCHRA, J. ROTH and W. WALSH: *Astrophys. J.*, **383**, 467 (1992).
[4] See, *e.g.*, M. ROWAN-ROBINSON: *The Cosmological Distance Ladder* (Freeman, San Francisco, Cal., 1985).
[5] M. FUKUGITA, C. J. HOGAN and P. J. E. PEEBLES: *Nature (London)*, **366**, 309 (1993).

[6] J. C. MATHER, E. S. CHENG, D. A. COTTINGHAM, R. E. EPLEE jr., D. J. FIXSEN, T. HEWAGAMA, R. B. ISAACMAN, K. A. JENSEN, S. S. MEYER, P. D. NOERDLINGER, S. M. READ, L. P. ROSEN, R. A. SHAFER, E. L. WRIGHT, C. L. BENNET, N. M. BOGGESS, M. G. HAUSER, T. KELSALL, S. H. MOSELEY jr., R. F. SILVERBERG, G. F. SMOOT, R. WEISS and D. T. WILKINSON: *Astrophys. J.*, **420**, 439 (1994).

[7] H. GUSH, M. HALPERN and E. H. WISHNOW: *Phys. Rev. Lett.*, **65**, 537 (1990).

[8] R. DAVIS, H. HODGES, G. SMOOT, P. J. STEINHARDT and M. S. TURNER: *Phys. Rev. Lett.*, **69**, 1856 (1992).

[9] E. L. WRIGHT: *Phys. Rev. Lett.*, **396**, L3 (1992).

[10] S. S. MEYER, E. S. CHENG and L. A. PAGE: *Astrophys. J.*, **371**, L1 (1991); K. GANGA, S. S. MEYER, E. S. CHENG and L. A. PAGE: *Astrophys. J.*, **410**, L57 (1993).

[11] P. J. E. PEEBLES, D. N. SCHRAMM, E. TURNER and R. KRON: *Nature (London)*, **352**, 769 (1991).

[12] G. F. SMOOT, C. L. BENNETT, A. KOGUT, E. L. WRIGHT, J. AYMON, N. W. BOGGESS, E. S. CHENG, G. DEAMICI, S. GULKIS, M. G. HAUSER, G. HINSHAW, P. D. JACKSON, M. JANSSEN, E. KAITA, T. KELSALL, P. KEEGSTRA, L. ROKKE, R. F. SILVERBERG, L. TENORIO, R. WEISS and D. T. WILKINSON: *Astrophys. J.*, **396**, L1 (1992); D. J. FIXSEN, E. S. CHENG, D. A. COTTINGHAM, R. E. EPLEE jr., R. B. ISAACMAN, J. C. MATHER, S. S. MEYER, P. D. NOERDLINGER, R. A. SHAFER, R. WEISS, E. L. WRIGHT, C. L. BENNETT, N. M. BOGGESS, T. KELSALL, S. H. MOSELEY jr., R. F. SILVERBERG, G. F. SMOOT and D. T. WILKINSON: *Astrophys. J.*, **420**, 445 (1994).

[13] G. F. SMOOT: in *First Course in Current Topics in Astrofundamental Physics*, edited by N. SANCHEZ and A. ZICHICHI (World Scientific, Singapore, 1992), p. 192.

[14] G. F. SMOOT, C. L. BENNETT, A. KOGUT, E. L. WRIGHT, J. AYMON, N. W. BOGGESS, E. S. CHENG, G. DEAMICI, S. GULKIS, M. G. HAUSER, G. HINSHAW, P. D. JACKSON, M. JANSSEN, E. KAITA, T. KELSALL, P. KEEGSTRA, L. ROKKE, R. F. SILVERBERG, L. TENORIO, R. WEISS and D. T. WILKINSON: *Astrophys. J.*, **396**, L1 (1992); E. L. WRIGHT: *Astrophys. J.*, **396**, L3 (1992).

[15] T. P. WALKER, G. STEIGMAN, D. N. SCHRAMM, K. OLIVE and H. S. KANG: *Astrophys. J.*, **376**, 51 (1991).

[16] E. W. KOLB, M. S. TURNER, A. CHAKRAVORTY and D. N. SCHRAMM: *Phys. Rev. Lett.*, **67**, 533 (1991).

[17] For recent reviews of dark matter see, *e.g.*, M. S. TURNER: *Phys. Scr. T*, **36**, 167 (1991); P. J. E. PEEBLES: *Nature (London)*, **321**, 27 (1986); V. TRIMBLE: *Annu. Rev. Astron. Astrophys.*, **25**, 425 (1987); J. KORMENDY and G. KNAPP: *Dark Matter in the Universe* (Reidel, Dordrecht, 1989); K. ASHMAN: *Proc. Astron. Soc. Pac.*, **104**, 1109 (1992); S. FABER and J. GALLAGHER: *Annu. Rev. Astron. Astrophys.*, **17**, 135 (1979).

[18] S. FABER and J. GALLAGHER: *Annu. Rev. Astron. Astrophys.*, **17**, 135 (1979).

[19] J. S. MULCHAEY, D. S. DAVIS, R. F. MUSHOTZKY and D. BURSTEIN: *Astrophys. J.*, **404**, L9 (1993).

[20] M. ROWAN-ROBINSON, A. LAWRENCE, W. SAUNDERS, J. CRAWFORD, R. ELLIS, C. S. FRENK, I. PARRY, X. XIAOYANG, J. ALLINGTON-SMITH, G. EFSTATHIOU and N. KAISER: *Mon. Not. R. Astron. Soc.*, **247**, 1 (1990); N. KAISER, G. EFSTATHIOU, R. ELLIS, C. FRENK, A. LAWRENCE, M. ROWAN-ROBINSON and W. SAUNDERS: *Mon. Not. R. Astron. Soc.*, **252**, 1 (1991); M. STRAUSS, M. DAVIS, A. YAHIL and J. P. HUCHRA: *Astrophys. J.*, **385**, 444 (1992).

[21] E. BERTSCHINGER and A. DEKEL: *Astrophys. J.*, **336**, L5 (1989); A. DEKEL, E. BERTSCHINGER, A. YAHIL, M. A. STRAUSS, M. DAVIS and J. P. HUCHRA: *Astrophys. J.*, **412**, 1 (1993); M. STRAUSS, A. YAHIL, M. DAVIS, J. P. HUCHRA and K. FISHER: *Astrophys. J.*, **397**, 395 (1992).

[22] A. SANDAGE: *Astrophys. J.*, **133**, 355 (1961); *Phys. Scr. T*, **43**, 7 (1992); ref.[1].
[23] E. LOH and E. SPILLAR: *Astrophys. J.*, **307**, L1 (1986); M. FUKUGITA, F. TAKAHARA, K. YAMASHITA and Y. YOSHII: *Astrophys. J.*, **361**, L1 (1990).
[24] See, *e.g.*, E. W. KOLB and M. S. TURNER: *Annu. Rev. Nucl. Part. Sci.*, **33**, 645 (1983); A. DOLGOV: *Phys. Rep.*, **222**, 309 (1993); A. COHEN, D. KAPLAN and A. NELSON: *Annu. Rev. Nucl. Part. Sci.*, **43**, 27 (1993).
[25] S. WEINBERG: *Gravitation and Cosmology* (Wiley, New York, N.Y., 1972).
[26] For a more complete pedagogical discussion of structure formation see, *e.g.*, ref.[1]; P. J. E. PEEBLES: *The Large-Scale Structure of the Universe* (Princeton University Press, Princeton, N.J., 1980); G. EFSTATHIOU: in *The Physics of the Early Universe*, edited by J. A. PEACOCK, A. F. HEAVENS and A. T. DAVIES (Adam Hilger, Bristol, 1990), p. 361.
[27] For a pedagogical discussion of CBR anisotropy see, *e.g.*, G. EFSTATHIOU: in *The Physics of the Early Universe*, edited by J. A. PEACOCK, A. F. HEAVENS and A. T. DAVIES (Adam Hilger, Bristol, 1990), p. 361. Also see J. R. BOND and G. EFSTATHIOU: *Mon. Not. R. Astron. Soc.*, **226**, 655 (1987); J. R. BOND, G. EFSTATHIOU, P. LUBIN and P. MEINHOLD: *Phys. Rev. Lett.*, **66**, 2179 (1991).
[28] R. K. SACHS and A. M. WOLFE: *Astrophys. J.*, **147**, 73 (1967).
[29] P. J. E. PEEBLES: *Nature (London)*, **327**, 210 (1987); *Astrophys. J.*, **315**, L73 (1987); R. CEN, J. P. OSTRIKER and P. J. E. PEEBLES: *Astrophys. J.*, **415**, 423 (1993).
[30] See, *e.g.*, A. VILENKIN: *Phys. Rep.*, **121**, 263 (1985); A. ALBRECHT and A. STEBBINS: *Phys. Rev. Lett.*, **69**, 2615 (1992); D. BENNETT, A. STEBBINS and F. BOUCHET: *Astrophys. J.*, **399**, L5 (1992).
[31] See, *e.g.*, N. TUROK: *Phys. Rev. Lett.*, **63**, 2652 (1989); A. GOODING, D. SPERGEL and N. TUROK: *Astrophys. J.*, **372**, L5 (1991).
[32] See, *e.g.*, C. W. MISNER: *Astrophys. J.*, **151**, 431 (1968); R. PENROSE: in *General Relativity: An Einstein Centenary Survey*, edited by S. W. HAWKING and W. ISRAEL (Cambridge University Press, Cambridge, 1979), p. 581; R. H. DICKE and P. J. E. PEEBLES: in *General Relativity: An Einstein Centenary Survey*, edited by S. W. HAWKING and W. ISRAEL (Cambridge University Press, Cambridge, 1979), p. 504.
[33] J. FRY, C. T. HILL and D. N. SCHRAMM: *Comments Nucl. Part. Phys.*, **19**, 25 (1989); A. GUPTA, C. T. HILL, R. HOLMAN and E. W. KOLB: *Phys. Rev. D*, **45**, 441 (1992).
[34] J. PRESKILL: *Annu. Rev. Nucl. Part. Sci.*, **34**, 461 (1984).
[35] C. B. COLLINS and S. W. HAWKING: *Astrophys. J.*, **180**, 317 (1973).
[36] A. H. GUTH: *Phys. Rev. D*, **23**, 347 (1981).
[37] Y. HU, M. S. TURNER and E. J. WEINBERG: *Phys. Rev. D*, **49**, 3830 (1994).
[38] S. COLEMAN: *Phys. Rev. D*, **15**, 2929 (1977); S. COLEMAN and R. DE LUCCIA: *Phys. Rev. D*, **21**, 3305 (1980).
[39] R. WATKINS and L. WIDROW: *Nucl. Phys. B*, **374**, 446 (1992); S. W. HAWKING, J. STEWART and I. MOSS: *Phys. Rev. D*, **26**, 2681 (1982).
[40] A. D. LINDE: *Phys. Lett. B*, **108**, 389 (1982).
[41] A. ALBRECHT and P. J. STEINHARDT: *Phys. Rev. Lett.*, **48**, 1220 (1982).
[42] A. ALBRECHT, P. J. STEINHARDT, M. S. TURNER and F. WILCZEK: *Phys. Rev. Lett.*, **48**, 1437 (1982); L. ABBOTT and M. WISE: *Phys. Lett. B*, **117**, 29 (1982); A. D. LINDE and A. DOLGOV: *Phys. Lett. B*, **116**, 329 (1982).
[43] For a review of first-order inflation see, *e.g.*, E. W. KOLB: *Phys. Scr. T*, **36**, 199 (1991).
[44] D. LA and P. J. STEINHARDT: *Phys. Rev. Lett*, **62**, 376 (1989).
[45] F. ADAMS and K. FREESE: *Phys. Rev. D*, **43**, 353 (1991).

[46] E. W. KOLB, D. SALOPEK and M. S. TURNER: *Phys. Rev. D*, **42**, 3925 (1990).
[47] E. J. WEINBERG: *Phys. Rev. D*, **40**, 3950 (1989); M. S. TURNER, E. J. WEINBERG and L. WIDROW: *Phys. Rev. D*, **46**, 2384 (1992).
[48] Q. SHAFI and A. VILENKIN: *Phys. Rev. Lett.*, **52**, 691 (1984).
[49] S.-Y. PI: *Phys. Rev. Lett.*, **52**, 1725 (1984).
[50] L. KNOX and M. S. TURNER: *Phys. Rev. Lett.*, **70**, 371 (1993).
[51] K. A. OLIVE: *Phys. Rep.*, **190**, 307 (1990).
[52] R. HOLMAN, P. RAMOND and G. G. ROSS: *Phys. Lett. B*, **137**, 343 (1984).
[53] K. FREESE, J. FRIEMAN and A. OLINTO: *Phys. Rev. Lett.*, **65**, 3233 (1990).
[54] A. D. LINDE: *Phys. Lett. B*, **129**, 177 (1983).
[55] A. D. LINDE: *Phys. Lett. B*, **132**, 317 (1983); A. B. GONCHAROV and A. D. LINDE: *Phys. Lett. B*, **139**, 27 (1984).
[56] H. MURAYAMA, H. SUZUKI, T. YANAGIDA and J. YOKOYAMA: *Phys. Rev. Lett.*, **70**, 1912 (1993).
[57] Q. SHAFI and C. WETTERICH: *Phys. Lett. B*, **129**, 387 (1983); **152**, 51 (1985).
[58] A. A. STAROBINSKI: *Phys. Lett. B*, **91**, 99 (1980); M. B. MIJIC, M. S. MORRIS and W.-M. SUEN: *Phys. Rev. D*, **34**, 2934 (1986).
[59] M. S. TURNER and F. WILCZEK: *Phys. Rev. Lett.*, **65**, 3080 (1990); A. KOSOWSKY, M. S. TURNER and R. WATKINS: *Phys. Rev. Lett.*, **69**, 2026 (1992).
[60] R. D. REASENBERG, I. I. SHAPIRO, P. E. MACNEIL, R. B. GOLDSTEIN, J. C. BREIDENTAL, J. P. BRENKLE, D. L. CAIN, T. M. KAUFMAN, T. A. KOMAREK and A. I. ZYGIELBAUM: *Astrophys. J.*, **234**, L219 (1979).
[61] M. S. TURNER and L. WIDROW: *Phys. Rev. Lett.*, **57**, 2237 (1986).
[62] R. M. WALD: *Phys. Rev. D*, **28**, 2118 (1983); L. JENSEN and J. STEIN-SCHABES: *Phys. Rev. D*, **34**, 931 (1986).
[63] L. JENSEN and J. STEIN-SCHABES: *Phys. Rev. D*, **35**, 1146 (1987).
[64] A. A. STAROBINSKII: *JETP Lett.*, **37**, 66 (1983).
[65] J. A. FRIEMAN and M. S. TURNER: *Phys. Rev. D*, **30**, 265 (1994).
[66] D. GOLDWIRTH and T. PIRAN: *Phys. Rep.*, **214**, 223 (1992).
[67] See, *e.g.*, A. D. LINDE: *Inflation and Quantum Cosmology* (Academic Press, San Diego, Cal., 1990).
[68] A. H. GUTH and S.-Y. PI: *Phys. Rev. Lett.*, **49**, 1110 (1982); A. A. STAROBINSKII: *Phys. Lett. B*, **117**, 175 (1982); S. W. HAWKING: *Phys. Lett. B*, **115**, 295 (1982); J. M. BARDEEN, P. J. STEINHARDT and M. S. TURNER: *Phys. Rev. D*, **28**, 679 (1983).
[69] V. A. RUBAKOV, M. SAZHIN and A. VERYASKIN: *Phys. Lett. B*, **115**, 189 (1982); R. FABBRI and M. POLLOCK: *Phys. Lett. B*, **125**, 445 (1983); L. ABBOTT and M. WISE: *Nucl. Phys. B*, **244**, 541 (1984); B. ALLEN: *Phys. Rev. D*, **37**, 2078 (1988).
[70] See, *e.g.*, A. D. LINDE: *Phys. Lett. B*, **158**, 375 (1985); D. SECKEL and M. S. TURNER: *Phys. Rev. D*, **32**, 3178 (1985); M. S. TURNER, A. COHEN and D. KAPLAN: *Phys. Lett. B*, **216**, 20 (1989).
[71] At first sight, first-order inflation might seem very different from slow-rollover inflation, as reheating occurs through the nucleation of percolation of true-vacuum bubbles. However, such models can be recast as slow-rollover inflation by means of a conformal transformation, and the analysis of metric perturbations proceeds as in slow-rollover inflation. See, *e.g.*, E. W. KOLB, D. SALOPEK and M. S. TURNER: *Phys. Rev. D*, **42**, 3925 (1990).
[72] E. W. KOLB and M. S. TURNER: *The Early Universe* (Addison-Wesley, Redwood City, Cal., 1990), Chapt. 8.
[73] E. R. HARRISON: *Phys. Rev. D*, **1**, 2726 (1970); YA. B. ZEL'DOVICH: *Mon. Not. R. Astron. Soc.*, **160**, 1 (1972).
[74] P. J. STEINHARDT and M. S. TURNER: *Phys. Rev. D*, **29**, 2162 (1984).

[75] The material presented in this section is a summary of work completed during this school and will published elsewhere, M. S. Turner: *Phys. Rev. D*, **48**, 3502 (1993).
[76] M. S. Turner: *Phys. Rev. D*, **48**, 5539 (1993).
[77] J. M. Bardeen, J. R. Bono, N. Kaiser and A. Szalay: *Astrophys. J.*, **304**, 15 (1986).
[78] D. H. Lyth and E. D. Stewart: *Phys. Lett. B*, **274**, 168 (1992); E. D. Stewart and D. H. Lyth: *Phys. Lett. B*, **302**, 171 (1993).
[79] L. Abbott and M. Wise: *Nucl. Phys. B*, **244**, 541 (1984).
[80] M. White: *Phys. Rev. D*, **46**, 4198 (1992).
[81] M. S. Turner, M. White and J. E. Lidsey: *Phys. Rev. D*, **48**, 4613 (1993).
[82] L. Abbott and M. Wise: *Nucl. Phys. B*, **244**, 541 (1984); F. Lucchin and S. Matarrese: *Phys. Rev. D*, **32**, 1316 (1985); R. Fabbri, F. Lucchin and S. Matarrese: *Phys. Lett. B*, **166**, 49 (1986).
[83] V. Belinsky, L. Grishchuk, I. Khalatanikov and Ya. B. Zel'dovich: *Phys. Lett. B*, **155**, 232 (1985); L. Jensen: unpublished (1985).
[84] A. A. Starobinskii: *Sov. Astron.*, **11**, 133 (1985).
[85] A. D. Linde: *Phys. Lett. B*, **108**, 389 (1982); A. Albrecht and P. J. Steinhardt: *Phys. Rev. Lett.*, **48**, 1220 (1982).
[86] S. D. M. White, C. Frenk and M. Davis: *Astrophys. J.*, **274**, L1 (1983); **287**, 1 (1983); J. Centrella and A. Melott: *Nature (London)*, **305**, 196 (1982).
[87] J. P. Ostriker: *Annu. Rev. Astron. Astrophys.*, **31**, 689 (1993).
[88] L. Knox and M. S. Turner: unpublished (1993).
[89] V. De Lapparent, M. Geller and J. Huchra: *Astrophys. J.*, **302**, L1 (1986); **332**, 44 (1988); M. Geller and J. Huchra: *Science*, **246**, 897 (1989).
[90] N. Bahcall: *Annu. Rev. Astron. Astrophys.*, **26**, 631 (1988).
[91] M. Davis, G. Efstathiou, C. S. Frenk and S. D. M. White: *Astrophys. J.*, **292**, 371 (1985).
[92] T. Broadhurst, R. S. Ellis, D. C. Koo and A. S. Szalay: *Nature (London)*, **343**, 726 (1990).
[93] S. J. Maddox, G. Efstathiou, W. J. Sutherland and J. Loveday: *Mon. Not. R. Astron. Soc.*, **242**, 43p (1990).
[94] A. Dressler, S. M. Faber, D. Burnstein, R. L. Davies, D. Lynden-Bell, R. J. Terlevich and G. Wegner: *Astrophys. J.*, **313**, L37 (1987); E. Bertschinger, A. Dekel, S. M. Faber, A. Dressler and D. Burnstein: *Astrophys. J.*, **364**, 370 (1990), and references therein.
[95] K. Fisher, M. A. Strauss, M. Davis, A. Yahil and J. P. Huchra: *Astrophys. J.*, **389**, 188 (1992); also see M. S. Vogeley, C. Park, M. J. Geller and J. P. Huchra: *Astrophys. J.*, **391**, L5 (1992); W. Saunders, C. Frenk, M. Rowan-Robinson, G. Efstathiou, A. Lawrence, N. Kaiser, R. Ellis, J. Crawford, X-Y. Xia and I. Parry: *Nature (London)*, **349**, 42 (1991).
[96] J. Loveday, B. A. Peterson, G. Efstathiou and S. J. Maddox: *Astrophys. J.*, **390**, 338 (1992); **400**, L43 (1992).
[97] The Sloan Digital Sky Survey is a collaboration between The University of Chicago, Fermilab, Johns Hopkins University, Princeton University and the Institute for Advanced Study.
[98] N. Kaiser: *Astrophys. J.*, **284**, L9 (1984); in *Inner Space/Outer Space*, edited by E. W. Kolb, M. S. Turner, D. Lindley, K. Olive and D. Seckel (University of Chicago Press, Chicago, Ill., 1986), p. 258.
[99] G. Efstathiou, J. R. Bond and S. D. M. White: *Mon. Not. R. Astron. Soc.*, **258**, 1p (1992).

[100] R. Y. Cen and J. P. Ostriker: *Astrophys. J.*, **393**, 22 (1992); A. E. Evrard, F. J. Summers and M. Davis: *Astrophys. J.*, in press (1994); E. Bertschinger and J. Gelb: *Comput. Phys.*, **5**, 164 (1991); J. Gelb and E. Bertschinger: *Astrophys. J.*, in press (1994); N. Katz, L. Hernquist and D. Weinberg: *Astrophys. J.*, **399**, L109 (1992); G. Evrard: *Mon. Not. R. Astron. Soc.*, **235**, 911 (1988); C. Park and J. R. Gott: *Mon. Not. R. Astron. Soc.*, **249**, 288 (1991); C. Park: *Mon. Not. R. Astron. Soc.*, **242**, 59p (1990); C. Frenk, S. D. M. White, G. Efstathiou and M. Davis: *Astrophys. J.*, **351**, 10 (1990).

[101] For a synopsis of structure formation in the CDM scenario see, *e.g.*, G. R. Blumenthal, S. Faber, J. Primack and M. J. Rees: *Nature (London)*, **311**, 517 (1984).

[102] See, *e.g.*, J. P. Ostriker: *Annu. Rev. Astron. Astrophys.*, **31**, 689 (1993); F. Adams, J. R. Bond, K. Freese, J. Frieman and A. J. Olinto: *Phys. Rev. D*, **47**, 426 (1993); J. Gelb, J. Frieman and B. Gradwohl: *Astrophys. J.*, **403**, L5 (1993); R. Cen, N. Gnedin, L. Kofman and J. P. Ostriker: *Astrophys. J.*, **399**, L11 (1992).

[103] R. Davis, H. Hodges, G. Smooth, P. J. Steinhardt and M. S. Turner: *Phys. Rev. Lett.*, **69**, 1856 (1992); F. Lucchin, S. Matarrese and S. Mollerach: *Astrophys. J.*, **401**, L49 (1992); D. Salopek: *Phys. Rev. Lett.*, **69**, 3602 (1992); A. Liddle and D. Lyth: *Phys. Lett. B*, **291**, 391 (1992); J. E. Lidsey and P. Coles: *Mon. Not. R. Astron. Soc.*, **358**, 57 (1992); T. Souradeep and V. Sahni: *Mod. Phys. Lett. A*, **7**, 3541 (1992).

[104] M. S. Turner, G. Steigman and L. Krauss: *Phys. Rev. Lett.*, **52**, 2090 (1984); M. S. Turner: *Phys. Scr. T*, **36**, 167 (1991); P. J. E. Peebles: *Astrophys. J.*, **284**, 439 (1984); G. Efstathiou, W. J. Sutherland and S. J. Maddox: *Nature (London)*, **348**, 705 (1990); L. Kofman and A. A. Starobinskii: *Sov. Astron. Lett.*, **11**, 271 (1985).

[105] Q. Shafi and F. Stecker: *Phys. Rev. Lett.*, **53**, 1292 (1984); S. Achilli, F. Occhionero and R. Scaramella: *Astrophys. J.*, **299**, 577 (1985); S. Ikeuchi, C. Norman and Y. Zahn: *Astrophys. J.*, **324**, 22 (1988); A. van Dalen and R. K. Schaefer: *Astrophys. J.*, **398**, 33 (1992); M. Davis, F. Summers and D. Schlegel: *Nature (London)*, **359**, 393 (1992); J. Holtzman and J. A. Primack: *Astrophys. J.*, in press (1993); A. Klypin, J. Holtzman, J. Primack and Enikö Regös: *Astrophys. J.*, **416**, 1 (1993); D. Pogosyan and A. A. Starobinsky: DAMTP/IOA/MRAO preprint (1993).

[106] M. S. Turner and B. J. Carr: *Mod. Phys. Lett. A*, **2**, 1 (1987).

[107] T. Gaier, J. Schuster, J. Gundersen, T. Koch, M. Sieffert, P. Meinhold and P. Lubin: *Astrophys. J.*, **398**, L1 (1992); D. C. Alsop, E. S. Cheng, A. C. Clapp, D. A. Cottingham, M. L. Fischer, J. O. Gundersen, E. Kreysa, A. E. Lange, P. M. Lubin and P. R. Meinhold: *Astrophys. J.*, **395**, 317 (1992); A. C. S. Readhead, C. R. Lawrence, S. T. Myers, W. L. W. Sargent, H. E. Hardebeck and A. T. Moffet: *Astrophys. J.*, **346**, 556 (1989); P. DeBernardis, S. Masi, F. Melchiorri, B. Melchiorri and N. Vittorio: *Astrophys. J.*, **396**, L57 (1992); R. A. Watson, C. M. G. Delacruz, R. D. Davies, A. N. Lasenby, R. Rebolo, J. E. Beckman and S. Hancock: *Nature (London)*, **357**, 660 (1992); S. S. Meyer, E. S. Cheng and L. A. Page: *Astrophys. J.*, **371**, L1 (1991); J. O. Gundersen *et al.*: *Astrophys. J.*, in press (1993); P. R. Meinhold *et al.*: *Astrophys. J.*, in press (1993).

[108] R. Crittenden, J. R. Bond, R. Davis, G. Efstathiou and P. J. Steinhardt: *Phys. Rev. Lett.*, in press (1993).

[109] M. S. Turner and L. M. Widrow: *Phys. Rev. D*, **37**, 2743 (1988); B. Ratra: *Astrophys. J.*, **391**, L1 (1992).

[110] A. Abramovici, W. E. Althouse, R. W. P. Drever, Y. Gursel, S. Kawamura,

F. J. RAAB, D. SHOEMAKER, L. SIEVERS, R. E. SPERO, K. S. THORNE, R. E. VOGT, R. WEISS, S. E. WHITCOMB and M. E. ZUCKER: *Science*, **256**, 325 (1992); K. S. THORNE: in *300 Years of Gravitation*, edited by S. W. HAWKING and W. ISRAEL (Cambridge University Press, Cambridge, 1987), p. 330.

[111] See, *e.g.*, J. R. PRIMACK, D. SECKEL and B. SADOULET: *Annu. Rev. Nucl. Part. Sci.*, **38**, 751 (1988); D. O. CALDWELL: *Mod. Phys. Lett. A*, **5**, 1543 (1990); P. F. SMITH and J. D. LEWIN: *Phys. Rep.*, **187**, 203 (1990).

[112] U.-L. PEN, D. N. SPERGEL and N. TUROK: Princeton Observatory preprint 485 (1993); D. COULSON, D. SPERGEL and N. TUROK: work in progress (1993).

Cosmic Microwave Background.

P. L. RICHARDS

Department of Physics and Center for Particle Astrophysics
University of California - Berkeley, Cal. 94720

1. – Introduction.

The following is a summary of three lectures given at the International School of Physics at Varenna, in the Summer of 1992. These lectures were intended to give the students at the school a context for understanding recent results from the COBE satellite.

The cosmic microwave background (CMB), sometimes called the cosmic background radiation, was discovered by PENZIAS and WILSON in 1965[1]. Their direct microwave measurement detected an isotropic sky emission corresponding to $T = 3.5$ K at 4.08 GHz. This discovery was not driven by cosmological ideas, but rather by the availability of a radiotelescope designed for satellite communication which had very low sidelobe response. PENZIAS and WILSON realized that this telescope was appropriate for measurements of very diffuse radiation and constructed a Dicke-switched radiometer which compared the sky signal with a LHe-cooled cold load. The radiation that they discovered was more widely distributed than anticipated.

As is the case with many important scientific discoveries, previous experiments interpreted with hindsight give evidence for the CMB. The earliest such experiment appears to be a measurement of an excitation temperature of 2.3 K for interstellar molecular CN by MCKELLAR in 1940 which was reported in 1950[2]. In 1946, DICKE et al.[3] reported a direct measurement of the sky temperature giving a value < 20 K.

The first theoretical basis for expecting a CMB was given by GAMOW[4] in 1948 and by ALPHER and HERMAN also in 1948[5]. These early explorations of nucleosynthesis led naturally to the idea of a background temperature for the Universe. Unfortunately, their findings were not communicated to experimentalists who were capable of making such measurements. Motivated by cyclic universe models, DICKE, ROLL and WILKINSON[6] set up an experimental program which was a few months away from measuring the background when the Penzias and Wilson result was announced.

The existence of the CMB follows directly from the idea of a hot big bang. It is assumed that at a sufficiently early time the Universe was hot enough to ionize all matter. The resulting plasma interacted sufficiently strongly with electromagnetic radiation that the photons were in thermal equilibrium with the matter. This equilibrium was maintained until the Universe cooled sufficiently for recombination of the electrons onto the nuclei. At this time the photons became decoupled from the matter and pursued their separate evolution, which was adiabatic (constant entropy) expansion. According to elementary statistical physics, the number of photons per mode in thermal equilibrium at temperature T is $n = (\exp[h\nu/kT] - 1)^{-1}$. Since the entropy of the photon gas depends on n, this number is constant during adiabatic expansion. This can occur only if the temperature T is proportional to frequency ν, which in turn is inversely proportional to the scale factor of the Universe. The consequence is that the temperature of the CMB evolves as $1 + z$.

Many reviews of CMB measurements are available in the literature. An excellent early review is given by WEISS[7]. USON and WILKINSON have given a very readable review with extensive lists of experimental results[8]. READHEAD and LAWRENCE reviewed the isotropy of the CMB in 1992[9]. In addition, the papers describing results from the Cosmic Background Explorer Satellite (COBE) include careful summaries of previously published results[10-17].

This lecture will not attempt to duplicate these reviews. It will focus on experimental issues and attempt to provide the background that will help students to understand the recent experimental results and place them in context.

2. – Receivers for direct measurements.

It should be noted that CMB measurements are quite different from experiments done to measure higher-energy photons. The photon rates are very high, but such low-energy photons cannot be detected individually. Although there are limits to measurement accuracy set by fluctuations in the rate of arrival of photons, measurements are usually limited by receiver or environmental noise.

In the design of an experiment, a choice must be made of the throughput (Etendu) $A\Omega$. Here A is the area of the beam at focus and Ω is the solid angle of divergence of the beam. In single-mode systems, where the signal is coupled to a device which is small compared to a wavelength, the throughput is limited to one polarization with $A\Omega = \lambda^2$. This regime is referred to as diffraction limited. The power reaching the receiver in a single mode from a black-body source can be written $P_\nu \, d\nu = h\nu \, d\nu (\exp[h\nu/kT] - 1)^{-1}$. This function is flat at low frequencies and rolls off exponentially when $h\nu > kT$. In multimode optical systems the throughput $A\Omega$ is a constant set by the size of optical components. The power reaching such a receiver is equal to the number of modes $A\Omega/\lambda^2$ times the power per mode $P_\nu \, d\nu$ times 2 for polarization. Consequently it can be written as $2A\Omega\lambda^{-2}\{h\nu \, d\nu(\exp[h\nu/kT] - 1)^{-1}\}$. For constant $A\Omega$, this function increases as frequency squared at low frequencies, has a peak near $h\nu = kT$, and decreases at higher frequencies. This is the peaked Planck spectrum seen in textbooks.

It should be noted that in the low-frequency, or Rayleigh-Jeans, limit the power from a black-body is proportional to T. For a single mode, $P_\nu = kT$. For many modes, $P_\nu = 2A\Omega\nu^2 kT/c^2$. In this low-frequency limit, T is a measure of power. These equations relating P_ν and T are inverted to provide the definition of antenna temperature T_A at all frequencies. The antenna temperature T_A is equal to the thermodynamic temperature for $h\nu \ll kT$ but not for higher frequencies.

It is also necessary to choose the bandpass B that is measured. Because the spectrum of the CMB is a broad continuum, the measured signal typically varies in proportion to B. Many sources of system noise vary as $B^{1/2}$. Consequently, the ratio of signal to noise often varies as $B^{1/2}$, so a broad bandpass can be useful.

It is necessary to convert the radiofrequency signal to a d.c. output. This is either done by direct detection or by heterodyne down-conversion followed by direct detection. For purposes of calibration and removal of the effects of slow changes in receiver response, most receivers alternately view the sky and a cold reference. For absolute temperature measurements, this reference is a cold black-body or load. For anisotropy measurements, this reference is usually (but not always) another patch of sky. The final data correspond to the difference between the signals received from these two sources.

Receivers for measurements of the cosmic microwave background are made up of generic building blocks well known to radio and optical engineers. The components chosen for a given receiver depend upon the frequency, the noise requirements, etc. A list of these generic receiver components can be made as follows:

A modulator or Dicke switch is used to switch the receiver between the sky and the cold reference. Examples of modulators include a moving mirror in front of the antenna, a waveguide switch after the antenna, or a Michelson Fourier spectrometer. Also, the aperture synthesis techniques of radioastronomy can be used both to select the fields viewed and to switch between them, or differences between carefully matched bolometers can be measured directly.

An antenna is used as a spatial mode filter. This can be a parabolic reflector in combination with a feed horn for a narrow field of view, or one of several kinds of horn used alone for a wider field of view.

A linear amplifier can be used at the front end of a receiver for radiofrequencies (RF) below ~ 100 GHz. Examples include masers, parametric amplifiers and transistor amplifiers. High-electron-mobility transistor (HEMT) amplifiers now give the best combination of low noise and broad bandwidth.

A mixer can be used to down-convert the incoming RF signal to a lower, or intermediate (IF) frequency, where it is more easily amplified and filtered. The mixer multiplies the sky signal with that from a local oscillator. Such mixers are made with Schottky and superconductor-insulator-superconductor (SIS) diodes.

Bandpass filters are used to reject unwanted radiation.

Multimode square-law detectors such as bolometers or single-mode detectors such as Schottky diodes can be used to directly convert RF or IF signals to d.c.

Lock-in (phase sensitive) detectors multiply the signal by a reference signal derived from the Dicke switch. The steady-state lock-in output is then proportional to the difference between the sky signal and the signal from the cold reference. This step is often implemented by digital software.

Measurements made over extended periods of time are then averaged. This is equivalent to a low (frequency)-pass filter to reject high-frequency noise.

Measurements of the CMB have challenged the ingenuity of receiver developers, and many different types of receiver have been used. It is useful to consider several generic receiver types. Although receivers are made up from many components, they are traditionally named for the first active or nonlinear component, whether it be a heterodyne mixer, a HEMT amplifier, or a bolometric detector.

In a typical heterodyne receiver used for frequencies $\lesssim 90$ GHz, the sky and the reference are viewed through two single-mode scalar horn antennas and a waveguide Dicke switch is used to alternately connect the antennas with the mixer. A mixer then multiplies the switched sky signal with the signal from a local oscillator which produces an output at the sum and difference frequencies. The difference frequency, or IF, is then bandpass-filtered, amplified and rectified in a Schottky diode used as a square-law detector. As with all receivers, the diode output is then lock-in detected using the Dicke switch as reference and the output is low-pass filtered or averaged and stored.

The availability of radiofrequency amplifiers made with high-electron-mobility transistors (HEMT) with low noise makes an alternative single-mode receiver attractive. After the sky radiation is collected and Dicke-switched as described above, it is amplified and bandpass filtered at the RF without down-conversion to an IF. Finally, it is rectified in a diode and lock-in detected using the drive for the Dicke switch as the reference. In an aperture synthesis interferometer, switching of the field of view on the sky is achieved by adding two or more beams with a time-varying phase shift before rectification. The receivers described thus far are limited to a single mode $A\Omega = \lambda^2$.

Bolometric receivers are commonly used for frequencies ≥ 90 GHz. The sky signal is chopped against a reference with a moving mirror as a Dicke switch and collected by an antenna which defines the throughput $A\Omega$. This signal passes through an optical bandpass filter and falls on a bolometric (thermal) radiation detector. The output of the bolometer is then amplified and lock-in detected using the Dicke switch as the reference, and then averaged and stored. Bolometric receivers usually have multimode throughput, but can use single-mode optics such as scalar horns and waveguides. They typically have wide bandwidths. They require very low temperatures to obtain high sensitivity.

Many different sources of noise in receivers give a fluctuating contribution to the receiver output. The use of a Dicke switch in front of the receiver with a lock-in amplifier and low-pass filter after it means that receiver noise is important only in a narrow band around the switch frequency. Receiver noise is often characterized by a noise temperature T_N. This noise temperature is defined as the physical temperature of a matched load or black-body at the receiver input that contributes an amount of noise to the output equal to that from the receiver itself. In most receivers, the dominant contribution to the noise temperature comes from the first active, or nonlinear, component.

Heterodyne down-converters and linear amplifiers preserve phase while increasing the number of photons. Consequently, they are subject to quantum noise which can be expressed as $T_N \gtrsim h\nu/k$. In practice, the noise temperature of present HEMT amplifiers is significantly larger than quantum noise and its numerical value in kelvin is approximately equal to the frequency in GHz. Improved InP HEMT's may reduce this noise by a factor ~ 2.

Noise in bolometric receivers is a complicated subject. Because of the relatively wide bandwidths used and the relatively narrow dynamic range of bolometers, the sensitivity of the receiver depends on the level of background radiation in the band. This background can arise from warm parts of the apparatus, the atmosphere, or even the CMB. It causes both heating of the bolometer and fluctuations due to photon noise. Low emissivity or cold optics and cold bandpass filters to reduce the background are important features of experimental design. General considerations favor coherent single-mode receivers such as HEMT's for low frequencies and narrow bandwidths and direct detector receivers such as bolometers for high frequencies and broad bandwidths [18].

All of the receivers which we have described involve square-law detection. This process can be carried out in a bolometric detector or in a diode detector. Noise is processed in a characteristic way by square-law detectors, which is summarized by the well-known Dicke formula. In a system with noise temperature T_N, the antenna temperature difference which gives a signal level equal to the noise is $\Delta T = T_N(B\tau)^{-1/2}$. Here B is the bandwidth before detection and τ is the integration time after detection. Averaging after detection over time τ effectively introduces a low-pass filter with post-detection bandwidth $B_2 = (2\tau)^{-1}$. This noise-equivalent antenna temperature can be converted to a noise-equivalent power using the definition of antenna temperature given above.

3. – Sources of interference.

Compared with the CMB, all other signals are local interference. An analysis of extragalactic point sources which might confuse measurements of the CMB has been given by FRANCESCHINI et al. in 1989 [19]. These point sources are not predicted to be strong enough to interfere with measurements of the spectrum of the CMB, but can influence measurements of anisotropy at the level $\Delta T/T \lesssim 10^{-5}$. Unpublished lists of bright, flat-spectrum radiosources exist that have been compiled by major millimeter wave radio-observatories for use in telescope pointing and calibration.

Synchrotron radiation is produced by electrons spiraling in the magnetic field of our galaxy. The antenna temperature of this radiation scales approximately as $\nu^{-2.7}$. A map of sky brightness at 408 MHz produced by HASLAM et al. [20] is dominated by synchrotron radiation. It is a common practice to extrapolate these measurements to higher frequencies to estimate the contribution of synchrotron radiation to CMB anisotropy measurements. Synchrotron radiation is also a serious source of confusion for low-frequency CMB spectrum measurements.

Free-free emission, or thermal bremsstrahlung, is a little-measured component of the galactic emission which is expected to confuse CMB anisotropy measurements. This radiation arises from scattering events between electrons and protons. Its antenna temperature scales as $\nu^{-2.1}$ in the frequency band of CMB measurements. Its distribution can be estimated from measurements of H_α emission at 6563 Å. Measurements of H_α emission have been made on limited areas of the sky by REYNOLDS [21]. Estimates based on these limited data suggest that the free-free emission may be comparable to the synchrotron emission from our galaxy in the frequency range between 50 and 100 GHz. A conservative upper limit on free-free emission has been obtained by making the assumption that all of the brightness in the Haslam map is free-free emission and extrapolating to higher frequencies as $\nu^{-2.1}$.

Thermal emission from galactic dust is an important contaminant for CMB anisotropy measurements. The temperature of this dust is set by the balance between the absorption of starlight from the interstellar radiation field and the emission of infrared. Because of the shift of the spectrum to higher frequencies, hot spots radiate much more strongly than cooler regions, producing a relatively uniform temperature. Most of the galactic dust emission is surprisingly well characterized by a thermal spectrum with a single temperature of ~ 20 K and an emissivity which varies as $\nu^{+1.5}$. Some idea of the galactic dust emission can be deduced from the IRAS 100 μm galactic dust emission maps. Such maps are used to locate regions of the sky where intermediate-scale CMB anisotropy measurements are not strongly influenced by dust emission. On half-degree angular scales, the dust emission contrast observed at 400 GHz is well correlated with the IRAS map at a factor ~ 8 higher frequency [22].

There is evidence for the existence of a component of cold dust in our galaxy [11, 23]. There is concern that such dust may have a morphology which is unrelated to the IRAS dust, and may, therefore, be difficult to separate from anisotropy of the CMB. Thus far, no convincing direct measurements of the emission from cold dust have been made at high galactic latitudes where CMB anisotropy experiments are carried out.

Dust in the solar system has a much higher temperature than galactic dust but, due to the small column density, a lower antenna temperature in the frequency band of CMB measurements. Analysis of measurements from IRAS [24] and early results from the diffuse infrared background experiment (DIRBE) on COBE have shown that this dust emission is more uniformly distributed than galactic dust, but does have structure in the form of bands symmetrically disposed around the plane of the ecliptic. The effect of this radiation on high-frequency CMB measurements is minimized by avoiding the ecliptic plane. BENNETT et al. [14] give a careful account of galactic interference on the angular scales of the COBE DMR experiment. Each component, however, can be expected to have its own dependence on angular scale.

The transparency of the atmosphere is high enough that it is not a serious problem for CMB measurements, but molecular emission from the atmospheric lines is often a serious source of confusion. Atmospheric oxygen has emission lines at 60 and 120 GHz as well as at higher frequencies. The 60 GHz line is split by the Earth's magnetic field into numerous components and is difficult to model theoretically. Ozone emission occurs in bands of many weak lines beginning near the peak of the CMB spectrum and extending to higher frequencies. Atmospheric water has a smaller number of strong lines over the same frequency range.

Contributions of atmospheric emission to measurements of the CMB spectrum are subtracted by measuring the sky brightness as a function of angle from the zenith and extrapolating to zero atmosphere, or by subtracting a calculated spectrum. The molecular parameters of important atmospheric constituents are thoroughly studied. Calculations of atmospheric emission at ground-based sites have limited accuracy because of uncertainties with line shapes. They are more accurate for balloon elevations. In anisotropy measurements, the chopping is in azimuth to avoid any signal from a horizontally stratified stable atmosphere. Fluctuations of the atmosphere near the chopping frequency, however, contribute noise. Horizontal gradients in the emission due to weather systems can mimic CMB anisotropy unless measurements are available at several frequencies.

Measurements made below ~ 5 GHz are not seriously troubled by atmospheric emission and can be carried out wherever convenient. Observations of both the spectrum and the anisotropy of the microwave background are made up to ~ 45 GHz and in a window near 90 GHz from high dry sites. The South Pole which has a pressure altitude of 3.4 km has proved useful.

For frequencies above 90 GHz, balloons, rockets or spacecraft are required.

In summary, the most favorable frequency range for measurements of both the spectrum and the anisotropy of the CMB is close to the peak in the (constant throughput) Planck spectrum. The atmospheric emission becomes negligible at lower frequencies, but galactic synchrotron radiation grows very rapidly. Free-free emission may prove troublesome on the low-frequency side of the peak. For frequencies at or beyond the peak, the atmospheric emission is a serious problem for ground-based experiments. At sufficiently high frequencies, the rise in atmospheric emission and galactic dust emission with frequency will interfere with CMB anisotropy measurements.

Thermal emission from the apparatus or the Earth can cause serious problems for CMB measurements. Diffraction effects cause any antenna to receive some signal from sources at large angles from the center of the beam. This effect is usually minimized by nonuniform illumination, or apodization, of the antenna aperture. For example, the far-field antenna pattern resulting from a Gaussian illumination of the antenna aperture falls off much more rapidly with angle than the Airy pattern of the uniformly filled aperture. Reflecting ground is commonly used so that the signal observed in the «far sidelobes» comes from the relatively cold sky and not from the Earth.

In the Rayleigh-Jeans region of the CMB spectrum below 90 GHz, the brightness of the Earth is $\sim 10^2$ higher than that of the CMB. This factor increases exponentially with increasing frequency as the CMB spectrum begins to fall. Since the Earth subtends a much larger solid angle than the central lobe of the antenna, the antenna pattern of successful CMB measurements must fall from 10^{-6} to 10^{-9} at the horizon, depending on the frequency and kind of measurement. The antenna temperature of the Earth is independent of frequency. The spectral index is reduced by a factor ν for each time the radiation is diffracted. Since the effective number of diffractions is often not known at large angles, however, earthshine diffracted into the apparatus cannot be directly identified from its spectrum.

4. – Microwave measurements of the CMB spectrum.

Microwave heterodyne receivers are sufficiently sensitive to measure the spectrum of the CMB without serious problems from receiver noise. For a relatively poor receiver with noise temperature $T_N = 10^3$ K and bandwidth $B = 10^8$ Hz, the Dicke formula gives $\Delta T = T_R (B\tau)^{-1/2} = 0.1$ K for $\tau = 1$ s.

An important factor in the accuracy of CMB spectral measurements is the construction of cold black-body loads with near-unit emissivity and well-defined temperature. In many microwave experiments, the cold load is made from commercial microwave-absorbing foam immersed in liquid helium at the bottom of an upward-looking cryostat. Care must be taken to correct for the emission from any window over the cryostat, from the warm walls of the cryostat, and even from reflections at the surface of the liquid helium. Commercial absorbing foams are designed to have little surface reflection and are often configured in arrays of pyramids to further minimize reflection. Tests must be carried out to ensure that foam absorbers designed for ambient temperature are still effective at LHe temperatures.

Cold loads can be operated in a vacuum if they are cooled by conduction from a LHe bath. Such loads have the advantage that their temperature can be varied over a wide range. Since the thermal conductance of the absorbing material must be high enough to ensure temperature uniformity, foams are not appropriate for this application. The usual solution is to use a solid absorber such as an iron-filled epoxy in the form of a deep circular cone. A high-quality load can be made in this configuration even from materials with surface reflectance $\sim 10\%$. The surface of the cone must be smooth to avoid scattering and the geometry must be chosen to ensure a large enough number of reflections before a ray entering the cone returns to the entrance.

Successful microwave measurements of the spectrum of the CMB have been carried out from ~ 400 MHz to ~ 90 GHz. There are many detailed differences in the instrumental designs used over this frequency range at different times by different investigators. There is, however, a common pattern to the measurements which can be illustrated by considering the procedure for data analysis. When the instrument is chopped between the sky and a reference, the various contributions to the antenna temperature which characterizes the instrumental output can be written in the form [9]

$$T_{\text{sky}} - T_{\text{ref}} = T_{\text{CMB}} + T_{\text{atmosphere}} + T_{\text{galaxy}} + T_{\text{ground}} + T_{\text{horn}} + T_{\text{offset}} .$$

When the instrument is chopped between a cold load in the sky horn and the reference, the signal can be written in the form

$$T_{\text{CL}} - T_{\text{ref}} = T_{\text{He}} + T_{\text{wall}} + T_{\text{horn}} + T'_{\text{offset}} .$$

The difference between the signals is, therefore,

$$T_{\text{sky}} - T_{\text{CL}} = T_{\text{CMB}} + T_{\text{atmosphere}} + T_{\text{galaxy}} + T_{\text{ground}} - T_{\text{He}} - T_{\text{wall}} + \Delta T_{\text{offset}} .$$

In order to obtain T_{CMB}, it is necessary to obtain values for the other terms and subtract them. The atmospheric contribution is usually evaluated by measuring data as a function of zenith angle θ. Assuming that the beam is well defined and that the atmosphere is horizontally stratified, $T_{\text{atmosphere}}(0) = T_{\text{atmosphere}}(\theta)/\sec\theta$. This method of atmospheric correction is adequate at low frequencies, but becomes increasingly difficult above 30 GHz, even for the best ground-based sites.

Values of T_{galaxy} can be obtained by convolving the beam with the Haslam map. Another procedure is to make measurements at two or more frequencies and to eliminate that part of the data which varies with frequency as expected for synchrotron radiation or free-free emission. Subtraction of T_{galaxy} gets increasingly difficult at low frequencies where galactic synchrotron radiation becomes very large. The contributions from the walls of the cold load T_{wall} are minimized by good design and calculated corrections are applied. Ideally, the calibration loads should be placed in front of the sky horn so that horn emission is directly subtracted in the calibration. This can prove difficult at low frequencies because the size of the antenna increases with wavelength. Sometimes the calibration is carried out behind the sky horn and estimates of the sky horn contribution are subtracted. The signal from the ground T_{ground} is minimized by using a well-designed antenna and by surrounding the receiver with ground shields. Again this becomes very difficult at low frequencies because of the size of equipment required.

Microwave measurements of the spectrum of the CMB are typically limited by systematic errors. When nonstatistical systematic errors are dominant, the most difficult quantity to evaluate is the size of the error. Therefore, the practice of computing averages of different measurements of CMB temperature weighted according to their stated errors has been criticized as favoring the data of the overly optimistic experimentalist.

Because it is difficult to decide which experiments are accurate and which are not, the practice has developed of quoting more recent experiments and discarding older ones. This practice is unfair to the early measurements by WILKINSON and collaborators[8] which have proved in hindsight to be of high quality. It also leaves out the pioneering work by HOWELL and SHAKESHAFT[25], whose measurements at very low frequencies where the galactic contamination is very large are still important. It is interesting to note in this regard that a compilation of microwave measurements made by PEEBLES in 1971[26] gave a weighted average of 2.73 K which is extremely close to the presently accepted value.

More recently, a large number of experiments have been carried out by a US-Italian collaboration led by SMOOT[27]. A particularly successful measurement at 24 GHz was made by JOHNSON and WILKINSON in 1986[28]. The significant atmospheric contribution at this frequency from ground-based sites was eliminated by observing from a balloon platform.

5. – Direct measurements at frequencies beyond the peak.

Several factors hindered direct measurements of the temperature of the CMB beyond the peak. The ratio of the brightness of a 2.7 K black-body to one at 300 K is 10^{-2} at low frequencies, reaches 10^{-3} near the peak of the Planck curve, and decreases exponentially at higher frequencies. As a consequence, all parts of the apparatus in contact with the radiation must be cooled with liquid helium. The sidelobe and atmospheric problems are much worse than at lower frequencies. Successful measurements have been done from balloon, rocket and spacecraft platforms.

Because heterodyne or RF amplifier technology is less well developed at high frequencies, measurements beyond the peak are generally made with bolometric detectors. The metal film absorber for the widely used composite bolometric detector was invented with the requirements of CMB measurements in mind [29]. Bolometric receivers require spectral discrimination. Many experiments use fixed-frequency band-pass filters based on a series of Fabry-Perot etalons made from metal mesh. Other experiments have used versions of the Martin-Puplett polarizing Michelson interferometer [30] which was also invented for CMB measurements. This device is used as a Fourier-transform spectrometer to produce a spectrum of the detected power as a function of frequency and as a Dicke switch. The requirements of high-frequency CMB measurements have also contributed to novel antenna technologies, such as the Winston light concentrator with apodizing flare [31].

Building on the balloon technology developed by WEISS and collaborators, WOODY and RICHARDS [32] were able to provide definitive evidence that the spectrum of the CMB decreases with frequency beyond a peak near 6 cm^{-1} (180 GHz). This balloon experiment used a liquid-helium-cooled Winston cone antenna with a linear flare at the entrance and a polarizing Michelson Fourier spectrometer. The detector was the first ^{3}He-cooled bolometric detector to be used for astronomy. Calibration was done before the flight using a cold load inserted into the antenna. The detector response from 3 to 40 cm^{-1} showed a bump characteristic of a ~ 3 K black-body plus emission lines characteristic of atmospheric O_2, O_3 and H_2O. When an atmospheric model was subtracted from the data and the calibrations were applied, the estimate of the brightness of the night sky as a function of frequency clearly showed the peaked form of a Planck spectrum for $T \sim 3$ K.

A statistical analysis of the known errors in this experiment showed that the measured spectrum was not quite consistent with a Planck curve. The best-fit temperature of 2.96 K was significantly higher than the average of existing measurements at lower frequencies. Despite warnings about the possibility of undetected systematic errors, this indication of an excess generated considerable theoretical activity. The general conclusion was that the excess was difficult to reconcile with mainstream cosmological theories. The basic point is that, since the 10^9 CMB photons per baryon dominate the entropy of the Universe, it is very difficult to find a mechanism that can cause a $\sim 10\%$ distortion in the spectrum of the CMB.

PETERSON et al. [33] modified the experiment by replacing the Fourier spectrometer with band-pass filters and by using zenith-angle scans, rather than spectral fits to subtract the atmospheric interference. The results gave the clue that the upper lip of the antenna might be making a contribution to the signal.

Finally, BERNSTEIN et al. [34] were able to show that the Woody-Richards excess did arise from the warm lip of the antenna. The geometrical theory of diffraction used to analyze the antenna performance prior to flight was not sufficiently accurate at long wavelengths to reveal the existence of the effect.

The Woody-Richards experiment and its successors were limited by a fundamental problem with balloon measurements. There must be residual atmosphere to support the balloon and it is difficult to keep this atmosphere from condensing onto a cold antenna. The compromise solutions adopted were not effective enough to avoid a detectable excess signal. Many suggestions were made by the community about potential sources of systematic error in this experiment. In hindsight, none of these suggestions turned out to be correct.

6. – Rocket measurements of the spectrum of the CMB.

Because of the complete absence of atmosphere, small rockets which reach elevations of ~ 300 km for ~ 5 min are, in principle, good platforms for high-frequency spectral measurements. In practice, however, a very long development period was required before experimenters learned how to carry out such complex experiments within the size and time constraints of a sounding rocket flight. A very successful measurement of the high-frequency spectrum was announced by HALPERN and GUSH [35] very shortly after the first COBE results. Their success was the culmination of a ~ 20 year development program including two previous flights which encountered significant difficulties. The University of British Columbia instrument used a cooled Winston cone antenna with apodizing flare, a polarizing Michelson interferometer as both a spectrometer and a Dicke switch and two ^{3}He-cooled bolometric detectors. This instrumentation had many similarities to the Far Infrared Absolute Spectrophotometers (FIRAS) experiment on COBE. The measured spectrum was an excellent fit to a Planck curve from 1 to 30 cm^{-1} with a temperature of 2.736 K.

Another series of rocket experiments were carried out by a collaboration between Nagoya University and Berkeley [36]. These experiments used dichroic beam dividers to measure six frequency bands from 8 to 100 cm^{-1}. The second flight produced six measurements of sky brightness including what appeared to be a valid measurement of the spectrum of the CMB at three frequencies beyond the peak. The measurement at 9 cm^{-1} gave a value consistent with expectations. However, measurements at 14 and 21 cm^{-1}, where the Planck brightness is very small, showed a significant excess.

This Nagoya-Berkeley excess stimulated intensive theoretical work because it was superficially consistent with the expectation that the CMB photons might be distorted by Compton scattering from a hot intergalactic medium, or that intense starlight from an early generation of stars might have been thermalized by dust and redshifted to the frequency of the observed excess. Detailed calculations showed that the measurements were difficult, but not impossible, to reconcile with standard big-bang cosmologies.

The Nagoya-Berkeley group flew a third time just prior to COBE to check their result. The data were contaminated by radiofrequency interference, but were consistent with no excess. Further analysis of the data from the second flight carried out after the COBE announcement revealed the possibility of coherent electrical pickup from the light chopper.

7. – CMB spectral measurements from molecular lines.

Important measurements of the temperature of the CMB were made by observing the excitation temperature of interstellar CN. These gave the first evidence that the CMB spectrum does not continue to increase as frequency squared, and that the CMB temperature is the same in the remote molecular cloud as at the Earth [37]. The technique used is to measure the strength of visible-UV absorption lines due to CN in the spectrum of starlight passing through a molecular cloud. The absorption lines arise from transitions from three ground-state levels to higher vibrational levels of the molecule. They can be analyzed to give relative populations of the ground-state levels and thus, through the Boltzmann factor, values of the excitation temperature. The measurements are difficult because they depend upon estimates of the strengths of partially saturated lines which are not fully resolved. Excitation temperatures are deduced which correspond to ground-state splittings of 3.79 and 7.58 cm^{-1}. The contribution of local excitations to the temperature must in principle be subtracted from the measurements. Estimates now suggest that this correction is relatively small. Recent work is described by MEYER and JURA [38] and by CRANE et al. [39].

8. – The Far Infrared Absolute Spectrophotometer experiment.

The FIRAS experiment on COBE [10] used a polarizing Michelson Fourier spectrometer and ^{4}He-cooled composite bolometers to cover the 1 to 100 cm^{-1} range in two bands. The 7° field of view was defined by a Winston cone antenna with an apodizing flare which was cooled to 1.5 K. The bolometers at the two outputs of the interferometer view the difference between the signals from the two inputs, which were the sky and the variable-temperature cold load. The temperature of the cold load was adjusted to cancel the sky signal. The fact that cancellation occurred at all frequencies showed that the spectrum of the sky was very close to the same as that of the black-body load.

The results of the FIRAS experiment are being released in a series of papers.

The spectrum from the first 9 min of data were published by MATHER et al. [10]. WRIGHT et al. [15] report on a map of galactic dust emission and the observation of [C I], [C II], [N II] and CO spectral lines. Additional results include the observation of the spectrum of the dipole anisotropy of the microwave background and very tight limits on spectral distortions. An upper limit of $6 \cdot 10^{-5}$ is placed on the y parameter which measures the heating of the CMB photon by individual Compton scattering events at relatively late times. A limit of $2 \cdot 10^{-3}$ is placed on the chemical potential μ/kT, which measures heating of the CMB at relatively early times when multiple Compton events would be expected.

9. – Anisotropy of the CMB.

Early observations of the CMB used its isotropy to distinguish it from galactic emission. This isotropy was crucial in identification of the radiation as cosmological in origin. It was recognized quite early, however, that the CMB is expected to be isotropic only in the co-moving reference frame. Observers in other reference frames should see a dipolar Doppler shift varying as $\Delta T(\theta)/T = 1 + (v/c)\cos\theta$, where v is the peculiar velocity of the observer plus a small (kinematic) quadrupole. Furthermore, the existence of structure in the Universe today shows that the matter distribution could not have been completely isotropic at the time of decoupling. All models predict that this structure created anisotropy in the CMB at smaller angular scales.

Apparatus used to measure the anisotropy of the CMB differs from apparatus used to measure the spectrum in several ways. First, a selection must be made of the angular scale on which the anisotropy is to be observed. Early measurements focused on the dipole and quadrupole anisotropies, so they typically used horn antennas to produce beams with full width $\sim 10°$. More conventional techniques of radioastronomy including paraboloidal reflectors and aperture synthesis arrays are useful for smaller scales. Experiments have now been carried out with beam sizes down to a few arcseconds. Anisotropy measurements are typically carried out with a switching technique which rapidly compares two or more sky positions to reduce the effects of gain drift. Comparison of the sky with a cold load has also been used. Calibration remains an important requirement and is carried out by switching between loads with different temperatures.

Attempts to measure anisotropies of $\Delta T/T = 10^{-5}$ or less place very stringent requirements on receiver sensitivity or, alternatively, on measurement time. Many anisotropy experiments have been carried out using heterodyne radiometers. The current generation of experiments below ~ 50 GHz typically uses HEMT amplifiers. Bolometric receivers are used for frequencies beyond ~ 80 GHz.

10. – Anisotropy measurements.

One motivation for measurements of the anisotropy of the CMB at large angular scales was to observe the predicted Doppler shift due to the peculiar motion of the observer. Many groups attacked this problem using heterodyne techniques with gradually improving results. The first credible measurement of the dipolar anisotropy was made from a U2 high-altitude aircraft by SMOOT et al. [40] using an apparatus consisting of two scalar horns, a microwave Dicke switch and a 33 GHz heterodyne radiometer.

Other noteworthy large-scale experiments in the pre-COBE era include a JPL-Princeton balloon experiment at 24 GHz by FIXEN et al. [41] which used a maser amplifier for low noise. This experiment has evolved into a single-horn balloon maser experiment which is chopped against an internal cold load [42].

LUBIN and NETO [43] used an optically chopped single-horn SIS quasi-particle heterodyne mixer receiver for anisotropy measurements at 90 GHz. This balloon experiment was able to see the dipolar anisotropy in the raw data and provided stringent limits of the quadrupole component.

The Soviet Relict experiment (KLYPINE et al. [44], and STRUKOV and SKULACHEV [45]) was the first measurement of the anisotropy of the CMB from space. It produced an all-sky map, but significant antenna sidelobe response greatly complicated the data analysis.

Groups at the University of Rome and at MIT produced important upper limits from bolometric balloon experiments with a variety of chopping schemes (MEYER et al. [46], MELCHIORRI et al. [47] and DE BERNARDIS et al. [48]).

DAVIES and LASENBY [49] are doing ground-based HEMT experiments from Tenerife at 5°. TIMBIE and WILKINSON [50] use an SIS heterodyne correlation receiver for (0.5-5)° measurements from Saskatoon. MEINHOLD and LUBIN [51] made 90 GHz SIS heterodyne measurements from a balloon at 0.5°. GAIER [52] did 90 GHz HEMT measurements from the South Pole at 1°. READHEAD et al. [53] did 20 GHz HEMT observations from Owens Valley at 2 arcmin.

In summary, this work showed [9] that the large-scale anisotropy of the cosmic microwave background was dominated by a dipole anisotropy of Doppler origin with $\Delta T/T = 10^{-3}$. This anisotropy is thought to arise from the peculiar velocity of the observer. In combination with redshift surveys it gives a picture of velocity fields in our region of the Universe. These experiments also placed useful upper limits on the amplitude of any quadrupole component in the CMB and upper limits in the range $\Delta T/T \leq (1-5) \cdot 10^{-5}$ on the fluctuations in the CMB on angular scales down to 2 arcmin.

11. – COBE Differential Microwave Radiometer experiment.

The COBE DMR experiment contained three pairs of relatively conventional two-horn microwave Dicke-switched heterodyne receivers with 7° FWHM beams 60° apart on the sky. The center frequencies of the receivers were at 31, 53 and 90 GHz. The satellite is in a polar orbit and is oriented at ~ 90° to the Sun. It spins about its axis to explore a 120° circle on the sky and obtains full-sky coverage after one year of observation. The data will be reported year by year as they become available.

The DMR data [12, 13] are analyzed by subtracting the instrumental offset, calibrating and rejecting data which are contaminated by the Earth or the Moon. Corrections are then applied to the data for contributions from the Moon, Jupiter, the Galaxy, the Doppler dipole and the effect of the Earth's magnetic field on the Dicke switch [17]. A matrix inversion procedure is then used to obtain a 6144 pixel single-beam map of the sky from the two-beam differential data. This map is then smoothed by a Gaussian function equal in width to the antenna beam. The reported mean dipole is $\Delta T = (3.36 \pm 0.1)$ mK in the direction $l = 264.7 \pm 0.08$, $b = 48.2 \pm 5$. This dipole is then subtracted from the map and the kinematic, that is noncosmological, quadrupole term is also subtracted. These maps show no obvious features away from the galactic plane except for receiver noise. The contribution of noise to the maps was evaluated by comparing the maps made from the difference between the outputs of receivers a and b, which subtracts the sky and leaves only noise, with the maps made from the sum of the outputs from receivers a and b which includes both sky signal and noise. Several procedures are reported which average over most of the sky to improve the signal-to-noise ratio. The sky r.m.s. for a 10° beam was computed by deleting the galactic plane. The result is independent of the galactic cut if $|b| > 15°$ for 53 and 90 GHz and $|b| > 30°$ for 31 GHz. Correlation functions are calculated directly from the temperature map, assuming various galactic cuts. Their form is not Gaussian but is consistent with a flat power spectrum of primordial fluctuations with an uncertainty of ~ ±0.7 in the spectral index. All six maps show significant quadrupole contributions, but the quadrupole components depend somewhat upon the galactic cut used. The values of the 10° sky r.m.s. and the galactic signals are consistent in all three bands.

Comparison of the three frequencies clearly shows at least two contributions to galactic confusion. Several different galactic subtraction procedures were used to obtain reliable CMB anisotropy maps and cosmic quadrupole components [13]. One procedure was to construct model maps from other data including FIRAS results and the Haslam map. A second procedure was to fit the DMR data to functions with assumed spectral indices. The third procedure was to take linear combinations of the DMR maps to cancel the galactic plane. The result is a cosmic quadrupole component $Q_{\text{r.m.s.}} = (13 \pm 4)$ µK which corresponds to $\Delta T/T = 4.8 \cdot 10^{-6}$.

12. – Power spectrum of fluctuations.

Measurements of the anisotropy of the cosmic microwave background are important because they can constrain cosmological models. Such models must be consistent with the snapshot of the matter distribution on the surface of last scattering provided by the microwave background, and with the large-scale structure observed in the Universe today. Calculations of the CMB anisotropy from a given model involve an average over an ensemble of equivalent universes. Since we have only one universe to measure, we do not expect to obtain the ensemble average value even for the correct cosmological model. The best that we can do is to evaluate statistical quantities by averaging over the sky. A discrepancy is expected, especially at low spatial frequencies where few pixels are available. This discrepancy is known as the cosmic variance. A second source of discrepancy arises when the sky coverage is limited. This can result either from galactic obscuration or from experimental limitations. Since the number of independent pixels measured is important, measurements at large angles require more complete sky coverage such as that supplied by the DMR experiment.

The predictions of a given cosmological model can conveniently be represented in the form of a power spectrum of anticipated fluctuations as a function of spatial frequency on the surface of last scattering. Alternatively, the power spectrum can be related to angular frequency on the observable sky by a somewhat model-dependent conversion. A distance of 100 Mpc corresponds to a present value in degrees of $\Omega h \sim 1$. Here, Ω is the normalized density and h is the normalized Hubble constant. To account for the spherical nature of the sky, these plots conventionally use the quantum number l of the spherical harmonic as a measure of angular frequency. For high angular frequencies, this is equivalent to 2π divided by the angle. It should be noted that many different power spectra are discussed by cosmologists. They include one-, two- and three-dimensional representations of the fluctuations of matter and radiation at various epochs. The discussion here refers only to a one-dimensional prediction on the present sky of $l^2 C(l)$ as a function of l.

Inflationary cosmologies generally predict a flat or nearly flat primordial power spectrum consistent with the early hypothesis of Zel'dovich. This power spectrum is then modified by evolutionary effects. The predicted power spectrum remains independent of l for $l < 50$. Such small values of l corresponded to points on the surface of last scattering which have not been in causal contact since inflation. The fluctuations in the microwave background at these angular scales occur because a photon emitted from a clump of matter must climb out of the gravitational potential well before reaching us [54]. At larger values of l, gravitational interactions between matter density fluctuations have occurred. In cold-dark-matter models the power spectrum increases for $l > 50$ to a peak near $l = 200$ due to Doppler shifts arising from gravitationally induced flows. For larger values of l, the power spectrum contains subsidiary peaks and then decreases rapidly because of a statistical averaging effect arising from the thickness of the surface of last scattering. This characteristic power spectrum associated with cold-dark-matter models is sensitive to the assumed value of the baryon density. Other cosmological models give radically different power spectra. One major function of CMB anisotropy measurements is to measure the power spectrum for our observable sky and thereby provide information about cosmological models by constraining the permitted range of theoretical power spectra.

The results of the COBE/DMR experiment can be restated in terms of this power spectrum viewpoint. Most important, it has demonstrated the existence of fluctuations in the microwave background which are probably cosmological. Second, it has provided a value for the amplitude of the power spectrum for small l (large angular scales). This is sometimes described as providing a normalization for cosmological models. Third, the COBE/DMR experiment provides some information about the slope of the power spectrum up to angular frequencies corresponding to the inverse of the 7° beam size. The accuracy of this last result is significantly limited by the cosmic and sample variances at small l, but the results are consistent with a flat power spectrum as expected for scales larger than the horizon.

It is a matter of great importance to be able to compare experiments made on different angular scales with each other and with cosmological models. These comparisons are in fact quite difficult and have led to much confusion. Any experiment can be characterized by a window function as a function of l. For experiments which produce a temperature map such as COBE, including experiments which are chopped against an internal load [46], this window function has the form of a Gaussian cut-off at a value of l which corresponds to the beam size. For chopped experiments which compare two adjacent regions of the sky, the window function has a peak. The low-frequency side of the peak is controlled by the amplitude of the chop and the high-frequency side by the beam size. The r.m.s. average of the experimental signal can be understood as a convolution between the power spectrum of the sky fluctuations and the window function of the experiment. Consequently, r.m.s. values of $\Delta T/T$ for different experiments cannot be directly compared if they correspond to different window functions.

At a time when most anisotropy experiments were producing only upper limits, the convention arose of assuming that the sky fluctuations have a Gaussian autocorrelation function with a single correlation angle. Using this assumption, it is possible to compute an unambiguous limit to $\Delta T/T$ for each assumed value of the correlation angle. The result is a U-shaped upper limit on a plot of $\Delta T/T$ vs. correlation angle. This assumption of a Gaussian autocorrelation function with a single correlation angle is equivalent to the assumption of a specific form for the power spectrum for each angular scale, namely l^2 times a Gaussian roll-off. This approach is useful only if the power spectrum of the real sky has such a peak. Otherwise, it can introduce significant errors. For example, the Gaussian assumption is not too bad a representation of the cold-dark-matter spectrum near the peak, but is a poor representation for small values of l.

This fact led to confusion when the first attempts were made to compare the COBE measurement with upper limits at smaller angular scales by WRIGHT *et al.* [15]. Using this representation, it appeared that the COBE detection was significantly larger than upper limits set by other experiments. This apparent discrepancy was an artifact of assuming a Gaussian form for an autocorrelation function which is decidedly non-Gaussian.

The ideal method of data analysis would be to invert the convolution of the window function with the power spectrum and thus obtain measure points on the power spectrum. In principle, this can be done only if the width of the window function can be neglected. In practice, experimental window functions are wide compared to the peak in the power spectrum predicted by cold-dark-matter models. Consequently, it is not possible to convert a measured r.m.s. to a point on the power spectrum in a model-independent way. The situation is quite analogous to using broad-band photometry to estimate the properties of a rapidly varying line spectrum. Comparisons between experiments can always be made using model power spectra such as the one predicted by cold-dark-matter cosmologies.

It is worth pointing out that the information content of potential experiments is larger than a single point on the power spectrum. Experiments with a wide window function, such as those which produce a temperature map, or experiments with a chopping angle significantly larger than the beam size, contain information about the slope and curvature of the power spectrum. This information can be extracted by considering a correlation matrix of the experimental points rather than a simple r.m.s. It also should be pointed out that experiments will eventually map CMB features on the real sky. The shape of individual features will provide the best indication of whether there are topological defects such as those predicted to arise from cosmologies containing cosmic strings.

13. – Present and future measurements of CMB anisotropy.

The power spectrum viewpoint highlights the motivation for measuring the fluctuations in the CMB over a wide range of angular scales. Measurements of the anisotropy on the quadrupole to 10° angular scale provide information on regions of the Universe which have not been in causal contact since inflation. The physics that is being probed includes the Sachs-Wolfe effect, primordial gravitational waves, and the possibility that the primordial Universe was anisotropic or inhomogeneous. Measurements on the 1° angular scale probe the origin of the presently observed large-scale structure in the Universe, adiabatic fluctuations, bulk flows and CDM models. Measurements on smaller angular scales probe Compton distortions associated with possible late reionization, bulk motions and high-Z dust.

Enormous effort is now going into the search for anisotropy of the CMB on a wide range of angular scales. This discussion does not do justice to this work. Many current experiments emphasize angular scales close to 1° where a peak in the power spectrum is predicted by cold-dark-matter models. Such experiments can be divided into two classes. Low-frequency experiments using HEMT amplifier receivers are being carried out at frequencies below 50 GHz from high dry sites including the South Pole. The frequencies are low enough that atmospheric contamination is manageable and long integration times are available. As the frequency of the experiment is increased, the receiver noise, receiver bandwidth and atmospheric interference increase, but the interference from galactic synchrotron and free-free emission decreases. Other experiments are being carried out at frequencies above ~ 80 GHz from balloon platforms using the higher sensitivity available from broad-band bolometric receivers. The balloon platform reduces the importance of atmospheric noise at high frequencies but limits the available observing time. Experiments are also being carried out using including bolometric techniques from the South Pole in the 90 GHz atmospheric window.

There is an active debate regarding which technology is most powerful for measuring intermediate-scale anisotropies. The answer will depend upon many experimental factors as well as the nature of the galactic contamination. Although COBE has evaluated the galactic contamination in the statistical sense at large angular scales [14], there is evidence that the various constituents depend strongly on angular scale. For example, the 100 μm IRAS dust emission has been Fourier analyzed [55]. The power spectrum is shown to fall as l^{-3} over the range from 7° to 4 arcmin. Most likely, both kinds of experiment will be necessary for a high degree of confidence.

The first generations of intermediate-scale anisotropy experiments produced only upper limits. Measurements at a single spectral frequency are sufficient to produce such a limit. More recently, however, experiments are beginning to observe structure which cannot be argued away as sidelobe or other spurious phenomena. Once structure is seen, it is extremely valuable to have measurements at a number of well-separated frequencies in order to interpret the results with confidence. The present generation of multimode bolometric experiments have 3-4 bands over a factor ~ 5 in frequency with a nearly constant beam size. Current single-mode experiments, by contrast, typically extend over a factor ~ 1.5 in frequency. Most of these experiments do not make a 2-dimensional map, but measure a single line on the sky. In this case, accurate spectral information can be deduced only if the beam sizes are the same in all of the bands.

14. – The MAX experiment.

This discussion of CMB anisotropy will conclude with a description of the millimeter wave anisotropy experiment (MAX) which is well known to the author. The MAX experiment is a collaboration between groups at the University of California at Berkeley and the University of California at Santa Barbara. It is a balloon experiment which measures the sky with a 0.5° beam and a 1.3° sinusoidal chop at frequencies of 3, 6, 9 and 12 cm^{-1}. The telescope is an off-axis Gregorian design with an underfield 1 m primary mirror, a nutating ellipsoidal secondary mirror and a ^{3}He-cooled dichroic photometer[56].

Measurements from the second flight of MAX[51] on 6/90 and the third flight on 6/91 show fluctuations in sky brightness which are candidates for intermediate-scale anisotropy of the CMB. Measurements near the star γ Ursae Minoris[57,58] show essentially no evidence for galactic dust emission which would be most significant in the 12 cm^{-1} band. There is correlated structure in the 6 and 9 cm^{-1} bands whose relative amplitudes are consistent with CMB fluctuations and correspond to $\Delta T/T = (4-5) \cdot 10^{-5}$ for a Gaussian autocorrelation function with coherence angle $\theta_c = 25'$. Although the spectrum of these fluctuations is also consistent with free-free emission and synchrotron radiation, the magnitude of these galactic contaminants is expected to be one or more orders of magnitude smaller at the frequencies of the measurement.

Measurements of the sky in a region with higher dust emission near the star μ Pegasi[59] show a clear contribution from galactic dust emission in all bands, especially at 12 cm^{-1}. The morphology of this dust correlates very well with the IRAS 100 μm dust map. This observation proves that the experiment was capable of measuring a sky signal. Analysis of the data at 6 and 9 cm^{-1} also shows evidence for a second component whose morphology is not well defined. If it is assumed that this second component is entirely CMB fluctuations, then the conservative upper limit of $\Delta T/T < 2.5 \cdot 10^{-5}$ can be computed for a Gaussian autocorrelation function with coherence angle $\theta_c = 25'$.

Whenever an anisotropy in sky brightness is observed, the possibility must be considered that it arises from the sidelobe response of the telescope picking up variations in earthshine. One limitation of balloon observations is on the size of the ground shields which can be effectively carried. MAX has addressed this question by redesigning the telescope baffles between flights and by analyzing the data into bins which correspond to different elevation angles. The observations near μ Pegasi were carried out over an extended period of time when the elevation of the source varies from well below that of γ Ursae Minoris to well above. These observations place an upper limit on the sidelobe response which is significantly smaller than the structure observed near γ Urase Minoris on the same flight.

We believe that the MAX experiment has now observed the long-sought CMB fluctuation on intermediate angular scales. These experiments can be extended to make maps of $\sim 10^3$ pixels using arrays of more sensitive bolometers.

Future plans for the MAX experiment include a new photometer with more sensitive bolometers cooled below 100 mK by an adiabatic demagnetization refrigerator for higher sensitivity. There are also plans to modulate the primary mirror so as to avoid chopping the spillover around the primary mirror. This new design will eliminate chopping aberrations and permit the use of a multiarray photometer in the future. The goal is to map $\sim 10^3$ pixels so as to make the effect of sampling variance unimportant. The correlation observed between the millimeter wave dust emission and the IRAS 100 µm dust map shows that regions of the sky exist where the CMB anisotropy can be mapped without serious dust contamination at the level of $\Delta T/T \sim 10^{-6}$.

15. – Summary.

Measurements of the cosmic microwave background provide our earliest look at the Universe. Until recently, the numbers arising from these measurements were restricted to a temperature and upper limits on both the deviations of the spectrum from the Planck curve and upper limits on cosmological anisotropies. Although very useful, the information content in these observations was limited. The recent observation of anisotropy on several angular scales and the expectation that similar observations will soon be available at many angular scales promise a rich new era in microwave background cosmology.

* * *

The author is most grateful to the International School of Physics at Varenna for the invitation to speak and to N. VITTORIO for his patience during the preparation of this manuscript.

REFERENCES

[1] A. A. PENZIAS and R. W. WILSON: *Astrophys. J.*, **142**, 419 (1965).
[2] G. HERZBERG: in *Molecular Spectra and Molecular Structure* (D. Van Nostrand Co. Inc., New York, N.Y., 1950), p. 496.
[3] R. H. DICKE, R. BERYNGER, R. L. KYHL and A. V. VANE: *Phys. Rev.*, **70**, 340 (1946).
[4] G. GAMOW: *Phys. Rev.*, **74x**, 505 (1948).
[5] R. A. ALPHER and R. C. HERMAN: *Nature (London)*, **162**, 774 (1948).
[6] R. H. DICKE, P. J. E. PEEBLES, P. G. ROLL and D. T. WILKINSON: *Astrophys. J.*, **142**, 414 (1965).
[7] R. WEISS: *Annu. Rev. Astron. Astrophys.*, **18**, 489 (1980).
[8] J. M. USON and D. T. WILKINSON: in *Galactic and Extragalactic Radio Astronomy*, 2nd edition, edited by G. L. VERCHUR and K. I. KELLERMAN (Springer-Verlag, Berlin, 1988), p. 603.
[9] A. C. S. READHEAD and C. R. LAWRENCE: *Annu. Rev. Astron. Astrophys.*, **30**, 653 (1992).
[10] J. C. MATHER, E. S. CHENG, R. E. EPLEE jr., R. B. ISAACMAN, S. S. MEYER, R. A.

Shafer, R. Weiss, E. L. Wright, C. L. Bennett, N. W. Boggess, E. Dwek, S. Gulkis, M. G. Hauser, M. Janssen, T. Kelsall, P. M. Lubin, S. H. Moseley jr., T. L. Murdock, R. F. Silverberg, G. F. Smoot and D. T. Wilkinson: *Astrophys. J. Lett.*, **354**, L37 (1990).

[11] E. L. Wright, J. C. Mather, C. L. Bennett, E. S. Cheng, R. A. Shafer, D. J. Fixsen, R. E. Eplee jr., R. B. Isaacman, S. M. Read, N. W. Boggess, S. Gulkis, M. G. Hauser, M. Janssen, T. Kelsall, P. M. Lubin, S. S. Meyer, S. H. Moseley jr., T. L. Murdock, R. F. Silverberg, G. F. Smoot, R. Weiss and D. T. Wilkinson: *Astrophys. J.*, **381**, 200 (1991).

[12] G. F. Smoot, C. L. Bennett, A. Kogut, J. Aymon, C. Backus, G. De Amici, K. Galuk, P. D. Jackson, P. Keegstra, L. Rokke, L. Tenorio, S. Torres, S. Gulkis, M. G. Hauser, M. A. Janssen, J. C. Mather, R. Weiss, D. T. Wilkinson, E. L. Wright, N. W. Boggess, E. S. Cheng, T. Kelsall, P. Lubin, S. Meyer, S. H. Moseley, T. L. Murdock, R. A. Shafer and R. F. Silverberg: *Astrophys. J. Lett.*, **371**, L1 (1991).

[13] G. F. Smoot, C. L. Bennett, A. Kogut, E. L. Wright, J. Aymon, N. W. Boggess, E. S. Cheng, G. De Amici, S. Gulkis, M. G. Hauser, G. Hinshaw, C. Lineweaver, K. Loewenstein, P. D. Jackson, M. Janssen, E. Kaita, T. Kelsall, P. Keegstra, P. Lubin, J. Mather, S. S. Meyer, S. H. Moseley, T. Murdock, L. Rokke, R. F. Silverberg, L. Tenorio, R. Weiss and D. T. Wilkinson: *Astrophys. J. Lett.*, **396**, L1 (1992).

[14] C. L. Bennett, G. F. Smoot, M. Janssen, S. Gulkis, A. Kogut, G. Hinshaw, C. Backus, M. G. Hauser, J. C. Mather, L. Rikke, L. Tenorio, R. Weiss, D. T. Wilkinson, E. L. Wright, G. De Amici, N. W. Boggess, E. S. Cheng, P. D. Jackson, P. Keegstra, T. Kelsall, R. Kummerer, C. Lineweaver, S. H. Moseley, T. L. Murdock, J. Santana, R. A. Shafer and R. F. Silverberg: *Astrophys. J. Lett.*, **396**, L7 (1992).

[15] E. L. Wright, S. S. Meyer, C. L. Bennett, N. W. Boggess, E. S. Cheng, M. G. Hauser, A. Kogut, C. Lineweaver, J. C. Mather, G. F. Smoot, R. Weiss, S. Gulkis, G. Hinshaw, M. Janssen, T. Kelsall, P. M. Lubin, S. H. Moseley jr., T. L. Murdock, R. A. Shafer, R. F. Silverberg and D. T. Wilkinson: *Astrophys. J. Lett.*, **396**, L13 (1992).

[16] N. Boggess, J. C. Mather, R. Weiss, C. L. Bennett, E. S. Cheng, W. Dwek, S. Gulkis, M. G. Hauser, M. A. Janssen, T. Kelsall, S. S. Meyer, S. H. Moseley, T. L. Murdock, R. A. Shafer, R. F. Silverberg, G. F. Smoot, D. T. Wilkinson and E. L. Wright: *Astrophys. J.*, **397**, 420 (1992).

[17] A. Kogut, G. F. Smoot, C. L. Bennett, E. L. Wright, J. Aymon, G. De Amici, G. Hinshaw, P. D. Jackson, E. Kaita, P. Keegstra, C. Lineweaver, K. Loewenstein, L. Rokke, L. Tenorio, N. W. Boggess, E. S. Cheng, S. Gulkis, M. G. Hauser, M. A. Janssen, T. Kelsall, J. C. Mather, S. Meyer, S. H. Moseley, T. L. Murdock, R. A. Shafer, R. F. Silverberg, R. Weiss and D. T. Wilkinson: *Astrophys. J.*, **401**, 1 (1992).

[18] P. L. Richards and L. Greenberg: in *Infrared and Millimeter Waves*, Vol. 6, edited by K. J. Button (Academic Press, New York, N.Y., 1982), p. 149.

[19] A. Franceschini, L. Toffolatti, L. Danese and G. De Zotti: *Astrophys. J.*, **344**, 35 (1989).

[20] C. G. Haslam, U. Klein, C. J. Salter, H. Stoffel, W. E. Wilson, M. N. Cleary, D. J. Cooke and P. Thomasson: *Astron. Astrophys.*, **100**, 208 (1981).

[21] R. J. Reynolds: *Astrophys. J. Lett.*, **392**, L35 (1992).

[22] Most recent IRAS maps obtainable from The Infrared Processing and Analysis Center (Jet Propulsion Laboratory, Pasadena, Cal.).

[23] E. L. WRIGHT: *Astrophys. J.*, **320**, 818 (1987).
[24] F. BOULANGER and M. PERAULT: *Astrophys. J.*, **330**, 964 (1988).
[25] T. F. HOWELL and J. R. SHAKESHAFT: *Nature (London)*, **210**, 1318 (1966).
[26] P. J. E. PEEBLES: *Physical Cosmology* (Princeton University Press, Princeton, N.J., 1971).
[27] G. SMOOT, S. M. LEVIN, C. WITEBSKY and G. DE AMICI: *Astrophys. J.*, **331**, 653 (1988).
[28] D. G. JOHNSON and D. T. WILKINSON: *Astrophys. J.*, **313**, L114 (1987).
[29] J. CLARKE, G. I. HOFFER, P. L. RICHARDS and N.-H. YEH: *J. Appl. Phys.*, **48**, 4865 (1977).
[30] D. H. MARTIN and E. PUPLETT: *Infrared Phys.*, **10**, 105 (1969).
[31] J. C. MATHER: *IEEE Trans. Antennas Prop.*, **AP-29**, 967 (1981).
[32] D. W. WOODY and P. L. RICHARDS: *Astrophys. J.*, **248**, 18 (1981).
[33] J. B. PETERSON, P. L. RICHARDS and T. TIMUSK: *Phys. Rev. Lett.*, **55**, 332 (1985).
[34] G. M. BERNSTEIN, M. L. FISCHER, J. B. PETERSON, P. L. RICHARDS and T. TIMUSK: *Astrophys. J. Lett.*, **377**, L1 (1989).
[35] H. P. GUSH, M. HALPERN and E. H. WISHNOW: *Phys. Rev. Lett.*, **65**, 537 (1991).
[36] T. MATSUMOTO, S. HAYAKAWA, H. MURAKAMI, S. SATO, A. E. LANGE and P. L. RICHARDS: *Astrophys. J.*, **329**, 567 (1988).
[37] P. THADDEUS: *Annu. Rev. Astron. Astrophys.*, **10**, 305 (1972).
[38] D. M. MEYER and M. JURA: *Astrophys. J.*, **297**, 119 (1985).
[39] P. CRANE, D. J. HEGYI, M. L. KUTNER and N. MANDOLESI: *Astrophys. J.*, **346**, 136 (1989).
[40] G. F. SMOOT, M. V. GORENSTEIN and R. A. MULLER: *Phys. Rev. Lett.*, **39**, 898 (1977).
[41] D. J. FIXEN, E. S. CHENG and D. T. WILKINSON: *Phys. Rev. Lett.*, **50**, 620 (1983).
[42] S. P. BOUGHN, E. S. CHENG, D. A. COTTINGHAM and D. J. FIXEN: *Astrophys. J. Lett.*, **391**, L49 (1992).
[43] P. M. LUBIN, T. V. NETO, G. EPSTEIN and G. SMOOT: *Astrophys. J. Lett.*, **298**, L1 (1985).
[44] A. A. KLYPINE, M. V. SAZHIN, I. A. STRUKOV and D. P. SKULACHEV: *Sov. Astron. Lett.*, **13**, 104 (1987).
[45] I. A. STRUKOV and D. P. SKULACHEV: *Sov. Astron. Lett.*, **10**, 1 (1984).
[46] S. S. MEYER, E. S. CHENG and L. A. PAGE: *Astrophys. J. Lett.*, **371**, L7 (1991).
[47] F. MELCHIORRI, B. O. MELCHIORRI, C. CECCARELLI and L. PIETRANERA: *Astrophys. J. Lett.*, **250**, L1 (1981).
[48] P. DE BERNARDIS, S. MASI, F. MELCHIORRI, B. MELCHIORRI and N. VITTORIO: *Astrophys. J. Lett.*, **396**, L57 (1992).
[49] R. D. DAVIES, A. N. LASENBY, R. A. WATSON, R. A. DAINTREE, E. J. HOPKINS, J. BECKMAN, J. SANCHEZ-ALMEIDA and R. REBOLO: *Nature (London)*, **326**, 462 (1987).
[50] P. TIMBIE and D. T. WILKINSON: *Astrophys. J.*, **353**, 140 (1990).
[51] P. R. MEINHOLD and P. M. LUBIN: *Astrophys. J. Lett.*, **370**, L11 (1991).
[52] T. GAIER, J. SCHUSTER, J. GUNDERSEN, T. KOCH, M. SEIFFERT, P. R. MEINHOLD and P. LUBIN: *Astrophys. J. Lett.*, **398**, L1 (1992).
[53] A. C. S. READHEAD, C. R. LAWRENCE, S. T. MYERS, W. L. W. SARGENT, H. E. HARDBECK and A. T. MOFFET: *Astrophys. J.*, **346**, 566 (1989).

[54] R. K. Sachs and A. M. Wolfe: *Astrophys. J.*, **147**, 73 (1967).
[55] T. N. Gautier III, F. Boulanger, M. Perault and J. L. Puget: *Astron. J.*, **103**, 1313 (1992).
[56] M. L. Fischer, D. C. Alsop, E. S. Cheng, A. C. Clapp, D. A. Cottingham, J. O. Gundersen, T. C. Koch, E. Kreysa, P. R. Meinhold, A. E. Lange, P. M. Lubin, P. L. Richards and G. F. Smoot: *Astrophys. J.*, **388**, 242 (1992).
[57] D. C. Alsop, E. S. Cheng, A. C. Clapp, D. A. Cottingham, M. L. Fischer, J. O. Gundersen, E. Kreysa, A. E. Lange, P. M. Lubin, P. R. Meinhold, P. L. Richards and G. F. Smoot: *Astrophys. J.*, **395**, 317 (1992).
[58] J. O. Gundersen, A. C. Clapp, M. Devlin, W. Holmes, M. L. Fischer, P. R. Meinhold, A. E. Lange, P. M. Lubin, P. L. Richards and G. F. Smoot: *Astrophys. J. Lett.*, **413**, L1 (1993).
[59] P. Meinhold, A. Clapp, D. Cottingham, M. Devlin, M. Fischer, J. Gundersen, W. Holmes, A. Lange, P. Lubin, P. Richards and G. Smoot: *Astrophys. J. Lett.*, **409**, L1 (1993).

Cosmic Background Radiation: Present Knowledge and Future Programs.

B. MELCHIORRI

Istituto di Fisica dell'Atmosfera-CNR - piazzale L. Sturzo, Roma, Italia

L. BRANCA and F. MELCHIORRI

Dipartimento di Fisica dell'Università «La Sapienza» piazzale Aldo Moro 2, Roma, Italia

1. – Cosmic background radiation: how Planckian is it?

About 25 years ago YA. B. ZEL'DOVICH[1] noted that the «*cosmic background radiation is like a 100 year old virgin—being virgin just because nobody wished to violate her!*». This picturesque statement has not been taken too seriously by cosmologists: year after year, theoreticians have accumulated papers filled by nice formulae describing possible distortions of the CBR spectrum, if a huge amount of energy injection had occurred some time in the past[2-4]. Year after year, experimental physicists have claimed the discovery of some spectral feature the explanation of which required energy injections even larger than those considered by theoreticians[5-7]. We are now excited by recent results of two space experiments, COBE[8] and COBRA[9]; they provide clean CBR spectra, where no distortions are detectable, within 0.5 percent of the peak. Do we have to be surprised by these results? This is exactly what ZEL'DOVICH predicted and what every theoretician should anticipate.

In the framework of big-bang cosmology the CBR has in fact very few chances to be disturbed. The first possibility was at the time of primordial nucleosynthesis, but SUNYAEV[10] has shown that any energy injection at these early times cannot produce effects detectable today, no matter how large it was. The second possibility is located between the time of matter-radiation equivalence and the recombination time, *i.e.* around $10^4 \geq Z \geq 10^3$, when adiabatic waves of comoving dimensions smaller than those of a typical galaxy were erased by the «photon viscosity». A tiny *y-distortion*[10] is expected in this case, well below the present level of sensitivity of COBE or COBRA. Its detection would represent a support to the internal consistency of the idea that CBR anisotropies are the footprints of the process of galaxy formation.

The last possibility is represented by the re-ionization of the Universe after the recombination. A high-temperature plasma would again produce a *y-distortion*, certainly detectable if the entire Universe was re-ionized at large redshifts. This possibility is, however, excluded by the fact that CBR anisotropies have been observed at angular scales smaller than the horizon [11, 12]. A soft, low-temperature re-ionization would produce a plasma radiating via a free-free mechanism, but this distortion is detectable in the radio region only.

In conclusion, unless we introduce exotic phenomena, the CBR spectrum has to be Planckian and what one can expect to detect in COBE or COBRA experiments is a very low *y-distortion*. The natural question is: are these experiments optimized to this purpose? There is the possibility that some systematic effect can prevent us from detecting this distortion, at least below some level?

Both COBRA and COBE consist of a polarizing interferometer which compares the sky brightness with that emitted by an on-board reference blackbody. Care has been dedicated to make this black-body really black. The final accuracy of these systems is described in terms of minimum detectable spectral distortion as a fraction of the peak intensity. In fig. 1 we have plotted the expected temperature error bars as a function of the frequency. They are at minimum around the peak of 2.7 K black-body, as expected. It is unfortunate that the *y-distortion* becomes also close to zero around the same wave number. There is a fundamental limitation in such kind of instruments: when the brightness level in which we are interested is so low that only one photon is arriving within the time constant of the detector (typically 10^{-2} s), then the efficiency of the instrument goes rapidly to zero in the Wien region of the spectrum, due to the particlelike behaviour of the photons, which prevents interference from occurring: only the wavelike fraction is, therefore, detected, and this fraction is a decreasing function of the frequency. For instance, at 20 cm^{-1} in a frequency interval of 0.1 cm^{-1} and in the sampling frequency of 10^2 Hz we expect about 10^4–10^5 photons arriving from a 2.7 K black-body: at the same wave number the instrumental efficiency is reduced more than a factor ten if we are searching for a spectral distortion $y \leq 10^{-4}$–10^{-5}. This fact seems not to have been considered by the COBE team.

Another not negligible effect of interferometers is that of shifting the true frequency of operation by an amount which depends on the solid angle. Let us assume, for the sake of simplicity, that a detector at very low temperature is collecting the sky radiation $S(\nu)$ through an interferometer having an internal beam divergence Ω (measured in steradians). As the moving mirror scans its path we get an interferogram

$$(1) \qquad W(x) = \frac{1}{2}\pi a^2 \int_0^\infty S(\nu)\left[1 + \cos 4\pi\nu x\left(1 - \frac{\Omega}{4\pi}\right)\right] d\nu\,.$$

If Ω is exactly known, the true spectrum is recovered via the Fourier in-

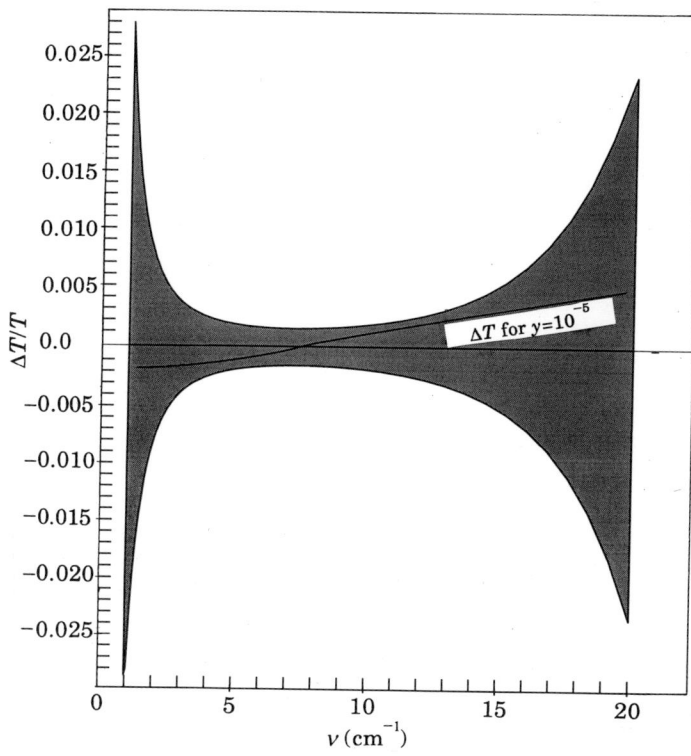

Fig. 1. – Errors on the CBR temperature due to a statistical uncertainty of 1% of the peak of a 2.7 K black-body. A Compton distortion with a Comptonization parameter $y = 10^{-5}$ is also shown. Both distortion and errors increase for frequencies away from the peak.

verse transformation. In the practical case, however, not only Ω is affected by errors, but especially it is frequency dependent: its measurement at a given frequency, for instance, by means of interstellar lines, does not allow us to predict its value at other frequencies. In fig. 2 we have shown the type of spectral distortion induced by an uncertainty on Ω and compared it with a *y-distortion*: both are overamplified to make the similarity in the two behaviours evident.

This problem is made worse in the COBE experiment due to the presence of three solid angles: that of the instrument, that of the internal reference, that of the external black-body. It is hard to believe that the solid angle remains the same in the three cases, *i.e.* when the instrument is pointing at the free sky, when it is coupled with the small internal reference and finally when it is observing the larger external black-body. Therefore, an Ω distortion is expected at some level and its amplitude is shown in fig. 3 as a function of the frequency. It means that this kind of instrument can provide a zero output even if the two

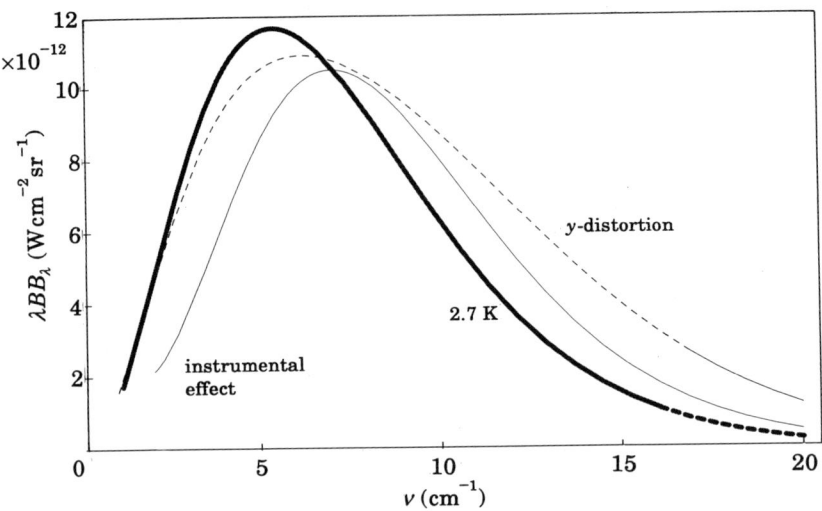

Fig. 2. – Distortions with respect to a Planckian spectrum due to a Comptonization (y-distortion) and due to an error in the measurement of the solid angle (which affects the inverse Fourier transform). Both distortions are greatly exaggerated in order to show their similarity.

reference black-bodies are perfect black-bodies but the sky *has* an intrinsic *y-distortion*, which is simply compensated by a small mismatch among the three solid angles involved in the observations. On the other hand, if a *y-distortion* is detected, one has to prove that it is not due to the above effect.

We may conclude this section by confirming that ZEL'DOVICH was right and

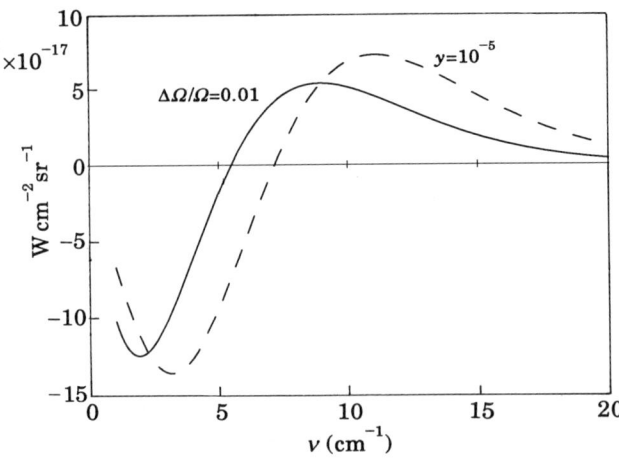

Fig. 3. – Comparison between a Compton distortion of $y = 10^{-5}$ and a possible instrumental effect due to a mismatch between reference black-bodies and sky angles $\Delta\Omega/\Omega \simeq 0.01$.

that the CBR spectrum is just Planckian: at the level of accuracy already acquired it would be difficult to convince the scientific community of a possible detection of some distortion. Future prospects in this field are in the search for a free-free distortion in the long-wavelength region.

2. – Considerations about the dipole anisotropy.

The dipole anisotropy has been measured both in amplitude and direction for the first time by the Princeton Group in the radio region and by the Florence Group in the Wien region [13]. In order to explain its significance, let us hypothesize that the last-scattering surface has a temperature distribution which is not uniform, but is represented by a dipolelike pattern:

$$(2) \qquad T_{\mathrm{LS}} = T_0(1 + \beta \cos \Theta),$$

where β has here the meaning of a small coefficient. An observer would attribute this dipole anisotropy to a relative motion between himself and the LS surface, *i.e.* by interpreting β as v/c, where v is the velocity of the observer with respect to the LS surface. How can we disentangle a true motion, *i.e.* a *kinematic dipole anisotropy*, from such a *thermal* effect? To this purpose it is worthwhile to remember that the instruments are sensitive to the brightness of the radiation and not to its temperature. Only in the R.J. region of the spectrum the two quantities coincide. Let us insert the temperature behaviour into the Planck formula and develop it in series as done by BOTTANI, DE BERNARDIS and MELCHIORRI [14]:

$$(3) \quad I_{\mathrm{obs}}(\nu_{\mathrm{obs}}) = I_{\mathrm{BB}}(\nu_{\mathrm{obs}}) \left[1 + \frac{1}{2}(3-\alpha)\beta^2 + (3-\alpha)\beta \cos \Theta \right] +$$

$$+ \left[\left[3 - 2\alpha + \frac{1}{2} \frac{\nu_{\mathrm{obs}}^2}{I_{\mathrm{BB}}(\nu_{\mathrm{obs}})} I_{\mathrm{BB}}''(\nu_{\mathrm{obs}}) \right] \beta^2 \cos^2 \Theta \right],$$

where $\alpha = \dfrac{\nu_{\mathrm{obs}}}{I_{\mathrm{BB}}} \dfrac{\mathrm{d}I_{\mathrm{BB}}}{\mathrm{d}\nu_{\mathrm{obs}}} = 3 - \dfrac{xe^x}{e^x - 1}$ is the spectral index of a black-body.

Therefore, there is a quadrupole term, proportional to $\cos^2 \Theta$, given in amplitude by

$$(4) \qquad I_{\mathrm{qw}} = \frac{1}{2}\beta^2 I_{\mathrm{BB}}(\nu_{\mathrm{obs}})(3-\alpha)(2-\alpha-x/2).$$

This quadrupole is present in the Wien region of the spectrum and tends to zero in the radio region: it is due to the nonlinear dependence of the brightness on the temperature in the Planck formula and it has nothing to do with the relative motion between the observer and the LS surface.

Let us now consider the case of a real motion between the observer and the last-scattering surface. The observer will see a temperature pattern given by

(5) $$T_{\text{obs}} = T_0 \gamma (1 + \beta \cos \theta^*),$$

where θ^* is the angle between the line of sight and the direction of motion, *as seen by the LS surface*. To get a consistent formula one has to change θ^* by means of the aberration formula

(6) $$\cos \theta = \frac{\cos \theta^* + \beta}{1 + \beta \cos \theta^*},$$

where θ is the angle as seen by the observer. We get

(7) $$T_{\text{obs}} = \frac{T_0}{\gamma (1 - \beta \cos \theta)}.$$

The reader is invited to note the difference with respect to eq. (2): it is due to the use of the aberration formula (6). This new expression for T_{obs} when inserted in the Planck formula gives rise to additional terms in the quadrupole component: we have called it «aberration quadrupole I_A» in order to underline its physical origin

(8) $$I_A = I_{\text{BB}}(\nu_{\text{obs}})(3 - \alpha)\beta^2 \cos^2 \theta.$$

I_A tends to $\beta^2 \cos^2 \theta$ in the R.J. region of the spectrum. Equation (8) when compared with the previous equation (4) proves that it is possible to disentangle the kinematic nature of the observed dipole anisotropy through measurements carried out in the R.J. region of the spectrum, where $I_{\text{qw}} = 0$ while $I_A = \beta^2 \cos^2 \theta$. Therefore, it is I_A the quantity one has to measure in order to prove that the dipole anisotropy has a kinematic origin and not some kind of «thermal» origin. Unfortunately, I_A is quite small and in any case definitely smaller than the quadrupole observed by COBE. This means that the analogue of the Bradley experiment for the motion of the Earth around the Sun, *i.e.* the measurement of the aberration of the CBR due to the observer's motion, is an extremely difficult experiment.

Let us return to eq. (3) by limiting ourselves to the dipole term

(9) $$\Delta I = I_{\text{BB}} \beta (3 - \alpha).$$

It has been pointed out by several authors that the dipole term is very sensitive to spectral distortions, being proportional to the first derivative of the spectrum, as expressed by the spectral index α. Let us introduce the quantity

(10) $$3 - \alpha_x = \frac{\Delta I_{\text{obs}} - \Delta I_{\text{BB}}}{\beta (I_{\text{obs}} - I_{\text{BB}})},$$

where $_{\text{obs}}$ refers to the observed brightness spectrum and dipole anisotropy

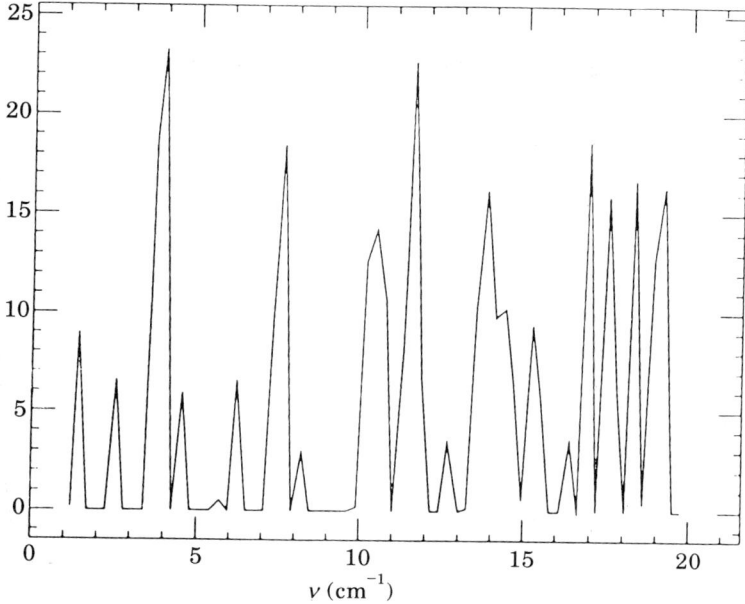

Fig. 4. – Possible deviations from a pure black-body spectrum as limited by dipole observations. The upper limits due to the FIRAS dipole spectrum on the quantity $3 - \alpha$ indicative of the spectral shape are plotted vs. frequency: only sharp lines are allowed.

spectrum of COBE. This quantity provides information on the amplitude of the distortion spectral index allowed by the observations within the experimental uncertainties. It is plotted in fig. 4.

It is clear that COBE observations exclude the presence of continuous emission but not that of lines: this point has been discussed by MELCHIORRI and TANZILLI[15]. Generally speaking, the dipole anisotropy could be a powerful method for detecting lines of cosmological origin. The sharpest the line, the larger the ratio between the dipole amplitude and the total brightness would be. The advantage of this method with respect to the conventional spectroscopic searches depends on the level of galactic contamination, which can be significantly reduced by fitting the data with a dipole distribution.

It has been shown that the dipole anisotropy can be used to discriminate between cosmological and local lines; in fact narrow lines can produce huge dipole signals and a peculiar shape. Therefore, one can hope that the deviations from the best-fit dipole observed by COBE-FIRAS are, at least in part, due to cosmological lines.

Several cosmological processes are expected to produce emission and absorption lines. The first candidates are electronic transitions between hydrogen and helium energetic levels occurring before and during the recombination epoch. Considerable deviations of the CBR spectrum from that of a black-body can derive from the nonequilibrium state between radiation and matter. Obviously the strength of these lines depends on the energy release which occurred in the early Universe. The first group of spectral features arises when the Universe becomes optically thick in the Lyman lines (z_{ot}): in this case the absorption of hard photons is accompanied by emission at lower frequencies. LYUBARSKY and SUNYAEV have calculated the dependence of the optical depth in the Lyman lines on cosmological parameters; they have also drawn the shape of these spectral distortions, but have used too high values for the Comptonization parameter y and the chemical potential μ.

We have reanalysed the situation considering more consistent values for these parameters as provided by the COBE-FIRAS results of the CBR spectrum. We restrict our attention to the case of y-distortions, because this is the situation that causes higher deviations from a Planckian curve. If we suppose $\Omega = 1$, $\Omega_b = 0.3$ and $h = 1$, we obtain $z_{ot} \simeq 4000$ for the L_α and L_β transitions. Using $y = 0.001$, the highest distortion takes place in absorption at $\nu \simeq$ $\simeq 24.5$ cm^{-1} with $\Delta I_\nu \simeq -6 \cdot 10^{-15}$ W/cm^2 sr cm^{-1}. Inside the analysed band, distortions are characterized by lower values:

$$\Delta I_\nu(\nu_{32}) \simeq 1.5 \cdot 10^{-16} \text{ W/cm}^2 \text{ sr cm}^{-1},$$

$$\Delta I_\nu(\nu_{43}) \simeq 2.8 \cdot 10^{-17} \text{ W/cm}^2 \text{ sr cm}^{-1}.$$

It is possible to calculate distortions in the dipole anisotropy spectrum, but ΔD_ν cannot be greater than ΔI_ν. By a simple analysis, that is by solving the radiation transport equation, we can place upper limits to the quantity ΔD_ν:

$$\Delta D_\nu(\nu_{13}) \lesssim 10^{-16} \text{ W/cm}^2 \text{ sr cm}^{-1},$$

$$\Delta D_\nu(\nu_{32}) \lesssim 3 \cdot 10^{-17} \text{ W/cm}^2 \text{ sr cm}^{-1};$$

these are beyond the present instrumental sensitivity of COBE-FIRAS.

A second group of spectral features could arise during the recombination period. However, the situation is worse; in fact $y < 0.01$ gives rise to a distortion in the CBR spectrum $\Delta I_\nu \leq 10^{-15}$ W/cm^2 sr cm^{-1} at $\nu \simeq 14.5$ cm^{-1}, due to the Balmer series. So we arrive at the conclusion that there is no possibility of explaining the distortions in the dipole anisotropy as observed by COBE by means of the above-mentioned lines.

Other candidates are molecules formed by gas phase reactions in the post-recombination Universe. Due to the small number of atomic species, the chemistry is rather simple during this epoch. Molecular processes can be very important in the evolution of the Universe and yield distortions in the CBR spectrum. Before protoclouds form, when the gas temperature is lower than the radiation one, molecular absorption can increase the gas temperature if the collisional disexcitation rate is greater than the radiative one. In the following evolution, molecular processes are very effective in the cooling and triggering of the collapse of primordial gas clouds. The analysis of the LiH molecule seems to be very promising for understanding the early phase of the development of protoclouds.

We have concentrated our attention to the case of the H_2 molecules. For the sake of simplicity we calculate the distortions in the CBR dipole anisotropy spectrum in a very peculiar situation: we suppose that the Universe is still homogeneous and isotropic at the moment of H_2 formation. This situation is fairly simple and lets us avoid making assumptions about the physical conditions of the cloud; furthermore it can be easily generalized if the cloud's parameters are known. The study of the roto-vibrational cascade following molecular formation should be necessary for an accurate understanding of the problem; however, due to the preliminary character of this work, we will neglect it.

In accordance with LEPP and SHULL, we assume H_2 formation via the gas phase mechanism commonly known as the H^- process; the abundance fraction of H_2 molecules (f_{H_2}) is $1.4 \cdot 10^{-6}$ and $z_b = 64$, where z_b is the epoch when the H^- process is effective and the emission process begins. Distortions in the CBR spectrum observed today at the frequency ν_0 are calculated using the following equation:

$$(11) \qquad \Delta I_\nu(\nu_0) = \frac{c}{4\pi H_0} \int_0^\infty \frac{L(\nu, z) E(z)}{(1+z)^5 (1+\Omega z)^{1/2}}.$$

$E(z)$ is an empirical weighting function, given by a combination of classical error functions, which takes into account the evolution of the sources; it depends on z_b as well as on the *time interval* Δz required to stabilize the emission or absorption process. Assuming a Lorentzian shape for a generic line in the intrinsic luminosity density $L(\nu, z)$ at the epoch z and using the quadrupole transition probabilities, it is possible to estimate the distortions in the CBR spectrum. The large-scale anisotropy due to the Doppler effect can be obtained from eq. (9). In particular ΔD_ν can become 25% of the corresponding ΔI_ν if $\Delta z_b/z_b \simeq$ $\simeq 10^{-4}$. The real conditions taking place in protoclouds can significantly increase these distortions. In fact, in a collapsing cloud, not only does density increase, but also f_{H_2}: when $n_H \gtrsim 10^8$ cm^{-3}, three-body reactions involving H and H_2 species are so efficient that all the atomic hydrogen can be converted into molecules before dissociation.

3. – Considerations about the CBR anisotropy.

3`1. *Problems in detecting the CBA: one-channel experiments.* – When a radiometer is observing two sky regions $\Delta\Theta$ degrees apart with a beam σ degrees wide (it is customary to use the value of the standard deviation of a Gaussian shape to identify the beam width) the output ΔS can be written as

(12) $$\Delta S = \Re[\Delta I_{\text{inst}} + \Delta I_{\text{atm}} + \Delta I_{\text{gal}} + \Delta I_{\text{RJ}} + \delta\Delta I_{\text{dipole}} + \Delta I_{\text{CBR}}],$$

where $\Re$ is the response of the system to the incoming power, and

ΔI_{inst} refers to spurious signals arising from radiation emitted by the instrument itself;

$\Delta I_{\text{atm}} = \Delta I_{O_3} + \Delta I_{O_2} + \Delta I_{H_2O} + \ldots$ is due to the atmospheric fluctuations both in composition and temperature of the various components at balloon altitude;

$\Delta I_{\text{gal}} = \Delta I_{\text{dust}} + \Delta I_{\text{synchrotron}} + \Delta I_{\text{free-free}}$ is due to galactic anisotropies produced by dust clouds of different temperature T_d, emissivity ε_d depending on the observed sky region and to synchrotron and free-free contributions;

$\Delta I_{\text{RJ}} = \Delta I_{\text{Earth}} + \Delta I_{\text{Sun}} + \Delta I_{\text{Moon}} + \ldots$ is due to the various celestial objects radiating in the Rayleigh-Jeans region of the spectrum, entering the field of view via scattering, reflection and diffraction;

$\delta\Delta I_{\text{dipole}}$ is the gradient of the CBR dipole anisotropy, which depends on the amplitude of modulation: for small amplitudes $\Delta\theta$ we have $\delta\Delta I_{\text{dipole}} \simeq -\Delta\theta \times \Delta I_{\text{dipole}} \sin\Psi$, Ψ being the angle between the direction of the dipole maximum and that of modulation, and $\Delta\theta$ is in radians.

To disentangle true ΔI_{CBR} anisotropies from the listed spurious contributions one is forced to adopt various more or less efficient procedures. We discuss here what can be done for a single-channel experiment, *i.e.* when spectral information is not available, as for the 1978 ULISSE flight. A short review of the instrument follows.

The instrument consisted of a cryogenic dewar (fig. 5) containing 1.5 litres of liquid helium at ambient pressure (42 km of altitude), *i.e.* at an operating temperature of $T_{\text{bath}} \leq 1.2$ K. A composite Ge-Ga bolometer (2.5 mm in diameter) + parabolic cone + a set of filters + a TPX lens defined a field of view of 2.2° (Gaussian standard deviation), or about 5° FWHM, with a spectral response from 450 to 3000 μm, and a NET (thermodynamic) of 0.5 mK Hz$^{-1/2}$.

The NEP of a bolometer depends on the operating temperature and the incoming background as shown in fig. 6. In our case the background is due to the emission from the optical components and from the residual atmosphere. It is clear from fig. 6 that an operating temperature much smaller than 1 K is not useful in our case. Our large background is due to the choice of a very large spectral bandwidth and throughput $A_b \times \Omega_b \simeq 0.3$ cm^2 sr, where A_b, Ω_b are the bolometer area and solid angle, respectively.

The choice of the bolometer dimensions, operating temperature and spectral bandwidth is usually constrained by two opposite requirements: in order to

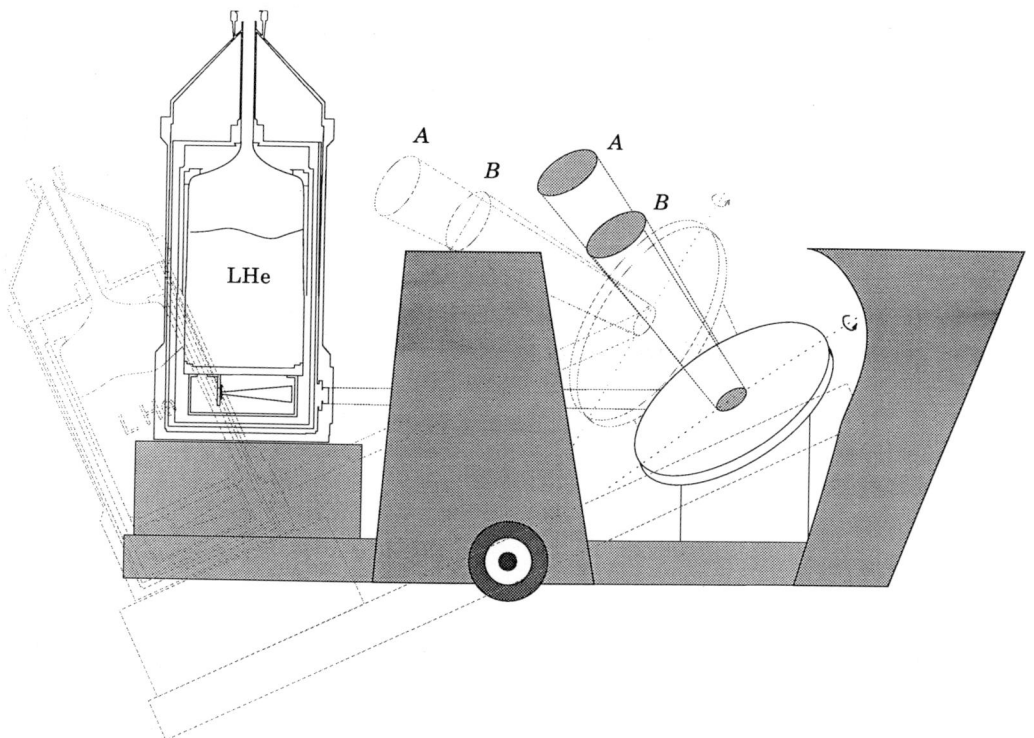

Fig. 5. – Scheme of the 1978 flight. Radiation coming from the sky at an elevation of 40° from two regions about 5° large, and 6° apart, is reflected by the wobbling mirror into the line of sight of a TPX lens which concentrates the power into a ^{4}He cryostat. The entire system is shielded from ground radiation and mounted on a platform which can be tilted up to ± 30°[13].

minimize galactic and atmospheric signals one is led toward low frequencies and small bandwidth; the corresponding loss in sensitivity is compensated by increasing the bolometer area and decreasing its operating temperature. On the other hand, large bolometers operating at low temperatures are very sensitive to cosmic rays which thereby increase the low-frequency noise. The choice we made in our first ULISSE flight turned out to be satisfactory, the galactic and atmospheric contaminations being just comparable with the final instrumental sensitivity at high galactic latitudes. Other groups have decided for the alternative solution of large-area, (0.1-0.3) K bolometers: their quoted NEP are comparable to or larger than ours.

The beam was directed to the sky by a large flat metallic mirror, oversized by a factor three with respect to the optical beam at 2 mm wavelength, and modulated mechanically with an amplitude $\Delta\Theta \simeq 6°$ in the sky at a sinusoidal frequency of 10 Hz.

The switched signals undergo a 10^4 amplification, synchronous demodulation

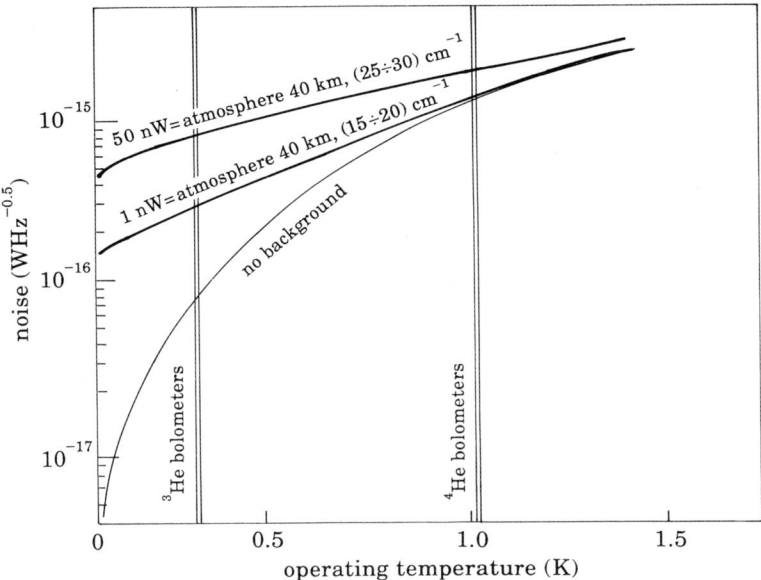

Fig. 6. – Noise-equivalent power for an ideal bolometer under different backgrounds: when the background is larger than 10 nW, there is a small advantage in using ^{3}He bolometers rather than ^{4}He.

and 0.1 s integration by two, 90° out of phase, lock-ins. The two outputs provide information, respectively, about signal + noise and noise alone. This technique is similar to that employed by COBE, where two separate instruments A and B have been used and the quantity $(S_A + S_B)/2 \propto$ signal + noise, while $(S_A - S_B)/2 \propto$ noise.

The dewar, the optical system and the wobbling mirror were mounted on a magnesium platform, which can be tilted $\pm 20°$, via telecommand. It follows that the direction of modulation can be made parallel to the horizon by observing the instrumental offset and consequently tilting the platform. This procedure allowed us to correct for small disalignments due to the evaporation of cryogenic liquids. The entire system was surrounded by metallic shields. The upper edge of the flat mirror (the largest optical component) stands about 10 cm below the edge of the shield.

In the following we discuss the calibration procedure adopted for the ULISSE experiment.

The calibration can be divided into two steps: i) measureament of the angular response and ii) absolute spectral calibration of the radiometer.

The angular response of the system is of great relevance in CBA experiments, where one should minimize spurious signals from strong sources outside the nominal field of view. The angular response has been measured in laboratory by a mercury lamp + a mechanical chopper in the focal plane of one metre

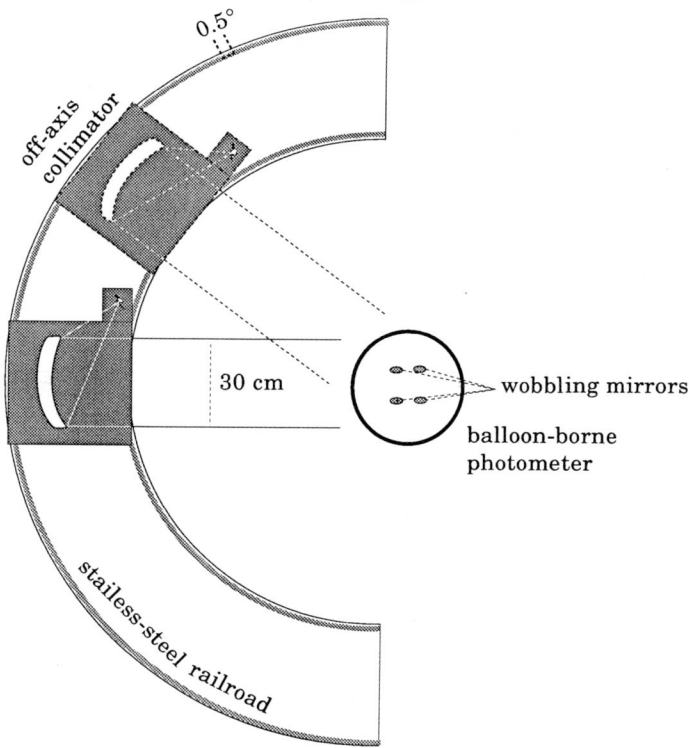

Fig. 7. – Laboratory set-up employed in measurements of the angular response: a Hg lamp is employed at the focal plane of a 30 cm off-axis collimator, which is mounted on a semi-circular railroad.

off-axis parabola: as shown in fig. 7, the entire system was mounted on a stainless-steel railroad, to explore the sidelobes of the instrument. The results have been corrected for the lock-in efficiency to sinusoidal modulation: two effects have to be considered:

1) In the case of sources lying in the line of sight the response depends on the shape of the source, being greater for pointlike sources, as schematically illustrated in fig. 8a.

2) In the case of sources lying outside the line of sight a sinusoidal modulation sharpens the angular response of the system, as schematically illustrated in fig. 8b.

The first point is of interest for a correct calibration, while the second provides information on the effective sidelobes of the instrument. To explore this second point during the flight, we have used the Sun and the Moon as strong, pointlike sources outside the geometrical field of view: laboratory measurements and in-flight observations are combined in fig. 9.

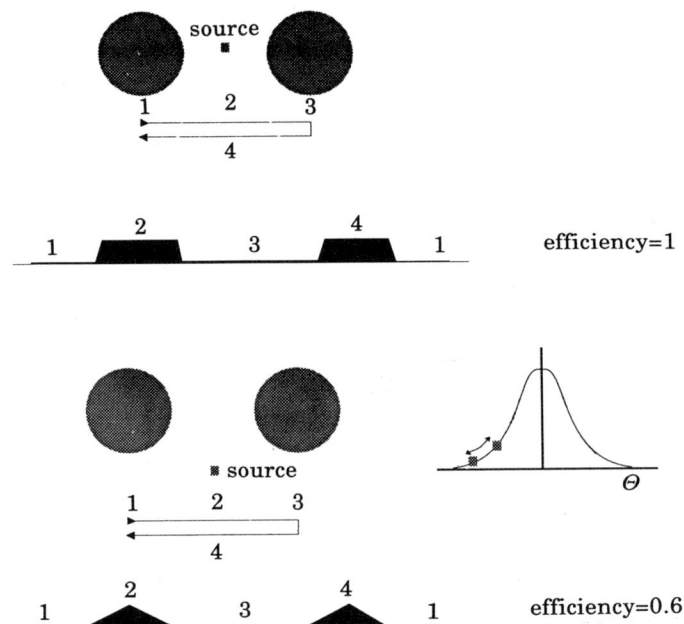

Fig. 8a. – Modulation efficiency in the case of sinusoidal modulation for a pointlike source in the line of sight and outside the nominal beam size. Since the lock-in is sensitive to the time spent by the source in the beam (dark area), the main effect of a sinusoidal modulation is that of sharpening the angular response.

We employed the laboratory and in-flight measurements of the angular response in order to derive information on the out-of-axis response of our system. Despite our efforts, we were unable to measure the instrumental response at a level well below 10^{-5} of the peak signal: this means that the Sun can be still detectable at angles larger than 30° as a faint source in the range of (10–30) μK. The situation becomes much worse for extended sources like the Earth and the balloon itself. The corresponding problems will be discussed later.

The absolute spectral calibration of our instrument was a rather difficult task: it was divided into two separate problems.

1) The measurement of the spectral response: this task was accomplished for by means of a lamellar grating interferometer + a reticle spectrometer covering the region 1 μm–5 mm. Great care has been taken to search for «leaks» in the region where the system is assumed to be opaque to the radiation. Since the bolometers are sensitive in the entire electromagnetic spectrum, a rejection factor greater than 10^9 is required in the near infrared. To check for this value of rejection several identical filters have been realized and inserted sequentially in the beam, while the instrument was illuminated by a strong UV and visible source.

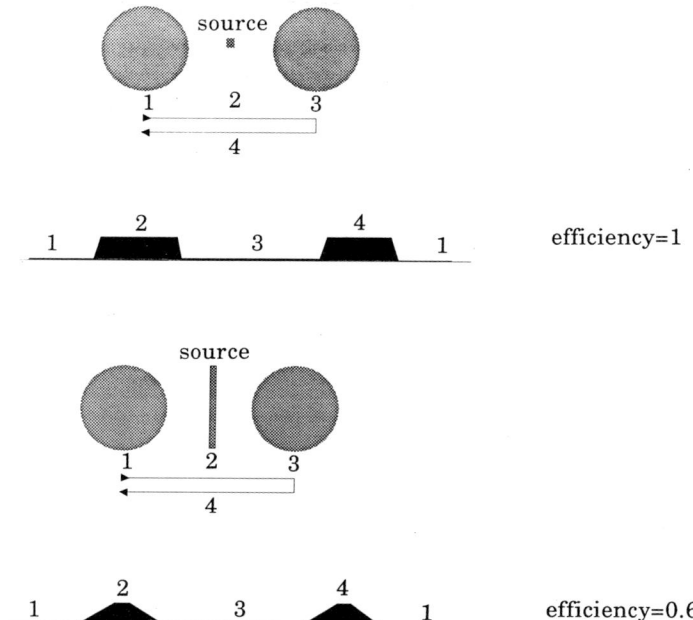

Fig. 8b. – Modulation efficiency for a pointlike and an extended source: the efficiency for a sinusoidal modulation decreases in the case of a source ⩾ the field of view.

It would be quite cumbersome to convert these pieces of information into an absolute laboratory calibration for our instrument. As a matter of fact, our experiments suffered a significant uncertainty in calibration until precise meas-

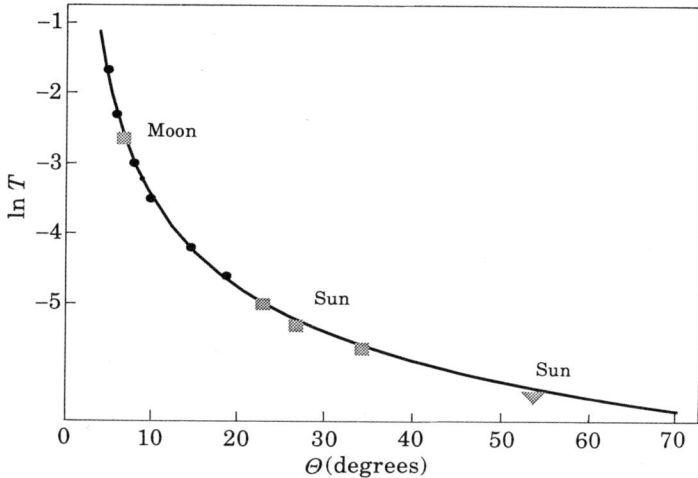

Fig. 9. – The Sun and the Moon have been used as strong sources to explore the off-axis rejection of the balloon-borne radiometer: here we plotted the angular response of the instrument of fig. 5[13]. T = transmittance.

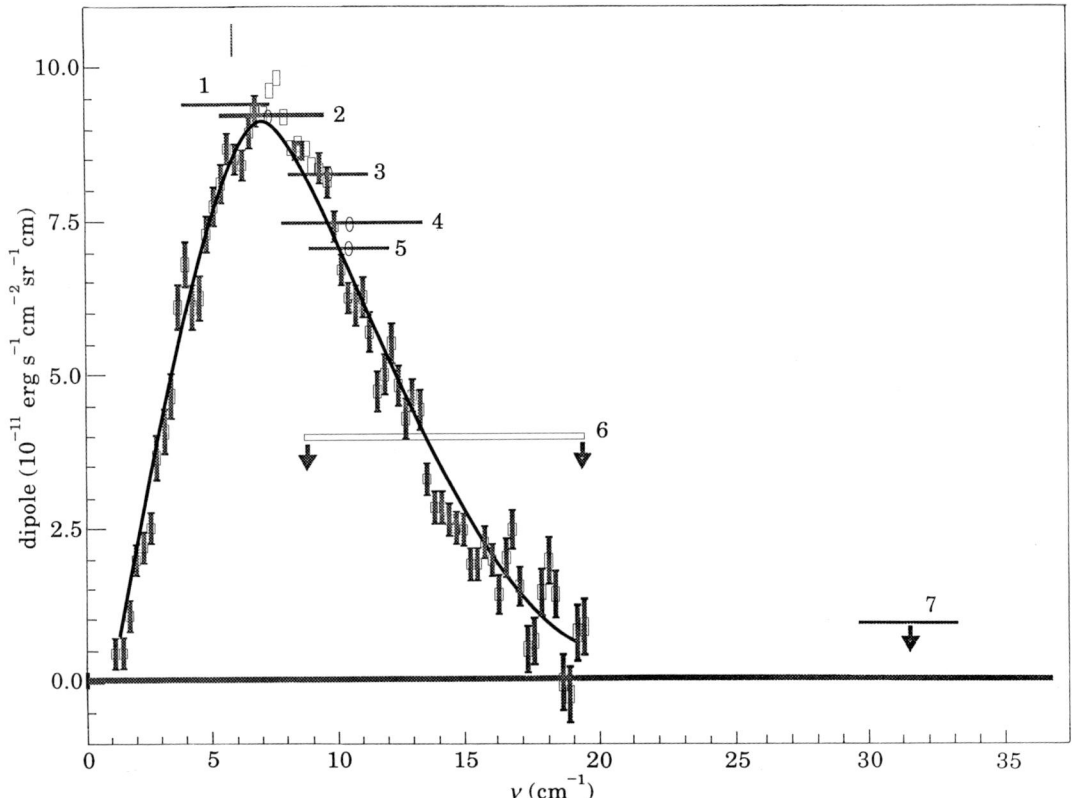

Fig. 10. – The dipole anisotropy spectrum as measured by COBE-FIRAS compared with the broad-band values (horizontal bars with numbers) obtained in our flights. Bar 2 refers to the amplitude of the dipole as measured in the 1978 balloon flight. The agreement is within a factor 1.5 and we have employed FIRAS data to re-calibrate our results [14].

urements of the CBR dipole anisotropy were available. We will not re-discuss here the problems encountered in calibration, because all our results have been re-calibrated after COBE-FIRAS has produced a precise spectrum of the dipole anisotropy. In fig. 10 we have collected all our data and compared them with the FIRAS dipole spectrum. From this figure one can derive that the largest difference is in the 1980 flight, the old calibration being a factor 1.5 lower than that derived by comparison with FIRAS data.

To discuss the various sources of spurious signals let us underline that our system is sensitive to gradients only: a diffuse uniform background will be not detected at all. Besides, a gradient that is constant in time would produce an offset, easily removed by data analysis. A gradient with a time constant different from that associated with the observation of celestial sources can be identified by a time analysis.

In the study of various spurious contributions we adopted, where possible, the following procedure:

1) choose the experimental set-up that maximizes the spurious effect, to make it easily detectable;

2) measure the dependence of the spurious effect on the experimental parameters;

3) extrapolate the results to the values of the in-flight parameters.

This procedure was suggested by the consideration that many spurious effects were at the threshold of our instrumental sensitivity.

a) The plane wobbling mirror. The emissivity ε of a flat metallic mirror is estimated to be smaller than 1% in the far infrared. For unpolarized radiation and for incidence angles far from 90° we have [16]

$$(13) \qquad \varepsilon = 1 - \frac{R_\parallel + R_\perp}{2} \simeq \frac{n_{\text{air}}}{n_{\text{mirror}}} \times \cos\vartheta \simeq \left(\frac{4\pi}{\mu\sigma\lambda c}\right)^{1/2} \times \cos\vartheta,$$

where ϑ is the incidence angle (45° in our case), $R_\parallel$ and $R_\perp$ are the two values for the reflectivity of polarized radiation, n_{air} and n_{mirror} are the refractive indices: in the case of an aluminium mirror $n_{\text{mirror}} \simeq (4\pi/\mu\sigma\lambda c)^{1/2}$ with σ the electrical conductivity and μ the magnetic permeability. By introducing the values appropriate for Al we get $\varepsilon \simeq 3 \cdot 10^{-2} \lambda^{-1/2} \cos\vartheta$, with λ in micrometres. For a gold-plated aluminium mirror, as in our case, we anticipate a lower emissivity, of the order of 10^{-4} around 1 mm of wavelength.

Since the mirror is wobbling, its angle with respect to the optical axis of the detector is changing sinusoidally up to a maximum amplitude $\Delta\Theta \simeq 3°$: correspondingly, the emissivity changes by an amount $\Delta\varepsilon \simeq 3 \cdot 10^{-3} \lambda^{-1/2} \Delta\cos\vartheta \simeq$ $\simeq 10^{-4} \lambda^{-1/2}$. For the sake of simplicity let us consider the case of a uniform background at temperature T_s, while the mirror is at temperature T_m: the radiation coming into the detector when the mirror is in the two extreme positions $\Theta_1 =$ $= \vartheta - 1.5°$ and $\Theta_2 = \vartheta + 1.5°$ is given by

$$(14) \qquad \begin{cases} T_1 = T_s(1 - \varepsilon_1) + T_m \varepsilon_1, \\ T_2 = T_s(1 - \varepsilon_2) + T_m \varepsilon_2, \end{cases}$$

where T_1 and T_2 are the antenna temperatures in the two cases.

The output of our system is sensitive to the difference

$$(15) \qquad \Delta T = T_2 - T_1 = (T_s - T_m)(\varepsilon_1 - \varepsilon_2).$$

This last equation can be rewritten as

$$(16) \qquad \Delta T = \varepsilon(T_s - T_m)\Delta\cos\vartheta.$$

or

(17) $$\Delta T_{\text{thermodynamic}} \simeq 3 \times 3.7\,(T_s - T_m)\,\mu\text{K},$$

where the factor 3 arises from the difference between the antenna and the thermodynamic temperature, and $\varepsilon \simeq 10^{-4}$, $\Delta\cos\vartheta \simeq 3.7\cdot 10^{-2}$.

Unfortunately $T_m \simeq 240$ K is much larger than the sky temperature $T_s \leq$ $\leq (10\text{--}20)$ K: therefore, we anticipate an offset of (1–3) mK during the flight. As already stated, a constant offset is easily removed from the data: we have to worry about the fluctuations of this offset.

In order to limit the fluctuations in $\Delta T_{\text{thermodynamic}}$ within 30 μK, T_m has to be stable within 0.2 K, while $\delta\Delta\Theta$ has to be smaller than 50 microradians, $\simeq 10$ arcseconds.

These are stringent requirements. T_m has been monitored during the flight: it appeared to follow the ambient temperature with a time constant of about 15 min. After a rapid cooling down to $-40\,°$C during the first two hours of flight, it started to heat up very slowly and gradually to the final value of $-30\,°$C attained at the end of the flight. The corresponding offset was uncorrelated with the gondola movements and easily removed from the data as a polynomial fit to the time series of the data.

Short-term fluctuations in $\delta\Delta\Theta$ arise from mechanical slacks. They represent a source of noise excess in the detector. We monitored them by measuring the r.m.s. of the output of 90° out-of-phase lock-in when the wobbling motor was turned off and on: we found an increase in the noise by a factor 1.3, when the mirror was wobbling.

Long-term fluctuations are possible for electronically controlled modulations, due to the instability in the electronic offset. Therefore, we preferred a mechanical modulation system, where a stainless-steel cam of sinusoidal profile drove the motion of the mirror.

The measurement of long-term fluctuations in the laboratory turned out to be a rather difficult task: the main problem is represented by the fact that eq. (16) assumed T_s to be uniform over the region explored by the wobbling mirror. If not, an offset is produced proportional to the temperature gradient. We attempted to realize this ideal condition of uniform background in our laboratory by pointing at a flat surface made by Eccosorb, oversized with respect to the beam path, the temperature of which was found to be uniform within the sensitivity of our thermometers ($\leq 10^{-3}$ K over a 50×50 cm^2 surface), and stable within 10^{-3} K/h. To check for temperature gradients over the Eccosorb surface, we mounted the panel over a platform which was rotated by 180° every few minutes, so that the regions intercepted by the two beams are exchanged. The experiment was carried out in the underground of the Physics Building, where the ambient temperature was rather constant with a maximum change of about 0.3 °C around the sunset. The metallic mirror was heated up to a maximum of 200 °C by an electronically controlled resistor. The temperature of the mirror was monitored by electronic transducers attached to the backside of the mirror with Thermocoat RS Heat Sink glue (thermal conductivity $1.7 \cdot 10^{-3}$ cal cm^{-1} s^{-1} °C^{-1}). The same thermometers have been employed to monitor the temperature distribution across the Eccosorb panel.

One should note that the procedure we listed previously is not fulfilled here: the signals anticipated in our laboratory set-up are smaller than those expected in flight.

Moreover, for temperatures higher than 50 °C the surface of the mirror is rapidly oxidized. Therefore, we were forced to use smaller temperature differences in our long-time integrations. For a $\Delta T \simeq 20$ °C we observed an offset of the order of 100 µK slightly changing with a 12 h time constant. Inserted in eq. (16) it would correspond to a mirror emissivity lower than 10^{-4}. More interesting for our purposes was the fact that the offset was rather constant in time. Once the offset was removed, the data did not show any deviation from a typical Gaussian noise distribution. Therefore, we concluded that the effects of the wobbling mirror are ≤ 100 µK for time scales of the order of 30 min. In a linear extrapolation this corresponds to less than 30 µK per pixel (typical time scale $\simeq 5$ min).

b) The shields. The emission by the shields can be diffracted and modulated by the wobbling mirror and, therefore, detected as a signal. The shields are reflecting aluminium surfaces and they have been oriented to reflect the sky at 60° of elevation. The temperature of the shields was monitored by several thermometers and found constant within 0.03 K in night time. In day time much care has to be taken to avoid a strong temperature rise due to the Sun radiation. We have employed a system already successfully used in cryogenic superinsulation that is now standard in balloon experiments. A mylar foil has been aluminized over one side. The clean surface is exposed to the Sun, so that visible rays pass it and are reflected by the aluminized surface with a great efficiency. The aluminized surface heats up, due to its bad near-i.r. emissivity, but the mylar substrate has a very good emissivity in the infrared and cools down the surface efficiently. Usually we employed three of these mylar foils, separated by high-density polystyrene panels, each three centimetres thick. Finally an aluminium foil was glued to the inner part to realize the reflecting inner shield. Temperature measurements showed a rise of about 60 °C for the outside of the system, when exposed to Sun radiation, an increase of about (1-5) °C for the inner part of a single aluminized mylar foil, while the temperature of the inner fourth aluminium foil remained unchanged within 0.03 °C.

Even if the temperature is perfectly uniform and constant, we can get signals from zones of different emissivity. A cleaning procedure was, therefore, established before the flight to have an emissivity as uniform as possible.

The final budget due to shield emissivity is difficult to assess. We believe that it is irrelevant in night time. We have checked for any spurious effect by inserting an electrical heater approximately in the centre of each panel. No signals have been observed when the balloon was pointing in a fixed direction in the sky and the temperature of each shield was sequentially raised by (1–3) °C. From this test we derived the conclusion that the shields cannot contribute to the signal for more than 100 μK within time scales of 30 min.

c) Earth's radiation. The Earth is shielded by our system of metallic shields, but its radiation can be diffracted by the shields themselves and modulated, if not uniform. A rough estimate of the Earth's spurious signals can be done by schematizing the shields and the metallic mirror as two circular apertures through which the Earth's radiation is diffracted: we anticipate a diffracted power of the order of 3 mK for a 300 K black-body. Again, we are interested in fluctuations of this uniform background. Unfortunately, temperature gradients as large as 30 K are expected in our case, because both the hot African coast and the colder Mediterranean Sea are in the horizon of the gondola. Gradients as large as 300 μK are possible, after diffraction over the shields. Since they are distributed over angular scales of about 180°, our 6° modulation should produce signals of about 30 μK, just at the limit of our per-pixel sensitivity. Larger fluctuations are expected if the gondola is rapidly rotating, however. We have observed them during the ascent, when the pointing system was not operating.

In order to check for the efficiency of the shields, we have removed one of them by telecommand few minutes before the end of the flight. Therefore, it was possible to observe the effect of the Earth radiation directly diffracted by the metallic mirror even in the case of slow gondola movements (the time constant was of the order of one turn per hour): we observed signals at a level of (100–300) μK synchronous with the gondola motion.

Similar fluctuations seem to have plagued several experiments in the past: they have been observed by the MIT group as well as by the MAX group. These groups measured the spectrum of the fluctuations and found it substantially different from that of a black-body, marginally consistent with the stratospheric ozone spectrum. This fact led them to attribute the observed fluctuations to atmospheric effects. We believe that they were more probably due to the Earth's radiation entering via diffraction: since the diffracted power is strongly wavelength dependent, it is not surprising that the spectrum is not that of a black-body. One should stress that the spectrum of this diffracted radiation is close to that of the CBR in the (3–10) cm^{-1} region: therefore, it can easily mimic the CBA. Since it is synchronous with the gondola movements but not with the sky rotation, we can disentangle it by comparing signals observed at the same azimuth but separated in time more than $\simeq$ 30 min. Unfortunately, the time constant for the Earth's radiation is not very different from this number, both because the balloon is moving at a relatively high speed ($\simeq$ 100 km/h) and the ground temperature is rapidly changing along the day. The only solution to this problem is that of avoiding fast gondola rotations.

d) Atmospheric background. The atmospheric background was measured in flight by tilting the optical platform, so that the direction of modulation was no longer parallel to the horizon. One should note that an atmospheric offset could be present even in the case of modulation parallel to the horizon, if the modulation amplitude is not symmetric with respect to the zero reference. In fact, the mean elevations in the sky at the extreme angles of modulation are slightly higher than the elevation corresponding to the zero reference. In first approximation we get

$$\Delta T \simeq 3 \cdot 10^{-2} T_s \Delta \Psi,$$

where $\Delta \Psi$ is the asymmetry in the modulation. It imposes an accuracy in the modulation amplitude comparable with that of eq. (16).

The instrumental offset due to various different causes has been set to zero in the ULISSE 1978 flight by appropriately tilting the platform via telecommand. We found that gondola pendulations up to several arcminutes were induced by turning on the pointing system: in our case this pendulation has the same effect as tilting the platform and produces a large variable offset. To avoid this problem we pointed the system approximately toward the chosen sky direction and turned off the pointing system, so that the gondola was moving freely around the selected direction.

The presence of fluctuations in the atmospheric background has been monitored in the following way:

Fluctuations with somewhat short time scales (few minutes) have been tested by observing the structure of the signal when the gondola was pointing at the same sky zone (this excludes true sky anisotropies). For a field of view of $\sigma = 2.2°$ essentially the same sky region is observed for about 10 min, if the gondola is at rest. To disentangle atmospheric induced fluctuations from the intrinsic detector noise, the straightforward way is that of using two 90° out-of-phase lock-in channels: the channel in phase with the wobbling mirror provides the signal + noise, while the second channel is sensitive to the noise alone.

No evidence for atmospheric fluctuations has been found at the level of 100 μK, at time scales of 5 min. Much longer time scales can be investigated by re-observing the same atmospheric region at time intervals of about 30 min. A slightly worse upper limit has been found at these time scales, of the order of 300 μK. Moreover, we cannot disentangle spurious atmospheric effects from the diffracted Earth's radiation, as discussed previously.

e) Balloon surface reflections. An $8 \cdot 10^5$ m³ balloon subtends a total angle of about 40° for a gondola suspended 100 m below. If the line of sight of the instrument is directed toward the balloon, we detect the power radiated by the balloon itself and, more worse, that radiated by the Earth and reflected from the balloon surface. A simple optical ray tracing shows that the Earth's radiation, reflected by the balloon, is in the line of sight if the instrument is pointing at elevations greater than 80°. It is possible to avoid this radiation directly entering into the instrument by covering the entire optical system, but it always illuminates the shields. The estimated reflectivity of the balloon surface is of the order of 0.01 at millimetric wavelengths. The contribution of this effect can be estimated by substituting the rejection factor due to the shields with the factor 0.01. To check the effective rejection of the system, we made it possible to change the elevation angle: at an angular elevation of about 70° large fluctuations of (200–400) μK have been observed when the gondola was rotating: by measuring the dependence of this effect on the elevation angle we can conclude that it is well below 30 μK for an elevation angle of 45°.

Also the Moon and the Sun can be reflected by the balloon surface: a software has been developed which automatically takes care of this, by removing the data corresponding to Moon and Sun closer than 50° to the optical axis.

f) Galactic background. For years the question of galactic contamination has remained unsolved. It is still believed that large-bandwidth systems, like ULISSE, are affected in a dramatic way by patchy dust emission. The first clear answer to this problem has been provided by FIRAS: the measured dipole spectrum, obtained by comparing two wide sky regions 180° apart at high galactic latitudes, shows that the dust contribution is ≤ 50 μK if integrated over the (3–20) cm^{-1} region. Since the dust anisotropy increases with the beam size and the beam separation, we are led to the conclusion that in the case of the ULISSE experiment the signals at high galactic latitudes are «essentially» not contaminated by dust: an assumption which does not exclude that a few hot spots are still present. The FIRAS dipole allows us to estimate the contribution of dust to various spectral regions: by limiting the observations to the far-infrared side, *i.e.* (3–10) cm^{-1}, we get a decrease of dust contribution by a factor 1.6 only, with respect to the wider (3–25) cm^{-1} band employed in our experiment.

Also in the case of the ULISSE 1978 flight, the first check for the cleanness of the observed sky regions consisted in searching for the dipole anisotropy. For a sky modulation of 6° the instrument would detect gradients ≤ 0.3 mK. Therefore, a clean detection of the dipole anisotropy made us confident for the absence of disturbances at a similar level. In fig. 11 we quoted few hours of data as received at ground and directly plotted on the chart recorder, prior to any manipulation. We note the presence of the dipole gradient, while the gondola was freely rotating, as well as large signals when the instrument was crossing the galactic plane (GP). The ULISSE 1978 flight for the first time clearly detected the dipole anisotropy in the far infrared [13].

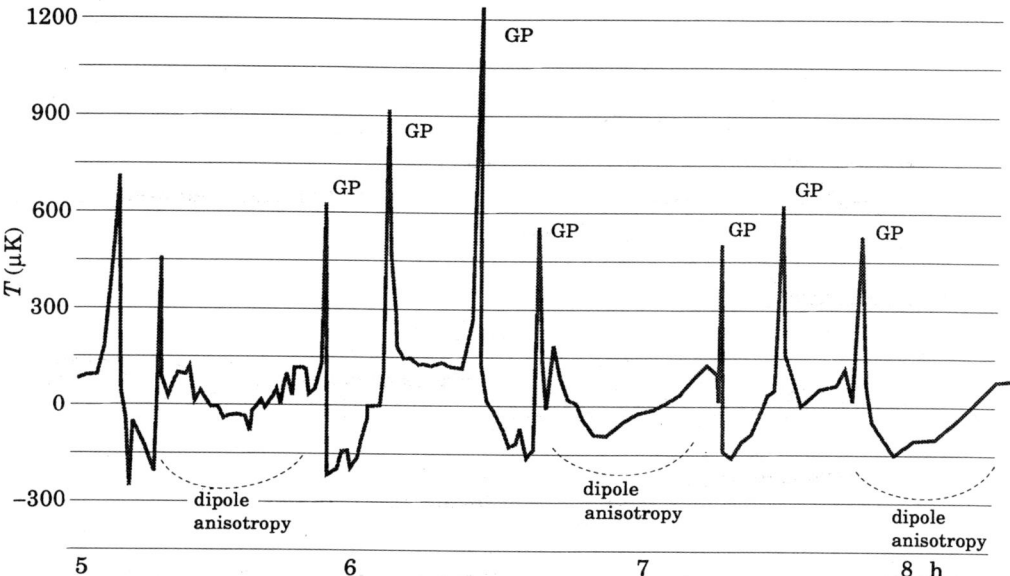

Fig. 11. – Signal sent to the chart recorder by the instrument of fig. 5, while the gondola was freely rotating. GP is the galactic plane crossed by the instrument (or approached). The gradient due to the dipole anisotropy is also shown [13].

Obviously, one cannot exclude the presence of some residual dust emission: in 1978 the data of IRAS were not available and we attempted a correlation with Heiles map on the H I, 21 cm line, being H I a good but not perfect dust indicator [17]. The only safe way we had to test for dust contamination was just that of cutting the data at various galactic latitudes and observing the r.m.s. value of the residuals: a technique similar to that employed in the analysis of the COBE data. We found that no significant changes in the r.m.s. of fluctuations were present for $b_{\text{cut}} \geqslant 30°$ [13].

In fig. 12 we plotted the observed sky region in the ULISSE 1978 flight as well as the selected strip on which the search for CBA was carried out. Two different reasons motivated the choice of that region: a) it is at high galactic latitude, b) it is the sky region where the instrument spent the largest time. The sky region at least 20° away from the galactic plane has been observed several times: in fig. 13 only one scan, or string, is shown, after subtraction of the dipole term. The gray area corresponds to the r.m.s. value of the 90° out-of-phase lock-in and provides information on the detector noise. The various levels of this noise correspond to the different time spent by the instrument in the various observed regions. The same analysis has been done for the remaining strings. The sky roughness, measured by subtracting from the variance of $(SA + SB)/2$ that of $(SA - SB)/2$, is 45 µK: in performing this analysis we disregarded the experimental point at the left of the plots, which was well above

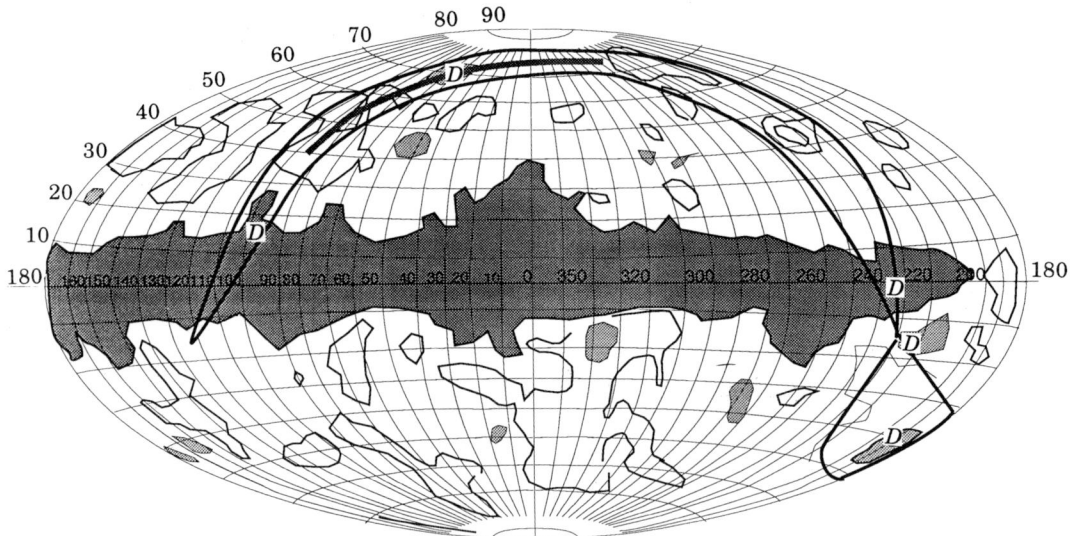

Fig. 12. – Sky region observed in the course of the 1978 balloon flight. The sky coverage was not at constant rate, the gondola spent the largest part of the time on the left-hand side. Some of the hottest regions observed by COBE are shown for comparison, but only the Galaxy has to be considered as confirmed, up to now. D indicates hot spots detected in our experiment at a signal-to-noise ratio larger than 3. The strip at high galactic latitude was that employed for the search of the CBA [12, 13].

the mean. The variance of the data is comparable within the experimental errors with the variance of the detector noise of 32 μK.

In fig. 14 we have quoted all the data relative to a sky region shown in fig. 12 as a dashed zone. If one excludes the largest bump, we get a «signal» of

$$\Delta I_{\text{CBR}} = (70 \pm 40) \, \mu\text{K},$$

where the errors are determined by a reduction of a factor ten in the maximum likelihood.

It is interesting to note that the observed sky variance is comparable with that of COBE at the angular scale of 10°.

The data collected by the ULISSE 1978 flight have been used to search for large-scale gradients, like dipole and quadrupole patterns. The results are resumed in the map of fig. 15. While the dipole anisotropy was consistent in direction and amplitude with that observed in the radio region, we found evidence for a possible quadrupolelike distortion [13].

The observed sky region was insufficient to compute all the quadrupole components: Q_2 and Q_5 (in right ascension and declination coordinates) have been found to be highly correlated and the r.m.s. amplitude Q_r has been estimated to be of the order of (0.5–0.8) mK.

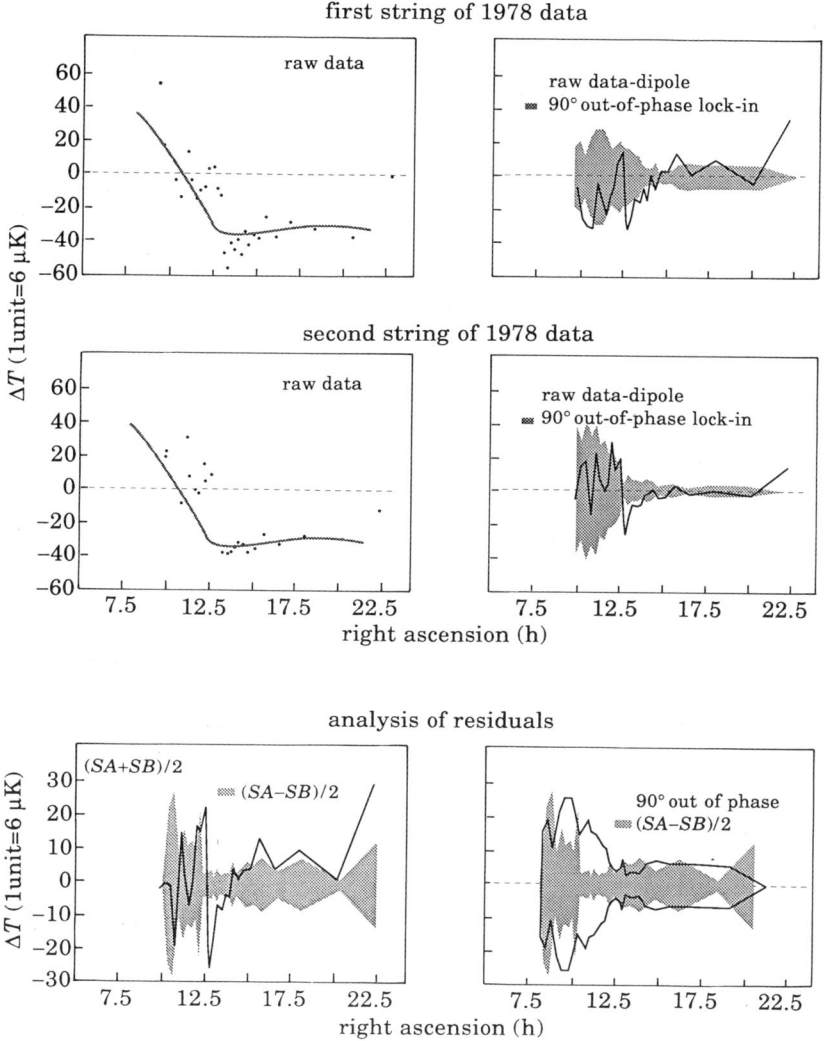

Fig. 13. – Analysis of a string of data of the 1978 flight: signal observed in the strip after removal for the dipole anisotropy. The shaded area corresponds to the noise in the 90° out-of-phase lock-in.

It is now clear that the quadrupolelike distortion we observed is not intrinsic to CBA, since radio observations have set much more stringent upper limits. However, if it is due to dust and not to some instrumental artefact or atmospheric hot spot, it should be present at a level 10–10^2 times lower in the 53 GHz COBE quadrupole. Therefore, it is interesting to see if our data are at least consistent with the COBE quadrupole. In fig. 16 we have plotted the signal we expected from the COBE quadrupole in the region we explored, multiplied by 80:

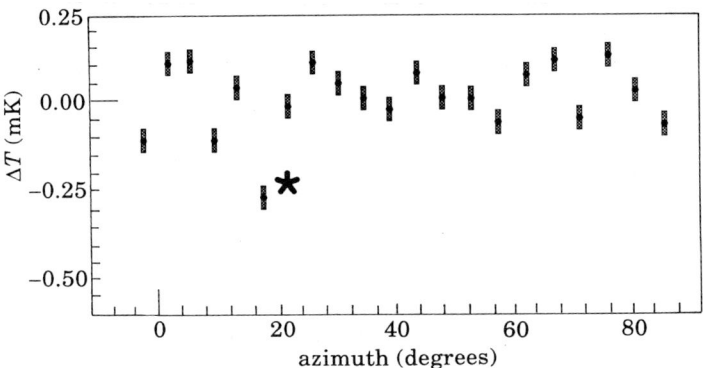

Fig. 14. – Final data collected in the shaded region of fig. 12 after removal of the dipole anisotropy. The asterisk corresponds to the region indicated as D in fig. 12.

the shaded area takes into account the uncertainty in the various Q_i (which are relative to galactic coordinates in the case of COBE data). We have also plotted as a continuous line the quadrupolelike distortions we observed, as due to the

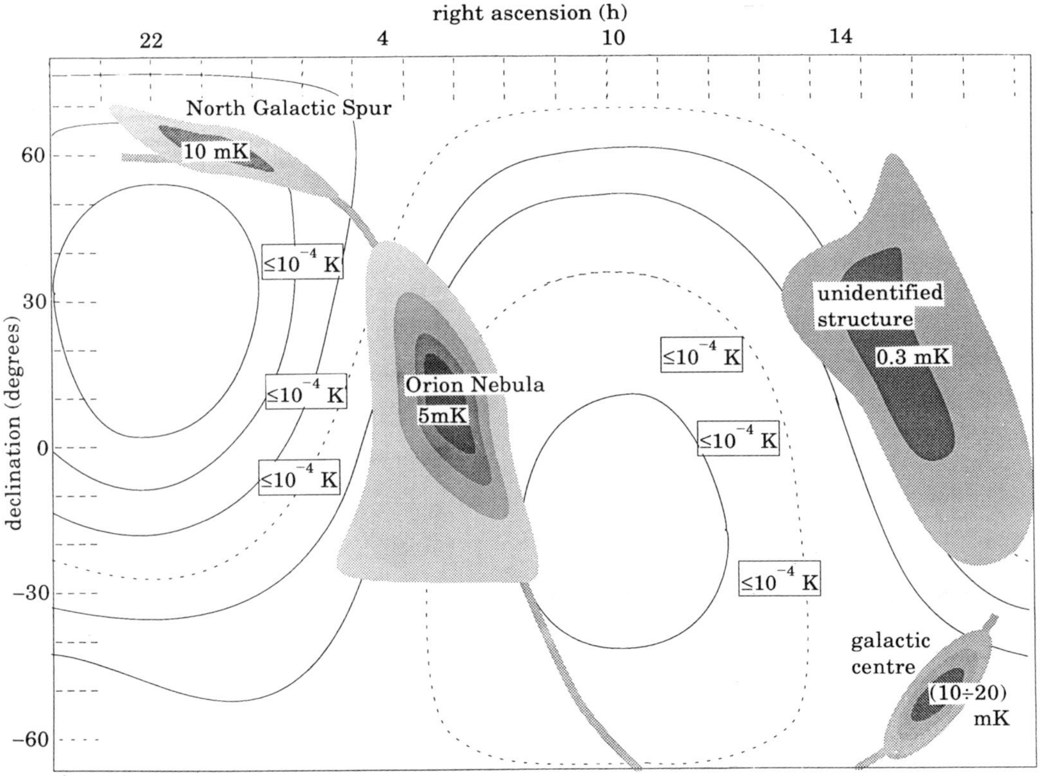

Fig. 15. – Map of the region explored by the 1978 balloon flight. The unidentified region on the right was the major source contributing to the quadrupolelike distortion.

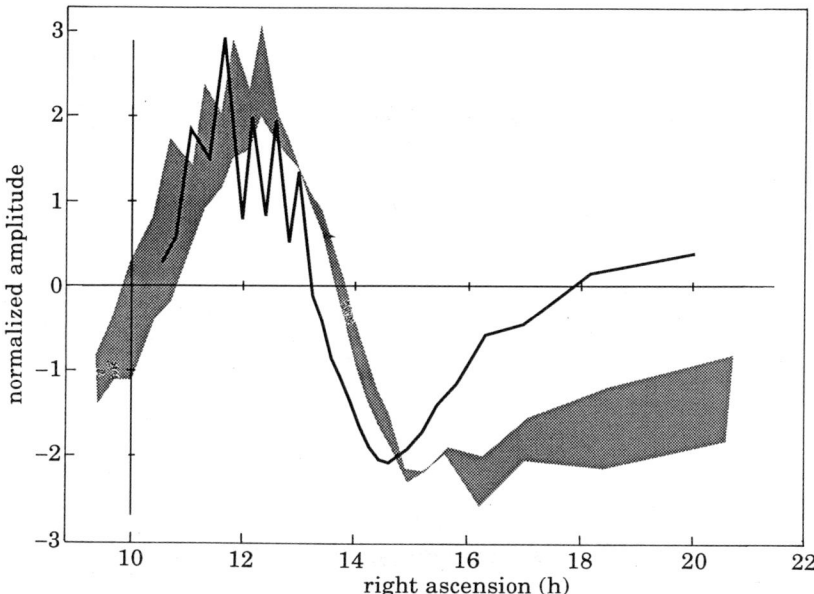

Fig. 16. – The quadrupolelike distortion observed in the 1978 flight (continuous line) compared with the COBE quadrupole amplified by a factor 80.

Q_5 component only (in right ascension and declination coordinates). Due to our limited sky coverage, the agreement could well be fortuitous, but it suggests a bit of caution in interpreting the COBE quadrupole as not contaminated by dust.

The main scientific results of this flight were the detection, for the first time, of the dipole anisotropy in the infrared and a stringent upper limit of about 70 µK to CBR anisotropies at angular scales of 6° which can be interpreted as a detection if dust emission is negligible.

The main technical information we have derived by the ULISSE 1978 experiment was the following: many local effects are strongly sensitive to the gondola movements. Anisotropies in the Earth's radiation, diffracted or reflected from the balloon surface, and anisotropies in atmospheric emission are both large-scale effects: they are reduced with the beam size and if gondola motion is avoided. On the other hand, the pointing systems easily induce gondola oscillations at a level of arcminutes or fraction of arcminutes: it follows that the best procedure consists in pointing toward a clean sky region and subsequently disactivating the pointing system, to let the gondola free of reaching a stable attitude.

3˙2. *Problems in detecting the* CBA: *multichannel experiments*. – Equation (12) can be rewritten several times for several different photometric channels.

There is no reasonable hope, however, to be able to set up a complete system of equations from which the various terms can be extracted. It would require too many channels, increasing the final noise to an unacceptable amount.

It follows that essentially two ways have been followed: in the first one, three or four channels have been employed to build up a colour-colour diagram, like that recently published by the MAX team. In this way one does not correct for spurious anisotropies, but limits himself to check if the observed signals are compatible to some of the estimated spectra of the spurious backgrounds.

We have followed a second way, in which two carefully selected channels have been employed. Let us rewrite eq. (12) for these two channels as

$$\Delta S_1 = \Re[\Delta I^1_{\text{atm}} + \Delta I^1_{\text{gal}} + \Delta I^1_{\text{RJ}} + \delta \Delta I^1_{\text{dipole}} + \Delta I^1_{\text{CBR}}], \tag{18}$$

$$\Delta S_2 = \Re[\Delta I^2_{\text{atm}} + \Delta I^2_{\text{gal}} + \Delta I^2_{\text{RJ}}]. \tag{19}$$

The absence of CBA terms in eq. (19) means that the wavelength of this channel is short enough to make the CBR contribution negligible. To disentangle the CBA one has to select the central wavelengths and the bandwidth of the two channels in such a way to have

$$\frac{P_1}{P_2} = \frac{\Delta I^1_{\text{atm}}}{\Delta I^2_{\text{atm}}} = \frac{\Delta I^1_{\text{gal}}}{\Delta I^2_{\text{gal}}} = \frac{\Delta I^1_{\text{RJ}}}{\Delta I^2_{\text{RJ}}}. \tag{20}$$

In attempting to fulfil the requirements (20) we considered the scheme illustrated in fig. 17. One channel has a filter extending from 3 cm^{-1} up to an appropriate frequency ν^*, while the second channel is centred at the frequency $\kappa\nu^*$, where κ is an appropriate constant, practically ranging from 1 to 2. For instance, in fig. 18 we have plotted the three ratios (20) as a function of ν^* for $\kappa = 1.2$ and 1.6 and for a galactic-emission model as given by the COBE-DIRBE results (spectral index 1.3). One can see that there are two regions where the three ratios are similar, namely around 8 cm^{-1} and 38 cm^{-1}. The first frequency is not acceptable, because the CBR is still

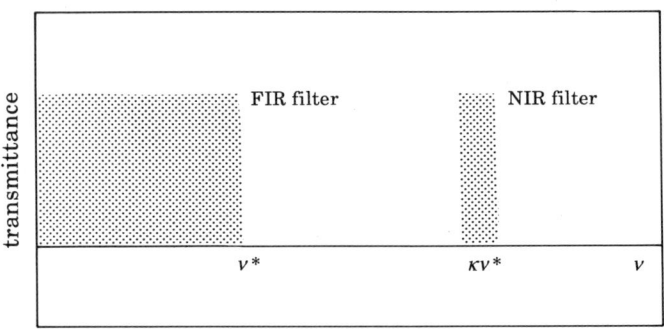

Fig. 17. – Scheme of the two-filter radiometer: a wide far-infrared filter is followed by a narrow mid-infrared filter.

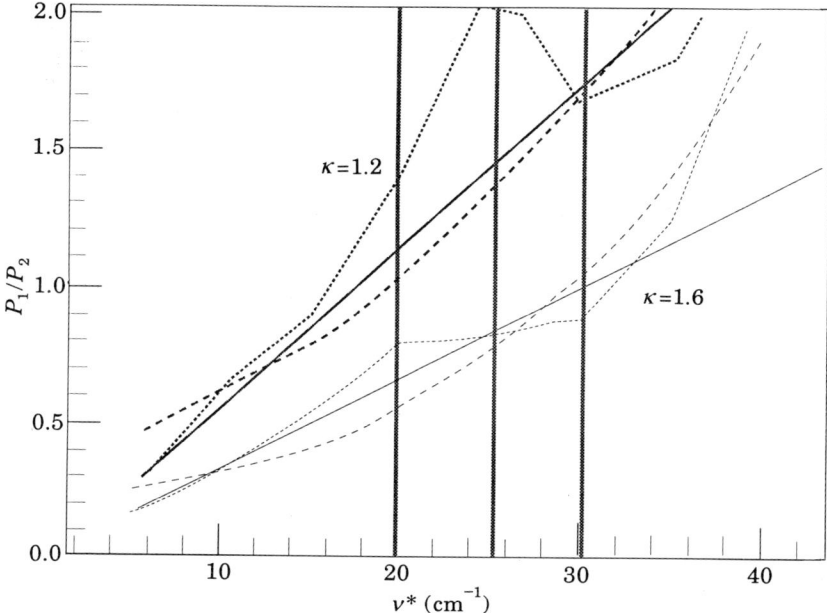

Fig. 18. – Ratios of eq. (20) plotted against the cut-off frequency ν^* for two values of the ratio between the short-wavelength cut-off of the first filter and the central wavelength of the second filter κ. Continuous lines, R.J. sources; dashed lines, galactic emission; dotted lines, atmosphere. The three vertical lines indicate the three cut-off frequencies selected for both the values of κ in the 5 balloon flights to reach the optimum of correlation. Practically the maximum of correlation was found for $\nu^* \simeq 32$ cm^{-1}, $\kappa = 1.6$.

present in both channels, while the second one provides an interesting possibility.

Unfortunately, the precise choice of the best ν^* and κ can be determined only through experiments, the spectral properties of the atmosphere and of the galaxy being not exactly known. This was the goal of the ULISSE program, where anisotropy searches have been alternated to studies of atmospheric and galactic contaminations. Practically, we explored the three possibilities indicated in fig. 18 by the vertical lines.

The procedure was the following:

First we analysed the effect of atmospheric fluctuations by inducing a spurious atmospheric noise: the direction of modulation was made not parallel to the horizon, by that introducing an atmospheric offset. This was made fluctuating by random changing the amplitude of modulation. In fig. 19 we report the correlation between two channels when this procedure was activated: the slope of the correlation is then compared with that found in exploring the dust clouds at high galactic latitudes, like l 134. Finally, a R.J. source (a calibrator at ambient temperature) is inserted in the field of view and the next slope is determined.

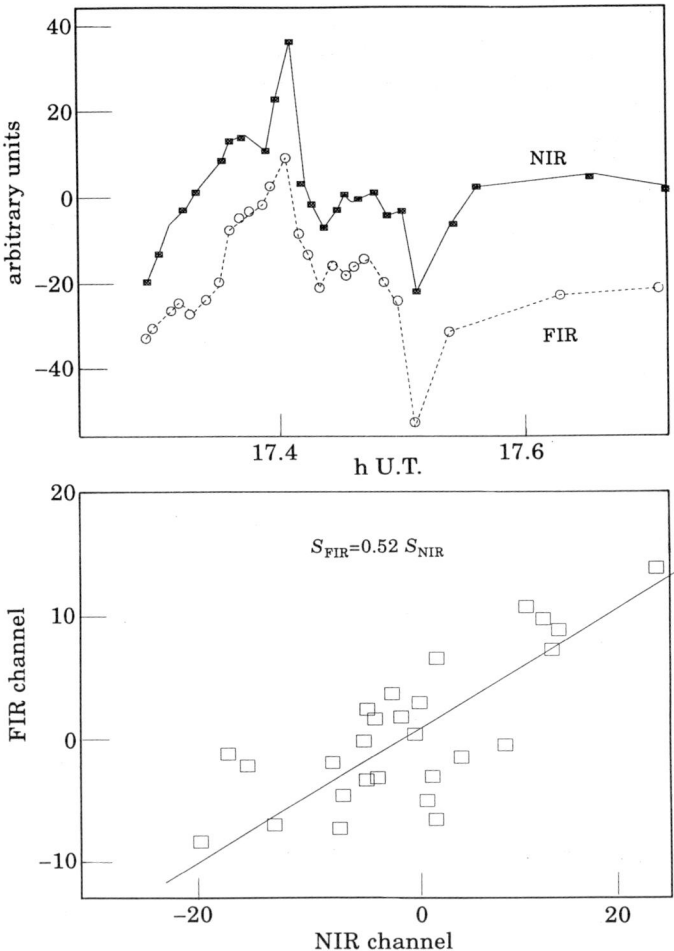

Fig. 19. – Fluctuations induced in the FIR channel by the transit of the l 134 cloud (upper panel). Similar fluctuations have been induced in the NIR channel and the correlation between the two is studied in the lower panel.

In this way, by trials and errors, we finally arrived at the best compromise. The spectral properties of the two selected channels are shown in fig. 20.

The selected sky strip was determined by a compromise between galactic latitude, time of observation, direction of modulation parallel to the galactic plane. The data are plotted in fig. 21 both for the B and C channels (infrared and far infrared, respectively).

In the upper panel of this figure we have plotted the signals evaluated by simulating the ULISSE experiment on IRAS data, taking into account all the experimental details, like the antenna profile and the sinusoidal modulation. The r.m.s. from IRAS is of the order of 1 MJ/sr: the dust contamination is,

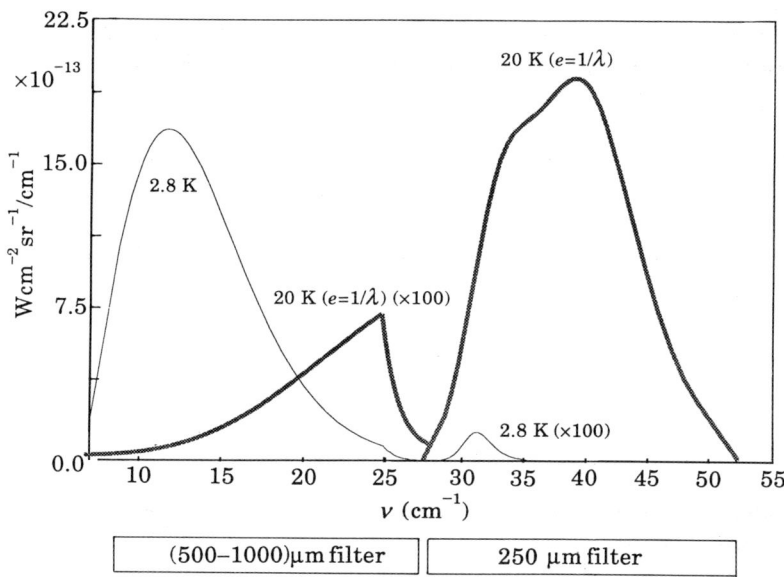

Fig. 20. – The main properties of the two filters optimized for galactic-dust correlation: the power entering in the photometer is shown for the two cases of CBR and a dust cloud.

therefore, comparable with that observed by the MAX experiment. The influence of dust in our data is clearly shown by a comparison with channel C signal. The residual discrepancies between IRAS and C channel cannot be immediately ascribed to different dust components or CBA, due to the presence of the various spurious effects previously discussed.

These effects are significantly removed by comparing directly the two on-board channels B and C. A «clean» data set has been obtained using the linear regression

$$\Delta S_c = \Delta S_L^i - A \Delta S_H^i - B,$$

where the constants A and B were obtained minimizing the quantity

$$\chi^2 = \sum_i (\Delta S_L^i - A \Delta S_H^i - B)^2.$$

The value of the constant A found in this way is in good agreement with the one expected from the COBE-FIRAS dust spectrum, a fact that shows a predominance of dust contamination. B turned out to be null within the errors. The clean data are plotted in panel D). A large-scale behaviour is evident and it is well fitted by the expected gradient from the dipole anisotropy (continuous line).

As previously stated, the best source for calibration in our experiments was always represented by the CBR dipole anisotropy, whose spectrum is

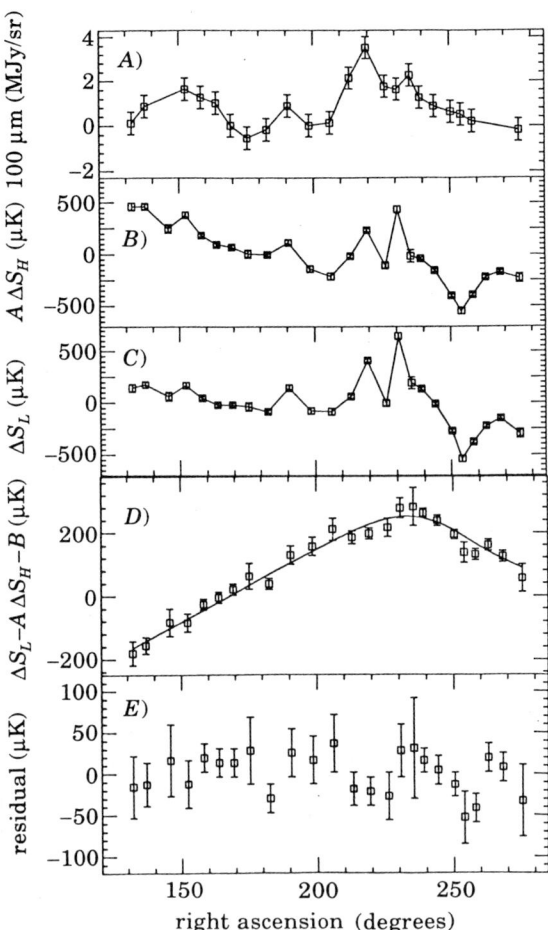

Fig. 21. – Panel A) simulation on the IRAS sky of the ULISSE experiment. Panel B) data collected in the channel with the filter NIR of fig. 17. Panel C) data collected by the channel FIR of fig. 17. Panel D) subtraction from C to B: the dipole anisotropy is fitted with the direction and amplitude of FIRAS and DMR results. Panel E) residuals after removal of dipole anisotropy[18].

well established in the full wavelength range of our low-frequency channel.

To obtain an unambiguous in-flight calibration we proceeded in this way: we fitted the anisotropy data using the relation

(21) $$\Delta S_c = \Re \frac{2}{\Im} \int_0^\Im D[\phi(t), \delta(t)] R(t) \, dt + \text{offset},$$

where $\Re$ is the system responsivity (i.e. μK/telemetry units), $\Im$ is the period of the sinusoidal beam switching, ϕ and δ are the right ascension and declination of the beam during the beam switching around the direction i; $R(t) = \pm 1$ is the demodulation function in phase with the sinusoidal modulation and D is the dipole anisotropy of the CBR. We used for D the direction given by COBE-DMR and the amplitude provided by fig. 10. Our detector responsivity that best fits the previous relation is $\Re = (2.58 \pm 0.13)$ μK/telemetry units, in good agreement with the one measured in the laboratory: the offset remains negligible. The best fit to the dipole is shown as a continuous line. The uncertainties in DMR and FIRAS data introduce an uncertainty of 10% in the responsivity. Since the intermediate-scale anisotropy of the CBR is expected to have the same spectrum of the dipole anisotropy, it is not needed to apply any spectral correction and we can use $\Re$ to convert our telemetry units into thermodynamic temperature fluctuations of the CBR. The residuals from the dipole fit are plotted in panel E). The standard deviation of these data is 22 μK, while the mean error is 24 μK. These are the lowest fluctuations never observed by an instrument devoted to the search of CBR anisotropies. For comparison, COBE has a per-pixel sensitivity of 75 μK, while MAX is around 50 μK and Lasenby and Davies' experiment is around 30 μK.

3˙3. *Data analysis and discussion.* – While the standard deviation of the data provides direct information on the performances of the detectors, the comparison among various results is hampered by our ignorance about the real distribution of the CBA. As a rule a large primordial power index conveys power on large-scale structures: a differential measurement with a small beam separation becomes unable to fully detect the gradients due to these large-scale structures. It follows that the experiment is less sensitive than one with a larger beam separation. In the case of COBE the matrix of the data has been inverted, thereby presenting the final results as due to a single-beam instrument: the comparison with our differential results requires the knowledge of the power spectrum of the anisotropies.

Obviously, fluctuations at scales much smaller than the beam size are averaged and diluted by the instrument. This effect is similar in COBE and in our instrument due to the similar beam size.

Let us start by describing the distribution of the CBA in a way that is now standard in theoretical analysis: We introduce an autocorrelation function $C(\theta, \sigma)$ defined as

(22) $$C(\theta, \sigma) = [\langle W_\sigma (T(\gamma) - T_0)(T(\theta + \gamma) - T_0) \rangle],$$

where T_0 is the 2.73 K mean temperature of the CBR, while the average is carried out over all the possible directions, *i.e.* over all the possible γ. W_σ means that, before the final average, each product is appropriately weighted for the beam size σ.

We can formally describe $C(\theta, \sigma)$ as the product of a constant term C_0 and another term $\wp(\theta, \sigma)$ that depends on the real distribution of the anisotropies in the sky.

It is the goal of theoreticians to provide estimates of both terms. Once $\wp$ has been selected, one can compare the prediction for C_0 with the results arising from a given experiment to reject the theory. It is not the purpose of the present lecture that of discussing the constraints that our results provide for various theories. We will limit ourselves to a comparison between our results and those obtained by COBE and MAX in the framework of what could be considered a possible scenario for galaxy formation. We assume cold dark matter, $\Omega = 1$, no bias in galaxy formation. If the COBE results are true, we can normalize to them all the possible predictions with various power indices ($n = 1$ being the usually preferred Harrison-Zel'dovich, or inflation index).

The first conclusion is that it does not exist a value of n that is acceptable within one standard deviation. At two standard deviations $n = 1$ is marginally acceptable.

At our level of sensitivity COBE-DMR anisotropies should have been marginally detected. Therefore, our results are starting to constrain COBE data.

The fact that much larger signals have been observed in the 1978 flight at high galactic latitudes represents a serious problem. In this respect a similar situation seems to be present in the MAX experiment, where a factor two of difference in anisotropy amplitude has been found in two different sky directions.

There are indications that or a) the galactic contamination is still important in COBE results, or b) the distribution of CBR anisotropies is not Gaussian, or both.

3'4. *Considerations on CBR anisotropies and primordial molecules.* – The following discussion has been fully developed by MAOLI, MELCHIORRI and TOSTI in [19].

The potentiality of the cosmic-background-radiation (CBR) anisotropy studies lies on the hypothesis that the post-recombination Universe (*i.e.* $10 \leq\ \leq Z \leq 10^3$) was transparent to CBR photons, so that measurements of anisotropies at various angular scales can be directly compared with the predictions of theories of galaxy formation. It is, therefore, required that the post-recombination Universe was not re-ionized, because an ionized medium would scatter CBR photons, thereby changing their energy distribution and spatial direction. Strong evidence against a hot re-ionization (*i.e.* characterized by an electron temperature $\geq 10^6$ K) is provided by recent COBRA[9] and COBE-FIRAS[8] results: their spectral measurements of the CBR have set stringent upper limits to any spectral deviation from a pure Planckian curve as well as to the Comptonization parameter, excluding the existence of a hot ionized medium. If we neglect for a while the remaining possibility of a soft re-ionization ($T_e \leq\ \leq 10^5$ K) or a very late re-ionization (*i.e.* occurring at redshift $Z \leq 70$), we are led to the conclusion that CBR anisotropies, observed by COBE-DMR at angular scales from 90° to 10°, are directly testing primordial perturbations: unfortunately, as more data become available from new intermediate-scale experiments, it appears more and more evident that no one among the various available theories is fitting the data in a satisfactory way.

For instance, the Tenerife[20] and the South Pole experiments[21] seem to indicate upper limits well below the values expected by a simple extrapolation from COBE-DMR data on the basis of cold-dark-matter models. This situation has led some authors to revitalize the hypothesis of a moderate secondary ionization around $Z \simeq 100\text{--}50$ in order to appropriately reduce intermediate-scale anisotropies to values compatible with the CDM theory, while avoiding spectral distortions within the limits posed by the COBE-FIRAS results[22].

Even more disturbing is the fact that bolometric observations in the millimetric region at similar angular scales, like the ULISSE experiment[17], the MAX experiment[23] and the ARGO experiment[24], provide r.m.s. values more than 2σ higher than the radio results. The disagreement between radio and infrared measurements could well be fortuitous, the observations being carried out on different sky regions: in such a case, however, one is forced to consider non-Gaussian spatial distributions of the anisotropies.

Therefore, it appears worthwhile to re-analyse the general question of the cosmic transparency to CBR photons in order to explore the possibility that CBR anisotropies could be affected by some not jet considered mechanism, even in the case of a neutral Universe. In view of the problems raised by observations, such an unknown mechanism should present the following peculiarities:

a) It should not affect the CBR spectrum.

b) It should be able to erase, at least partially, CBR anisotropies at small angular scales and, possibly, in a such frequency-dependent way to explain the colour excess in the millimetric region.

c) Secondary anisotropies arising from the process should pass the severe upper limits already posed by several radio and infrared observations at small angular scales.

Points *a)* to *c)* may be fulfilled by the process of elastic Thomson scattering between CBR photons and primordial molecules: more specifically, we intend to study the influence of primordial molecules on CBR anisotropies in the redshift range $Z_f \Leftrightarrow Z_d$, where Z_f is the *formation redshift*, *i.e.* when the Universe becomes cold enough to stop the photodissociation of newly formed molecules ($Z_f \simeq 100\text{--}400$ depending on the molecular species), Z_d is the dissociation redshift, *i.e.* when the high-energy photons of the first formed objects start again to photodissociate them ($Z_d \simeq 5\text{--}70$ depending on the evolutionary models of the structures).

In the above redshift range a certain amount of primordial molecules is expected to be present. The possible mechanisms of interaction between these molecules and CBR photons are thermal emission and absorption and elastic Thomson scattering. Since molecular abundances are modest, collisional excitation and de-excitation are negligible: therefore, thermal emission and absorption are also negligible. PUY et al. [25] have estimated that emission from molecules could hardly reach 10^{-6} of the CBR brightness.

The situation could be different if the CBR spectrum had some intrinsic distortion, because molecules may act as heat pumps, exchanging energy between rotational and vibrational levels. One could even argue whether the perfectly Planckian shape of the CBR spectrum is due to this molecular action: in any case molecules tend to restore the black-body spectrum and not to add detectable spectral distortions.

The remaining effect is the Thomson scattering between photons and molecules. This is an elastic scattering during which a photon is first absorbed and then re-emitted at the same frequency, but not in the same direction. Obviously, the effect does not alter the CBR spectrum, while it could in principle smear out the primordial spatial distribution of the CBR: For a given molecular line ν_{ij} a CBR photon arriving from the last-scattering surface can be resonantly scattered if $\nu_{LS}(1+Z)/(1+Z_D) = \nu_{ij}$; taking into account the finite natural linewidth $\Delta\nu_{ij}$ (essentially due to the Doppler broadening), this condition will be fulfilled for a redshift range $\Delta Z/Z = \Delta\nu_{ij}/\nu_{ij} \simeq 10^{-5}$.

At first glance, the paucity of molecules and the limited path of interaction suggest that elastic scattering is also irrelevant. One should remember, however, that in the case of a harmonic oscillator the Thomson cross-section is several orders of magnitude greater than that for free electrons.

Due to this, DUBROVICH [26] considered the effect of Thomson scattering by H_2D^+ on CBR anisotropies, thereby concluding that a significant erase may occur, *if* the abundance of H_2D^+ is large enough: the abundance of H_2D^+ is rather uncertain, and the requested density ($\simeq 10^{-6}[H]$) hardly reached.

However, a variety of primordial molecules exist in a neutral or moderately ionized Universe, for a wide range of redshifts, their resonant frequencies being broadened into a continuum so that CBR anisotropies may be smeared out over a wide spectral region. The above facts guarantee that the requirements a) and b) are satisfied (*i.e.* the CBR spectrum is unchanged, while the CBR anisotropies may be erased). Also, the wavelength dependence of the effect is determined by the resonant frequencies of rotational and vibrational levels of the various molecules redshifted along their range of existence $Z_f - Z_d$.

Molecule formation in the early Universe starts when photons have not enough energy to dissociate them. The final abundance for each species depends on the initial atomic abundance and on the associative and dissociative rates of the various reactions. A complete list of possible reactions can be found in [25, 27-29].

Of particular relevance is the case of molecules containing lithium: LiH and

LiH$^+$. The abundances of these molecules are rather uncertain: for LiH the data available in the literature range from 1 to 10% of the primordial ^{7}Li abundance [27, 30] to 60% [25]. More recently, LIPOVKA [31] has suggested that almost all the available Li is trapped in LiH, if one correctly takes into account the temperature dependence of the photoassociation rate. For LiH$^+$ the situation is even more uncertain: since ionized Li recombines later than hydrogen, DALGARNO and LEPP [29] concluded that LiH$^+$ could even be more abundant than LiH and PALLA and FINK [32] estimated an abundance of about 50% of Li; more recently the abundance of LiH$^+$ has been estimated to be much lower than that of LiH [33]. The rates for the involved chemical processes are almost unknown, however.

In any case, if χ is the fraction of Li converted in LiH, the values theoretically expected for LiH range from $2 \times \chi 10^{-10}$ predicted by standard nucleosynthesis [34-36] up to a maximum of $\chi 10^{-8}$ in the case of inhomogeneous nucleosynthesis [37, 38]. We want to point out that the primordial Li abundance is related to the baryon-photon ratio: the uncertainties in this number allow a range from 10^{-10} to 10^{-9} even in the framework of the standard nucleosynthesis. Measurements toward population-II stars tend to a low value like $(1-2) \cdot 10^{-10}$ [39, 40], but these results have been criticized by various authors, since it is possible that a substantial fraction of Li was destroyed through some form of internal mixing [41, 42]. The most recent measurements of Li abundance is that of LEMOINE et al. [43] giving an abundance of $3.4 \cdot 10^{-9}$ toward ρ Ophiuchi.

The resonance cross-section of Thomson scattering has to be normalized on the line width and the optical depth τ is

(23) $$\tau = \int \sigma \, n_{\mathrm{mol}} \, \mathrm{d}l,$$

where

$$\sigma = \frac{\lambda^3 A_{ij}}{8\pi c} \frac{\nu}{\Delta \nu_{\mathrm{D}}},$$

$$n_{\mathrm{mol}} = \Omega \omega_{\mathrm{b}} \rho_c \alpha_{\mathrm{m}} n_{vj} (1+Z)^3,$$

$$\mathrm{d}l = \frac{c \, \mathrm{d}Z}{H_0 (1+Z)^2 \sqrt{1 + \Omega Z}},$$

where the notations mean

σ = cross-section,
$\Delta \nu_{\mathrm{D}}$ = Doppler broadening,
A_{ij} = Einstein coefficients,
Ω = total (i.e. baryonic and nonbaryonic) density relative to ρ_c,
ω_{b} = baryonic fraction of total density,
ρ_c = critical density,

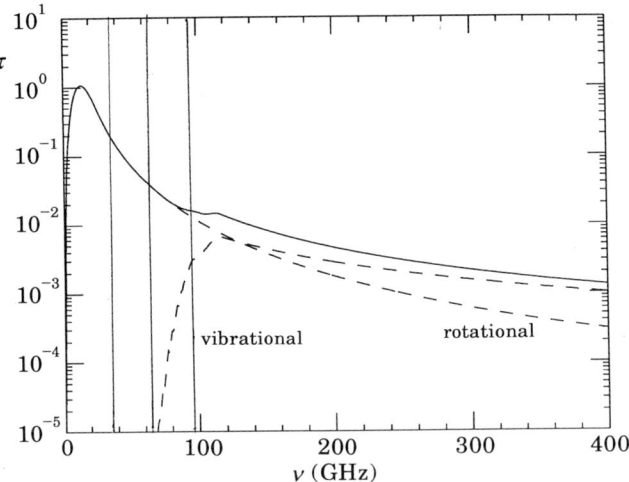

Fig. 22. – Expected optical depth from LiH molecules formed at $Z = 400$ with a mean abundance of $2 \cdot 10^{-10}$ [H]. The contributions of both the rotational and vibrational states are shown (from [19]).

α_m = molecular abundance to H,
n_{vj} = population of level with quantum numbers v, j.

The LiH optical depth is shown in fig. 22. We have considered the most significant cases, namely LiH formation at redshift 400 with a redshift of photodissociation, respectively, 5 and 70.

The optical depth is computed in the case of standard nucleosynthesis, *i.e.* $2 \cdot 10^{-10}$ [H]. The LiH fraction χ is assumed to range from 0.1 to 1. For comparison, the frequencies of the COBE-DMR radiometers are indicated.

The first important consequence of fig. 22 is that no significant effects are expected in the case of a Li abundance $\leq 2 \cdot 10^{-10}$ and a LiH abundance ≤ 0.1[Li]. Although these numbers are consistent with our knowledge about primordial nucleosynthesis and early-Universe chemistry, we cannot exclude values 10-100 times larger for the quantity χ[LiH]. In such a case primary CBR anisotropies will be affected by resonant scattering if $\theta \leq \theta_H$ and $\nu \leq 80$ GHz, where θ_H is the angular diameter of the horizon at the effective redshift of Thomson scattering. For a given power spectrum of primary CBR anisotropies, the effect of the Thomson scattering is that of affecting all the angular scales smaller than $\theta_H(\nu)$ by an amount of the order of $e^{-\tau}$. Therefore, we anticipate a possible decrease in CBR anisotropies which is wavelength dependent, as shown in fig. 23. In this figure we have plotted the anisotropies arising from some CDM models for galaxy formation. The quantity $(1/4\pi) l(2l + 1) C_l$ essentially represents the power of $\langle (\Delta T/T)^2 \rangle$ at the angular scale $\theta \simeq 1/l$, while the spatial wave number l roughly indicates the inverse in radians of the angular

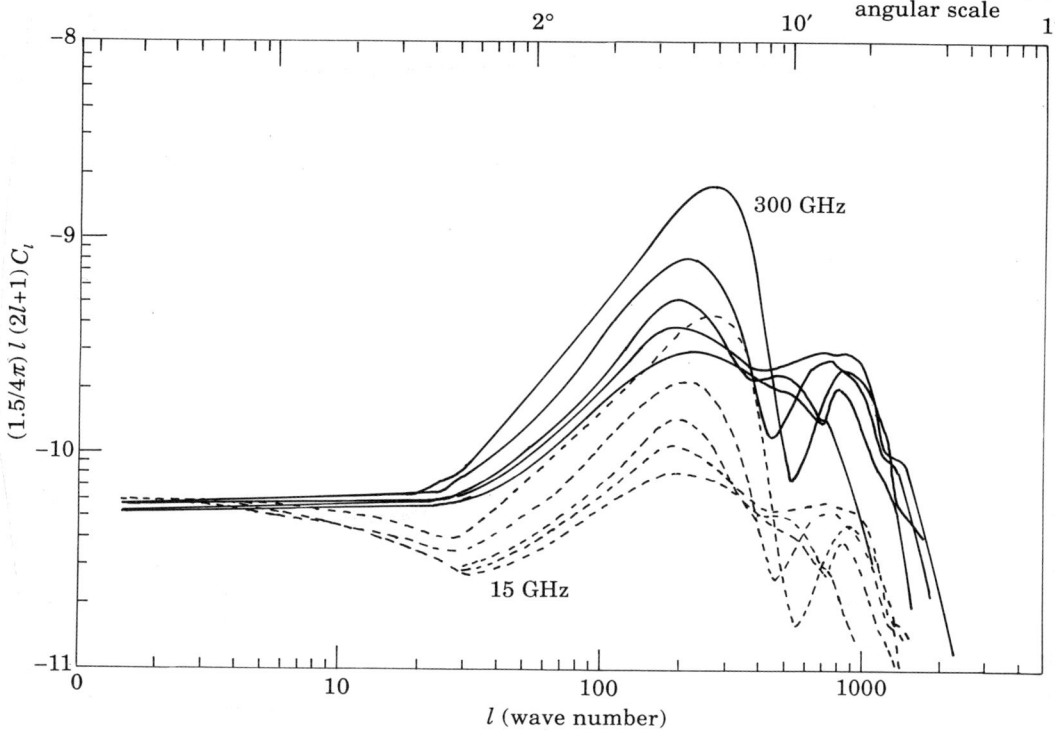

Fig. 23. – Power spectrum of CBR anisotropies, as expected in the framework of CDM theories: the effect of molecular scattering is indicated. The scattering is efficient at small angular scales and at low frequencies; LiH = 10^{-9}, $Z = 400$, $h = 0.5$, $\Omega = 0.5, 0.2, 0.05, 0.03, 0.01$ from top to bottom.

scale. The various curves refer to different contents of baryonic matter Ω: it has been suggested by several authors that a precise measurement of CBR anisotropies may lead to an independent and precise determination of Ω. For the sake of simplicity we have plotted the case of LiH $\simeq 10^{-9}$. Other values of [LiH] would produce a quadratic proportional change in the results.

The main conclusion is that low-frequency anisotropies ($\nu \leqslant 60$ GHz) may be significantly erased by Thomson scattering, while in the millimetric region the effect is negligible. Our poor knowledge of the primordial Li abundance does not allow us to evaluate the amplitude of the effect. On the other hand, any decrease of $\Delta T/T$ with the wavelength should suggest the presence of molecular scattering. *Vice versa*, any upper limit to the difference in amplitude between infrared and radio anisotropies at the same angular scale may be used to exclude at the given level of confidence any overproduction of Li and of LiH with respect to the *standard* values of $2 \cdot 10^{-10}$ and 10%.

As primary CBR anisotropies are erased, new, secondary anisotropies are expected. We may consider three classes of secondary anisotropies; they all arise from the primordial clouds evolving from the perturbations present at the last-scattering surface. To characterize them, let us introduce a peculiar velocity V_p of the cloud and a collapsing velocity V_c. The first one is related to the present-day peculiar-velocity field $V_p(0)$ by $V_p = V_p(0)(1 + Z)^{-1/2}$. The second one measures the slow-down in expansion of the cloud with respect to the cosmic expansion and is given (in linear regime) by $V_c = \dfrac{2A}{3} \dfrac{\Delta Z}{1+Z} (1+Z)^{-1}$, where A is an appropriate normalizing constant ($A \simeq 10$ for protoclouds of galaxy size). We have three cases:

a) $\Delta\rho/\rho < 1$, $V_p > V_c$: the anisotropy arises from the anisotropic scattering: the situation is similar to the case of Doppler effect on the last-scattering surface. At the resonant frequency in a bandwidth $\Delta\nu = \nu_0 (\Delta Z/(1 + Z))$ we get a signal

$$\text{(24)} \qquad \frac{\Delta I(\nu_0)}{I(\nu_0)} = (3 - \alpha)\beta_p \tau,$$

where $\beta_p = V_p/c$, τ is the optical depth, $I(\nu_0)$ is the CBR brightness and $\alpha = (\nu/I)(\partial I/\partial \nu)$ is the spectral index of the CBR. In the R.J. region of the spectrum this relation simplifies to

$$\text{(25)} \qquad \frac{\Delta T}{T} = \beta\tau.$$

b) $\Delta\rho/\rho < 1$, $V_p < V_c$: the motion of the two opposite sides of the cloud produces a frequency shift $\pm \beta_c = \pm V_c/c$, so that an additional anisotropy is present at the two shifted frequencies given in amplitude by

$$\text{(26)} \qquad \frac{\Delta T}{T} = \beta_c \tau.$$

c) $\Delta\rho/\rho \geq 1$, $V_p \ll V_c$: as $\Delta\rho/\rho$ approaches unity, the cloud collapse compensates the cosmic expansion and the bandwidth of the scattered radiation approaches the natural linewidth, which is several orders of magnitude smaller than $\Delta Z/Z$. Therefore, the lines become thinner and the resonant cross-section greater. We get a signal

$$\text{(27)} \qquad \frac{\Delta T}{T} = \beta_c$$

in a bandwidth of the order $\Delta\nu/\nu \simeq 10^{-5}$.

To get an idea of the peculiarities of these anisotropies let us first consider the spectrum of a single cloud located at different redshifts but with mass corre-

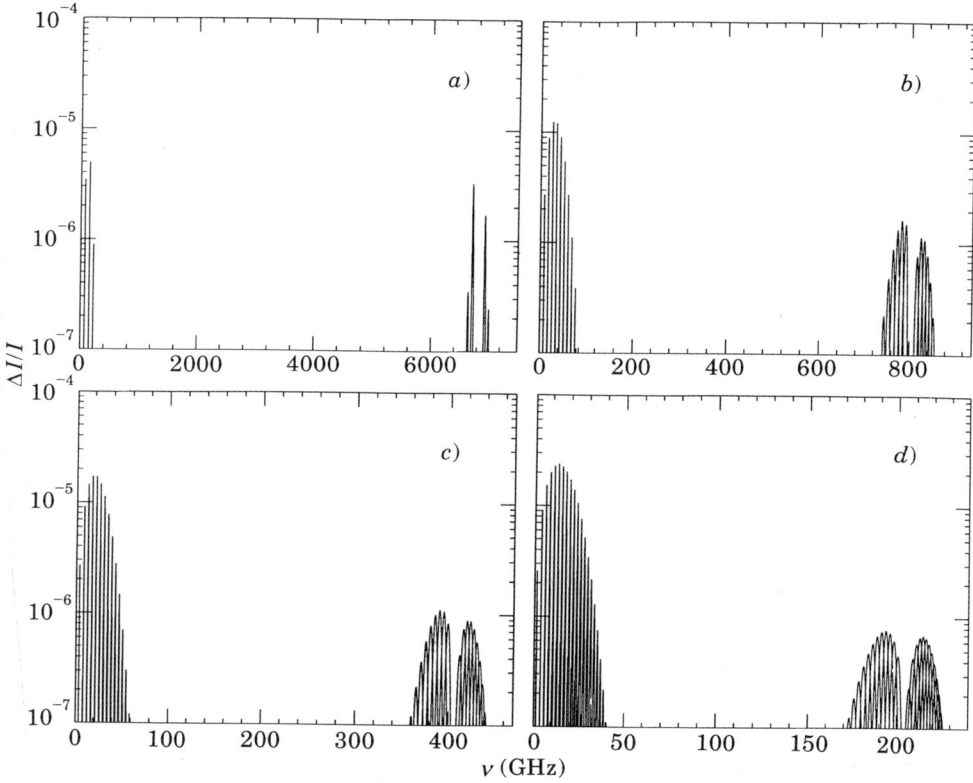

Fig. 24. – Theoretical spectrum for a typical isolated cloud of galaxies cluster size at $z = 5$ (a)), 50 (b)), 100 (c)), 200 (d)), moving at peculiar velocity $V_p = V_p(0)/\sqrt{1+Z}$. The resulting spectrum is essentially due to LiH ([LiH] = $3 \cdot 10^{-9}$).

sponding to a galaxy cluster today (this fixes the peculiar velocity at about 600 km/s now). In fig. 24a)-d) we have plotted the corresponding spectrum. In order to isolate a single cloud along the line of sight, the observer should use a telescope with an adequate spatial and spectral resolution. For a cloud with dimensions comparable with a protogalaxy, the spatial resolution would range around (10–50) arcseconds for a redshift $Z \simeq 100$, while the spectral resolution is given by $\Delta \nu \leqslant \nu(\Delta Z/Z)$ in cases a), b) and about 10^{-5} in case c). Even in these conditions one should take into account that other clouds along the line of sight at appropriate redshifts would radiate inside the selected bandwidth and, also, would scatter the radiation coming from the selected cloud. In order to estimate the real level of anisotropy for a given frequency of observation in cases a) and b), one has to take into account the effects of other N clouds along the line of sight having lines redshifted at the same frequency. The signal from each cloud is partially scattered by other clouds and the fluctuations observed in the sky by a radiotelescope will be smeared out by a factor $\sqrt{N}$, weighted over $\tau_i v_i/c$.

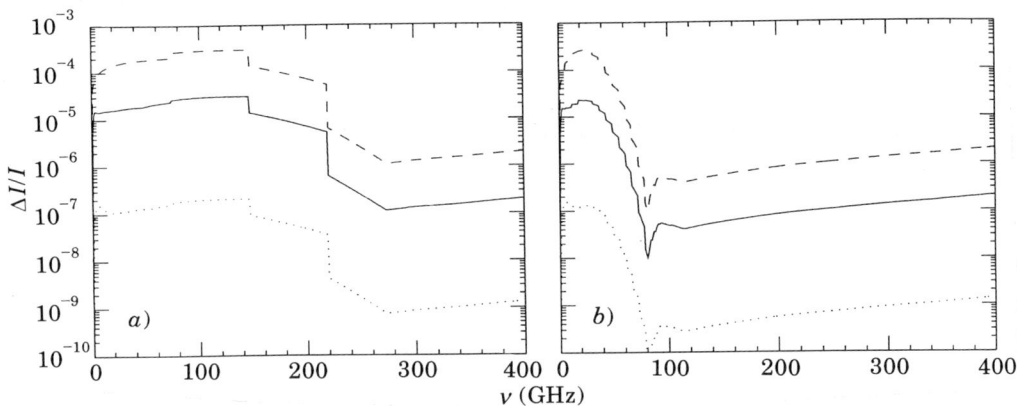

Fig. 25. – For a given frequency we considered the molecular lines giving the highest signal that we plotted after considering the action of other lines: a) $Z_d = 5$, b) $Z_d = 70$. The maximum signal is given by LiH rotational transitions from 10 GHz till (100–200) GHz (this last value depending on the re-ionization redshift Z_d). Adopted values of [LiH] are the same as in fig. 24.

The final result is

$$(28) \quad \left(\frac{\Delta T}{T}\right)_i = \frac{(3-\alpha)\tau_i\beta_i\left[1 - \sum_{(z<z_i)}^{j}\tau_j\right]}{\left[\sum_{0<Z<1000}^{J}\left(\tau_j\beta_j\left[1 - \sum_{(z<z_i)}^{k}\tau_k\right]\Big/\tau_i\beta_i\left[1 - \sum_{(z<z_i)}^{k}\tau_k\right]\right)\right]^{0.5}} .$$

In fig. 25a), b), the expected anisotropy is plotted vs. the frequency of observation.

We should note that this is a rough estimate of the real anisotropy, to which clouds of different size contribute. It is, however, adequate to provide an idea of the level of the expected signals for clouds in the linear regime. The angular size of these anisotropies would be that of protogalactic objects, i.e. $10''$–$100''$. The main conclusion is that secondary anisotropies have a very peculiar spectrum: In the range 20 GHz $\leq \nu \leq$ 250 GHz they may acquire detectable amplitudes. Moreover, in the region $\nu \geq$ 60 GHz, the Universe is rather transparent and $\tau \ll 1$, and one should be able to observe the effect of a single cloud. As pointed out before, this would require a spectral resolution $\Delta\nu/\nu \simeq \Delta Z/Z \simeq 10^{-3}$ for the typical dimensions of clouds corresponding to galaxies today ($\simeq$ 10 arcseconds). Moreover, the spectral distribution of the roto-vibrational lines would allow one to check the nature of the observed signals, the roto-vibrational signature being unique in this range of frequencies.

A rather different situation is expected in the case of clouds at the end of the linear regime: as $\Delta\rho/\rho$ approaches unity, the cloud collapse compensates the expansion of the Universe. The collapse velocity is close to $c\Delta Z/(1+Z)$ and the expected signals are remarkably large, even if concentrated into narrow lines.

Our model predicts a partial or total erase of anisotropies in the radio region for angular scales smaller than the horizon at the redshift for which the optical depth is unity. This effect is similar to the case of a partial secondary ionization but differs in being strongly frequency dependent. If the primordial LiH abundance is large (*i.e.* $\geq 10^{-9}$), we predict a substantial difference between the amplitude of the COBE-DMR 31 GHz channel and the 90 GHz channel. Therefore, one may use the COBE-DMR data to set upper bounds to the LiH abundance. Unfortunately, the amplitude of the 31 GHz channel is the most noisy in COBE data and the observed anisotropy could even be consistent with zero at this frequency. Much more sensitive to any possible erase are the various 30 GHz experiments carried out with HEMT's at the South Pole and at Saskatoon at angular scales of (0.5-1.5) degrees. The results of these experiments, although very preliminary, seem to indicate an anisotropy at a level of $1.5 \cdot 10^{-5}$ K, significantly (2σ) smaller than the value quoted by the MAX experiment [44] at 200 GHz. If this discrepancy will be confirmed in the same sky region, then the effect of primordial molecules would be proved. At the present level of knowledge, however, the difference could well be fortuitous.

We may tentatively resume the observational situation as follows:

1) Tenerife experiment: beam size = 5.6°, 15 GHz radiometer $\Rightarrow (60 \pm 12)$ μK, 37 GHz radiometer $\Rightarrow (86 \pm 20)$ μK. For a redshift of formation of $Z_F = 400$ we get [LiH] $\leq 2 \cdot 10^{-9}$.

2) SP and MAX experiments: beam size $\simeq 1°$, 30 GHz radiometer $\Rightarrow \leq 50$ μK, 200 GHz bolometer $\Rightarrow (100-150)$ μK. For a redshift of formation $Z_F = 400$ we get [LiH] $\leq 2 \cdot 10^{-9}$.

3) ULISSE and COBE: beam size $\simeq 6°$, 300 GHz radiometer $\Rightarrow (110 \pm 50)$ μK, 31 GHz receiver, 10° beam size $\Rightarrow \leq 50$ μK. For a redshift of formation $Z_F = 400$ we get [LiH] $\leq 4 \cdot 10^{-9}$.

Secondary anisotropies may be tested at small angular scales, as in the radio observations of FOMALONT and PARTRIDGE [45] at 8 GHz and 50″, although their spectral resolution of 50 MHz is a bit too low. Their upper limits of $2 \cdot 10^{-5}$ impose a LiH upper limit of $3 \cdot 10^{-9}$. Observations at 200 GHz have been carried out by DE BERNARDIS *et al.* [24] and recently improved in sensitivity, by means of the IRAM radiotelescope at 10 arcseconds resolution. The sensitivity of the system (about 1 mK) is close to the values predicted by inhomogeneous nucleosynthesis.

In conclusion, primordial molecules may play a significant role in altering the amplitude and power spectrum of CBR anisotropies: the effect depends essentially on the Li abundance.

Conversely, accurate measurements of anisotropies at large and small angular scales will provide precious information on the abundance of Li: the upper limits we may derive from the available data already exclude LiH values much greater than 10^{-9}: a further increase in accuracy of observations would possibly rule out inhomogeneous nucleosynthesis.

Secondary anisotropies arising from anisotropic scattering, if detected, would represent the only way to study the spatial distribution of perturbations after the decoupling and well before the formation of nonlinear structures. At the proposed level of sensitivity of various future space experiments ($\delta T \leqslant$ $\leqslant 1\ \mu K$) there are few doubts that the effect of primordial molecules would be detected, even in the case of *standard* Li and LiH abundances.

REFERENCES

[1] As quoted by R. A. SUNYAEV: in *Confrontation of Cosmological Theories with Observational Data*, IAU Symposium N. 63, edited by M. S. LONGAIR (Reidel, Dordrecht, 1974), p. 167.
[2] L. DANESE and G. DE ZOTTI: *Riv. Nuovo Cimento*, **7**, 277 (1977).
[3] J. R. BOND: in *The Early Universe*, edited by W. G. UNRUH and G. W. SEMENOFF (Reidel, Dordrecht, 1988), p. 283.
[4] P. SALATI: in *The Infrared and Submillimeter Sky after COBE*, edited by M. SIGNORE and C. DUPRAZ, NATO-ASI 359 (1991), p. 143.
[5] D. P. WOODY and P. L. RICHARDS: *Phys. Rev. Lett.*, **42**, 925 (1979).
[6] H. P. GUSH: in *Gamow Cosmology, Proc. S.I.F.*, Course LXXXVI, edited by F. MELCHIORRI and R. RUFFINI (North Holland, Amsterdam, 1982), p. 260.
[7] J. B. MERCER and S. WILSON: in *Far Infrared Astronomy: a Supplement to Vistas in Astronomy*, edited by M. ROWAN ROBINSON (Pergamon Press, London, 1975), p. 103.
[8] J. C. MATHER: *Astrophys. J. Lett.*, **37**, 354 (1990).
[9] H. P. GUSH, M. HALPERN and E. H. WISSHNOW: *Phys. Rev. Lett.*, **65**, 537 (1990).
[10] R. A. SUNYAEV: in *Confrontation of Cosmological Theories with Observational Data*, IAU Symposium N. 63, edited by M. S. LONGAIR (Reidel, Dordrecht, 1974), p. 167.
[11] M. J. DEVLIN, A. C. CLAPP, M. L. FISHER, A. E. LANGE, P. R. MEINHOLD and P. L. RICHARDS: in *Relativistic Astrophysics and Particle Cosmology, Ann. N.Y. Acad. Science*, Vol. 688 (1992), p. 809.
[12] F. MELCHIORRI, B. MELCHIORRI, C. CECCARELLI and L. PIETRANERA: *Astrophys. J.*, **250**, L1 (1981).
[13] R. FABBRI, I. GUIDI, F. MELCHIORRI and V. NATALE: *Phys. Rev. Lett.*, **44**, 1563 (1980).
[14] S. BOTTANI, P. DE BERNARDIS and F. MELCHIORRI: *Astrophys. J. Lett.*, **384**, L1 (1992).
[15] F. MELCHIORRI and P. E. TANZILLI: *Int. J. Mod. Phys. D*, **1**, 605 (1993).
[16] J. P. KRAUS: *Radioastronomie* (McGraw-Hill, New York, N.Y., 1966), Chapt. 7.
[17] F. MELCHIORRI, B. OLIVO, C. CECCARELLI and L. PIETRANERA: *Astrophys. J. Lett.*, **227**, L129 (1981).

[18] P. DE BERNARDIS, S. MASI, B. MELCHIORRI and F. MELCHIORRI: in *Observational Tests of Cosmological Inflation*, NATO-ASI, Vol. 348, edited by T. SHANKS (1962), p. 443.
[19] R. MAOLI, F. MELCHIORRI and D. TOSTI: *Astrophys. J.*, **425**, 372 (1994).
[20] A. N. LASENBY, R. D. DAVIES, R. REBOLO and J. E. BECKMAN: in *Present and Future of CMR, Proceedings Santander*, edited by J. L. SAUT, E. MARTINEZ and L. CARGON, *Lecture Notes in Physics*, Vol. **429** (Springer-Verlag, Berlin, 1994), p. 91.
[21] J. SCHUSTER, G. TODD, J. GUNDERSEN, P. MEINHOLD, T. KOCH, M. SEIFFERT, C. A. WUENSCHE and P. LUBIN: *Astrophys. J.*, **412**, L47 (1993).
[22] J. SILK, D. SCOTT and H. WAYNE: submitted to *Phys. Rev. D* (1993).
[23] M. DEVLIN, D. ALSOP, A. CLAPP, D. COTTINGHAM, M. FISHER, J. GUNDERSEN, W. HOLMES, A. LANGE, P. LUBIN, P. MEINHOLD, P. RICHARDS and G. SMOOT: *Proc. Nat. Acad. Sci. USA*, **688**, 809 (1992).
[24] P. DE BERNARDIS, V. DUBROVICH, P. ENCRENAZ, R. MAOLI, S. MASI, G. MASTRANTONIO, B. MELCHIORRI, F. MELCHIORRI, M. SIGNORE and P. E. TANZILLI: *Astron. Astrophys.*, **269**, 1 (1993).
[25] D. PUY, G. ALECIAN, J. LE BOURLOT, J. LE ORAT and G. PINEAU: *Astron. Astrophys.*, **267**, 337 (1992).
[26] V. K. DUBROVICH: *Pis'ma Astron. Ž*, **1**, 10 (1975).
[27] S. LEPP and M. SHULL: *Astrophys. J.*, **280**, 465 (1984).
[28] K. KIRBY and A. DALGARNO: *Astrophys. J.*, **224**, 444 (1978).
[29] A. DALGARNO and S. LEPP: in *Astrochemistry*, edited by S. P. TARAFDAR and M. P. VARSHNI (Reidel, Dordrecht, 1985), p. 109.
[30] V. K. DUBROVICH: *Izv. Spet. Observatorii*, **13**, 40 (1981).
[31] P. LIPOVKA: personal communication (1993).
[32] F. PALLA and E. FINK: personal communication (1992).
[33] M. BELLINI, P. DE NATALE, M. INGUSCIO, E. FINK, D. GALLI and F. PALLA: *Astrophys. J.*, **424**, 507 (1994).
[34] J. YANG, M. S. TURNER, G. STEIGMAN, D. N. SCHRAMM and K. OLIVE: *Astrophys. J.*, **281**, 493 (1984).
[35] L. KAWANO, D. SCHRAMM and G. STEIGMAN: *Astrophys. J.*, **327**, 750 (1988).
[36] M. ARNOULD and M. FORESTINI: in *Proceedings of La Rabida School on Nuclear Astrophysics* (Springer-Verlag, Berlin, 1989), p. 170.
[37] H. REEVES, P. DELBOURGO-SALVADOR, J. AUDOUZE and P. SALATI: *J. Phys.*, **9**, 179 (1988).
[38] H. KURKI-SUONIO and R. A. MATZNER: *Phys. Rev. D*, **39**, 1046 (1989).
[39] F. SPITE and M. SPITE: *Astron. Astrophys.*, **115**, 357 (1982).
[40] R. REBOLO, P. MOLARO and J. E. BECKMAN: *Astron. Astrophys.*, **192**, 192 (1988).
[41] S. VAUCLAIR: in *IAU Symposium 132, The Impact of Very High S/N Spectroscopy on Stellar Physical*, edited by G. CAYREL DE STROBLE and M. SPITE (Kluver, Dordrecht, 1987), p. 432.
[42] S. VAUCLAIR, in *Dark Matter*, edited by J. AUDOUZE and J. TRAN THANH VHAN (Editions Frontières, Gyf-sur-Yvette, 1988), p. 269.
[43] M. LEMOINE, R. FERLET, A. VIDAL-MAJAR, C. EMERICH and P. BERTIN: submitted to *Astron. Astrophys* (1992).
[44] P. RICHARDS, A. C. CLAPP, M. J. DEVLIN and A. E. LANGE: submitted to *Astrophys. J. Lett.* (1994).
[45] E. B. FOMALONT, R. B. PARTRIDGE, J. D. LOWENTHAL and R. A. WINDHORST: *Astrophys. J.*, **404**, 8 (1993).

The Large-Scale Structure of the Universe in Non-Gaussian CDM Models.

F. LUCCHIN and L. MOSCARDINI

Dipartimento di Astronomia dell'Università
vicolo dell'Osservatorio 5, I-35122 Padova, Italia

S. MATARRESE

Dipartimento di Fisica G. Galilei dell'Università
via Marzolo 8, I-35131 Padova, Italia

A. MESSINA

Dipartimento di Fisica A. Righi dell'Università
via Irnerio 46, I-40126 Bologna, Italia

1. – Introduction.

One of the bases of modern cosmology and of current theories of structure formation is the assumption that some kind of nonbaryonic dark matter accounts for most of the mass density in the Universe. In particular, the scenario based on cold dark matter (CDM; see, *e.g.*, ref.[1,2]) has been thoroughly investigated in the last years because of its high predictive power and capability of explaining many properties of the structures observed in the Universe, both on galactic scales and on very large ones. The standard version of this model is characterized by Gaussian adiabatic primordial perturbations with a Harrison-Zel'dovich spectrum, $P(k) \propto k^n$, with spectral index $n = 1$, in the context of a flat universe with zero cosmological constant. An important parameter of the model is the *linear bias factor* $b = 1/\sigma_8$, σ_8 being the r.m.s. mass fluctuation at $8h^{-1}$ Mpc (h is the Hubble constant H_0 in units of $100 \text{ km s}^{-1} \text{ Mpc}^{-1}$). The detection of large-angular-scale cosmic microwave background (CMB) anisotropies by the COBE DMR experiment[3] yields a normalization of the amplitude of the CDM power spectrum, giving $b \approx 0.8$.

However, the standard CDM scenario has recently met increasing problems, essentially represented by the high ratio of small- to large-scale power. In particular, using the COBE normalization, the CDM model predicts excessive velocity dispersion on scales of order $1h^{-1}$ Mpc, while, on larger scales, it is unable to reproduce the slope of the galaxy angular-correlation function obtained from the APM survey[4]. Although the COBE data fix the normalization of mass fluctuations in a given model (*i.e.* the value of b) up to the uncertainty on the gravitational-wave contribution, there still remains a large uncertainty in how to translate this information into the galaxy distribution. This is due to the nonlinear biasing mechanism of galaxy formation and on its possible scale and/or environment dependence (see, *e.g.*, ref.[5,6]).

In order to overcome the difficulties of the standard CDM model, various alternative models have been proposed. An *ad hoc* value for the relic cosmological constant Λ, filling the gap between the dynamical and the inflation predicted value of Ω_0, *i.e.* $\Lambda = 3(1 - \Omega_{0,\text{dyn}})H_0^2$, alleviates the large-scale problem (see, *e.g.*, ref.[7]). Hybrid (*e.g.* hot plus cold) dark-matter models may overcome both the small- and large-scale problem by suitably balancing the power content on the opposite sides[8-10]. A more natural way to reduce the high ratio of small- to large-scale power of the standard CDM model is to lower the spectral index of primordial perturbations: «tilted», *i.e.* $n < 1$, models boost power from small to large scales[10-16]. A primordial spectrum with $n < 1$ can be originated during the inflationary dynamics in models such as power law[17], extended[18] and natural inflation (see, *e.g.*, ref.[13]). The COBE DMR experiment renewed the interest in these models: its results are in fact consistent with a scale-free spectrum with $n = 1.15^{+0.45}_{-0.65}$ on scales $\gtrsim 10^3 h^{-1}$ Mpc. When normalized to fit the COBE data, the Harrison-Zel'dovich spectrum, modulated by the CDM transfer function, leads to excessive small-scale power, while spectra with $n < 1$ imply reduced power on all scales below $\sim 10^3$ Mpc. Moreover, it has been shown by many authors that a consistent normalization of tilted CDM models to the COBE data, properly accounting for the gravitational-wave contribution to the Sachs-Wolfe effect, leads to a relevant enhancement of the linear biasing factor[19].

Here we consider a more radical change of the standard CDM cosmology obtained by dropping the hypothesis that the primordial perturbations have random phases. This alternative is possible either in the context of suitable inflationary scenarios or in more specific models based on phase transitions in the early universe. The former possibility has been widely discussed by many authors[20-25]: one can obtain phase correlations on cosmologically relevant scales provided that a model driven by a multiple scalar field is assumed[26-29]. In the latter case non-Gaussian perturbations are produced during phase transitions by primordial seeds, relic cosmological defects, such as monopoles, cosmic strings or textures, which, later on, will trigger and drive the gravitational instability on the surrounding matter[30-33].

We report here the results of a long-term program to simulate the evolution of the large-scale structure of the Universe from non-Gaussian initial conditions[34-38]. To select our primordial non-Gaussian statistics we followed two simple guiding principles: i) some sort of scale invariance, in the same vein which, in the early Seventies, allowed to single out the Harrison-Zel'dovich power spectrum, much before inflation could provide this prediction; ii) a multiplicative process acting nonlinearly on an underlying Gaussian field, a kind of generalization of the origin of the Gaussian distribution from a discrete Bernoulli process via the central-limit theorem. In particular, we chose two simple non-Gaussian statistics for the primordial peculiar gravitational potential Φ before the modulation of the CDM transfer function: the scale-invariant *chi-squared* statistics (χ^2), one of the simplest multiplicative distributions, and the *lognormal* statistics (LN), representing the extreme case of multiplicative process. The Poisson equation formally splits each model into two distributions for the linear mass fluctuation δ_M, characterized by opposite sign for the skewness $\langle \delta_M^3 \rangle$: we termed respectively χ_p^2 and LN_p the models with positive skewness and χ_n^2 and LN_n those with negative skewness. Our models belong to the wider class of *skewed* non-Gaussian distributions, which are noninvariant under the transformation $\delta_M \to -\delta_M$ leading to an excess probability of overdensities compared to underdensities (skew-positive) or *vice versa* (skew-negative).

An initially nonvanishing skewness is a typical (though not mandatory) signature of cosmological non-Gaussian mass density fluctuations. Recently, SAUNDERS *et al.* [39] found a positive skewness up to large scales in the QDOT data. There is a wide debate on whether such a positive skewness is to be ascribed to primordial processes or just to the nonlinear gravitational evolution of a primordially Gaussian field [40, 41]. In any case, our simulations have shown that an initial skewness in the particle distribution strongly affects the subsequent dynamics and leads to a different texture of the present universe. In general, seed-mass models and, in particular, texture-seeded ones [31, 42] present relevant similarities with our skew-positive models. On the opposite, cosmic explosions (see, *e.g.*, ref. [43]) and cosmic bubbles, relics of a period of extended inflation [18, 44], originate a skew-negative density field.

Our simulations of the above skewed models represent the Universe on a cube of $260 h^{-1}$ Mpc side (h was fixed to 0.5); periodic boundary conditions were assumed as usual. Calculations were performed on a CRAY-YMP432 at the CINECA Computer Center in Bologna using a particle-mesh (PM) code (see, *e.g.*, ref. [45]) with $N_p = 128^3$ particles on $N_g = 128^3$ grid points: each particle, representing a fluid element of dark matter, carries a mass of $4.7 \cdot 10^{12} M_\odot$. We ran two realizations for each of the four models. We then performed a statistical analysis of the present texture of our models using some relevant tests. A deeper statistical investigation of these simulations, mainly devoted to the study of the peculiar velocity field (bulk motions, pairwise velocities, abundance of «great attractors»), will be the subject of a future work [46].

The plan of this lecture is as follows. In sect. 2 we describe our skewed non-Gaussian models. Section 3 presents the results of the statistical analysis on the present spatial distribution obtained by the N-body simulations. Section 4 contains a brief discussion of the results.

2. – Skewed CDM models.

In order to obtain our skewed CDM models, we start from realizations of the peculiar gravitational potential Φ, the gauge-invariant variable whose statistics is generally determined by primordial processes. Unlike the density perturbation field δ_M, this quantity is neither forced to have zero mean, because only its gradients are physically meaningful, nor needs it to satisfy the positive-mass constraint. The peculiar gravitational potential can be used to move the particles according to the Zel'dovich algorithm [47], leading to the following phase-space particle distribution [48]:

$$f_{\text{Zel}}(x, p; t) = \bar{\rho}(t)\delta(p + \nabla_q \Phi(q)), \qquad q = x - b(t)p, \tag{1}$$

$\bar{\rho}(t) \propto a(t)^{-3}$ being the average mass density, $a(t)$ the universal scale factor, $p = dx/db$ the particle «momentum» per unit mass and $b(t) \propto a(t)$ the growing mode of linear perturbations. Before orbit mixing the Zel'dovich approximation can be used to compute the mass density and peculiar velocity fields in the quasi-linear regime.

The fully nonlinear dynamics of our collisionless dark-matter fluid is governed by the following set of equations. The Poisson equation relates the potential Φ to the mass fluctuation δ_M:

$$\nabla^2 \Phi = 4\pi G a^2 \bar{\rho} \delta_M. \tag{2}$$

The Euler equation can be used to follow the evolution of the comoving peculiar velocity field $v \equiv dx/dt$:

$$\frac{\partial v}{\partial t} + 2\frac{\dot{a}}{a}v + v \cdot \nabla v = -\frac{1}{a^2}\nabla\Phi. \tag{3}$$

Finally the mass density obeys the continuity equation

$$\frac{\partial \rho}{\partial t} + 3\frac{\dot{a}}{a}\rho + \nabla \cdot (\rho v) = 0. \tag{4}$$

The numerical techniques used in the PM code solve this set of differential equations. Alternatively, it is possible to obtain approximate solutions using less accurate methods in the weakly nonlinear regime [47, 49-51].

Our distributions are built up in such a way that Φ always has the power spectrum

$$\mathcal{L}_\Phi(k) \propto k^{-3} T^2(k), \tag{5}$$

$T(k)$ being the transfer function of the CDM scenario [2],

$$T(k) = [1 + 6.8k + 72.0k^{3/2} + 16.0k^2]^{-1}. \tag{6}$$

We assume that the Φ statistics is primordially determined. This statistics is not altered during the linear evolution, independently of the physical processes which modify the wave number content of the gravitational potential, only provided one deals with scales not affected by acoustic oscillations. These considerations led us to the following procedure to obtain our models.

i) The potential $\Phi(x)$ is given by the convolution of a real function $\tau(x)$

with a homogeneous and isotropic random field $\varphi(\boldsymbol{x})$,

(7) $$\Phi(\boldsymbol{x}) = \int \mathrm{d}\boldsymbol{y}\, \tau(\boldsymbol{y}-\boldsymbol{x})\, \varphi(\boldsymbol{y}).$$

ii) The field $\varphi(\boldsymbol{x})$ is related to a Gaussian random process $w(\boldsymbol{x})$ by some nonlinear mapping, while the function $\tau(\boldsymbol{x})$ is determined by its Fourier transform,

(8) $$\tilde{\tau}(\boldsymbol{k}) \equiv \int \mathrm{d}\boldsymbol{x}\, \exp[-i\boldsymbol{k}\cdot\boldsymbol{x}]\, \tau(\boldsymbol{x}) = T(k)\, F(k);$$

$F(k)$ is a suitable, positive, correction factor needed to obtain the exact CDM power spectrum in our finite simulation box.

iii) $w(\boldsymbol{x})$ is a Gaussian field with power spectrum

(9) $$\mathscr{L}_w(k) = \frac{2\pi^2 k^{-3}}{\ln(k_M/k_m)},$$

where k_m and k_M, respectively, refer to the minimum and maximum wave number of our simulation box. According to the above equation w has unit r.m.s. value $w_{\mathrm{r.m.s.}} \equiv \langle w^2 \rangle^{1/2} = 1$. The *flicker-noise* assumption $\mathscr{L}_w(k) \propto k^{-3}$ maintains the non-Gaussian features up to the largest simulation scale, corresponding to the w coherence length $\sim 1/k_m$. Thanks to the equipartition property in Fourier space, this spectrum is essentially stable against the nonlinear mapping $w \to \varphi$.

Let us describe in more detail the two different classes of our skewed non-Gaussian models.

2˙1. *Chi-squared models.* – The first class of non-Gaussian distributions is obtained by considering a chi-squared statistics with a single degree of freedom,

(10) $$\varphi(\boldsymbol{x}) = A_\varphi w^2(\boldsymbol{x}).$$

The φ power spectrum is

(11) $$\mathscr{L}_\varphi(k) \propto A_\varphi^2 k^{-3} [\beta(k) + 2\ln(1+k/k_m) - 1],$$

where $\beta(k) = (1-k/k_m)^2$, for $k_m \leq k \leq 2k_m$, and $\beta(k) = 1 + 2\ln(-1+k/k_m)$, for $2k_m \leq k \ll k_M$. Due to the fact that the constant A_φ can be either positive or negative, the original model formally splits into two distributions for δ_M. The quantities Φ and δ_M are related via the Poisson equation, so the sign of A_φ implies the opposite sign for the corresponding density skewness $\langle \delta_M^3 \rangle$. We refer to these distributions as *negative chi-squared* model (χ_n^2), if $\langle \delta_M^3 \rangle < 0$, and *positive chi-squared* model (χ_p^2), if $\langle \delta_M^3 \rangle > 0$. One can argue that the sign of A_φ, which is nothing but a phase, has been fixed by the primordial physical mechanism giving rise to the perturbation field.

FAN and BARDEEN (ref. [52]; see also ref. [27]) proposed an inflationary model where adiabatic perturbations are described on suitable scales by a squared Gaussian process, our χ_p^2, whose w field has, however, a non-scale-free spectrum. LYTH proposed a similar model for isocurvature axion perturbations [53].

2'2. *Lognormal models.* – The second class of our non-Gaussian models is obtained assuming a lognormal statistics for φ,

(12) $$\varphi(\boldsymbol{x}) = A_\varphi \exp[w(\boldsymbol{x})].$$

As in the previous case, the freedom for the sign of A_φ splits the model into two distributions for the density perturbation δ_M: the *negative lognormal* model (LN_n) and the *positive lognormal* model (LN_p), according to the sign of their skewness. The φ power spectrum is approximately given by

(13) $$\mathscr{P}_\varphi(k) \propto A_\varphi^2 k^{-3}.$$

Different realizations of the lognormal statistics would be obtained by varying the amplitude of the underlying Gaussian process w: however, if $w_{r.m.s.}$ were too small, the non-Gaussian character of φ would only manifest in rare high peaks; if, on the contrary, $w_{r.m.s.}$ were too high, the φ power spectrum would largely deviate from the flicker-noise shape. The lognormal statistics has been considered in the cosmological context by many authors (see, *e.g.*, ref. [54,55]).

It is interesting to note that the two above statistics belong to the same general class

(14) $$\varphi(\boldsymbol{x}) \propto |1 + \alpha^{-1} w(\boldsymbol{x})|^\alpha:$$

for $\alpha = 2$ and $w_{r.m.s.} \gg 1$ the chi-squared distribution is recovered, while the limit $\alpha \to \infty$ yields the lognormal model. It is also clear that, in the general non-Gaussian case, a statistical distribution for Φ does not imply the same distribution for δ_M.

The chi-squared model represents just an example of scale-invariant non-Gaussian statistics [55]. On large scales, where the transfer function is not effective $(T(k) \to 1)$, the gravitational potential of scale-invariant distributions obeys the scaling relation

(15) $$\langle \widetilde{\Phi}(\mu\boldsymbol{k}_1) \ldots \widetilde{\Phi}(\mu\boldsymbol{k}_N) \rangle \approx \mu^{-3N} \langle \widetilde{\Phi}(\boldsymbol{k}_1) \ldots \widetilde{\Phi}(\boldsymbol{k}_N) \rangle$$

for any N, up to logarithmic corrections arising from the wave number cut-offs on the underlying Gaussian field. This represents a straightforward generalization of the Harrison-Zel'dovich criterion of scale invariance to non-Gaussian statistics. As a simple consequence, the large-scale skewness $\langle \delta_M^3 \rangle$ scales as the power $3/2$ of the variance $\sigma_M^2 \equiv \langle \delta_M^2 \rangle$, as opposed to the scaling by the power of 2, induced by gravitation on smaller scales [40, 41]. This property is shared by many non-Gaussian models, such as the texture-seeded scenario [31, 42], and could provide a signature for the existence of primordial phase correlations on large scales. In any case, the process of nonlinear gravitational clustering on small scales always leads to a positive sign for the skewness.

For distributions obeying eq. (15) the spatial particle distribution deriving from the Zel'dovich algorithm is the same at time t_2 as it was at time t_1, provided that all lengths are scaled by the factor $\mu = (t_2/t_1)^{1/3}$.

Before concluding this section we have to discuss the criteria adopted to select the «galaxies» and to define the «present time» in our simulations.

The definition of galaxies in dissipationless N-body simulations is a widely debated problem with relevant consequences on the final results for the formation of large-scale structures. Galaxies are usually identified as a suitable subset of the particles in the simulation or by using the so-called peak-background scheme (see, e.g., ref. [56]). Some authors have stressed the relevant role of merging in the galaxy formation processes: for instance, COUCHMAN and CARLBERG [57] adopted a galaxy-finding algorithm which seems to mimic adequately the merging process and to allow for a low-«bias» model. To identify the «galaxies» in our simulations we adopt a simple bias criterion assuming that the galaxies preferentially form in the peaks of the underlying density field. We filter the initial density field with a Gaussian window function of radius $0.5 h^{-1}$ Mpc and pick up as «galaxies» the particles closest to each peak, defined as the grid point with a positive density contrast larger than 26 nearest grid points: this results in a galaxy catalog formed by $\sim 60\,000$ per each simulation. Due to the exceedingly high mass of our particles, following from the large box size and low resolution, and to the rather simplified «galaxy» identification criterion, we can only assume that our peak regions roughly trace the actual galaxy distribution. Low-resolution simulations are in fact quite inadequate to properly describe the clustering of galaxy-sized objects [58].

We stopped our simulations at the «present time» t_0, defined as the time when the galaxy-galaxy correlation function is best fitted by the power law $\xi(r) = (r/r_g)^{-\gamma}$, with $\gamma = 1.8$ in the interval $0.3 \leq \xi \leq 30$, in agreement with observational data. The quantity r_g is called the galaxy correlation length. A parametrization of different epochs may be obtained by using the bias factor b, defined as the inverse r.m.s. linear mass fluctuation on a sharp-edged sphere of radius $8 h^{-1}$ Mpc. Our present times t_0 correspond to $b = 1.5$ for the skew-positive models and $b \approx 0.5$ for the skew-negative ones.

Note that the method to define the galaxies used in this work is different from the «excursion regions» technique used in ref. [38], where a larger galaxy number density, $3 \cdot 10^{-2} h^3$ Mpc^{-3}, was necessary in order to generate simulated Lick catalogs with $\approx 530\,000$ galaxies in the whole simulation box. A consequence of this change is, for example, that the present epoch, i.e. the slope $\gamma = 1.8$ for the correlation function, is reached later here than in ref. [38].

3. – Results.

The main result of our simulations is the evidence that the sign of the initial skewness strongly affects the evolution and the present texture of the Universe. This fact is clearly displayed by the different visual appearance: in fig. 1 we show the present-time galaxy distributions in slices of thickness $65h^{-1}$ Mpc. Skew-positive models show a lumpy structure, characterized by isolated knots surrounded by large regions of quasi-uniform density; skew-negative models, on the contrary, give rise to a cellular structure with long filaments, extended sheets and large voids.

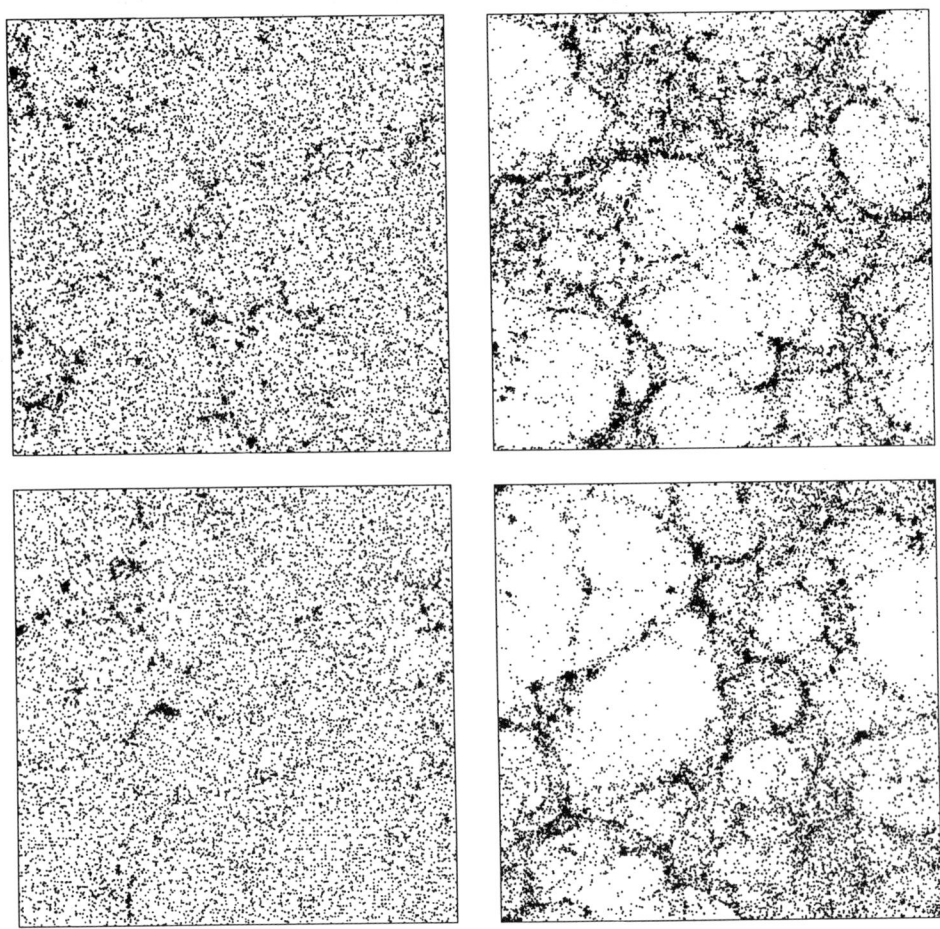

Fig. 1. – Projected slices of depth $65h^{-1}$ Mpc of the galaxy distribution at the present time. The slices refer to different models: χ_p^2 (top left), χ_n^2 (top right), LN_p (bottom left) and LN_n (bottom right).

3'1. *Probability distributions.* – Figure 2 shows the probability distribution P for the density contrast obtained using our procedure to give the initial conditions for N-body simulations. We adopt a log-linear plot of P vs. δ/σ (where δ and σ, respectively, refer to the density contrast and its r.m.s. value): in this plot a Gaussian distribution, shown for reference, would appear as an inverted parabola, while a skew-positive (negative) distribution has a long tail of positive (negative) fluctuations. Of course, there is symmetry with respect to zero between positive and negative models of the same statistics. Lognormal models appear to be more extreme than χ^2 models, leading to higher probability for high (in absolute value)-sigma events.

The subsequent dynamical evolution strongly changes these distributions. Figure 3 shows P at the present time for the mass and for the distribution of the «galaxies», defined according to ref.[38]. In this case we adopt a smoothing with Gaussian windows of radii $6h^{-1}$ Mpc and $12h^{-1}$ Mpc. At this time the tail lies on the right also for the skew-negative models because the aggregating action of gravity eventually shifted the power of the mass density field from underdense to overdense regions; this feature is less relevant for the larger filtering radius where this nonlinear effect had smaller time to act. We find that the skewness of the galaxy distribution, calculated in cubic cells of side L, vanishes beyond a scale $L \sim 35h^{-1}$ Mpc, even for the skew-negative models. Of course, the mass distribution behaves differently: $\langle \delta_M^3 \rangle$ for the skew-negative models changes sign on a scale $L \sim (40\text{–}50)h^{-1}$ Mpc. This result is not in contradiction with the analysis of the QDOT survey [39].

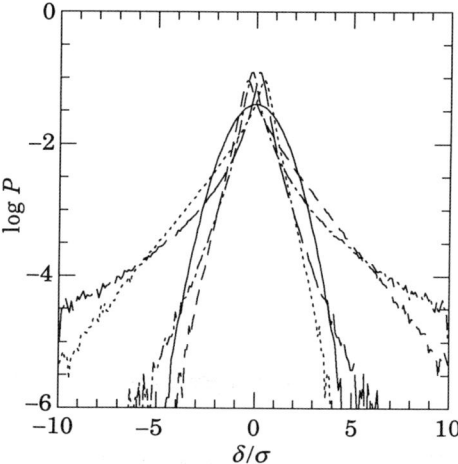

Fig. 2. – The probability distribution of the density contrast from the initial conditions of N-body simulations. The χ_p^2, χ_n^2, LN_p and LN_n models are represented by short-dashed, dotted, dot-dashed and long-dashed lines, respectively. Solid line refers to a Gaussian distribution, shown as reference.

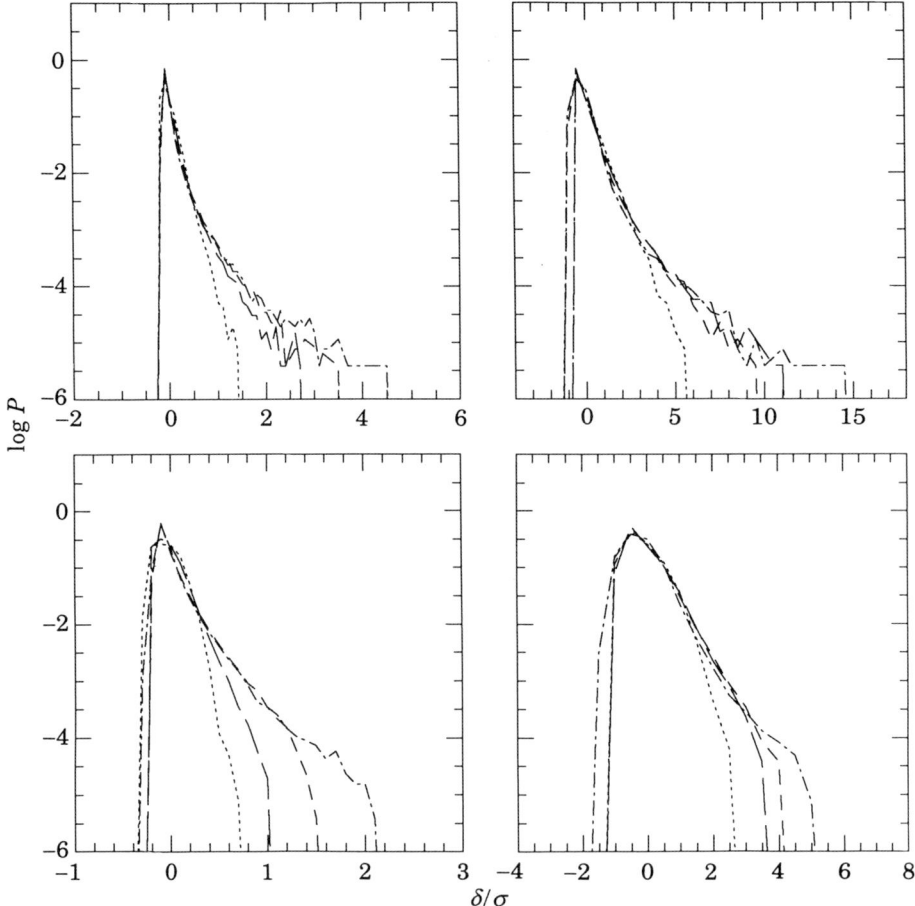

Fig. 3. – The probability distribution of the mass (left) and galaxy (right) density at the present time. The density is smoothed with a Gaussian window of radius $6h^{-1}$ Mpc (top) and $12h^{-1}$ Mpc (bottom). The density contrast is plotted in units of the r.m.s. fluctuation at the filter scale. The models are represented as in fig. 2.

3'2. *Correlation function.* – The two-point correlation function $\xi(r)$ is the typical test in the statistical analysis of the spatial distribution both in observational data samples and in numerical simulations. To calculate the correlation function we chose to follow the procedure described in ref. [59], which gives $1 + \xi(r) = (N_p / \bar{n} N_c \Delta V)$, where N_p is the number of pairs with distance between r and $r + \Delta r$, N_c the number of randomly chosen centres, $\bar{n}$ the mean particle density and ΔV the volume of the considered spherical shell. Figure 4 shows the galaxy-galaxy correlation function at the present time, *i.e.* when its slope is $\gamma \approx$ ≈ 1.8. The corresponding correlation length r_g is $3.6 \pm 1.0, 3.2 \pm 1.8, 7.3 \pm 0.8$ and 8.8 ± 0.9, in units of h^{-1} Mpc, for the χ_p^2, LN_p, χ_n^2 and LN_n models, respectively;

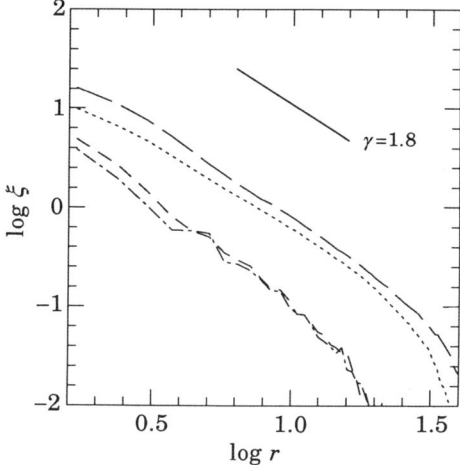

Fig. 4. – The galaxy two-point correlation function $\xi(r)$ at the present time vs. the logarithm of the distance r (in units of h^{-1} Mpc). The models are represented as in fig. 2. For reference the solid line with slope $\gamma = 1.8$ is reported.

the errors are obtained considering both the standard deviation of the fit and the scatter between the two realizations of each model. The difference between the two skew-negative models, not so evident in our previous simulations[35], can be ascribed to the present box size which allows the more extreme LN_n model to develop more strongly its large-scale coherence, leading to an exceedingly high correlation length. For comparison, the measured correlation length of the galaxy distribution is $r_g = (5-8) h^{-1}$ Mpc. The skew-negative models seem the most interesting ones: their correlation length is consistent with the analysis of the CfA survey[60]: decreasing the slope of $\xi(r)$ (which defines t_0) from 1.8 to 1.6, as suggested by the CfA analysis, would slightly increase the value of b and reduce the correlation length. Note that the choice of the galaxy definition used here reduces the discrepancy between observations and the results for LN_n model, noted in ref.[38].

3'3. *Topology of isodensity contours.* – The genus statistics proved to be a very useful test to describe the topology of the isodensity contours $\delta_M(x) = \nu \sigma_M$ of the matter distribution. This test has been widely applied both to data samples and to N-body simulations (see, *e.g.*, ref.[61]). The genus of a surface is the difference between the number of holes and isolated regions it contains and gives the maximum number of cuts that can be made on a contour surface without dividing it into two distinct pieces. For the Gaussian distribution the mean genus per unit volume shows symmetry between high- and low-density regions, displaying the so-called «spongelike» topology. Departures from this symmetry arise in the presence of nonvanishing skewness in the matter distri-

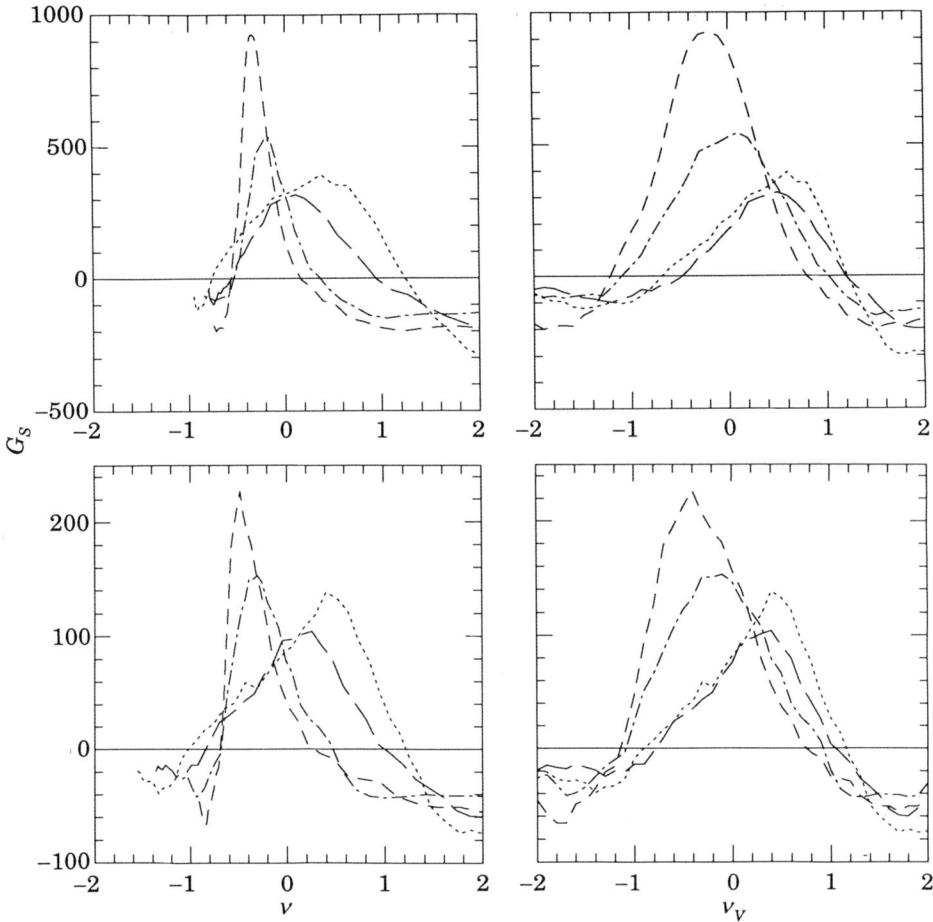

Fig. 5a. – The total genus G_S vs. the threshold ν (left) and ν_V (right) for the mass distribution at the present time. The mass density is smoothed with a Gaussian window of radius $6h^{-1}$ Mpc (top) and $12h^{-1}$ Mpc (bottom). The models are represented as in fig. 2.

bution. Figure 5 shows the total genus curve $G_S(\nu)$ in our simulation box for the present galaxy and mass distributions filtered with Gaussian windows of radii $6h^{-1}$ Mpc and $12h^{-1}$ Mpc vs. the relative threshold ν. We also calculated G_S as a function of the more traditional effective threshold ν_V, defined so that $V_{>\nu} \equiv (1/\sqrt{2\pi}) \int_{\nu_V}^{\infty} dy \exp[-y^2/2]$, $V_{>\nu}$ being the volume fraction enclosed by the contour $\nu \sigma_M$; $\nu_V = \nu$ for a Gaussian distribution (see also ref. [62]).

Let us first comment on the $G_S(\nu)$ curves. Both the mass and the galaxies in skew-positive models display the «meatball» topology (G_S curve shifted towards negative ν), typical of a lumpy distribution; the effect is less relevant for LN_p. This is indeed the kind of topology displayed by composite data obtained

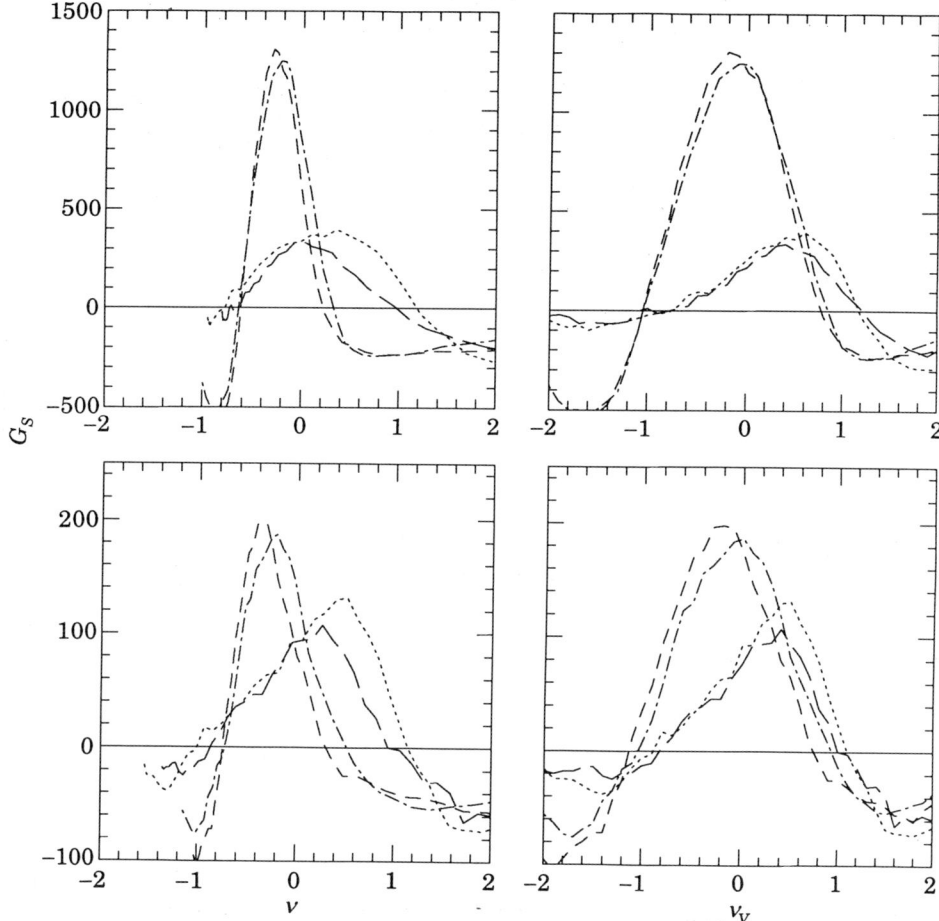

Fig. 5b. – The total genus G_S vs. the threshold ν (left) and ν_V (right) for the galaxy distribution at the present time. The galaxy density is smoothed with a Gaussian window of radius $6h^{-1}$ Mpc (top) and $12h^{-1}$ Mpc (bottom). The models are represented as in fig. 2.

from different galaxy catalogs [63]. A cellular or «swiss-cheese» topology (G_S curve shifted towards positive ν) results in skew-negative models: the algorithm adopted to select the galaxies does not lead to a change of the galaxy topology compared to the mass. For the same skew-negative models, however, we found [36] that the groups, selected by a «friends-of-friends» algorithm, display a meatball topology. The topology derived from the $G_S(\nu_V)$ curves does not show relevant differences.

In a recent work [38], we measured, using the Euler-Poincaré characteristic (EP) of the regions above a threshold density, the two-dimensional topology of projected galaxy catalogs generated, from the same simulations, with the selection criteria used in the Shane-Wirtanen galaxy counts, but adopting a different galaxy definition. We found that all our models produce a topology dominated by a meatball shift when normalized to the known small-scale clustering properties of galaxies. This makes it rather difficult to separate the effects of nonlinear gravitational clustering from phase correlations intrinsic to the primordial perturbations. It is, however, possible to discern differences in the topological behaviour of the same models at different stages of gravitational evolution using the projected catalogs. Models characterized by a positive skewness of the distribution of primordial density perturbations are found inconsistent with the Lick data. On the contrary, a CDM model with negative skewness of the primordial density field (chi-squared model) fits the cell count frequency distribution well and matches both the shape and the amplitude of the EP curve. These conclusions were shown to be rather insensitive to the form of the luminosity function used to construct the catalogs.

3`4. *Void probability function.* – Looking at fig. 1, it is easy to note the different importance of underdense regions in the considered models. In ref.[35], adopting the nearest-neighbour statistics[64], we found that the typical radius of the largest void in the simulations was strongly dependent on the initial skewness: skew-negative models have very large voids, formed by the percolation of smaller, initially underdense, regions, while in the skew-positive models the quasi-uniform distribution of particles results in very small voids. In order to obtain a more quantitative result, in this work we apply a different statistics, the void probability function (VPF), the probability $P_0(R)$ that a sphere of radius R, randomly placed in the simulations, contains no galaxies. This statis-

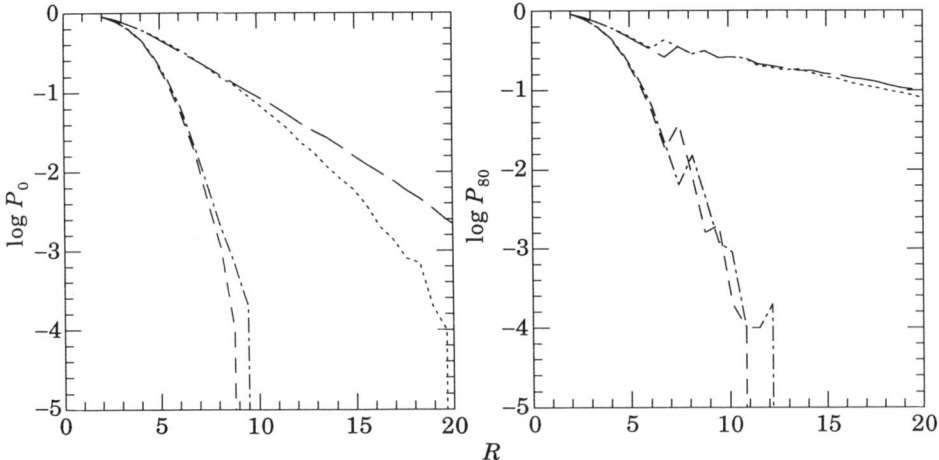

Fig. 6. – The void probability function, P_0 (left panel), and the probability that the average density in a randomly placed sphere is more than 80% below the global mean density, P_{80} (right panel), vs. the radius R (in h^{-1} Mpc) for the galaxy distribution at the present time. The models are represented as in fig. 2.

tics has been recently applied to the CfA2 survey [65]. In fig. 6 we show the results from the galaxy catalogs obtained in our simulations. Once again, we find a difference between skew-positive and skew-negative models: in the first case, P_0 falls to zero at $R = 10h^{-1}$ Mpc, while, in the second one, it is very high even at $R = 20h^{-1}$ Mpc. There are no particular differences between the two non-Gaussian models in the skew-positive case, while LN_n results to be once again more extreme than χ_n^2. In order to avoid the strong dependence of $P_0(R)$ on the number density of galaxies, WEINBERG and COLE [66] introduced the statistics $P_{80}(R)$, the probability that the average density in a randomly placed sphere is more than 80% below the global mean density. The resulting trend for our simulations (see fig. 6) is very similar to the $P_0(R)$ one: the underdense regions in the skew-negative models have a larger probability than in skew-positive ones.

4. – Discussion.

We have presented an analysis of the large-scale structure of the Universe as resulting from skewed non-Gaussian initial perturbations in a CDM cosmogony. Observations seem to require more structure on large scales than provided by the standard version of the CDM model. Higher-order moments of non-Gaussian statistics may, however, strongly affect the probability distribution of high peaks in the density field, giving rise to a richer amount of structures on large scales.

Our analysis confirms that the initial skewness of mass fluctuations is a strongly discriminating parameter. Actually, the models with the same skewness sign show a similar trend in most statistical tests. Many properties of our distributions can be summarized as follows. Skew-positive models appear lumpy on all scales with isolated rich clusters emerging from a nearly uniform density field. The analysis of the two-dimensional topology of mock catalogs drawn from our simulations [38] also revealed discrepancies with the properties of the Lick data. These conclusions agree with previous results [35, 66] and altogether suggest some difficulties for the texture-seeded CDM model [31, 42].

The structures in skew-negative models are characterized by a high level of connectivity with large voids surrounded by extended sheets and filaments. The galaxy correlation length for the χ_n^2 model, in particular, is in rather good agreement with observations. WEINBERG and COLE [66] recently evolved scale-free skew-negative models and concluded that these models present many attractive properties, but fail in reproducing the cluster mass-to-light ratios, the pairwise velocities of galaxies and the expected type of topology. A partial explanation of these difficulties could be ascribed to the normalization they used to single out the present epoch from scale-free initial distributions. The normalization of the linear mass fluctuation, resulting from our «galaxy»-finding algorithm and from the criterion to define the present time, implies a bias factor $b \approx$ ≈ 0.5, which may lead to large-scale CMB anisotropies larger than those detected by COBE [3]: these are in fact consistent with the unbiased (*i.e.* $b = 1$) standard CDM model. The apparent flaw of skew-negative models is likely to be overcome by improving the biasing scheme, for instance by taking into account cooperative mechanisms able to affect the galaxy clustering on large scales [6]. Moreover, the analysis of CMB anisotropies on very large scales for non-Gaussian models cannot be reduced to a simple calculation of the r.m.s. fluctuation. These problems will be analysed in more detail in a future work, where the debated issue of identifying the correct biasing scheme to select galaxies will be investigated [46].

In conclusion, phase correlations in the primordial perturbation field play a relevant role in the nonlinear evolution of cosmic structures. This kind of initial conditions may be motivated by physical models in the frame of the dynamics of phase transitions in the early universe. The present work has shown that the non-Gaussian alternative within the CDM scenario offers a viable solution to the large-scale problem.

* * *

This work has been partially supported by Ministero dell'Università e della Ricerca Scientifica e Tecnologica and by Consiglio Nazionale delle Ricerche (Progetto Finalizzato Sistemi Informatici e Calcolo Parallelo). The staff and the management of the CINECA Computer Center (Bologna) are warmly acknowledged for their assistance and for allowing the use of computational facilities.

REFERENCES

[1] G. BLUMENTHAL, S. FABER, J. PRIMACK and M. REES: *Nature (London)*, **311**, 517 (1984).
[2] M. DAVIS, G. EFSTATHIOU, C. S. FRENK and S. D. M. WHITE: *Astrophys. J.*, **292**, 371 (1985).
[3] G. F. SMOOT, C. L. BENNET, A. KOGUT, E. L. WRIGHT, J. AYMON, N. W. BOGGESS, E. S. CHENG, G. DE AMICI, S. GULKIS, M. G. HAUSER, G. HINSHAW, P. D. JACKSON, M. JANSSEN, E. KAITA, T. KELSALL, P. KEEGSTRA, C. LINEWEAVER, K. LOEWENSTEIN, P. LUBIN, J. MATHER, S. S. MEYER, S. H. MOSELEY, T. MURDOCK, L. ROKKE, R. F. SILVERBERG, L. TENORIO, R. WEISS and D. T. WILKINSON: *Astrophys. J. Lett.*, **396**, L1 (1992).
[4] S. J. MADDOX, G. EFSTATHIOU, W. J. SUTHERLAND and J. LOVEDAY: *Mon. Not. R. Astron. Soc.*, **243**, 692 (1990).
[5] A. BABUL and S. D. M. WHITE: *Mon. Not. R. Astron. Soc.*, **253**, 31P (1991).
[6] R. G. BOWER, P. COLES, C. S. FRENK and S. D. M. WHITE: *Astrophys. J.*, **403**, 405 (1993).

[7] G. EFSTATHIOU, W. J. SUTHERLAND and S. J. MADDOX: *Nature (London)*, **348**, 705 (1990).
[8] M. DAVIS, F. J. SUMMERS and D. SCHLEGEL: *Nature (London)*, **359**, 393 (1992).
[9] A. N. TAYLOR and M. ROWAN-ROBINSON: *Nature (London)*, **359**, 396 (1992).
[10] J. M. GELB, B. GRADWOHL and J. A. FRIEMAN: Fermilab preprint (1992).
[11] N. VITTORIO, S. MATARRESE and F. LUCCHIN: *Astrophys. J.*, **328**, 69 (1988).
[12] G. TORMEN, F. LUCCHIN and S. MATARRESE: *Astrophys. J.*, **386**, 1 (1992).
[13] F. C. ADAMS, J. R. BOND, K. FREESE, J. A. FRIEMAN and A. V. OLINTO: *Phys. Rev. D*, **47**, 426 (1993).
[14] R. CEN, N. Y. GNEDIN, L. A. KOFMAN and J. P. OSTRIKER: *Astrophys. J. Lett.*, **399**, L11 (1992).
[15] A. R. LIDDLE and D. H. LYTH: *Phys. Rep.*, **231**, 1 (1992).
[16] G. TORMEN, L. MOSCARDINI, F. LUCCHIN and S. MATARRESE: *Astrophys. J.*, **411**, 16 (1993).
[17] F. LUCCHIN and S. MATARRESE: *Phys. Rev. D*, **32**, 1316 (1985).
[18] D. LA and P. J. STEINHARDT: *Phys. Rev. Lett.*, **62**, 376 (1989).
[19] R. L. DAVIS, M. H. HODGES, C. F. SMOOT, P. J. STEINHARDT and M. S. TURNER: *Phys. Rev. Lett.*, **69**, 1856 (1992); A. DOLGOV and J. SILK: *Phys. Rev. D*, **47**, 2619 (1993); A. R. LIDDLE and D. H. LYTH: *Phys. Lett. B*, **291**, 391 (1992); J. E. LIDSEY and P. COLES: *Mon. Not. R. Astron. Soc.*, **258**, 57P (1992); F. LUCCHIN, S. MATARRESE and S. MOLLERACH: *Astrophys. J. Lett.*, **400**, L43 (1992); D. S. SALOPEK: DAMTP preprint (1992); T. SOURADEEP and V. SAHNI: *Mod. Phys. Lett. A*, **7**, 3541 (1993).
[20] A. ORTOLAN, F. LUCCHIN and S. MATARRESE: *Phys. Rev. D*, **38**, 465 (1988).
[21] S. MATARRESE, A. ORTOLAN and F. LUCCHIN: *Phys. Rev. D*, **40**, 290 (1989).
[22] J. D. BARROW and P. COLES: *Mon. Not. R. Astron. Soc.*, **244**, 188 (1990).
[23] L. KOFMAN, G. BLUMENTHAL, H. HODGES and J. PRIMACK: in *Proceedings of the Workshop on Large-Scale Structure and Peculiar Motions in the Universe*, edited by D. W. LATHAM and L. N. DA COSTA, ASP Conference Series, Vol. 15 (1991), p. 339.
[24] D. S. SALOPEK and J. R. BOND: *Phys. Rev. D*, **43**, 1005 (1991).
[25] S. MOLLERACH, S. MATARRESE, A. ORTOLAN and F. LUCCHIN: *Phys. Rev. D*, **44**, 1670 (1991).
[26] L. KOFMAN and A. LINDE: *Nucl. Phys. B*, **282**, 555 (1987).
[27] D. SALOPEK, J. R. BOND and J. M. BARDEEN: *Phys. Rev. D*, **40**, 1753 (1989).
[28] T. J. ALLEN, B. GRINSTEIN and M. B. WISE: *Phys. Lett. B*, **197**, 66 (1987).
[29] D. S. SALOPEK: *Phys. Rev. D*, **45**, 1139 (1992).
[30] N. TUROK: *Phys. Rev. Lett.*, **63**, 2625 (1989).
[31] C. PARK, D. N. SPERGEL and N. TUROK: *Astrophys. J. Lett.*, **372**, L53 (1991).
[32] R. J. SCHERRER and E. BERTSCHINGER: *Astrophys. J.*, **381**, 349 (1991).
[33] R. J. SCHERRER: *Astrophys. J.*, **390**, 330 (1992).
[34] A. MESSINA, L. MOSCARDINI, F. LUCCHIN and S. MATARRESE: *Mon. Not. R. Astron. Soc.*, **245**, 244 (1990).
[35] L. MOSCARDINI, S. MATARRESE, F. LUCCHIN and A. MESSINA: *Mon. Not. R. Astron. Soc.*, **248**, 424 (1991).
[36] S. MATARRESE, F. LUCCHIN, A. MESSINA and L. MOSCARDINI: *Mon. Not. R. Astron. Soc.*, **252**, 35 (1991).
[37] A. MESSINA, F. LUCCHIN, S. MATARRESE and L. MOSCARDINI: *Astroparticle Phys.*, **1**, 99 (1992).
[38] P. COLES, L. MOSCARDINI, M. PLIONIS, F. LUCCHIN, S. MATARRESE and A. MESSINA: *Mon. Not. R. Astron. Soc.*, **260**, 572 (1993).

[39] W. Saunders, C. Frenk, M. Rowan-Robinson, G. Efstathiou, A. Lawrence, N. Kaiser, R. Ellis, J. Crawford, X.-Y. Xia and I. Parry: *Nature (London)*, **349**, 32 (1991).
[40] P. Coles and C. S. Frenk: *Mon. Not. R. Astron. Soc.*, **253**, 727 (1991).
[41] J. Silk and R. Juszkiewicz: *Nature (London)*, **353**, 386 (1991).
[42] R. Y. Cen, J. P. Ostriker, D. N. Spergel and N. Turok: *Astrophys. J.*, **383**, 1 (1991).
[43] D. H. Weinberg, J. P. Ostriker and A. Dekel: *Astrophys. J.*, **336**, 9 (1989).
[44] A. R. Liddle and D. Wands: *Mon. Not. R. Astron. Soc.*, **253**, 637 (1991).
[45] R. W. Hockney and J. W. Eastwood: *Computer Simulations using Particles* (McGraw-Hill, New York, N.Y., 1981).
[46] F. Lucchin, S. Matarrese, A. Messina and L. Moscardini: in preparation.
[47] S. F. Shandarin and Ya. B. Zel'dovich: *Rev. Mod. Phys.*, **61**, 185 (1989).
[48] B. Grinstein and M. B. Wise: *Astrophys. J.*, **320**, 448 (1987).
[49] S. N. Gurbatov, A. I. Saichev and S. F. Shandarin: *Mon. Not. R. Astron. Soc.*, **236**, 921 (1989).
[50] A. Nusser, A. Dekel, E. Bertschinger and G. R. Blumenthal: *Astrophys. J.*, **379**, 6 (1991).
[51] S. Matarrese, F. Lucchin, L. Moscardini and D. Saez: *Mon. Not. R. Astron. Soc.*, **259**, 437 (1992).
[52] Z. H. Fan and J. M. Bardeen: University of Washington preprint (1992).
[53] D. H. Lyth: *Phys. Rev. D*, **45**, 3394 (1992).
[54] P. Coles and B. J. T. Jones: *Mon. Not. R. Astron. Soc.*, **248**, 1 (1991).
[55] F. Lucchin and S. Matarrese: *Astrophys. J.*, **330**, 535 (1988).
[56] C. S. Frenk: *Phys. Scr.*, **T36**, 70 (1991).
[57] H. M. P. Couchman and R. G. Carlberg: *Astrophys. J.*, **389**, 453 (1992).
[58] E. Bertschinger and J. M. Gelb: *Comput. Phys.*, **5**, 164 (1991).
[59] A. A. Klypin and S. F. Shandarin: *Mon. Not. R. Astron. Soc.*, **204**, 891 (1983).
[60] V. de Lapparent, M. J. Geller and J. P. Huchra: *Astrophys. J.*, **333**, 44 (1988).
[61] A. L. Melott: *Phys. Rep.*, **193**, 1 (1990).
[62] P. Coles and J. D. Barrow: *Mon. Not. R. Astron. Soc.*, **228**, 407 (1987).
[63] J. R. Gott, J. Miller, T. X. Thuan, S. E. Schneider, D. H. Weinberg, C. Gammie, K. Polk, M. Vogeley, M. S. Jeffrey, S. P. Bhavsar, A. L. Melott, R. Giovanelli, M. P. Haynes, R. B. Tully and A. J. S. Hamilton: *Astrophys. J.*, **340**, 625 (1989).
[64] B. S. Ryden and E. L. Tuner: *Astrophys. J.*, **287**, L59 (1984).
[65] M. S. Vogeley, M. J. Geller and J. P. Huchra: *Astrophys. J.*, **382**, 44 (1991).
[66] D. H. Weinberg and S. Cole: *Mon. Not. R. Astron. Soc.*, **259**, 652 (1992).

Nonlinear Evolution of Self-Gravitating Collisionless Matter.

S. Matarrese and O. Pantano

Dipartimento di Fisica G. Galilei dell'Università
via Marzolo 8, I-35131 Padova, Italia

F. Lucchin and L. Moscardini

Dipartimento di Astronomia dell'Università
vicolo dell'Osservatorio 5, I-35122 Padova, Italia

D. Saez

Departamento de Fisica Teorica de la Universidad - Burjassot, Valencia, España

1. – Introduction.

An important theoretical issue in cosmology is to understand the physical processes that occurred during the gravitational collapse of matter which gave rise to the observed large-scale structure of the Universe. A complementary issue is to reconstruct the initial conditions of the clustering process, *e.g.* the value of the cosmological parameters, the type of dark matter, the statistics of the primordial perturbations, starting from observational data such as the spatial distribution of galaxies or their peculiar velocities. Much work has recently focused in the latter direction, since more and more data on peculiar velocities of optical galaxies, as well as very large and complete galaxy redshift surveys, have become available both in the optical and in the infrared.

A widely applied and well-motivated approximation when dealing with the dynamics of dark matter, either cold or hot, is to treat it as a system of particles having negligible nongravitational interactions, a self-gravitating collisionless system. The dynamics of such a system is usually approached by different techniques, depending on the specific application. For instance, the evolution of small perturbations on a Friedmann-Robertson-Walker (FRW) background is followed by analytical methods. The nonlinear evolution in cases where some symmetries are present can also sometimes be followed analytically: typical examples being the spherical top-hat model for the Newtonian case[1] and the Tolman-Bondi (TB) solution[2] in general relativity (GR). There are also useful approximations valid in the mildly nonlinear regime, such as the Zel'dovich approximation[3]. Besides this classical approach, a number of variants have been proposed, all trying to overcome the inability to follow the development of structures beyond caustic formation. The most promising approach is adhesion theory[4], where artificial viscosity is introduced to mimic the gravitational sticking of particles around pancakes. An alternative method, called frozen-flow approximation, recently proposed by MATARRESE, LUCCHIN, MOSCARDINI and SAEZ[5], allows an extrapolation beyond the time when orbit crossing would have occurred according to the Zel'dovich formulation. This is based on the idea of extrapolating the growing mode of the linear velocity field beyond its actual range of validity, while solving exactly the continuity equation. Other approximations have been explored, such as the second-order perturbation expansion (e.g., ref.[6]), the algorithms proposed by GIAVALISCO et al.[7] and YAHIL[8] and the Lagrangian approach of ref.[9]. Different approximations apply in the highly nonlinear regime, such as the hierarchical closure ansatz for the BBGKY equations[1]. The most general problem of studying the fully nonlinear dynamics of a collisionless system in Newtonian theory can only be followed by numerical techniques, such as N-body codes (e.g., ref.[10]). Finally, a GR approach to the nonlinear evolution of scalar perturbations of a pressureless fluid has been proposed by MATARRESE, PANTANO and SAEZ[11], under the assumption of vanishing vorticity and negligible gravitational-wave interactions with the rest of the system.

Here we review various methods, trying to show their possible interconnections. Section 2 presents the Zel'dovich and frozen-flow approximations. Section 3 introduces the three-dimensional Burgers equation and discusses two limiting behaviours resulting from it. Section 4 contains a brief presentation of a GR algorithm. A final discussion is given in sect. 5.

2. – Newtonian dynamics of self-gravitating collisionless matter.

Let us start by writing the Newtonian equations for the evolution of collisionless particles in the expanding universe[1]. These can be written using suitably rescaled variables and in comoving coordinates. We shall assume that the Universe is spatially flat and matter dominated, so that the scale factor reads $a(t) = a_0 (t/t_0)^{2/3}$ (a subscript 0 will be used to define quantities at some «initial time» t_0). The Euler equations read

$$(2.1) \qquad \frac{d\boldsymbol{u}}{da} + \frac{3}{2a}\boldsymbol{u} = -\frac{3}{2a}\nabla\varphi,$$

where $\boldsymbol{u} \equiv d\boldsymbol{x}/da$ is a rescaled comoving peculiar velocity field. The symbol d/da stands for the total (convective) derivative $d/da = \partial/\partial a + \boldsymbol{u} \cdot \nabla$.

The continuity equation can be written in terms of the comoving matter density $\eta(\mathbf{x}, t) \equiv \rho(\mathbf{x}, t) a^3(t)/\bar{\rho}_0 a_0^3$ (where $\bar{\rho}_0$ is the mean mass density at t_0)

$$\text{(2.2)} \qquad \frac{d\eta}{da} + \eta \nabla \cdot \mathbf{u} = 0,$$

while the rescaled local (or peculiar) gravitational potential $\varphi \equiv (3t_0^2/2a_0^3) \cdot \phi(\mathbf{x}, t)$ is determined by local density inhomogeneities $\delta(\mathbf{x}, t) \equiv \eta(\mathbf{x}, t) - 1$ through the Poisson equation

$$\text{(2.3)} \qquad \nabla^2 \varphi = \frac{\delta}{a}.$$

We can restrict the analysis to initially irrotational flow. According to Kelvin's circulation theorem, in the absence of dissipation, vorticity is conserved along each fluid trajectory; in particular, a fluid with vanishing initial vorticity will forever remain irrotational. However, for a collisionless fluid such a property breaks down after caustic formation: a vorticity component is created in multi-stream regions, simply because the local Eulerian velocity field takes contributions from different Lagrangian fluid elements at the same position. Outside the regions of orbit mixing and/or after applying a suitable low-pass filter one can define a velocity potential by $\mathbf{u}(\mathbf{x}, a) = \nabla \Phi(\mathbf{x}, a)$. It is then easy to derive from the Euler equation a Bernoulli equation relating the velocity potential Φ to the gravitational one φ [12]:

$$\text{(2.4)} \qquad \frac{\partial \Phi}{\partial a} + \frac{1}{2}(\nabla \Phi)^2 = -\frac{3}{2a}(\Phi + \varphi).$$

By integrating along the particle trajectory $\mathbf{x}(a)$ one finds a formal solution of the continuity equation:

$$\text{(2.5)} \qquad \eta(\mathbf{x}, a) = \eta_0(\mathbf{q}) \exp\left\{ -\int_{a_0}^{a} da' \, \nabla \cdot \mathbf{u}[\mathbf{x}(\mathbf{q}, a'), a'] \right\}.$$

This can be compared with the formula obtained from mass conservation, $\eta(\mathbf{x}, a) d^3x = \eta_0(\mathbf{q}) d^3q$, where $\mathbf{q}$ is the initial (Lagrangian) position of the particle which has reached the (Eulerian) position $\mathbf{x}$ by the time $a(t)$. One has either the well-known Lagrangian form $\eta[\mathbf{q}(\mathbf{x}, a), a] = \eta_0(\mathbf{q}) \|\partial x/\partial q\|^{-1}$, where $\|\partial x/\partial y\|$ is the Jacobian determinant of the transformation $x \to y$, or the Eulerian one $\eta(\mathbf{x}, a) = \eta_0(\mathbf{q}) \|\partial q/\partial x\|$, which, however, requires inverting the trajectory to find $\mathbf{q}(\mathbf{x}, a)$. These solutions are only valid as long as no shell crossing (caustic) has occurred, so that there is a one-to-one correspondence between Lagrangian and Eulerian positions. Before caustic formation all these forms are identical.

Zel'dovich approximation. The simplest approximation is, of course, the linear one, which consists in neglecting the terms $\mathbf{u} \cdot \nabla \mathbf{u}$ in the Euler equations and $\nabla \cdot (\delta \boldsymbol{u})$ in the continuity equation. The resulting expressions read $\boldsymbol{u}_{\mathrm{lin}}(\boldsymbol{x}, a) = - \nabla \varphi_0(\boldsymbol{x})$, $\varphi_{\mathrm{lin}}(\boldsymbol{x}, a) = \varphi_0(\boldsymbol{x})$ and $\eta_{\mathrm{lin}}(\boldsymbol{x}, a) = 1 + a\nabla^2 \varphi_0(\boldsymbol{x})$, having neglected the contribution of decaying modes.

The next step is the Zel'dovich approximation (ZEL), based on the ansatz of extrapolating the equation $\boldsymbol{u} = -\nabla \varphi$ ($\Phi = -\varphi$) beyond linear theory; replacing this ansatz into the Euler equations gives

$$(2.6) \qquad \frac{\mathrm{d}\boldsymbol{u}}{\mathrm{d}a} = 0,$$

which has to be solved together with eq. (2.2). The resulting system can be taken as the definition of ZEL [3]. In this approximation a particle initially placed in $\boldsymbol{q}$ moves along a straight line with constant «speed» $\boldsymbol{u}$ determined by the value of the initial peculiar gravitational potential in $\boldsymbol{q}$, namely $\boldsymbol{x}(\boldsymbol{q}, \tau) = \boldsymbol{q} - \tau \nabla_{\boldsymbol{q}} \varphi_0(\boldsymbol{q})$, with $\tau \equiv a - a_0$. The velocity field is conserved along each particle trajectory: $\boldsymbol{u}_{\mathrm{ZEL}}(\boldsymbol{x}, \tau) = \boldsymbol{u}_0(\boldsymbol{q}) = -\nabla_{\boldsymbol{q}} \varphi_0(\boldsymbol{q})$. It is, of course, possible to define a velocity potential by $\boldsymbol{u}_{\mathrm{ZEL}}(\boldsymbol{x}, \tau) = \nabla_{\boldsymbol{x}} \Phi_{\mathrm{ZEL}}(\boldsymbol{x}, \tau) = \nabla_{\boldsymbol{q}} \Phi_0(\boldsymbol{q})$, which obeys the Hamilton-Jacobi equation [12]

$$(2.7) \qquad \frac{\partial \Phi}{\partial \tau} + \frac{1}{2}(\nabla_{\boldsymbol{x}} \Phi)^2 = 0.$$

The integration of the Zel'dovich-Bernoulli equation (2.7), in Eulerian space, has been considered by NUSSER and DEKEL [13] to reconstruct the primordial density field from the present large-scale velocity or density field. The solution of eq. (2.7) is $\Phi_{\mathrm{ZEL}}(\boldsymbol{x}, \tau) = \Phi_0(\boldsymbol{q}) + (\boldsymbol{x} - \boldsymbol{q})^2/2\tau$. The density field is usually represented in the Lagrangian form $\eta_{\mathrm{ZEL}}(\boldsymbol{q}, \tau) = \eta_0(\boldsymbol{q}) \|\mathbf{1} + \tau \boldsymbol{D}_0(\boldsymbol{q})\|^{-1}$, where $\mathbf{1}$ is the unit matrix and $\boldsymbol{D}_0$ the deformation tensor, with components $D_{0,ij}(\boldsymbol{q}) = \partial^2 \Phi_0(\boldsymbol{q})/\partial q_i \partial q_j$. The deformation tensor can be locally diagonalized, by going to principal axes X_1, X_2, X_3, with eigenvalues $\alpha_1, \alpha_2, \alpha_3$. We thus have

$$(2.8) \qquad \eta_{\mathrm{ZEL}}(\boldsymbol{q}, \tau) = \frac{\eta_0(\boldsymbol{q})}{(1 + \tau \alpha_1(\boldsymbol{q}))(1 + \tau \alpha_2(\boldsymbol{q}))(1 + \tau \alpha_3(\boldsymbol{q}))}.$$

According to the latter expression a singularity (caustic) in Lagrangian space would form at every point $\boldsymbol{q}$ where at least one eigenvalue, say α_1, is negative. A Lagrangian region where only one eigenvalue is negative will undergo collapse along one axis to a pancake configuration, one for which two eigenvalues are negative and one positive will collapse to a filament, finally one where all α_i are negative will collapse to a knot, or clusterlike configuration. The actual shape of these configurations depends on the precise ratio of the α_i.

Besides being, by construction, consistent with the growing mode of linear perturbations at early times, ZEL provides a good approximation up to the time of first shell crossing. The inconsistency of ZEL can be seen as follows. Inserting the ansatz $\Phi = -\varphi$ into the Poisson equation one gets an expression for the density fluctuation, $\delta_{\mathrm{dyn}} = -a\nabla \cdot \boldsymbol{u}$, which is nothing but the linear-theory relation between peculiar velocity and density fluctuation. This point has been discussed by NUSSER et al. [14], who refer to this determination of the density, $\eta_{\text{dyn}} = 1 + \delta_{\text{dyn}}$, as dynamical density, to distinguish it from the continuity density obtained from eq. (2.8). It is possible to go further this way by replacing this expression into eq. (2.2): one gets the mass density in Lagrangian form $\eta_{\text{dyn}}(\boldsymbol{q}, \tau) = \eta_0(\boldsymbol{q})/(1 - \tau\delta_+(\boldsymbol{q}))$, where $\delta_+(\boldsymbol{q}) = -(\alpha_1(\boldsymbol{q}) + \alpha_2(\boldsymbol{q}) + \alpha_3(\boldsymbol{q}))$, and the Poisson equation was used to relate φ_0 to δ_0 and we defined the (scaled) initial growing mode $\delta_+ \equiv \delta_0/a_0$.

It is then clear that the Zel'dovich ansatz is only exact for one-dimensional perturbations, where the two above expressions for the density coincide; in the general three-dimensional case it fails. As shown in ref.[14], the dynamical density contrast tends to overestimate the correct result, whereas δ_{ZEL} typically underestimates it.

Frozen-flow approximation. The frozen-flow approximation (FFA)[5] can be defined as the exact solution of the linearized Euler equations, where in the r.h.s. the growing mode of the linear gravitational potential is assumed. Such an equation is solved by $\boldsymbol{u}_{\text{FFA}}(\boldsymbol{x}, \tau) = \boldsymbol{u}_0(\boldsymbol{x}) = -\nabla_x \varphi_0(\boldsymbol{x})$, plus a negligible decaying mode. In this approximation the peculiar velocity field $\boldsymbol{u}(\boldsymbol{x}, a)$ is frozen at each point to its initial value, that is

(2.9) $$\frac{\partial \boldsymbol{u}}{\partial \tau} = 0,$$

which is the condition for steady flow[15]. Such an equation can be used, together with the continuity equation to define the FFA. Particle trajectories in the FFA are described by the integral equation $\boldsymbol{x}(\boldsymbol{q}, \tau) = \boldsymbol{q} - \int_0^\tau d\tau' \nabla_x \varphi_0[\boldsymbol{x}(\boldsymbol{q}, \tau')]$: particles during their motion update at each infinitesimal step their velocity to the local value of the linear velocity field, without memory of their previous motion, *i.e.* without inertia. This would be the case of a particle moving under the influence of a force in a medium with very large viscosity: the damping here is determined by the Hubble drag while the force is the gravitational one.

Also in this case a velocity potential can be defined, $\boldsymbol{u}_{\text{FFA}}(\boldsymbol{x}, \tau) = \nabla \Phi_{\text{FFA}}(\boldsymbol{x}, \tau)$, obeying the evolution equation $\partial \Phi/\partial \tau = 0$, which is solved by $\Phi_{\text{FFA}}(\boldsymbol{x}, \tau) = \Phi_0(\boldsymbol{x}) = -\varphi_0(\boldsymbol{x})$.

As we shall see shortly, no caustics are formed at finite time in the FFA, so that all expressions for the density can be used interchangeably. It is nevertheless interesting to write the comoving mass density as given by eq. (2.5), namely $\eta_{\text{FFA}}(\boldsymbol{x}, \tau) = \exp\left[\int_0^\tau d\tau' \delta_+[\boldsymbol{x}(\boldsymbol{q}, \tau')]\right]$, having assumed $\eta_0(\boldsymbol{x}_\star) \approx 1$. The logarithm of the density in $\boldsymbol{x}$ is given by the integral of the linear density field over

the trajectory of the particle which has arrived to x at time τ, starting from the Lagrangian position q. This is different from the lognormal model [16], where it is assumed that η is the exponential of the linear density field in the same Eulerian position. In both cases, however, the nonlinear density field tends to exponentiate the initial density fluctuation in the neighbourhood of high peaks: precisely the same effect predicted by the nonlinear biasing approach to cosmic-structure formation (*e.g.*, ref. [17] for the Gaussian case and ref. [18] for the general non-Gaussian one). While this is only approximately true for initial density peaks, it is a definite prediction for those points $x_\star$ that correspond to local maxima or minima of the initial velocity potential, *i.e.* great attractors or great repellers, respectively [19], in which case the FFA yields $\delta_{\text{FFA}}(x_\star, \tau) =$
$= \exp[\delta_+(x_\star)\tau] - 1$.

The FFA is, by construction, consistent with linear theory, so it follows correctly the evolution at early times (precisely as it happens for ZEL). The assumption of keeping the linear approximation for the velocity potential beyond the linear regime is justified by the fact that this quantity is more sensitive to large-wavelength modes than the density, it is, therefore, less affected by strongly nonlinear evolution. Streamlines are frozen to their initial shape, so multistream regions cannot form, unless they were already present in the initial velocity field. The FFA, therefore, avoids the formation of caustics at finite τ, so one can try to extrapolate the approximation after the time at which the first shell crossing would have appeared according to ZEL. A particle moving according to the FFA has zero component of the velocity in a place where the same component of the initial gravitational force is zero, it will then slow down its motion in that direction approaching such a position: particles in the FFA need infinite time to reach those places where a pancake, a filament or a knot will occur. Moreover, since, unlike ZEL, these particles move along curved paths, once they come close to pancake configurations they curve their trajectories, moving almost parallel to them, trying to reach the position of filaments. Again they cannot cross it, so they modify their motion, asymptotically approaching them, to finally fall, for $\tau \to \infty$, into the knots corresponding to the minima of the initial gravitational potential. Altogether, this type of dynamics implies an artificial thickening of particles around pancakes, filaments and knots, which mimics the real gravitational clustering around this type of structures. The physical thickening of the particle density around pancakes, filaments and knots, caused by the damped oscillations around these structures, is replaced by an approximately exponential slowing-down of particle motions, which, however, overestimates the actual particle deceleration. In the specular process of evacuation of initially underdense regions, the FFA overshoots the actual dynamics. Provided one gives up resolving the trajectories of individual particles, these effects produce a density field which looks roughly similar to the real one; in this sense the method should be considered intrinsically Eulerian.

MATARRESE et al. [5] have considered the evolution of structures on large scales as described by the FFA and compared it with the results of ZEL and of an N-body simulation. The structure of the algorithm used in the FFA simulation is very simple. At each time step the initial gravitational potential is used to calculate, by interpolation, the peculiar velocity in the position of each particle. The updated positions are then obtained by the standard leapfrog scheme to solve the (first-order) equations of motion. Unlike particle mesh (PM) codes, simulations based on the FFA do not need at each time step calculating the density field and double Fourier transforming to solve the Poisson equation. It is then possible to obtain a remarkable speed-up in the computational time. Using the same number of time steps to follow the growth of the gravitational instability, one can estimate that the FFA is more than 30 times faster than PM. Moreover, in the FFA, the number of time steps can be largely reduced. The calculation of the final particle positions in the Zel'dovich approximation avoids the updating of particle velocities: ZEL corresponds to a single FFA time step and the difference in the required CPU time is just the number of time steps.

The algorithm was tested assuming a standard cold-dark-matter (CDM) model, with Hubble constant $H_0 = 50 \text{ km s}^{-1} \text{ Mpc}^{-1}$. If the primordial perturbations are Gaussian, the only information one needs is the power spectrum of the primordial gravitational potential, which in the standard CDM cosmogony takes the form $\mathscr{P}_\varphi(k) = Ak^{-3}[1 + 6.8k + 72.0k^{3/2} + 16.0k^2]^{-2}$ [20], where the normalization factor A is fixed so that the linear mass variance at the present time is unity in a sphere of radius 800 km s^{-1}. MATARRESE et al. [5] have numerically evolved simulations with 64^3 particles on a 64^3 cubic grid with box size $L = 16\,000 \text{ km s}^{-1}$. The particles were uniformly distributed on the grid at the initial redshift $z_0 = 18$ and then moved according to ZEL and FFA up to the «present time», $z = 0$; a PM code was used to check the performance of the two approximations, with the same initial conditions. Particle positions and velocities were extracted at $z = 0.5$ and $z = 0$; the earliest of these times can also be interpreted as giving the mass distribution at the «present time» in a simulation with linear bias parameter $b = 1.5$.

Figure 1 shows projected particle positions in slices of depth 1000 km s^{-1} at the present time ($z = 0$), all drawn from the same initial conditions. Compared to the PM results, the FFA simulation recovers all the main structures in the correct places, even though they look thicker and the voids appear more empty and conspicuous. The FFA leads to an excess of substructure, which is left on the way during the evolution instead of being erased by the hierarchical clustering process as in the true dynamics. The structures obtained by the Zel'dovich approximation, instead, are less prominent and more fuzzy, as the particles have diffused away from the caustic positions after shell crossing.

The numerical results suggest that the FFA is able to reproduce the mass density distribution more accurately than ZEL. One can also check this by considering cell count statistics. One counts the number N of particles inside cubic

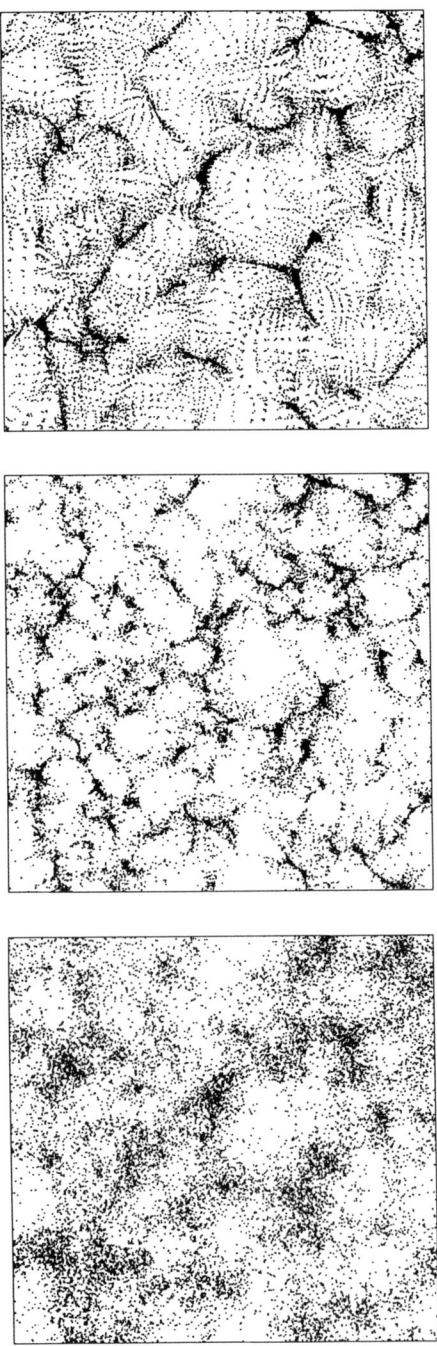

Fig. 1. – Slices with thickness $1000 \, \text{km s}^{-1}$, at the present time ($z = 0$), from simulations evolved from the same initial conditions. The particle distributions are taken from an N-body (PM) simulation (top panel), frozen-flow approximation (central panel) and Zel'dovich approximation (bottom panel).

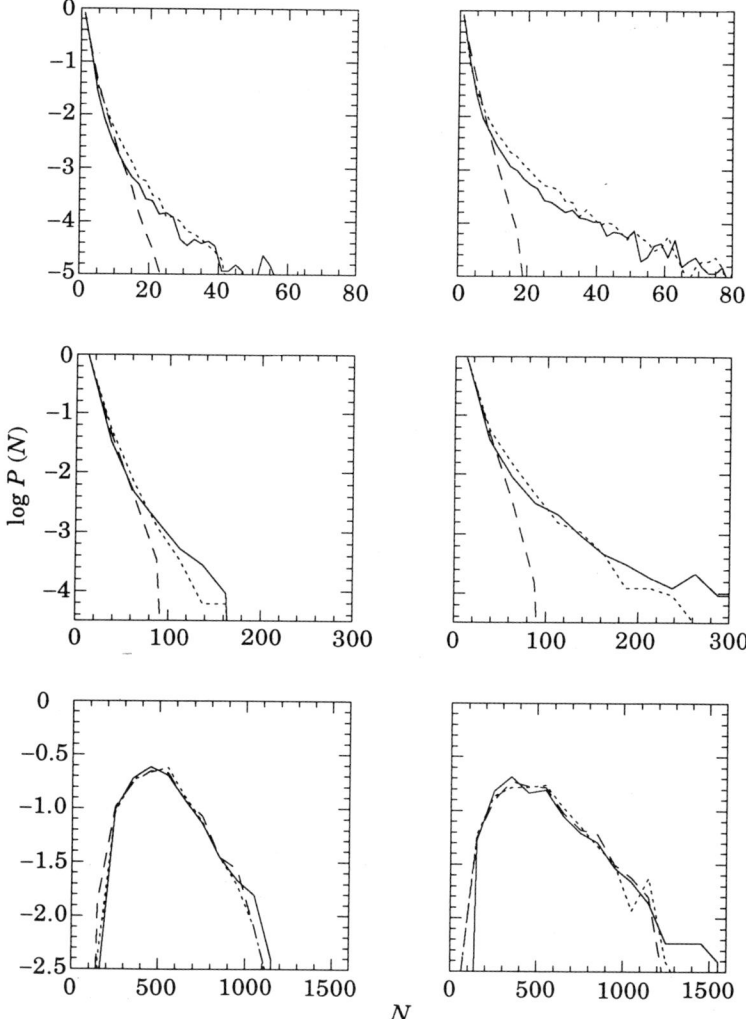

Fig. 2. – The cell count distribution $P(N)$ using different cubic cell sides R_b, at different times: from top to bottom, $R_b = 250, 500$ and 1000 km s^{-1}; left column $z = 0.5$, right column $z = 0$; —— PM, ----- FFA, --- ZEL.

cells with varying sides $R_b = 250, 500, 1000$ km s^{-1}. The cell count distribution $P(N)$ is shown in fig. 2 at $z = 0.5$ and $z = 0$. Even though the initial conditions are Gaussian, the clustering dynamics, assumed to be accurately described by the PM curve, produces a skew-positive distribution. Unlike ZEL, the FFA mimics reasonably well the long high-N tail; the two approximations become equivalent and similar to the PM results when the largest cell size is used. The advantages of the FFA become more evident when small scales and late times are considered. Further confirmation of these results comes from the values of

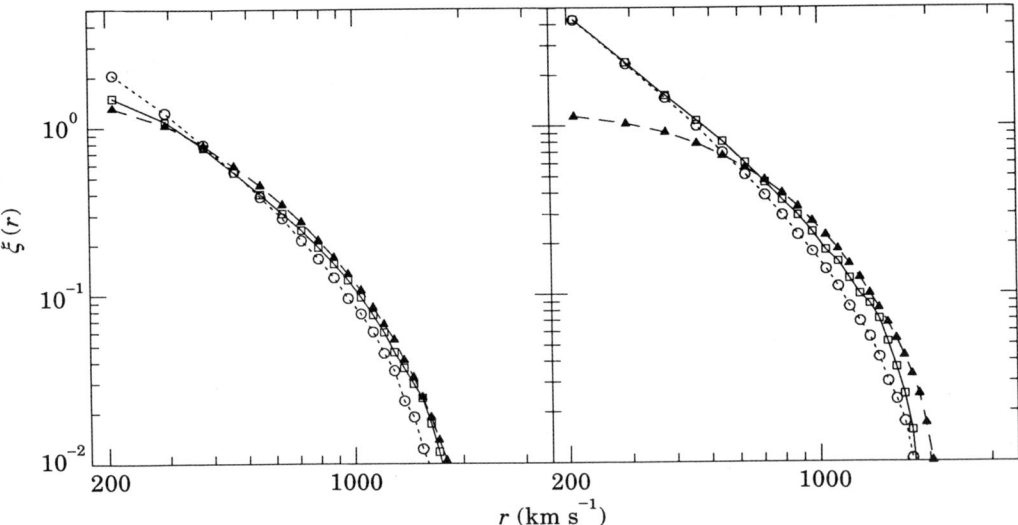

Fig. 3. – The particle two-point correlation function ξ vs. the distance r at redshifts $z = 0.5$ (left) and $z = 0$ (right) for the N-body simulation (PM, solid line with open squares), frozen-flow approximation (FFA, dotted line with open circles), Zel'dovich approximation (ZEL, dashed line with filled triangles).

the r.m.s. cell count fluctuation, $\delta N/N$. The values obtained by the PM simulation at the present time are 3.07, 1.83 and 0.51 with $R_b = 250$, 500 and 1000 km s^{-1}, respectively. The corresponding values from the FFA are 3.15, 1.83 and 0.47; the values from ZEL are 1.51, 1.08 and 0.46. The two-point correlation function $\xi(r)$ at $z = 0.5$ and $z = 0$ is shown in fig. 3. At the earliest time ZEL reproduces the PM results quite well. The FFA slightly overestimates and underestimates the correlation at small and large distances, respectively. In fact, the excess of substructure in the FFA increases the lumpiness, raising the small-scale correlation; on large scales, on the contrary, the structures are broken by the percolation of the voids and a depression of ξ results. The situation changes drastically at $z = 0$: in ZEL the absence of clustering after shell crossing inhibits the growth of the correlation function on small scales; the FFA instead gives a very accurate fit of the PM result even at small distances. These results suggest that the FFA is able to reproduce, at the statistical level, the clustering properties of the Universe even on scales reached by the nonlinear action of gravity.

3. – Adding artificial viscosity.

The FFA assumes that the velocity potential is linearly related at any time to the local value of the initial gravitational potential. One is, therefore, disregarding the nonlinear effects caused by the back-reaction of the evolving mass density on the peculiar velocity field itself (via the nonlinear evolution of the gravitational potential). This would result in a number of physical processes such as merging of pancakes, fragmentation and disruption of low-density bridges, which the FFA cannot describe. The slices of fig. 1 show that the results obtained by the FFA differ from the true nonlinear dynamics, displayed by the N-body simulation, mainly because of the lack of merging of protostructures. It would be interesting in this respect to test the approximation starting from non-Gaussian initial conditions, where the nonlinear process of merging can play a relevant role in determining the final texture of the Universe [21]. One can try to improve the FFA by allowing for some «artificial» evolution of the velocity potential in the vein of the adhesion model (ADM) [4], which aims to overcome the difficulties met by ZEL after caustic formation.

In the ADM one modifies the Zel'dovich approach by adding an artificial viscosity term to eq. (2.6), which is thus replaced by

$$(3.1) \qquad \frac{d\boldsymbol{u}}{da} = \nu \nabla^2 \boldsymbol{u}.$$

The viscosity is introduced to mimic the actual sticking of particles around pancakes, caused by the action of gravity even in a collisionless medium. The parameter ν plays the role of a coefficient of kinematical viscosity, which controls the thickness of pancakes.

The previous equation is the vector generalization of the well-known nonlinear diffusion or Burgers equation of strong turbulence [22]. One can still define a velocity potential through $\boldsymbol{u} = \nabla \Phi$, which can be determined through the Hopf-Cole substitution $\Phi = -2\nu \ln U$; the scalar field U satisfies the linear diffusion or Fokker-Planck equation, $\partial U / \partial \tau = \nu \nabla^2 U$, with the initial condition $U_0(\boldsymbol{x}) = \exp[-\Phi_0(\boldsymbol{x})/2\nu]$. The resulting velocity potential reads

$$(3.2) \qquad \Phi(\boldsymbol{x}, \tau) = -2\nu \ln \left[\frac{1}{(4\pi\nu\tau)^{3/2}} \int d^3q \exp\left[-\frac{1}{2\nu} S(\boldsymbol{x}, \boldsymbol{q}, \tau)\right]\right],$$

where one defines the action $S(\boldsymbol{x}, \boldsymbol{q}, \tau) \equiv \Phi_0(\boldsymbol{q}) + (\boldsymbol{x} - \boldsymbol{q})^2 / 2\tau$, satisfying the Hamilton-Jacobi equation (2.7). The corresponding velocity field is easily obtained by differentiation; the Eulerian positions of the particles are found by direct integration of the integral equation [23] $\boldsymbol{x}(\boldsymbol{q}, \tau) = \boldsymbol{q} + \int_0^\tau d\tau' \boldsymbol{u}[\boldsymbol{x}(\boldsymbol{q}, \tau'), \tau']$, while the density field can be obtained from eq. (2.5).

Adhesion model. The Burgers equation is usually considered in the limit of small (but nonvanishing) ν, which corresponds to the limit of large Reynolds numbers, $\mathcal{R}_0 = u_0 l_0 / \nu$, u_0 and l_0 being the characteristic amplitude and scale of the initial velocity field [24]. The product $u_0 l_0$ can be estimated either from the r.m.s. initial velocity potential smoothed on some scale R, $\langle \Phi_0^2(R) \rangle^{1/2}$, if this is

convergent, or from the square root of the structural function of Φ_0 [4] $D(r) =$
$$= \langle |\Phi_0(\boldsymbol{x}) - \Phi_0(\boldsymbol{x}+\boldsymbol{r})|^2 \rangle = (1/\pi^2) \int_0^\infty \mathrm{d}k\, k^2 \mathscr{L}_\varphi(k) W^2(kR)[1 - j_0(kr)], \quad \text{evaluated}$$
at a suitable lag, e.g. $r \approx R$. Here $\mathscr{L}_\varphi$ is the power spectrum of the initial gravitational potential and $W(kR)$ a suitable low-pass filter.

In the small-ν case the solution takes a simplified form which can be obtained from eq. (3.2) through a saddle point approximation,

$$(3.3) \qquad \Phi_{\mathrm{ADM}}(\boldsymbol{x}, \tau) \approx -2\nu \ln\left[\sum_\alpha \mathscr{T}(\boldsymbol{q}_\alpha)^{-1/2} \exp\left[-\frac{1}{2\nu} S(\boldsymbol{x}, \boldsymbol{q}_\alpha, \tau)\right]\right],$$

where $\mathscr{T}(\boldsymbol{q}) = \|1 + \tau \boldsymbol{D}_0(\boldsymbol{q})\|$, $\boldsymbol{D}_0$ is the deformation tensor and $\boldsymbol{q}_\alpha$ are the Lagrangian points which minimize the action S at given $\boldsymbol{x}$ and τ. The Zel'dovich approximation is recovered in the limit $\nu \to 0$. This model has been applied to perform numerical simulations of the large-scale structure of the Universe [23] or to obtain some physical insight into the structure formation process in simplified cases [25]. The model allows one to obtain the skeleton of the large-scale matter distribution by a geometrical technique based on the insertion of osculating paraboloids into the hypersurface $\varphi_0(\boldsymbol{q})$ [26]. The asymptotic properties of the solutions of Burgers equation [27] have been used to compute the velocity distribution and the mass function of knots [25, 28].

Viscid-flow approximation. Along the line of the ADM, one can modify the FFA by adding an artificial viscosity term to the r.h.s. of eq. (2.10), leading to the viscid-flow approximation (VFA):

$$(3.4) \qquad \frac{\partial \boldsymbol{u}}{\partial \tau} = \nu \nabla^2 \boldsymbol{u}$$

and, for the velocity potential, $\partial \Phi/\partial \tau = \nu \nabla^2 \Phi$. The problem is reduced to the solution of the linear diffusion equation, which yields

$$(3.5) \qquad \Phi_{\mathrm{VFA}}(\boldsymbol{x}, \tau) = \frac{1}{(4\pi\nu\tau)^{3/2}} \int \mathrm{d}^3 q\, \Phi_0(\boldsymbol{q}) \exp\left[-\frac{(\boldsymbol{x}-\boldsymbol{q})^2}{4\nu\tau}\right]$$

and similarly for the velocity. It is immediately clear that these results can be obtained from Burgers' theory in the limit of small Reynolds numbers, $\Phi_0 \ll 2\nu$ (or in the $\nu \to \infty$ limit at fixed $\nu\tau$). This allows one to connect the ADM to the VFA and, though indirectly, ZEL to the FFA: the Zel'dovich algorithm corresponds to the $\mathscr{R}_0 \to \infty$ limit of the adhesion model, whilst $\mathscr{R}_0 \ll 1$ yields the viscid-flow approximation, from which the frozen-flow one is finally obtained for $\nu \to 0$. Finally, all these approximations reduce to the linear one in the limit $\tau \to 0$.

The physical meaning of eq. (3.5) is evident: the time evolution acts as a smoothing of the velocity field which erases small-scale motions up to the filtering scale $R_G(\tau) = \sqrt{2\nu\tau}$, therefore freezing the dynamics on larger and larger scales. These points become even more clear if the velocity potential is written in Fourier space, $\hat{\Phi}_{\text{VFA}}(\boldsymbol{k}, \tau) = \Phi_0(\boldsymbol{k}) \exp[-k^2 \nu \tau]$. If the initial velocity potential does not contain contributions on scales smaller than a certain wave number cut-off $k_{\min}$, the dynamics is completely frozen and the Universe enters a stationary regime after a time given by $\tau \sim 1/\nu k_{\min}$.

In general, stable pancakes, filaments and knots form in the same positions dictated by the FFA: their density and thickness are now monitored by the value of the viscosity coefficient ν. Also here as in the ADM, however, the physically relevant case is obtained for small ν, because only in this case the thickness of the pancakes becomes negligible compared to their relative distance. Moreover, this is the condition to have an appreciable overdensity within these configurations. A high value of ν would imply a premature freezing of the dynamics before the actual nonlinear collapse could actually start. Preliminary numerical results seem to suggest that the VFA provides an improvement over the FFA. The possible application of this approximation to current cosmological scenarios of structure formation clearly needs further investigation.

4. – Relativistic dynamics of a self-gravitating collisionless fluid.

A GR method to follow the development of structures in a collisionless medium has been recently proposed in ref.[11]; we shall here briefly describe the basic properties of the method. Relativistic effects could be relevant either in connection with structures extending over very large scales or with high bulk motions. Although, at present, the biggest coherent structures that have been observed have size of about one order of magnitude below the horizon scale and typical peculiar velocities are largely nonrelativistic, it can be useful to have a general formalism able to correctly describe the formation of structures on even larger scales.

Let us introduce the equations which govern the dynamics of a collisionless perfect fluid in GR. A complete treatment of the problem and a full derivation of the equations presented here can be found in the review by ELLIS[29]. We use the signature $(-, +, +, +)$; Latin indices refer to space-time coordinates, $(0, 1, 2, 3)$, Greek indices to spatial ones, $(1, 2, 3)$. The relativistic dynamics of a collisionless (*i.e.* with vanishing pressure) self-gravitating perfect fluid is determined by Einstein's equations and by the continuity equations for the matter stress-energy tensor $T_{ab} = \rho u_a u_b$, where ρ is the energy density and u^a the four-velocity of the fluid ($u^a u_a = -1$). It is also useful to define the spatial projection tensor $h^{ab} \equiv g^{ab} + u^a u^b$ ($h_{ab} u^b = 0$). Differentiation of the velocity field yields the tensor $v_{ab} \equiv h_a{}^c h_b{}^d u_{c;d}$, for which $v_{ab} u^b = 0$, and the acceleration vector $\dot{u}^a \equiv u^a{}_{;b} u^b$, which is also spacelike, $\dot{u}^a u_a = 0$. An overdot denotes convective differentiation with respect to the proper time t of fluid elements, namely $\dot{A}_{a_1 a_2 \ldots a_n} = A_{a_1 a_2 \ldots a_n; b} u^b$. The antisymmetric part of the tensor v_{ab} is the vorticity tensor $\omega_{ab} \equiv v_{[ab]}$ (the symbol $_{[\ldots]}$ stands for antisymmetrization and $_{(\ldots)}$ for symmetrization), describing rigid rotations of fluid elements with respect to a locally inertial rest frame. Assuming irrotational motions, $\omega_{ab} = 0$, one still has two relevant quantities: the volume expansion scalar $\theta \equiv v^a{}_a$, and the symmetric and traceless shear tensor, $\sigma_{ab} \equiv v_{(ab)} - (1/3)\theta h_{ab}$. From the volume expansion scalar, giving the local rate of isotropic expansion, one can define a length scale l through $\theta = 3\dot{l}/l$, which reduces to the scale factor $a(t)$ in the homogeneous and isotropic FRW models; in that particular case $\theta = 3H$, where $H(t)$ is Hubble's constant. The shear tensor, on the other hand, describes a pure straining in which a spherical fluid volume is distorted into an ellipsoid with axis lengths changing at rates determined by the three $\sigma^a{}_b$ eigenvalues, σ_1, σ_2 and $\sigma_3 = -(\sigma_1 + \sigma_2)$. The vanishing-trace condition implies that this deformation leaves the fluid volume invariant, while, in the absence of vorticity, the principal axes of the shear keep their direction fixed during the evolution, in a locally inertial rest frame.

The fluid acceleration is only caused by pressure gradients, so in our case $\dot{u}^a = 0$: in the absence of pressure each fluid element moves along a geodesic. The continuity equation reads $\dot{\rho} = -\rho\theta$. The expansion scalar satisfies Raychaudhuri's equation

$$(4.1) \qquad \dot{\theta} = -\frac{1}{3}\theta^2 - 2\sigma^2 - 4\pi G\rho,$$

where G is Newton's constant and $\sigma^2 \equiv (1/2)\sigma^{ab}\sigma_{ab}$. In the FRW case, $\sigma_{ab} = 0$ and the latter equation reduces to the familiar Friedmann one, $3(\dot{H} + H^2) = -4\pi G\rho$. The shear is determined by the evolution equation

$$(4.2) \qquad \dot{\sigma}_{ab} = -\sigma_{ac}\sigma^c{}_b + \frac{2}{3}h_{ab}\sigma^2 - \frac{2}{3}\theta\sigma_{ab} - E_{ab},$$

where $E_{ac} \equiv C_{abcd}u^b u^d$ is the electric part of the Weyl tensor C_{abcd} (the latter being the part of the Riemann curvature not determined by local sources); $E^a{}_b$ is also called the tidal force field, for it contains that part of the gravitational field which describes tidal interactions; it is symmetric, traceless and flow orthogonal, $E_{ab}u^b = 0$. Tidal forces act on the fluid flow by inducing shear distortions. The tensor $E^a{}_b$ can be diagonalized by going to its principal axes (which do not generally coincide with those of the shear tensor), with eigenvalues E_1, E_2 and $E_3 = -(E_1 + E_2)$.

From the Weyl tensor one defines its magnetic part $H_{ac} = \frac{1}{2}\eta_{ab}{}^{gh}C_{ghcd}u^b u^d$ (η_{abcd} is the completely antisymmetric four-index tensor), which is also symmetric, traceless and flow orthogonal and contains, among other things, the part of the gravitational field which describes gravitational waves. Actually, gravitational waves are represented by the transverse traceless parts of E_{ab} and H_{ab}, satisfying $h^{bc}E_{ab;c} = 0$ and $h^{bc}H_{ab;c} = 0$ (*e.g.*, ref. [30]). While the tidal force field has a straightforward Newtonian analogue, which can be written in terms of derivatives of the gravitational potential, H_{ab} has no Newtonian counterpart. The important point is that, while in Newton's theory the gravitational potential is usually determined through a constraint equation, namely Poisson's one, in GR both E_{ab} and H_{ab} can be calculated through evolution equations.

A useful approximation [11] is to neglect the influence of H_{ab} on the evolution of E_{ab}: for initially scalar perturbations of an irrotational perfect fluid, this amounts to neglecting the interaction of gravitational waves (tensor modes) with the system. In such a case the tidal force field evolves according to

$$\dot{E}_{ab} = -h_{ab}\sigma^{cd}E_{cd} - \Theta E_{ab} + 3E_{c(a}\sigma_{b)}{}^{c} - 4\pi G\rho\sigma_{ab}. \tag{4.3}$$

Besides these evolution equations, there are many constraint ones that our variables have to satisfy; these will be automatically satisfied at any time during the evolution if one consistently sets up the initial conditions, *e.g.* by building up all initial values within linear theory.

The above approximation, together with the absence of pressure in the fluid, implies that no explicit spatial gradients occur in the evolution equations. There are, in general, spatial derivatives hidden in the convective time differentiation of tensors, because of the presence of the affine connection in covariant derivatives. In fact, while for a scalar quantity convective differentiation and simple differentiation can be made to coincide, provided one uses a reference frame comoving with the fluid, this no longer holds for a vector like u^a or for two-index tensors like σ_{ab} and E_{ab}.

No spatial gradients appear in all the above equations provided one refers to the rest frame of observers comoving with the fluid; the absence of nongravitational interactions implies that these observers freely fall in the gravitational field created by the fluid, while the equivalence principle ensures that they do not feel gravity locally. If $H_{ab} = 0$, GR proves as an economic way to account for the mutual gravitational interactions among different fluid elements without the need of simultaneously evolving all of them. However, as soon as the first caustics form, multistream regions appear and nonlocal effects start to play a relevant role in the subsequent evolution of these regions. There is a strong similarity between this method and ZEL: in both cases the evolution of each fluid element is completely determined by the local initial conditions and can be independently followed up to the time when it enters a multistream region. However, the present method is exact (except for having disregarded the role of H_{ab}) in the most general three-dimensional case, while ZEL is only exact for one-dimensional perturbations. Because of the choice of fluid variables and reference frame the method is a Lagrangian one: at the end of the calculations physical observables are known in the rest frame of each fluid element. The next step is then to reconstruct the Eulerian density and peculiar velocity fields on comoving spacelike hypersurfaces. This is indeed possible by integrating additional first-order equations to follow the relative displacement of neighbouring elements, which are represented by infinitesimal spacelike vectors ξ^a, evolving according to $\dot\xi^a = (1/3)\theta\xi^a + \sigma^a{}_b \xi^b$ [29].

The equations that govern the dynamics of our system form a set of twelve coupled partial differential equations involving twelve independent variables: σ_{ab} (five), E_{ab} (five), ρ and θ. This can be, however, reduced to a set of six equations for six unknowns by going to the simultaneous local principal axes of the shear and tidal force field. Three supplementary equations can be solved next to compute the components of each vector ξ^a and reconstruct the final grid. For a cubic grid with N_g^3 nodes, one would need to solve $6N_g^3$ first-order equations to obtain the dynamical variables plus $3(N_g - 1)(N_g^2 + N_g + 1)$ extra ones (still of first order) for the relative position vector components.

A numerical procedure based on this GR method was developed in ref. [11]; its accuracy was tested by integrating the nonlinear evolution of suitable spherical perturbations in an otherwise spatially flat FRW universe and comparing the results with the exact TB solution for the same initial profiles. Starting from an initial redshift $z_{in} = 50$ the system was evolved up to the present time for two profiles corresponding to a) a case of mildly nonlinear evolution (final central density contrast $\delta_c \simeq 3$) and b) a case of fully nonlinear evolution (final central density contrast $\delta_c \simeq 650$). The numerical integration was carried out over $N_g = 1000$ suitably spaced nodes. The results for the density contrast and the radial peculiar velocity from the centre are shown in fig. 4.

Let us finally mention that an exact solution for planar symmetry was also found in ref. [11], which turns out to be locally identical to the corresponding Zel'dovich solution.

5. – Conclusions.

We have considered various approximations designed to follow the nonlinear dynamics of collisionless matter. The FFA method [5] is based on solving exactly the continuity equation while extrapolating the linear approximation for the peculiar velocity field beyond its actual range of validity: this allows one to prevent the occurrence of orbit crossing, which represents the main drawback of ZEL. Thanks to this property, one obtains an approximate description of the Eulerian density field at later times and/or on smaller scales compared to ZEL. The application of the FFA to follow the evolution of structures within the standard CDM model gives a fairly accurate representation of the density pattern from a resolution scale of ~ 500 km s^{-1}, while the two-point correlation function fits quite well the true nonlinear result on even smaller scales. An advantage of the FFA over more advanced techniques, such as the ADM or a full N-body code, is a strong reduction in the computational time, without a relevant loss of

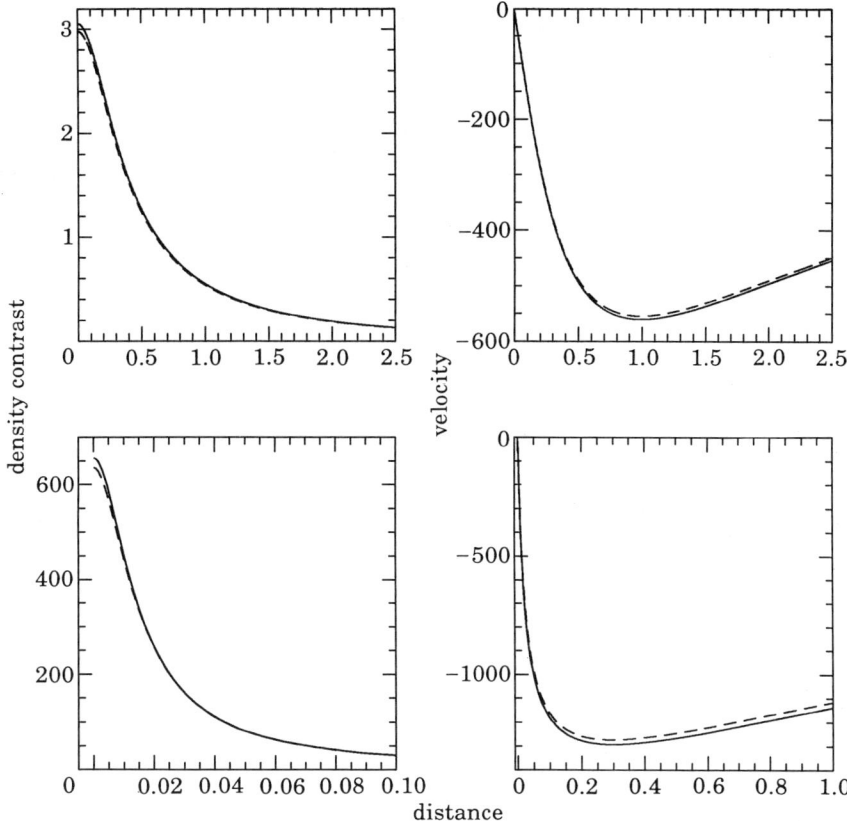

Fig. 4. – The results of the numerical integration (dashed lines) for the density contrast and the radial peculiar velocity (in suitable units) relative to the centre are compared with those given by the exact TB solution (continuous lines). Both quantities are plotted as functions of the proper distance from the origin given in suitable units. Cases a) and b) refer to different values for the central density contrast.

accuracy. The FFA could be applied to obtain fast simulations of the evolution of structures in the Universe on a wider range of scales and with a larger number of particles. Moreover, the FFA could be a reliable tool in reconstruction procedures of the initial density field from present-day data.

We then described a GR approach[11] to the nonlinear evolution of a self-gravitating collisionless fluid up to the epoch of caustic formation. The method relies on the approximation of neglecting the interaction of gravitational waves with the rest of the system and assuming irrotational motions. Under these assumptions a simple Lagrangian picture is obtained which, after self-consistent initial conditions have been assigned to a grid, allows one to follow the evolution of each fluid element separately in its own locally inertial rest frame. This method has two evident advantages: first, being Lagrangian, it automatically guarantees enhanced resolution in regions of higher density, *i.e.* where it is more needed; second, it allows one to follow each fluid element as being completely independent of the others, which obviously reduces the amount of computer storage needed to evolve the system. Only at the initial time the conditions ought to be specified simultaneously on the whole grid, which requires an amount of computer memory comparable to the one required to construct the initial conditions in an N-body code. For instance, if one considers an initial random (*e.g.*, Gaussian) realization of the peculiar gravitational potential in momentum space, $\hat{\varphi}(\boldsymbol{k})$ (*e.g.*, for a given choice of power spectrum), initial conditions on the grid should be obtained by inverse Fourier transforming $k_\alpha k_\beta \hat{\varphi}(\boldsymbol{k})$. An advantage of this method is that it allows one to concentrate on the evolution of a given region, where structures are forming, while its gravitational interaction with the rest of the Universe has been already included by the set-up of the initial conditions.

It would be important to understand how to follow the evolution of the system in the multistream regions which occur after caustic formation. Several methods have been developed to follow the evolution in the early nonlinear phases or to reconstruct the initial conditions of the clustering process, most of them trying to circumvent caustic formation: this can be obtained either by suitable smoothing procedures [31], or by artificially sticking the particles as their orbits first cross [4] or else by asympotically slowing down particle motions, so that orbit mixing never occurs [5]. Finally, one could consider the possibility of replacing the *fluid* approximation with the more realistic picture of a system of noninteracting particles.

* * *

This work has been partially supported by Italian MURST. DS thanks the Conselleria de Cultura Educacio i Ciencia de la Generalitat Valenciana and the Spanish DGICYT project PB90-0416 for financial support.

REFERENCES

[1] P. J. E. PEEBLES: *The Large Scale Structure of the Universe* (Princeton University Press, Princeton, N.J., 1980).
[2] R. C. TOLMAN: *Proc. Natl. Acad. Sci.*, **20**, 169 (1934); H. BONDI: *Mon. Not. R. Astron. Soc.*, **107**, 410 (1947).
[3] YA. B. ZEL'DOVICH: *Astrophysica*, **6**, 160 (1970); *Astron. Astrophys.*, **5**, 84 (1970).
[4] S. N. GURBATOV, A. I. SAICHEV and S. F. SHANDARIN: *Mon. Not. R. Astron. Soc.*, **236**, 385 (1989); S. F. SHANDARIN and YA. B. ZEL'DOVICH, *Rev. Mod. Phys.*, **61**, 185 (1989).
[5] S. MATARRESE, F. LUCCHIN, L. MOSCARDINI and D. SAEZ: *Mon. Not. R. Astron. Soc.*, **259**, 437 (1992).

[6] N. MAKINO, M. SASAKI and Y. SUTO: *Phys. Rev. D*, **46**, 585 (1992).
[7] M. GIAVALISCO, B. MANCINELLI, P. MANCINELLI and A. YAHIL: *Astrophys. J.*, **411**, 9 (1993).
[8] A. YAHIL: preprint (1991), unpublished.
[9] T. BUCHERT: *Astron. Astrophys.*, **223**, 9 (1989); *Mon. Not. R. Astron. Soc.*, **254**, 729 (1992); F. R. BOUCHET, R. JUSZKIEWICZ, S. COLOMBI and R. PELLAT: *Astrophys. J. Lett.*, **394**, L5 (1992).
[10] R. W. HOCKNEY and J. W. EASTWOOD: *Computer Simulations using Particles* (McGraw-Hill, New York, N.Y., 1981).
[11] S. MATARRESE, O. PANTANO and D. SAEZ: *Phys. Rev. D*, **47**, 1311 (1993).
[12] L. KOFMAN: in *Proceedings of the IUPAP Conference on Primordial Nucleosynthesis and Evolution of the Early Universe*, edited by K. SATO (Kluwer Academic, Dordrecht, 1991).
[13] A. NUSSER and A. DEKEL: *Astrophys. J.*, **391**, 443 (1992).
[14] A. NUSSER, A. DEKEL, E. BERTSCHINGER and G. R. BLUMENTHAL: *Astrophys. J.*, **379**, 6 (1991).
[15] G. K. BATCHELOR: *An Introduction to Fluid Dynamics* (Cambridge University Press, Cambridge, 1967).
[16] P. COLES and B. J. T. JONES: *Mon. Not. R. Astron. Soc.*, **248**, 1 (1990).
[17] H. D. POLITZER and M. B. WISE: *Astrophys. J. Lett.*, **285**, L1 (1984); N. KAISER and M. DAVIS: *Astrophys. J.*, **297**, 365 (1985).
[18] B. GRINSTEIN and M. B. WISE: *Astrophys. J.*, **310**, 19 (1986); S. MATARRESE, F. LUCCHIN and S. A. BONOMETTO: *Astrophys. J. Lett.*, **310**, L21 (1986).
[19] A. A. STAROBINSKY: unpublished (1988); A. F. HEAVENS: *Mon. Not. R. Astron. Soc.*, **251**, 267 (1991).
[20] M. DAVIS, G. G. EFSTATHIOU, C. S. FRENK and S. D. M. WHITE: *Astrophys. J.*, **292**, 371 (1985).
[21] L. MOSCARDINI, S. MATARRESE, F. LUCCHIN and A. MESSINA: *Mon. Not. R. Astron. Soc.*, **248**, 424 (1991); A. MESSINA, F. LUCCHIN, S. MATARRESE and L. MOSCARDINI: *Astroparticle Phys.*, **1**, 99 (1992); D. H. WEINBERG and S. COLE: *Mon. Not. R. Astron. Soc.*, **259**, 652 (1992); F. LUCCHIN, L. MOSCARDINI, S. MATARRESE and A. MESSINA: this volume, p. 141.
[22] J. M. BURGERS: *The Nonlinear Diffusion Equation* (Reidel, Dordrecht, 1974).
[23] A. NUSSER and A. DEKEL: *Astrophys. J.*, **362**, 14 (1990); D. H. WEINBERG and J. E. GUNN: *Mon. Not. R. Astron. Soc.*, **247**, 260 (1990).
[24] S. N. GURBATOV, A. I. SAICHEV and I. G. YAKUSHKIN: *Sov. Phys. Usp.*, **26**, 857 (1983).
[25] B. G. WILLIAMS, A. F. HEAVENS, J. A. PEAKOCK and S. F. SHANDARIN: *Mon. Not. R. Astron. Soc.*, **250**, 458 (1991).
[26] L. KOFMAN, D. POGOSYAN and S. SHANDARIN: *Mon. Not. R. Astron. Soc.*, **242**, 200 (1990); L. KOFMAN, D. POGOSYAN, A. MELOTT and S. SHANDARIN: *Astrophys. J.*, **393**, 437 (1992).
[27] S. KIDA: *J. Fluid Mech.*, **93**, 337 (1979); S. N. GURBATOV and A. I. SAICHEV: *Sov. Phys. JETP*, **53**, 347 (1981).
[28] S. N. GURBATOV, A. I. SAICHEV and S. F. SHANDARIN: ref. [4]; A. G. DOROSHKEVICH and T. V. KOTOK: *Mon. Not. R. Astron. Soc.*, **246**, 10 (1990).
[29] G. F. R. ELLIS: in *General Relativity and Cosmology*, edited by R. K. SACHS (Academic Press, New York, N.Y., 1971), p. 104.
[30] M. BRUNI, P. K. S. DUNSBY and G. F. R. ELLIS: *Astrophys. J.*, **395**, 34 (1992).
[31] A. DEKEL, E. BERTSCHINGER and S. M. FABER: *Astrophys. J.*, **364**, 349 (1990).

The Cosmological History of the Baryons.

A. BLANCHARD

Observatoire de Strasbourg, U.L.P.
11, rue de l'Université, 67000 Strasbourg, France

1. – Introduction.

The information that astronomers can receive from the Universe primarily comes from light which is emitted (or scattered) by ordinary matter, *i.e.* baryonic matter. The purpose of this lecture is to discuss what we may reasonably know about the baryonic content of the Universe, and then to introduce some more speculative aspects of the question.

1`1. *The quantity of seen baryons.* – Baryons are detected at different wavelengths, but most of the detected baryons appear in stars located in galaxies. The luminosity function of galaxies can be fitted by a Schechter function:

$$\phi(L) = \Phi_*(L/L_*)^\alpha \exp[-L/L_*].$$

The APM-Stromlo redshift survey [1] provides a good determination of the luminosity function:

$$\Phi_* = 1.4 \cdot 10^{-2} \, h^3 \, \text{Mpc},$$

$$L_* = 1.1 \cdot 10^{10} \, h^{-2} L_\odot,$$

$$M_* = -19.5 + 5 \log h,$$

$$\alpha = -0.97,$$

the luminosity is in the b_j band, close to the blue band. The density of light is then given by

$$\rho_{\text{light}} = \int \phi(L) L \, \mathrm{d}L = \Phi_* L_* \Gamma(2+\alpha) \approx 1.5 \cdot 10^8 \, h^{-2} L_\odot h^3 \, \text{Mpc}^{-3},$$

this last number may be uncertain by (30–50)%. The M/L corresponding to an Einstein-de Sitter universe then is

$$(M/L)_c \approx 1900\, h$$

(notice that this number depends on the band!).

The observed baryonic material inside galaxies is essentially in the form of stars and gas. Within disk galaxies, the maximum-disk hypothesis provides a robust upper limit on the mass-to-light ratio of disks of galaxies:

$$M/L = 6 \pm 4h(M/L)_\odot\,,$$

while in the dynamical mass-to-light ratio in the central part of ellipticals one has [2]

$$M/L = 15 \pm 5h(M/L)_\odot\,.$$

Ellipticals and spirals do have a slightly different luminosity function, but this can be neglected here. We can then infer a mean value for the mass-to-light ratio of the baryonic content of galaxies, and deduce the contribution of observed baryons in galaxies to the density of the Universe in terms of the critical density:

$$\Omega_b \approx 0.005$$

with an uncertainty of at least a factor of two. Actually the above estimate might include some nonbaryonic component as some nonbaryonic dark material can exist in disks as well as in the central regions of ellipticals. However, the above M/L is basically of the order of the value expected for the associated stellar population, and could probably not differ by more than a factor of two.

Other detected forms of baryons, cold gas, hot gas and intergalactic stars contribute a completely negligible amount. This does not mean, however, that a large contribution from such baryons does not exist, but it is not motivated by any arguments (with the exception of a hot-gas component that will be discussed in the following).

1˙2. *Estimates of the mean density of the Universe.* – Dynamical measurements of Ω_0 are extensively used in cosmology. It is important to keep in mind that they hold on an assumption which might not be correct. Such measurements are performed by estimating the mass of an object like a galaxy or a cluster, leading to the dynamical mass-to-light ratio, M/L. Then, the density of the Universe is deduced by assuming this ratio to be universal:

$$\rho_{\text{mass}} = \rho_{\text{light}}\, M/L\,. \tag{1}$$

The existence of dark matter is unambiguously demonstrated by the rotation curves of spiral galaxies, the largest M/L usually measured is of the order of $30h$ leading to $\Omega_0 \approx 0.015$.

Dynamical measurements of groups and clusters reveal a larger fraction of dark matter. Typically, the M/L ratios are in the range $200h$–$600h$, however the use of eq. (1) is doubtful: the fraction of ellipticals in clusters is not the same as in the field and, therefore, the stellar content does not have the same mean M/L than for the field galaxies. One might think that the ratio of dynamical mass to stellar mass is universal, providing a better extrapolation than eq. (1). This leads to $\Omega_0 \approx 0.07$–0.2. A short but excellent review of this subject may be found in [3].

1'3. *A sensible extrapolated value for the observed amount of baryons.* – The hot gas present in clusters deserves special attention. The intrinsic total contribution of clusters to the density of the Universe is rather small, but we might extrapolate their baryonic content, assuming the mass of gas to mass of stars ratio is universal. The Coma cluster is probably the best-studied cluster for this purpose, but the situation might change rapidly with the results expected from ROSAT. The gas content of the central part of Coma represents (20–30)% of the total mass [4]. This means that, inside Coma, nearly 80% of the visible baryons are in the gas component. It is rather likely that this ratio holds in the field; the existence of an intergalactic medium is, therefore, very natural and its density could be evaluated:

$$\Omega_{gas} \approx 0.03 \, .$$

Although this is not a proof, it is actually a strong argument in favour of the existence of an IGM. The possible caveat in this argument comes from the fact that the ratio of gas to stars seems to depend on the temperature of clusters [5]. If this effect is real, it might reflect that galaxy formation in clusters was *less* efficient for increasing richness [6]. This is a completely heretic idea, as this would mean that the clusters are antibiased, but it may be correct after all (this would solve some of the discrepancies mentioned later and it is an interesting exercise to examine whether such a scenario is plausible...).

1'4. *Early history.* – The theory of the (not so) early universe tells us about the first minutes of the history of the baryonic content of the Universe. After the quark-hadron phase transition, electrons and protons should exist in nearly exactly the same amount to ensure electric neutrality:

$$n_e = n_p \, .$$

Later on, matter and radiation are in a thermal equilibrium and consequently do have the same temperature:

$$T_m = T_\gamma \, .$$

The most important event of the gas history after that period is probably the primordial nucleosynthesis. The hot-big-bang picture has been originally proposed in order to explain the element abundances [7]. The primeval nucleosynthesis remains one of the greatest successes of this model. The discovery of the cosmological radiation and the remarkable blackbody shape of its spectrum [8] certainly guarantee that no model will be able to compete with this picture for a rather long time.... The composition of the cosmological gas is determined by the primeval nucleosynthesis, and is probably not altered until the stars will produce a tiny amount of heavier elements. Of course, this does not necessarily mean that these heavier elements are not important: for instance, they allow for our existence, and, more directly related to this lecture, they might change the thermal history of the gas, and provide fruitful constraints on it. As is well known, around 25% of the baryons turn into helium, while other elements are produced in much smaller quantity. Details on nuclear processes during primeval nucleosynthesis may be found in recent papers on the subject [9]. Among the various elements which are produced at nucleosynthesis, the most important ones are helium-4, helium-3, lithium-7 and deuterium as these elements are actually observed and their abundance can be compared to the predictions of nucleosynthesis. The main parameter that governs these abundances is the ratio η of the number of photons per baryon. The observed abundances of these elements allow one to determine a baryon-photon ratio and, therefore, the density of the baryonic content of the Universe. This quantity is now believed to be known with a 25% accuracy [9]:

(2) $$\rho_b \approx 3 \cdot 10^{-31} \, \text{g cm}^{-3}$$

corresponding to a number density

$$n = \frac{\rho_b}{m_p} \approx 1.8 \cdot 10^{-7} \, \text{cm}^{-3}.$$

It follows that the main uncertainty on the cosmological density parameter of the baryonic gas lies in the uncertainty of the Hubble constant H_0 ($= 100h$ km s^{-1}/Mpc):

(3) $$\Omega_b \approx 0.015 \, h^{-2}.$$

Here, we are faced with an important problem of modern cosmology: as we have seen in the first subsection, the *observed* amount of baryons (in stars and cold gas in galaxies) is of the order of 0.005, appreciably smaller than predicted by nucleosynthesis. This means that, if primordial nucleosynthesis is right, *most of the baryons are in a dark form*. This reinforces the case for the existence of an IGM containing most of the baryons. One could advocate that, if the Hubble constant is close to 100 km s^{-1}/Mpc, the discrepancy is not significant. However, with such a high value of H_0, one would have to abandon one or several of the following basic «facts» of cosmology:

Primeval nucleosynthesis is correct.

Dynamical measurements of the density of the Universe show that $\Omega_0 \geq 0.1$.

The age of the Universe t_0 is ≈ 15 Gy.

The cosmological constant Λ is zero.

Dark matter is nonbaryonic.

The first three sets of constraints are represented in fig. 1. The consistency of the model would need to have the three regions overlap. Actually, the most conservative point of view is to assume that $\Omega_0 = \Omega_b \approx 0.1$, $h \approx 0.5$, $t_0 \approx 20$ Gy (but this model is only weakly consistent with above constraints), the dark matter being then baryonic, probably in the form of unseen stars (brown dwarfs or «jupiters» being the most appealing candidates). In such a pure baryonic model, the Hubble constant could not be much higher as the nucleosynthesis constraint would then imply a very low Ω.

However, inflation has provided a theoretical argument for $\Omega_0 = 1$ [10]. Actually, this value is probably the only one that could easily emerge from a theory of the very early universe. In addition there now exist few significant observations that comfort this theoretical prejudice [11]. In this model a value $h = 0.5$ is almost unavoidable in order to escape a dramatic age problem.

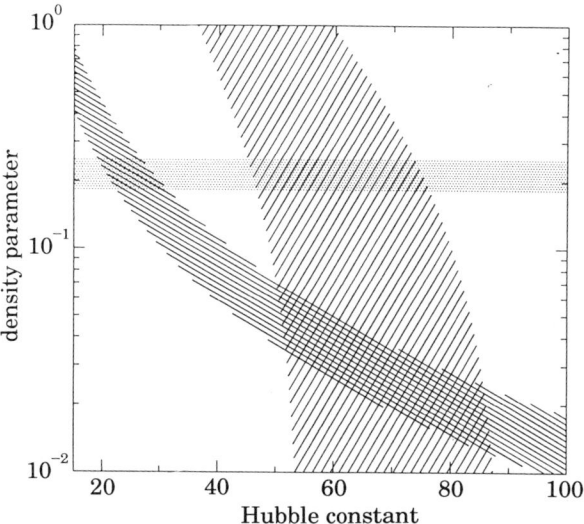

Fig. 1. – Various constraints can be set on the cosmological parameters Ω_0 and H_0. This diagram presents a few of them: the grey area corresponds to dynamical measurement in the Coma cluster [4] $0.18 \Omega \leq 0.25$, the first hatched area (narrow hatching) corresponds to the nucleosynthesis $0.018 \leq \Omega_b h^2 \leq 0.0092$ (this one applies only to the baryonic content of the Universe), the second hatched area corresponds to the constraint from the age of globular clusters: $11 \text{ Gy} \leq t_0 \leq 18 \text{ Gy}$.

Therefore, in each of the above models, which are able to reasonably well reproduce the observations, the value of Ω_b should be of the order of 0.05 (and the Hubble constant has to be of the order of 50 km s^{-1}/Mpc) in order to be consistent with the various constraints, and, therefore, 80 to 90% of the baryonic component should be dark, and this strongly advocates a detailed study of its history.

The implications of nucleosynthesis are quite important and so its validity must be addressed carefully. The quark-hadron phase transition could have produced inhomogeneities, leading to a nonstandard nucleosynthesis, allowing a larger value for the density of baryons. It seems that the observed abundances of light elements do not allow a value much larger than the standard one. The abundance of ^{4}He is of the order of 23%, with an uncertainty that could be of the order of 1%, leading to a large uncertainty in Ω_b. However, the observed abundances might not reflect the primeval abundances. The abundance of deuterium should be used with care as it could be destroyed in the chemical evolution of the Galaxy. Still it does provide a robust upper limit. The recent result from the space telescope has provided an accurate measurement of the interstellar deuterium[12], confirming and improving previous determinations. However, the observed value has to be corrected for evolution by an uncertain factor of the order of 3. Similar comments hold for the abundance of ^{3}He[13]. ^{7}Li is the last element useful to constrain primeval nucleosynthesis. The abundance of ^{7}Li from population-II stars with different temperatures is consistent with no variations from star to star (this is the famous Spite's plateau [14]). This is a strong argument for the observed value to be cosmological, *i.e.* unaffected by stellar evolution. However, it is known that stellar evolution could affect lithium abundance. Rotation, for instance[15], might deplete lithium. This would need a correcting factor, which would be difficult to determine precisely, as in the case of deuterium. If such effect is present, lithium would be less useful. Still, in the worst case, it is reasonable to think that the baryon density is known within a factor two, which is remarkable.

After the nucleosynthesis, the temperature of the Universe decreases. We will now pay attention to what happens when the temperature drops down to 4000 K. At that time, the electrons begin to recombine with protons to form H atoms, the ionization is becoming small (at $Z < 1500$) but is still high enough that photons diffuse on free electrons.

2. – Basic elements.

2˙1. *Time scale considerations.* – In the following we will meet several processes that may or may not play an important role in the evolution of the species which are involved. The first consideration for any possible process is to examine whether it is efficient or not. We will take as a first example the case of the diffusion of photons ($Z = 1100$) on the free electrons of the Universe. The process to consider is the diffusion of photons on cold electrons ($kT \ll h\nu$). The cross-section of this process at low energies ($h\nu \ll mc^2$) is (Thomson diffusion) $\sigma_T = 6.6 \cdot 10^{-25}$ cm^{-2} (the actual cross-section should take into account quantum correction, the Klein-Nishina formula should then be taken, but differs only in the relativistic case: $h\nu \geqslant mc^2$). The time scale for the diffusion on free electrons, therefore, is

$$t_T = \frac{1}{c\sigma_T n_e}.$$

This diffusion process will be important only if its time scale is shorter than the age of the Universe:

$$t = \frac{2}{3H_0}(1+z)^{-3/2} \approx 2h^{-1}10^{17}(1+z)^{-3/2} \text{ s}.$$

The electron density is

$$n_e \approx \chi \frac{\rho_b}{m_p} \approx \chi 1.8 \cdot 10^{-7}(1+z)^3 \text{cm}^{-3},$$

where χ is the ionization fraction (for simplicity we neglected the presence of helium). In the standard picture, the surface of last scattering is located at a redshift close to 1100. The opacity of the Universe to Thomson scattering is

(4) $$\tau_T = \int_0^{+\infty} c\sigma_T n_e \, dt \approx \int_0^{+\infty} 0.001 h^{-1} \chi(z) \frac{1+z}{(1+\Omega_0 z)^{1/2}} \, dz\,;$$

the fractional ionization is then found to be $\chi \approx 0.1$ by setting $\tau_T = 1$. A comprehensive discussion of the standard recombination may be found in the pioneering work [16]. A more recent analysis, with special care to the last-scattering surface, is given in [17].

2˙2. *The neutral period.* – After this period, matter is almost neutral. The temperature of the photon background evolves as

$$T_\gamma = 2.735(1+z).$$

Notice that increase with redshift means *decrease* with time. The temperature of the gas can be found from thermodynamic laws. Let u_m be the internal energy of a given comoving volume, and T_m its temperature:

(5) $$u_m = \frac{3}{2}nVkT_m,$$

where n is the total number density of particles. As the Universe expands, the pressure works and

$$du_m = -p\,dV = -nkT_m\,dV$$

leading to

$$\frac{dT_m}{T_m} = -\frac{2}{3}\frac{dV}{V},$$

therefore $T_m \propto V^{-2/3} \propto (1+z)^2$. This is the adiabatic cooling. After the recombination, the residual ionization is nonzero and allows some thermal contact, and matter actually decouples from radiation at a lower redshift $z \approx 200$ [18]. Notice that we have implicitly assumed that there were no other heating or cooling mechanisms. Thermodynamic laws are correct only if we are in a full-equilibrium situation where collisions are efficient enough. It is important, therefore, to verify this hypothesis. The time scale for collisions between neutral hydrogen atoms is

$$t_c \approx \frac{1}{n\sigma v},$$

where v is the thermal velocity, $\sigma \approx \pi a_0^2 \approx 10^{-16}$ cm^2. The ratio

$$\frac{t_c}{t} \approx 500\,h(1+z)^{-5/2}$$

indicates that, at small redshift, the gas, if completely neutral, is not in thermal equilibrium any more. However, we can still define its temperature by the relation

$$\frac{3}{2}kT_m = \frac{1}{2}mv^2;$$

the velocity of a freely moving particle in an expanding universe evolves according to

(6) $$v = v_0(1+z),$$

therefore the temperature decreases in exactly the same way as in the equilibrium case:

(7) $$T_m = T_0(1+z)^2.$$

2˙3. *A short comparison with observation.* – Actually, the picture we have described above is in complete disagreement with two basic observational facts: first, that dark matter is clumped (galaxies, clusters...) and there is no reason why most of the baryons will not be clumped in the dark-matter potentials. The second problem comes from the Gunn-Peterson test [19]. Observations of distant QSO's reveal no significant absorption even for Lyman-α photons, travelling in the IGM (apart from Lyman-α clouds). This implies a very ionized medium. The cross-section of such photons is $\sigma \approx 7.3 \cdot 10^{-18}\,\mathrm{cm}^2\,(Ry/h\nu)^3$. That is

$$\tau_{\mathrm{GP}}(z) \approx 1.5\,h^{-1}\,10^4\,(1-\chi)\,\frac{(1+z)\,\Delta z}{(1+\Omega_0 z)^{1/2}}\,dz\,;$$

the measurements imply that $\tau_{\mathrm{GP}} \leq 0.05$ at $z \approx 2.6$ [20]. Therefore, the neutral fraction is smaller than $4 \cdot 10^{-6}$. Strictly speaking it is possible that the IGM does not exist at all. Such a possibility, however, is very doubtful: as we have seen, the amount of observed baryons is much smaller than predicted by primeval nucleosynthesis. Even if the dark matter is baryonic, some baryons should remain after the process of galaxy formation. Furthermore, the existence of a substantial fraction of gas in the IGM is supported by the large amount of gas seen in clusters. Ionization of the IGM needs at some level energy injection in the medium (reheating), and eq. (5) has to be modified accordingly:

$$(8) \qquad u_{\mathrm{m}} = \frac{3}{2}\,nVkT_{\mathrm{m}} + \varepsilon_{\mathrm{h}}V\,dt - \varepsilon_{\mathrm{c}}V\,dt\,,$$

where ε_{h} and ε_{c} are, respectively, the rate of energy input and the total cooling rate of the gas. This leaves us with a considerable amount of freedom because of a variety of possible processes: the interaction with the CBR is able to cool (or heat) the gas, other background contributions might heat the gas, bulk flows and cosmic rays produced by stars might reinject energy in the IGM, and, at the same time, high-temperature gas is able to distort the CBR and to produce other backgrounds. In addition, the chemical composition modifies the cooling rates, and we have no serious reason to treat this problem as homogeneous... .

3. – Cooling and heating mechanisms.

Here, the different processes that could be taken into account will be briefly presented, but a full description of the possible scenarios will not be given. PEEBLES [21] earlier discussed these processes in the cosmological context. Basic elements can be found in [22]. The appendix of [23] gave the relevant formula to deal with this question, and has been widely used in the preparation of this lecture. No specific heating model will be discussed here. Such heating sources can be astrophysical (quasars, AGN, early galaxies, population-III stars), see, for instance, [24], or have a more exotic origin like decaying particles [23, 25].

I will now examine the consequences of the existence of cooling or heating mechanisms on the evolution of the gas. Hereafter, we will assume the medium to be homogeneous.

3'1. *Time evolution.* – Equation (8) can be written in the following form:

$$\frac{du_m}{u_m} = -\frac{4}{3}\frac{dt}{t} \pm \frac{dt}{\tau}, \qquad (9)$$

where

$$\tau = \left|\frac{u_m}{\varepsilon}\right|$$

is the cooling ($-$) or heating ($+$) time depending on the process under consideration. Again it is rather clear that only processes that have a characteristic time shorter than the age of the Universe are to be taken into account. Generally, the ratio τ/t will depend on t as a power law:

$$\frac{\tau}{t} = \left(\frac{t}{t_e}\right)^\alpha;$$

neglecting the adiabatic cooling in the above equation, the energy evolution can be found:

$$u_m \propto \exp\left[\pm\frac{1}{\alpha}(t/t_e)^\alpha\right],$$

which means that the temperature will relax very rapidly to an equilibrium value. Therefore, if the heating is a short burst of energy which is instantaneously injected into the medium, and if an efficient cooling is possible, the temperature will go down rapidly. In order to maintain this cooling, it is necessary to have source injection which provides energy over more than an expansion time. It is then not necessary to integrate the differential equations (which might easily be unstable). Rather, it is simpler to evaluate the final condition of the system by looking for the solution of the implicit equation

$$\varepsilon_h(n, T) - \varepsilon_c(n, T) = 0, \qquad (10)$$

which is satisfied when the equilibrium is achieved.

3'2. *Inverse Compton cooling.* – When photons diffuse on electrons by Thomson scattering the energy exchanged is of the order of

$$\Delta E \approx \frac{h\nu}{m_e c^2}(4T_e - h\nu).$$

The rate of energy transfer from the electrons to the photon gas is

$$\varepsilon_{cc} = \frac{4kT}{m_e c^2} c\sigma_T n_e aT_\gamma^4.$$

The associated time scale, therefore, is

$$t_{cc} \approx \frac{(3/2)nkT}{\varepsilon_{cc}} = \frac{3}{4} \frac{mc}{\sigma_T a T_\gamma^4} \text{ s},$$

leading to

$$\frac{t_{cc}}{t_H} = \frac{180}{(1+Z)^{5/2}}.$$

Therefore, at redshift greater than 10, the Compton mechanism will efficiently cool the gas, provided that it is ionized. It is easy to verify how efficient this is: if the gas is heated at $Z \approx 100$, at a redshift of 10 the temperature is dampened by a factor of 10^{-65}! The energy lost by the gas is transferred to the CBR, and produce distortion of the spectrum (the relaxation toward the Planckian shape cannot be achieved any more). The distortion is characterized by the y_c parameter

$$dy_c = \frac{4kT}{m_e c^2} \sigma_T n_e \frac{c}{H_0} \frac{dZ}{(1+Z)^2 (1+\Omega_0 Z)^{1/2}}.$$

Numerically

$$y_c = 4 \cdot 10^{-8} (1+Z_h)^{7/2} T_5 \quad \text{for adiabatic cooling},$$

$$y_c = 10^{-7} (1+Z_h)^{3/2} T_5 \quad \text{for } T = \text{const},$$

Z_h being the redshift at which the gas is heated, and T_5 the temperature in units of 10^5 K. This constraint provides a rather weak constraint on the history of the IGM. COBE did not detect any Compton distortion, implying $y_c \leq 10^{-3}$ [8]. This does imply that the temperature of the IGM could not have been hotter than 10^8 K for a long period [26]. Actually, it cannot be demonstrated that the Universe ever recombined [27].

3'3. *Bremsstrahlung cooling*. – Hot electron gas radiates energy via bremsstrahlung emission

$$\varepsilon_B = 1.4 \cdot 10^{-27} T^{1/2} n_e^2 \bar{g} \text{ erg cm}^{-3}/\text{s},$$

where $\bar{g}$ is the velocity average Gount factor. The associated time scale is

$$t_{cB} \approx 4 \cdot 10^{20} \frac{T_5^{1/2}}{(1+Z)^3} \text{ s}.$$

3'4. *Recombinative cooling*. – Recombinations on excited ground states lead to the emission of a photon, and the corresponding energy is lost for the gas. An accurate formula is given by

$$\varepsilon_R = \frac{64}{3}\sqrt{\frac{\pi}{3}}\,\alpha^4 a_0^2 c \sqrt{\frac{kT}{R_y}}\, n_e \left[-0.0713 + \frac{1}{2}\log\frac{R_y}{kT} + 0.6\sqrt[3]{\frac{kT}{R_y}}\right].$$

A good approximation, however, may be made by assuming that each recombination corresponds to an energy loss which is equal to one R_y when the mean kinetic energy of a particle of the gas is higher than 13.6 eV, and to kT when the kinetic energy is smaller. The recombination time is equal to

$$t_r = \frac{1}{n_e(\sigma v)_{\text{rec}}} = \frac{\sqrt{3\pi}}{n_e\,64\,\pi\alpha^4 a_0^2 c}\left(\frac{kT}{R_y}\right)^{2/3} = 5\cdot 10^{19}\,\frac{T_5^{2/3}}{\chi(1+z)^3}\,.$$

The energy lost is, therefore, approximated by

$$\frac{dE}{E} \approx \frac{dN_r}{kT}\left(\frac{1}{R_y} + \frac{1}{kT}\right) = \frac{dt}{t_{\text{cr}}},$$

leading to

$$t_{\text{cr}} \approx \frac{t_r}{1 + R_y/kT},$$

where t_r is the recombination time.

3'5. *Collisional-excitation cooling.* – When a free electron undergoes a collision with a neutral atom, the atom can be excited, and then de-excited radiatively to the ground state. The energy lost by this process is

$$\varepsilon_{\text{cec}} = 7.5\cdot 10^{-19}\, n_e^2 \chi(1-\chi)\exp[-R_y/kT].$$

The corresponding cooling time is

$$t_{\text{cec}} = 1.5\cdot 10^{14}\,\frac{1+\chi}{\chi(1-\chi)(1+z)^3}\,T_5\exp[+R_y/kT].$$

3'6. *The cooling function.* – When one is dealing with a cosmological problem in which cooling and heating processes compete, it is possible to use the cooling function to estimate the energy loss of the gas. This function is defined as

$$\varepsilon = n_e^2\,\Lambda(T),$$

and the function $\Lambda(T)$, taking all the above radiation mechanisms into account, has been tabulated. One should keep in mind that thermal equilibrium is assumed, which needs a density greater than the homogeneous value and the absence of any significant photoionization mechanism. In that case the global cooling time is just

$$t_c \approx \frac{3}{2} \frac{kT}{n_e \Lambda(T)}.$$

3˙7. *Ionization equilibrium.* – Actually, eq. (10) involves the ionization fraction, so that there are two unknown quantities T and χ. Therefore, a second equation is needed. Again we can write down the time evolution for the ionization fraction, but as long as the relevant time scales are shorter than the age of the Universe, we can assume an equilibrium situation for which the collisional ionization time t_I is equal to the recombination time t_R from

$$t_I = \frac{1}{n_{HI} \overline{\sigma_c v}} = \frac{1}{n_{HI} 7.2 a_0^2} \left(\frac{m_e}{kT}\right)^{1/2} \exp[R_y/kT] \approx$$

$$\approx \frac{2.2 \cdot 10^{14}}{(1-\chi)(1+z)^3} T_5^{-1/2} \exp[R_y/kT] \, s$$

and

$$t_R = \frac{1}{n_e \overline{\sigma_r v}} = \frac{3\sqrt{3\pi}}{n_e c \alpha^4 a_0^2} \left(\frac{kT}{R_y}\right)^{2/3} \approx \frac{5.8 \cdot 10^{19}}{\chi(1+z)^3} T_5^{2/3} \, s.$$

Therefore, the ionization fraction depends only on the temperature:

$$\chi \approx \frac{1}{1 + 2.8 \cdot 10^{-6} T_5^{-7/6} \exp[R_y/kT]}.$$

It follows that to satisfy the Gunn-Peterson test the temperature of the IGM needs to be of the order of $2 \cdot 10^5$ K. Actually, the equilibrium situation is not achieved at redshift $z \leq 5$ as the recombination time is larger than the age of the Universe. Another approximation in this calculation is that we neglect photoionization. However, photoionization is an efficient way to ionize the medium because of the long recombination time. If any energy heats the cosmological gas, a further equation for the ionization has to be solved:

$$d(\chi n) = n_{HI} \frac{dt}{t_I} - n_e \frac{dt}{t_R} + n_X \frac{dt}{t_i}.$$

When the gas is photoionized at some redshift and its temperature reaches a value above a few 10^3 K, it remains ionized because the recombination time is larger than the age of the Universe. If the IGM is photoionized, its temperature could be a few 10^4 K; quasars are the only well-known source of UV photons, but they are not numerous enough to achieve the observed level of ionization[28]. It is, therefore, not known whether this level has been achieved by energy injection or by photoionization by unknown sources. An interesting proposition is that the hard-X-ray background could heat the gas at a temperature close to 10^5 K [29], but this process seemingly cannot explain the Gunn-Peterson test.

4. – The history of the cosmological gas and galaxy formation.

Only baryons are observable in the Universe; on the other hand, as we have seen, dynamical measurements show that the main component of the mass is dark. Whether this dark matter is baryonic or not, it should be in a collisionless form. Thus for a long time cosmologists tried to understand galaxy formation by understanding the properties of this collisionless component, evolving under its own gravity. This is the standard gravitational-instability picture [30]. Numerical simulations have been widely used for this purpose. Nevertheless, it is rather clear that galaxy properties and their implications on galaxy formation theory need to include a description of the baryonic content of the Universe. The concept of «biasing» has revived the attention on dissipative processes in galaxy formation: galaxies may not be a random subset of the total content of the Universe, and their distribution may not reflect the overall distribution of the Universe, i.e. galaxies do not trace the mass. The difference between both distributions can be written

$$\delta_g = b \delta_\rho ,$$

where δ_ρ represents the density fluctuations of the mass, and δ_g the fluctuation of the galaxy number density. In the simplest case where the bias parameter is a constant, this implies that

$$\xi_g = b^2 \xi_\rho .$$

If such a difference does exist, it might alter the dynamical measurements of Ω_0. The simplest way to deal with this approach is to assume that b is a constant, but the truth may be much more complex. The fact that the stars inside a galaxy are more concentrated than the dark matter provides a direct evidence that «light does not trace the mass». This can be understood because the baryonic content of a galaxy has undergone dissipative processes. However, it is not obvious at all that the concept of biasing is relevant on larger scales, as suggested by the absence of luminosity segregation on large scales [31], as well as the absence of any significant difference between IRAS and optical galaxies. Actually, such a «bias» could arise even in the absence of dissipative processes [32]. On small scales, morphological segregation, the correlation function of clusters [33], the difference of clustering between IRAS galaxies and optical galaxies offer observational support to the concept of bias, but it is far from being clear that this may actually allow b to be of the order of 2-2.5 as seems necessary to solve the «Ω problem» [34].

Taking into account the dissipative processes that determine the cosmological evolution of the gas is a rather despaired tentative, but cosmologists have always been unrealistically optimistic, and this is actually the purpose of this lecture.

The most fundamental difference with previous sections is that the matter distribution is *a priori* inhomogeneous. Actually, the usual assumption of a homogenous IGM is unrealistic. There is a fundamental difficulty to describe the inhomogeneous state of the collisionless component and its evolution with time. This is already a nontrivial question. The possibility of deriving the mass function of structures that form inside the gravitational-instability picture has been handled by PRESS and SCHECHTER in 1974[35]. Numerical simulations[36] have greatly helped to understand what is going on, and at the end it seems that the Press and Schechter formula for the mass function that arose in the gravitational-instability picture is an amazingly good description of what actually arises. This recipe has long appeared as suspect to most cosmologists. Whether or not this approach is physically motivated is still a matter of debate[37, 38]. There are two basic assumptions necessary in order to derive the mass function. First, one has to assume that the nonlinear collapse of an overdense region is controlled by the linear amplitude of the mean density inside this region. The simplest way to express this is to assume that the regions that will collapse are exactly those which satisfy a sharp threshold criterium ($\delta \geq \delta_S$). The validity of this assumption seems rather natural, but is far from being obvious. Actually, it has been suggested that the condition of the collapse could be determined by the amount of kinetic energy in smaller structures already present in the collapsing object[39]. However, early numerical simulations[36] have suggested that such an effect is not present, and that the gravitational instability could be described in a simple scheme[40]. This has been confirmed directly by simulations addressing this point specifically[41]. The second assumption is that a region which is included in a nonlinear region after the density field has been smoothed with a scale length radius R will actually end up in an object with a typical mass of the order of $\bar\rho 4/3\pi R^3$. The mass function and the initial field could then be related by an *exact* equation:

$$(11) \qquad \mathscr{F}_{\mathrm{NL}}(> m) = \int_{\delta_S/\sigma(R)}^{\infty} s(\nu\sigma, \nu) F_R(\nu)\, d\nu,$$

where $\sigma(R)$ is the r.m.s. fluctuation of the smoothed field. The derivation of F_R seems, however, rather nasty. PRESS and SCHECHTER assumed that this function was Gaussian. The sharp-threshold assumption then leads to

$$(12) \qquad n(m) = -\left(\frac{1}{2\pi}\right)^{1/2} \delta_S \frac{\bar\rho}{m} \frac{1}{\sigma^2} \frac{d\sigma}{dm} \exp\left[-\frac{\delta_S^2}{2\sigma^2}\right].$$

They have then corrected it by a fudge factor of 2. Improvements on this formalism have been proposed since [37, 42], and have essentially tried to find a more robust derivation than the original one. However, a reasonable approximative derivation will necessarily lead to a formula close to the PS one [38].

The threshold value can be derived in the spherical model: the perturbation is assumed to be spherical. Its evolution can then be followed even when the density is becoming high. These solutions were originally used by LEMAÎTRE in the early thirties [43]. Such a solution cannot describe correctly the evolution of matter after the density becomes singular for the first time (orbit crossing). In an Einstein-de Sitter model, this occurs at a time where the linear constrast density is 1.68, therefore the nonlinear collapse is expected to be extremely fast. After orbit crossings, matter is expected to be in a more or less stationary state, *i.e.* virialized; this is an idealized picture, as accretion of matter is certainly a continuous process. The top-hat model allows us to follow the properties of collapsed objects: a positive linear fluctuation with a density contrast δ_i at some «initial epoch» z_i collapses at z_f:

$$1 + z_f = \frac{\delta_i(1 + z_i)}{1.68}.$$

The dynamical properties of the objects can then be estimated by assuming that the object is an isothermal sphere [44] and that its radius is twice as small as the maximum radius of expansion of the spherical region; the typical density of the object is of the order of 200 times the background. The circular velocity of the object can be derived:

(13) $$V_c^2 = \frac{GM}{R} = 127 \, M_{12}^{2/3} (1 + z_f) \, h_{50}^{2/3} \text{ km/s}.$$

The dispersion depends on the density profile:

(14) $$\sigma^2 = \frac{1}{\beta} V_c^2,$$

where β is the index of the density profile of the halo. The temperature of the baryonic gas inside the object can be inferred as well. The gas will be heated during the collapse, and reach a pressure equilibrium. The temperature of the gas can then be derived:

(15) $$T_{\text{gas}} = \frac{11.4 \cdot 10^5}{\Gamma} M_{12} (1 + z_f) \, h_{50}^{2/3} \text{ K};$$

this result comes from the sole assumption of hydrostatic equilibrium. The coefficient Γ depends on the properties of the object:

$$\Gamma = \frac{d \log \rho_{\text{gas}}}{d \log R} + \frac{d \log T}{d \log R}. \tag{16}$$

From observed properties of clusters, $\Gamma \approx 2.0$. The above formalism allows us to determine the evolution of the temperature distribution function of structures at any epoch. Numerical simulations by EVRARD [45, 46] greatly helped to check the validity of this simple scheme. The temperature he derived is only 20% lower than predicted by eq. (15). The above approach provides an interesting first step in understanding the evolution of the gas. As we have seen, the knowledge of the temperature and density of the gas is the basic ingredient necessary to determine its cooling time. If the gas can cool, say in a Hubble time, it will contract. As the density increases, the cooling time becomes even shorter. This instability would be stopped only if the gas turns into stars or into a cold disc of gas (when the gas is neutral, at $T \leq 10^4$ K it does not cool any more). Therefore, the cooling criterium can be understood as a criterium to distinguish the halos which lead to galaxies from those leading to clusters. This argument does provide a rough estimate of typical L_* galaxies [47]: galaxies correspond to halos for which the gas can cool, while clusters correspond to those for which the gas remains hot. The cooling runaway seems inevitable unless some process reheated the gas at a sufficient rate. In a hierarchical picture the structures present at one time are merged in a larger structure of the hierarchy. During this merging the gas will be reheated. The typical time scale of this process corresponds to the typical time scale for the temperature of halos to change. For the most massive halos, this time scale depends on the index of the fluctuation spectrum $\sigma \propto m^{-\alpha}$ [38]:

$$t_* = \frac{T_*}{\dot{T}_*} = \kappa_\alpha t_H(z), \qquad \text{with } \kappa_\alpha = \frac{9\alpha}{4 - 6\alpha},$$

where $t_H(z) = 0.66 H^{-1}(z)$, $H(z)$ being the Hubble constant at the time of formation of the object. At a given epoch all the objects have basically the same contrast density (actually, a density profile can be assumed; this does not change the following picture). The objects for which the cooling criterium

$$t_c = t_* \tag{17}$$

is satisfied can be delimited by the cooling curve in the density temperature (D-T) diagram as drawn in fig. 2. If all the baryonic content is in a gaseous form, the instantaneous total amount of gas able to cool in potentials can be ob-

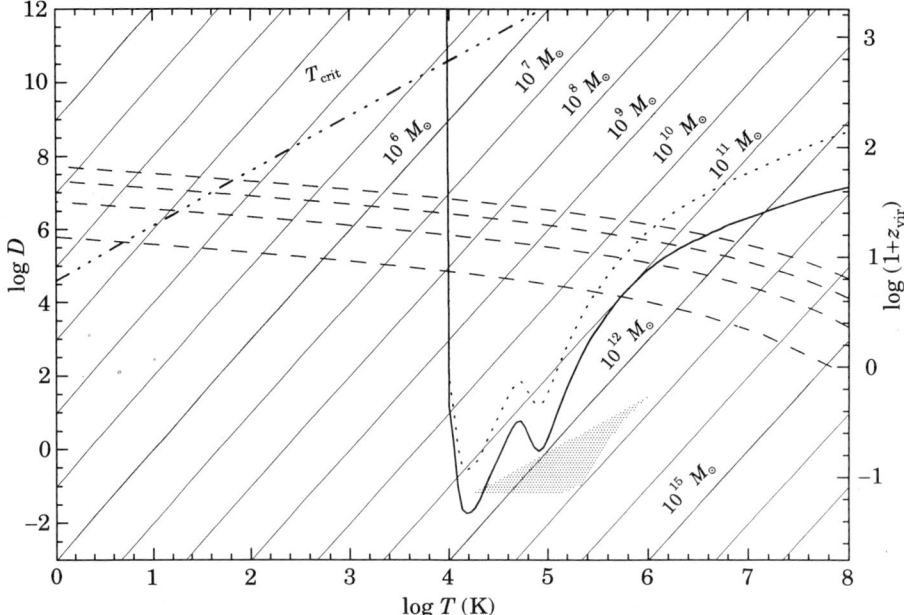

Fig. 2. – The density-temperature diagram for primordial gas (no metals). D is the dimensionless ratio of the local density to the present-day critical density, and $z_{\rm vir}$ gives the virialization redshift of the structure; $Z = 0$, $\Omega_{\rm b} = 0.06$, $h_{50} = 1.0$. Note that the redshift scale applies to virialized objects *only*. T is the temperature of the gas lying in the potential wells. Thin lines are constant *total* mass, while the *thick dash-three-dot line* gives the evolution of the cosmological gas after the recombination. The *thick dotted curve* is the D-T relation for the dynamical local cooling criterion ($t_{\rm cool} \lesssim \eta_{\rm d} t_{\rm dyn}$), while the *thick full line* corresponds to our criterion $t_{\rm cool} \lesssim \eta \kappa_\alpha t_{\rm H}$, with $\eta = 1$ (eq. (17)). Thin *dashed lines* represent the mass above which lie 50%, 10% and 1% of the structures, with a bias parameter of $b = 1.5$. The *shaded area* indicates the region where gas cools adiabatically once reheating terminates. This cooling curve is a useful tool to understand various regimes for the history of the cosmological gas.

tained by taking the sum of the gas present in halos that lies inside the cooling region:

$$g(z) = \frac{1}{\bar{\rho}} \int_{T_1}^{T_2} N(m)\, m\, {\rm d}m\,.$$

The derivation of the *total* amount of cooled gas seems more difficult to obtain because one has to take into account the full history of each halo. Extension of the Press and Schechter approach allows such a calculation [48, 49]. However, a simple argument can be used: any piece of gas lying in a structure with a temperature greater than 10^4 K has a history that can be represented by a line in the D-T diagram. At high redshift the gas is colder than 10^4 K, so that this curve *should cross* the cooling region. The total fraction can then be estimated just by estimating the amount of mass lying in objects with a virial temperature greater than 10^4 K. This means that, by now, most (around 80%) of the gas should have been cooled in some structures[38]. This is the overcooling problem: in the absence of reheating, most of the gas should have been cooled by now, while only 10% of this cooled gas is observable (as stars). It might be argued that most of this gas has actually been turned into stars, according to the previous calculation, but the fact that these stars are not shining any more at all is only possible if the IMF is strongly different now than in the past. The remnants can be in the form of small stars (jupiters, brown dwarfs) or high-mass stars. This latter case is less likely because high-mass stars produce metals which are ejected. However, the amount of gas in clusters indicates that a substantial fraction of the primordial gas should have escaped the cooling, while the above argument would predict that 100% has been cooled. The overcooling problem relies on several assumptions, and it is important to examine carefully its reality. Numerical simulations are limited by resolution. This is a striking problem: the cooling occurs in potentials with a virial temperature greater than 10^4 K and with a typical size of a few 10 kpc, which is generally out of the scope of 3D hydro-simulations. The fact that the overcooling has not been observed in most recent simulations is probably due to the resolution limit[50-52]. Interestingly, a 2D hydro-simulation benefits from a much better resolution. In this case, the overcooling problem has been observed[53].

Several solutions have been proposed to the overcooling problem. Basically, one has to assume the existence of some reheating. This could be due to the feedback of early stars, after the gas has cooled inside the potentials: as the first stars form, they heat the surrounding gas and prevent further star formation[54] and may even expel the gas of the potential[55]. In the CDM model, WHITE and FRENK[56] have given detailed calculations of the history of star formation and metal production. They conclude that galaxy formation is very late in such models, since the bulk of star formation occurs at $z \sim 0.5$; in addition, the metals are likely to be overproduced as in their calculation the amount of observed metals is already produced at $z = 1$. Furthermore, the faint end slope of the luminosity function is steeper than observed. Actually, WHITE and REES[57] have pointed out that the luminosity function of galaxies could be affected by the «feedback» mechanism and may, therefore, differ from the mass function of the dark matter, but the change in the slope is rather small. A different approach was proposed[38]: if the whole gas is maintained hot in a homogenous phase by some mechanism, then the picture is completely modified. If the gas is hot, the baryonic gas does not fall into potentials with virial temperature T (15) smaller than the temperature of the gas. Because most of the gas is not clumpy, the energy requirement to maintain it hot is minimal. In this model, star formation (and bright galaxies) do have a maximal formation rate at a relatively high redshift ($Z \approx 2-5$), subsequently most of the heating process stops, and the gas cools adiabatically. A heating mechanism might be the supernova winds [58] in first galaxies. Lower-temperature halos can then form. The number of such systems at their formation epoch is, therefore, determined by the number of halos at the epoch when the gas temperature was equal to their virial temperature (15). The galaxy luminosity function can then be estimated, assuming that halos are not destroyed by subsequent evolution of the clustering hierarchy. The slope of the mass function of cosmic structures is very close to -2. Therefore, the low-mass tail of the mass function of halos is

$$N(m, z) \sim \frac{1+z}{m^2}.$$

If the gas cools adiabatically, the velocity dispersion associated with the temperature of the gas is $\sigma = \sigma_0(1+z)$ and thus the mass-velocity dispersion relation gives the typical mass of the galaxies forming at the redshift z as $m \sim \sigma^3(1+z)^{-3/2} \sim (1+z)^{3/2}$ (see eq. (13)). The number density of galaxies then varies as

$$N(m) \sim m^{-4/3}.$$

Assuming an L-σ relation of the form $L \sim \sigma^\beta$, the present-day luminosity function can be derived: it has a power law dependence at faint luminosities

$$\phi(L) \sim L^{-(1+1/\beta)}.$$

From the observed values of $\beta \sim 2.7$ to 4.0, the resulting theoretical slope is interestingly close to the observed one. This model allows for recovery of the faint end slope for the luminosity function of galaxies.

As we have seen, the existence of hot IGM could efficiently suppress galaxy formation. The Gunn-Peterson test reveals that the gas is probably at a temperature higher than 10^4 K corresponding to a halo with circular velocity of the order of 20 km/s. Therefore, this suppression mechanism should work at some level. At low redshift, the gas is outside of the cooling region, therefore, if it does contract, its temperature will increase according to the adiabatic regime $T \propto \Delta^{2/3}$ meaning that the suppression mechanism would work on halos with circular velocity up to 100 km/s.

REFERENCES

[1] J. LOVEDAY, B. A. PETERSON, G. EFSTATHIOU and S. J. MADDOX: *Astrophys. J.*, **390**, 338 (1992).
[2] T. R. LAUER: *Astrophys. J.*, **292**, 104 (1985).
[3] S. D. M. WHITE: in *Proceedings of NATO Workshop on Observational Tests of Inflation*, edited by T. SHANKS, A. J. BANDAY, R. S. ELLIS, C. S. FRENK

and A. W. WOLFENDALE, Series C, Vol. 348 (Kluwer Academic Publishers, Amsterdam, 1991), p. 279.
[4] J. P. HUGUES: *Astrophys. J.*, **337**, 21 (1990); U. G. BRIEL, J. P. HENRY and H. BÖRINGHER: *Astron. Astrophys.*, **413**, L31 (1992).
[5] L. P. DAVID, K. A. ARNAUD, W. FORMAN and C. JONES: *Astrophys. J.*, **356**, 32 (1990).
[6] L. P. DAVID and G. BLUMENTHAL: *Astrophys. J.*, **389**, 510 (1992).
[7] G. GAMOV: *Phys. Rev.*, **74**, 505 (1948).
[8] J. C. MATHER, E. S. CHENG, R. E. EPLEE jr., R. B. ISAACMAN, S. S. MEYER, R. A. SHAFER, R. WEISS, E. L. WRIGHT, C. L. BENNETT, N. W. BOGGESS, E. DWEK, S. GULKIS, M. G. HAUSER, M. JANSSEN, T. KELSALL, P. M. LUBIN, S. H. MOSELEY jr., T. L. MURDOCK, R. F. SILVERBERG, G. F. SMOOT and D. T. WILKINSON: *Astrophys. J. Lett.*, **354**, L37 (1990); H. GUSH, M. HALPERN and E. WISHNOW: *Phys. Rev. Lett.*, **65**, 537 (1991).
[9] K. A. OLIVE, D. N. SCHRAMM, G. STEIGMAN and T. P. WALKER: *Phys. Lett. B*, **236**, 454 (1990); T. P. WALKER, G. STEIGMAN, D. N. SCHRAMM, K. A. OLIVE and K. A. KANG: *Astrophys. J.*, **376**, 51 (1991); M. S. SMITH, L. H. KAWANO and R. A. MALANEY: *Astrophys. J. Suppl.*, **85**, 219 (1993).
[10] A. H. GUTH: *Phys. Rev. D*, **23**, 247 (1981).
[11] E. BERTSCHINGER: in *Rencontres de Moriond in Astrophysics*, edited by J.-M. ALIMI *et al.* (Editions Frontières, Gif-sur-Yvette, 1990), p. 411.
[12] J. L. LINSKY, A. BROWN, K. GAYLEY, T. R. AYRES, W. LANDSMAN, S. N. SHORE and S. R HEAP: *Astrophys. J.*, **402**, 694 (1993).
[13] E. SHATZMAN: *Astron. Astrophys.*, **172**, 1 (1987).
[14] F. SPITE and M. SPITE: *Astron. Astrophys.*, **115**, 357 (1982).
[15] S. VAUCLAIR: *Astrophys. J.*, **335**, 971 (1988); M. H. PINSONNEAULT, C. P. DELIYANNIS and P. DEMARQUE: *Astrophys. J. Suppl.*, **78**, 179 (1992).
[16] P. J. E. PEEBLES: *Astrophys. J.*, **153**, 1 (1968).
[17] B. J. T. JONES and R. F. G. WYSE: *Astron. Astrophys. Lett.*, **149**, 144 (1985).
[18] D. PUY, G. ALECIAN, J. LE BOURLOT, J. LÉORAT and G. PINEAU DES FORÊTS: *Astron. Astrophys.*, **267**, 337 (1993).
[19] J. E. GUNN and B. A. PETERSON: *Astrophys. J.*, **142**, 1633 (1965).
[20] C. C. STEIDEL and W. L. W. SARGENT: *Astrophys. J.*, **318**, L11 (1987).
[21] P. J. E. PEEBLES: *Physical Cosmology* (Princeton University Press, Princeton, N.J., 1971).
[22] G. RYBICKI and A. LIGHTMAN: *Radiative Processes in Astrophysics* (John Wiley and Sons, New York, N.Y., 1979).
[23] A. STEBBINS and J. SILK: *Astrophys. J.*, **300**, 1 (1986).
[24] H. COUCHMAN: *Mon. Not. R. Astron. Soc.*, **214**, 137 (1985).
[25] X. ASSELIN, G. GIRARDI, P. SALATI and A. BLANCHARD: *Nucl. Phys. B*, **310**, 669 (1988).
[26] X. BARCONS, A. C. FABIAN and M. J. REES: *Nature (London)*, **350**, 685 (1991).
[27] J. BARTLETT and A. STEBBINS: *Astrophys. J.*, **371**, 8 (1991).
[28] P. R. SHAPIRO and M. L. GIROUX: *Astrophys. J.*, **321**, L107 (1987).
[29] S. COLLIN-SOUFFRIN: *Astron. Astrophys.*, **243**, 5 (1991).
[30] P. J. E. PEEBLES: *The Large-scale Structure of the Universe* (Princeton University Press, Princeton, N.J., 1980).
[31] J.-M. ALIMI, D. VALLS-GABAUD and A. BLANCHARD: *Astron. Astrophys.*, **180**, L5 (1988); A. J. S. HAMILTON: *Astrophys. J.*, **331**, L59 (1988).
[32] R. SCHAEFFER: *Astron. Astrophys.*, **180**, L5 (1987); S. D. M. WHITE, M. DAVIS, G. EFSTATHIOU and C. S. FRENK: *Nature (London)*, **330**, 451 (1987).

[33] N. KAISER: *Astrophys. J. Lett.*, **284**, L9 (1984).
[34] D. VALLS-GABAUD, J.-M. ALIMI and A. BLANCHARD: *Nature (London)*, **180**, L5 (1988).
[35] W. H. PRESS and P. L. SCHECHTER: *Astrophys. J.*, **187**, 425 (1974).
[36] G. EFSTATHIOU, C. S. FRENK, S. D. M. WHITE and M. DAVIS: *Mon. Not. R. Astron. Soc.*, **235**, 715 (1988).
[37] J. R. BOND, S. COLE, G. EFSTATHIOU and N. KAISER: *Astrophys. J.*, **379**, 440 (1991).
[38] A. BLANCHARD, D. VALLS-GABAUD and G. MAMON: *Astron. Astrophys.*, **264**, 365 (1992).
[39] P. J. E. PEEBLES: *Astrophys. J.*, **365**, 27 (1990).
[40] A. J. S. HAMILTON, P. KUMAR, E. LU and A. MATTHEWS: *Astrophys. J.*, **371**, L1 (1991).
[41] A. E. EVRARD and M. M. CRONE: *Astrophys. J.*, **394**, L1 (1992).
[42] J. A. PEACOCK and A. F. HEAVENS: *Mon. Not. R. Astron. Soc.*, **243**, 133 (1990).
[43] G. LEMAÎTRE: *Ann. Soc. Sci. Bruxelles A*, **53**, 51 (1933).
[44] J. BINNEY and S. TREMAINE: *Galactic Dynamics* (Princeton University Press, Princeton, N.J., 1987).
[45] A. EVRARD: *Astrophys. J.*, **341**, 26 (1989).
[46] A. EVRARD: in *Clusters of Galaxies*, edited by W. R. OEGERLE, M. J. FITCHETT and L. DANLY (Cambridge University Press, Cambridge, 1990), p. 287.
[47] J. BINNEY: *Astrophys. J.*, **215**, 483 (1977); M. J. REES and J. P. OSTRIKER: *Mon. Not. R. Astron. Soc.*, **213**, 75 (1971); J. SILK: *Astrophys. J.*, **211**, 638 (1977).
[48] R. G. BOWER: *Mon. Not. R. Astron. Soc.*, **248**, 332 (1991).
[49] S. COLE: *Astrophys. J.*, **367**, 45 (1991).
[50] R. CARLBERG and H. COUCHMAN: *Astrophys. J.*, **340**, 47 (1989).
[51] R. Y. CEN, A. JAMESON, F. LIU and J. P. OSTRIKER: *Astrophys. J. Lett.*, **362**, L41 (1990).
[52] N. KATZ, L. HERNQUIST and D. H. WEINBERG: *Astrophys. J. Lett.*, **399**, L109 (1992).
[53] R. E. KATES, E. V. KOTOK and A. A. KLYPIN: *Astron. Astrophys.*, **243**, 295 (1991).
[54] R. B. LARSON: *Mon. Not. R. Astron. Soc.*, **169**, 229 (1974).
[55] A. DEKEL and J. SILK: *Astrophys. J.*, **303**, 39 (1986).
[56] S. D. M. WHITE and C. S. FRENK: *Astrophys. J.*, **379**, 52 (1991).
[57] S. D. M. WHITE and M. J. REES: *Mon. Not. R. Astron. Soc.*, **183**, 341 (1978).
[58] M. E. TEGMARK, J. SILK and A. EVRARD: *Astrophys. J.*, **417**, 54 (1993).

ROSAT Observations of Clusters of Galaxies.

H. BÖHRINGER

Max-Planck-Institut für Extraterrestrische Physik - D-8046 Garching, B.R.D.

1. – Introduction.

In the hierarchical sequence of structures in the Universe clusters of galaxies constitute the next step of integration following the galaxies. They are in this hierarchical chain also the last and largest members which have clearly decoupled from the Hubble flow and are approaching a characteristic equilibrium configuration that allows a coherent modelling of these objects [1]. For the general understanding of galaxy formation a look at clusters of galaxies is, therefore, interesting for several reasons. In currently popular cosmological models galaxies and galaxy clusters have evolved from the same field of density fluctuations in the early Universe. Studying the statistics of galaxies and clusters in the present-day Universe we are probing the same power spectrum for the primordial fluctuations on slightly different scales, that is scales of the order of 1 Mpc for galaxies and of the order of 10 Mpc for clusters in comoving units. Also the high-density environment of clusters influences the evolution of galaxies and a different morphological population of galaxies is observed in the field and in clusters [2]. Moreover clusters can be used as targets to study a population of galaxies at fixed distance and epoch. In addition the gravitational potential of the cluster collects the gas expelled by galaxies in early evolutionary phases and makes it observable through X-rays so that one may obtain information on early starburst phases in the life of the cluster galaxies [3-6].

Clusters of galaxies are filled with a hot tenuous plasma which is emitting X-rays over quite extended regions in the cluster. Thus in X-rays clusters can be observed as a whole entity. The ROSAT X-ray Observatory which was launched on June 1st, 1990 [7] with its higher spatial resolution and increased sensitivity compared to previous X-ray telescopes is an ideal instrument for the study of clusters of galaxies. Firstly the hot intracluster gas that makes these galaxy clusters very prominent X-ray sources has its radiative emission maximum right in the soft-X-ray region partly covered by the detection band of the ROSAT X-ray telescope (0.1 to 2.4 keV). And secondly clusters of galaxies are very large objects that can nicely be imaged as extended objects out to large distances provided that the exposure is sufficiently deep.

The ROSAT mission is divided into two parts. Half a year at the beginning of the mission was devoted to an All Sky Survey that lasted from August 1990 to January 1991. A small area that was missed in this period due to instrument failures has been filled in later in 1991 and the sky coverage is now complete. The survey is now followed by pointed observations of which a fraction of about 15% is devoted to the detailed study of individual clusters. There is, therefore, a twofold interest in research on clusters of galaxies conducted with the ROSAT X-ray Observatory: statistical studies of a large sample of X-ray detected clusters and the detailed investigation of the physical properties of individual clusters.

Clusters of galaxies are—except for quasars—the most powerful X-ray sources in the sky with luminosities of 10^{43} to $3 \cdot 10^{45}$ erg s^{-1}. The emission originates from hot gas that fills the gravitational-potential well of the cluster close to hydrostatic equilibrium. Therefore, the X-ray emission is a good tracer of the gravitational potential of the cluster. It also provides direct information on the gas density distribution in the cluster. Therefore, X-ray observations are an invaluable tool for the study of the cluster morphology, the mass distribution in the cluster, the dynamical state of the cluster and the physical state of the intracluster medium. Two of the most impressive results obtained so far with ROSAT on galaxy clusters is the richness of substructure observed in nearby clusters and the large extent of the X-ray-emitting gas as well as the high gas mass fractions deduced.

In the ROSAT All Sky Survey several thousand clusters of galaxies are expected to be detected in X-rays. This is comparable to the number of clusters that have been catalogued up to date notably by ABELL [8] and ABELL, CORWIN and OLOWIN [9]. However, the selection criteria for the detection of clusters on photographic plates are poorly understood and it is even unclear how many of the objects detected as rich clusters on the plates are actually chance projections of a series of smaller galaxy groups in the line of sight. There is now evidence for these projection effects from spectroscopic studies [10], clustering analysis of ACO clusters [11] and N-body simulations of the growth of structure in the Universe [12]. In X-ray observations the galaxy cluster surface brightness is roughly proportional to the square of the galaxy density and can be even more peaked in clusters with cooling flows. Clusters of galaxies can, therefore, be detected in X-rays as three-dimensional, gravitationally bound entities.

Therefore, the galaxy cluster sample detected in the ROSAT All Sky Survey will be a good basis for several cosmological studies like the evolution properties of clusters, the mass distribution and the spatial distribution of clusters.

All this information can be used to test cosmological models and to obtain information on the primordial density fluctuation spectrum in the early Universe.

In the following sections we will give a short description of the ROSAT Observatory (sect. 2), discuss the observed X-ray morphology and structure of some nearby clusters (sect. 3), show results on the determination of the gas and gravitational mass in a sample of nearby clusters (sect. 4), and discuss the cosmological consequences of the large gas mass fractions observed in these clusters (sect. 5). The last section provides a conclusion. For all the relevant physical parameters that scale with distance we have assumed a value for the Hubble constant of $50 \text{ km s}^{-1} \text{ Mpc}^{-1}$.

The work presented here has been done in close collaboration with colleagues at MPE whose contributions I like to acknowledge: U. G. BRIEL, H. EBELING, G. HARTNER, D. NEUMANN, M. PIERRE, S. SCHINDLER, R. A. SCHWARZ, W. VOGES, R. G. CRUDDACE (NRL), A. EDGE (IoA Cambridge) and J. P. HENRY (IfA Hawaii).

2. – The ROSAT Observatory.

A description of the ROSAT Observatory can be found in [7, 13-16]. The main instrument on board of the ROSAT observatory is the X-ray telescope that consists of a fourfold nested Wolter-type mirror configuration. It covers an energy range from 0.07 to 2.4 keV. The second instrument is the Wide Field EUV Camera with a spectral window from 20 to 300 eV. Due to the strong absorption of the interstellar medium of the Galaxy in the EUV band the Wide Field Camera is only of limited application for extragalactic objects such as clusters and we will here concentrate on the results from the X-ray telescope.

The X-ray telescope carries two types of focal-plane detectors. The Position Sensitive Proportional Counter (PSPC) has a spatial resolution better than 30 arcsec within a radius of 20 arcmin around the telescope axis. It has a limited energy resolution which is roughly about 45% FWHM at 1 keV and the relative resolution varies roughly inversely proportional to the square root of the energy. The second type of detector, the high-resolution imager (HRI), has a better spatial resolution of less than 5 arcsec near the telescope axis but no energy discrimination.

The X-ray-emitting hot intracluster gas in galaxy clusters is optically thin and has a temperature roughly equivalent to the virial temperature of the cluster potential well. The expected and actually observed temperatures range from 2 to 10 keV. Figure 1 shows the spectra for hot plasma with temperatures in this range normalized to the same emission measure. One notes that in the temperature range relevant for clusters of galaxies the spectra are quite simi-

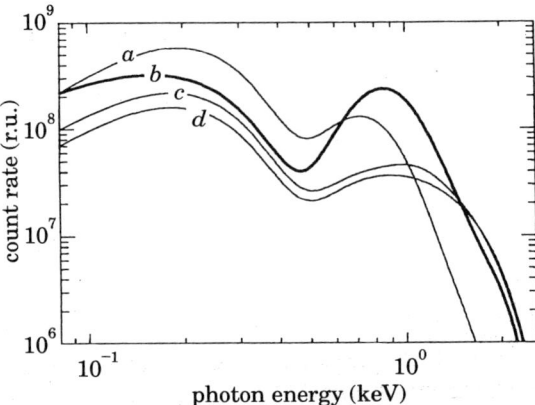

Fig. 1. – PSPC X-ray spectra for optically thin thermal plasma at temperatures of $3 \cdot 10^6$ K (a), $1 \cdot 10^7$ K (b), $3 \cdot 10^7$ K (c) and $1 \cdot 10^8$ K (d).

lar. The differences are becoming larger only for the lower temperatures. The ROSAT energy window is unfortunately too small to observe the exponential cut-off in the bremsstrahlung spectrum for the high cluster temperatures. Also the absolute energy flux integrated over the ROSAT window (defined as 0.1 to 2.4 keV) is not very sensitive for the gas temperature. Therefore, temperatures of the cluster gas can only be determined for spectra of very high quality (as shown below). This effect has the advantage, on the other hand, that the X-ray flux is directly related to the emission measure and thus gas densities can be determined without having precise temperature information.

During the ROSAT All Sky Survey [17] the sky was scanned by the ROSAT telescope in great circles in a plane perpendicular to the solar direction. Follow-

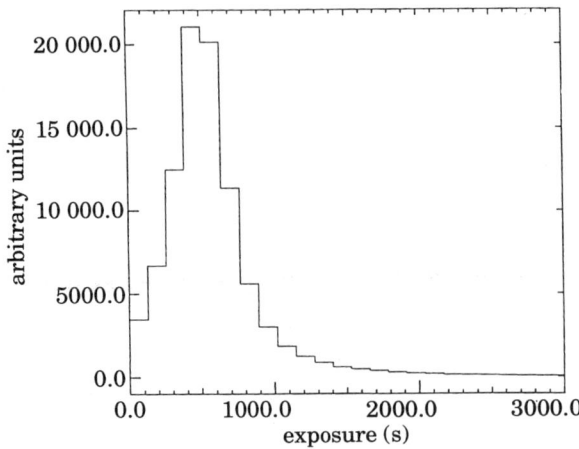

Fig. 2. – Exposure time histogram for the ROSAT All Sky Survey.

ing the Sun the whole sky is thus covered in half a year. Due to the overlap of the scan circles at the ecliptic poles the exposure is much greater at the poles. Figure 2 gives an exposure time histogram for the entire sky. In the first round of data analysis 49198 X-ray sources were detected (after removal of sources multiply detected in several strips) in the All Sky Survey.

3. – Observations of nearby clusters of galaxies.

The galaxy field of the nearby Virgo cluster extends over more than 10 degrees in the sky and, therefore, only selected fields were imaged with the EINSTEIN observatory [18]. In these observations X-ray emission in a halo of M 87 was detected extending out to about a degree. Scans with the nonimaging collimated detector on GINGA revealed extended emission on scales of several degrees [19].

In the ROSAT All Sky Survey the Virgo region was scanned almost homogeneously with an average exposure time of 460 s. Figure 3 shows a contour plot of the X-ray surface brightness in the energy band from 0.4 to 2.4 keV. The image has been smoothed with a variable Gaussian filter that provides a large-scale smoothing of up to 20 arcmin half-width for the low-surface-brightness regions but causes little distortion to the point sources. The most prominent fea-

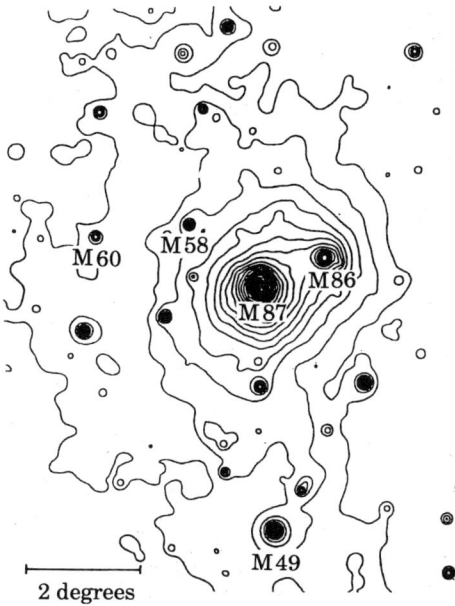

Fig. 3. – Contour plot of an X-ray image of the Virgo cluster from the ROSAT All Sky Survey. Only hard photons ($E \geq 0.4$ keV) and a variable Gaussian filter with a width of up to 20 arcmin were used to construct this image. The most prominent elliptical galaxies with observed X-ray emission are labelled in the plot.

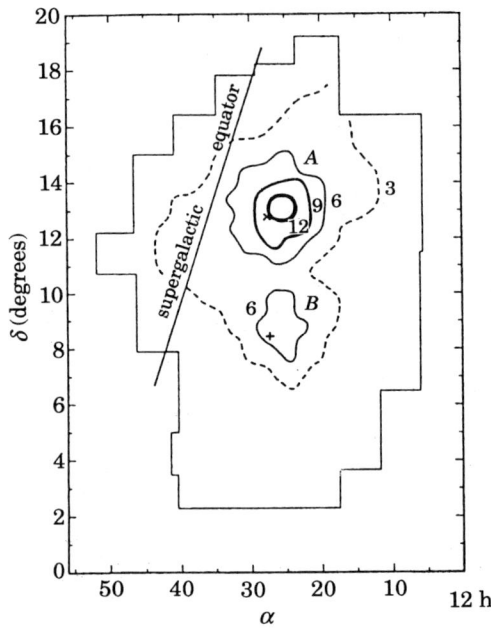

Fig. 4. – Contour plot of the galaxy density in the Virgo cluster from the photometric survey of Binggeli, Tammann and Sandage [20]. The large-scale structure is very similar to the X-ray image; × M 87, + M 49 (NGC 4472).

ture is the giant almost spherically symmetric X-ray halo around the bright elliptical galaxy M 87. The X-ray halos around the ellipticals M 86 and M 49 are also clearly visible and M 58 and M 60 show up as X-ray sources. Faint nonsymmetrical X-ray emission from the cluster extends over the whole region including both M 87 and M 49 with dimensions of about 7 and 4 degrees in north-south and east-west direction, respectively.

Figure 4 shows the density distribution of the galaxies in Virgo from the photometric survey of Binggeli, Tammann and Sandage [20]. The contour map of the galaxy column density is surprisingly similar to the X-ray image. In particular the asymmetric extension to the east in the direction of M 60 is visible in both displays. There are also interesting differences, however. While in the optical M 49 appears as the centre of a substantial subcluster in Virgo with an assumed mass of about one-third of the mass of the main substructure around M 87, the X-ray halo around M 49 is dwarfed by the halo of M 87 and it also has a very small core radius that is unresolved in these data and is quite untypical for a whole group or subcluster of galaxies.

The X-ray data of the Virgo cluster involve more than 30 000 photons, which allows one to obtain spectra of several regions throughout the cluster. In a first analysis the halo region of M 87 was subdivided into concentric rings with outer radii of 5, 10, 20, 60, 120 and 180 arcmin. Source spectra for the regions were

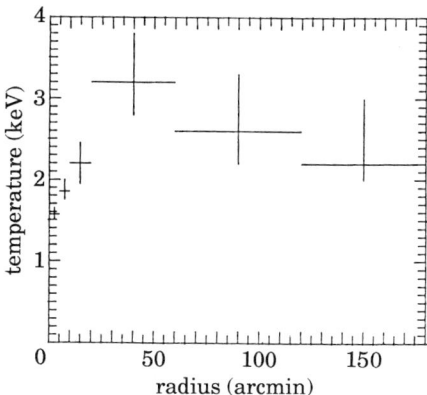

Fig. 5. – Temperature distribution of the X-ray-emitting gas in the halo of M 87 determined by fits of thermal model spectra to the PSPC data. The data were binned in concentric rings around M 87. A significant temperature decrease towards the centre is observed which is a signature of the cooling flow.

determined by subtracting the background emission obtained from two areas outside this halo region in the west and north of the cluster. The region to the east of the Virgo cluster has been found to be partly contaminated by enhanced soft foreground emission from the north polar spur region. The source spectra were then fitted by theoretical spectra from a Raymond-Smith model [21].

The result for the spectral analysis in all the regions is shown in fig. 5. There is a significant decrease in the derived temperature towards the centre in the inner region. This area is coincident with the cooling-flow region with a radius of about 100 kpc (17 arcmin). The temperature decrease is in agreement with the range of values one obtains by performing a one-phase cooling-flow structure analysis [22]. The current results lend good support to the cooling-flow picture, since the analysis of energy conservation without significant heat input of the surface brightness profile in the cooling-flow region leads to a similar temperature structure as it is observed with the ROSAT PSPC. To produce this result in the presence of a significant heat source would require a high degree of fine tuning.

A temperature determination of the overall region of the M 87 halo within 1 degree radius results in a value of $2.4^{+0.35}_{-0.2}$ keV which is in good agreement with the results from GINGA with a value of (2.38 ± 0.2) keV [23].

A deprojection analysis of the surface brightness profile (following [24]) yields a cooling-flow radius of about 120 kpc inside which the cooling time is shorter than the Hubble time. This cooling-flow radius of ~ 20 arcmin is consistent with the observed temperature drop inside a 15 arcmin radius and also its magnitude is within the limits of the predictions of a one-phase cooling-flow model. The flow rate in the cooling flow that is determined from the surface

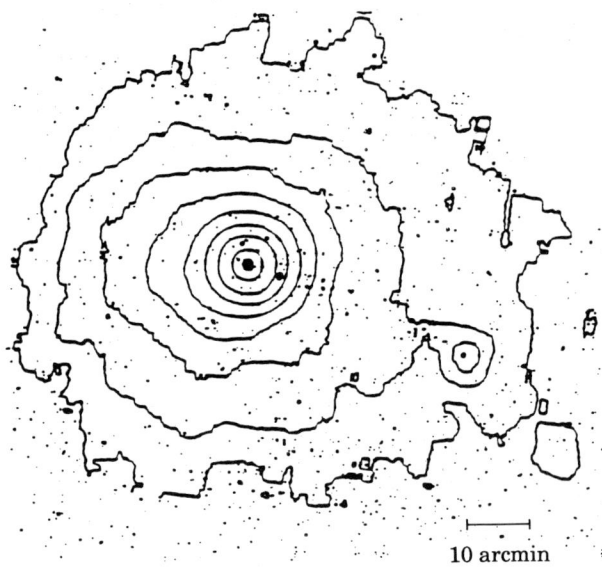

Fig. 6. – X-ray image of the Perseus cluster observed in the All Sky Survey. The contour levels are logarithmically spaced. The X-ray map is superposed onto the POSS plate. The main maximum coincides with NGC 1275, while the small western maximum coincides with IC 310.

brightness profile has a value of 20 (± 4) $M_\odot$ y^{-1} in good accord with the results by STEWART et al. [22].

The brightest cluster of galaxies detected in X-rays is the Perseus cluster. In the ROSAT All Sky Survey it was observed for ~ 505 s and detected with a count rate of 33.6 counts/s, where the counts were integrated out to a radius of 42 arcmin [25]. Figure 6 shows the X-ray image of Perseus from the All Sky Survey superposed on the POSS plate image. In previous observations with the EINSTEIN IPC [26] only the inner region of the cluster was imaged. Now with the «unlimited field of view» in the survey mode the X-ray halo was detected out to a radius larger than 2 Mpc. The asymmetric extension of the inner contour lines to the east which was already seen in the EINSTEIN image is also visible here. More surprisingly a trace of the large prominent chain of galaxies in Perseus is also clearly seen in the X-ray image on an even larger scale. An indication of this structure was also observed with the collimator experiment on Spartan [27]. The second maximum of the X-ray surface brightness within the western tail in the X-ray contours following the chain of galaxies coincides with the galaxy IC 310.

The Perseus cluster, which has a very strong cooling flow, shows a similar temperature decrease towards the cluster centre. Also here the temperature variation is consistent with the cooling-flow analysis that has also been performed previously with the imaging data of the EINSTEIN observatory by FABIAN et al. [24].

An analysis of the temperature distribution throughout Perseus on larger scale revealed another interesting result [25]. While the overall temperature is about $5.3(\pm 1.4)$ keV, one finds a significantly lower temperature of about $2.15[^{+0.9}_{-0.3}]$ keV on the eastern side where the excess emission reported above is observed. A plausible explanation for this phenomenon is the recent infall of a group or small cluster of galaxies into Perseus at the eastern side. If the gas from the group has not been thermalized and has a temperature in the range 1 to 2 keV, a small fraction of this gas superposed on the cluster emission can lead to the reduced temperatures observed. Thus the eastern asymmetry of Perseus and the temperature anomaly could be explained quantitatively by the same effect. In addition this shows that Perseus is a very unrelaxed cluster which also explains why there is such a large discrepancy in the mass determination from the optically observed galaxy velocity dispersion and from the X-ray data [24, 28, 29].

A signature for a cluster merger was observed even more prominently in the Coma cluster in the ROSAT survey [30]. A larger subclump is visible in the southwest of the main cluster and it is coincident with a subgroup in the galaxy density centred on NGC 4839 which was detected clearly in a photometric survey by MELLIER et al. [31].

An EXOSAT observation that was pointed near NGC 4839 revealed an X-ray gas temperature of $1.9[^{+1.7}_{-0.7}]$ keV for this subcluster. The X-ray luminosity in the ROSAT band for this subcluster is $3 \cdot 10^{43}$ erg s^{-1}. These values are typical for groups of galaxies (see [32]) and the temperature and luminosity value are well fitting with the typical temperature luminosity correlation for clusters (e.g., [33]), which supports the interpretation that this structure is a separate group of galaxies which is just about to merge with the main cluster.

A massive group of galaxies merging with a main cluster component has also been found in an early pointed observation of A 2256 [34]. In this case the signature of a merger event gives a nicely consistent picture in the optical, radio and X-ray image of the cluster [34-36]. A careful study of substructure in a statistical sample of clusters may provide interesting constraints on the mean density of the Universe, because clusters are only expected to be still growing if the mean density of the cluster environment is close to the critical density (e.g., [37]).

4. – Gas mass and gravitational mass of galaxy clusters.

Since the X-ray emissivity of the thermal gas in the ROSAT energy band is only weakly dependent on the temperature in the relevant temperature range, the observed X-ray surface brightness directly provides a value for the emission measure in the line of sight. Assuming that the cluster studied is spherically symmetric, the gas density profile can then readily be calculated.

For the ROSAT observations of clusters investigated so far we found that the β-model (*e.g.*, [38]) provides a good fit to the radial X-ray surface brightness profile even out to large radii. Therefore, the results of the fits where used for the further modelling with formulae of the following form:

$$S_x(R) = S_0 \left(1 + \frac{r^2}{r_c^2}\right)^{-3\beta + 1/2}. \tag{1}$$

The gas density is then given by

$$n_e(r) = n_{e0} \left(1 + \frac{r^2}{r_c^2}\right)^{3/2\beta} \tag{2}$$

with

$$n_{e0}^2 = S_0 \frac{(1+z)^4 \Gamma(3\beta)}{4\pi \Lambda_x r_c \Gamma(1/2) \Gamma(3\beta - 1/2)},$$

where S_0 is the central X-ray surface brightness in steradian, Λ_x is the emissivity of the thermal gas in the ROSAT band, r_c is the fitted core radius for the gas distribution, and n_e is the electron density in the intracluster medium.

The gas mass profile can then directly be obtained by integrating eq. (2). The errors introduced in this approach result from the assumption of isothermality which accounts for an uncertainty of less than 10%, the accuracy of the β-model fit which is good enough not to introduce a large error, and the assumption of spherical symmetry. The latter is accounted for by cutting out those regions that show local features of substructure as in Perseus and Coma. Another problem is the overall ellipticity that may introduce errors up to 10%.

The simultaneous knowledge of the gas density and temperature profile of a spherically symmetric cluster allows one to determine the gravitational-mass profile of the cluster on the assumption that the gas is in hydrostatic equilibrium. Since the sound travel distance in a Hubble time in the intracluster gas is of the order of 10 Mpc, large pressure gradients and large gas velocities are not expected except for local regions during a merger event. Therefore, the assumption of hydrostatic equilibrium should be a fair approximation. The cluster mass profile is then given by

$$M(r) = -\frac{kT_g(r) r^2}{m_h \mu G} \left(\frac{d \log T_g(r)}{dr} + \frac{d \log \rho(r)}{dr}\right), \tag{3}$$

where $\rho(r)$ and $T_g(r)$ are the density and temperature profile of the gas and the other parameters have their usual meaning.

The information on the temperature distribution is so sparse, however, that it is impractical to directly apply eq. (3) for the determination of the gravitational-mass profile. Therefore, we adopted two approaches to constrain the cluster masses. In one series of models we calculate the temperature profile (as well as the emission measure-weighted temperature profile integrated over the line of sight) for clusters with the given gas distribution and a large range of model potentials in a similar way as HUGHES[39]. The combination of all models that are consistent with the observed temperature and the actual gas density distribution define the range of allowed mass profiles. Alternatively we also explore the range of mass profiles for the allowed range of models with polytropic temperature distributions. Even though the model-dependent approach of the mass determination used here is, of course, somewhat unsatisfactory, we are convinced that the combination of the two methods allows us to cover the whole relevant parameter space for models with a reasonably smooth temperature profile.

The resulting curves for the gravitational and gas mass profiles for the Perseus clusters are shown in fig. 7. The observational data for the X-ray surface brightness profile extend only out to about 3 Mpc. At this radius the cluster mass reaches a value of $(1\text{--}2.6) \cdot 10^{15} h_{50}^{-1} M_\odot$, while the gas mass is $(3.2\text{--}4.3) \cdot 10^{14} h_{50}^{-2.5} M_\odot$. Thus one finds that the gas contributes $(13\text{--}43)\% \, h_{50}^{-1.5}$ to the mass of the cluster over this large volume that may comprise the total cluster.

In fig. 8, where the density profiles for the gas and the total matter are shown, one notes that the profiles for the gas and the gravitational matter touch for the mass models with steep gravitational-mass profiles. This effect is one of the interesting constraints that can be set by these model calculations. In the

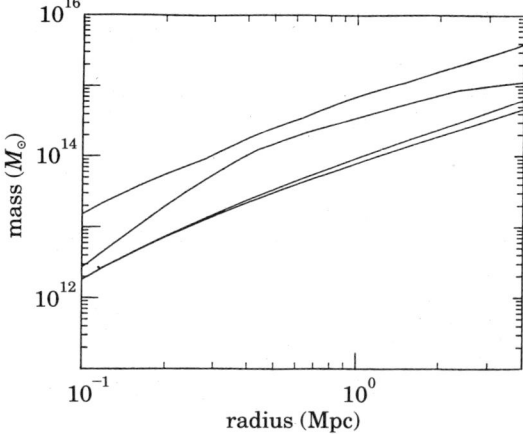

Fig. 7. – Integrated radial mass profiles for the X-ray-emitting gas and the gravitating matter in the Perseus cluster of galaxies as determined from X-ray data. The upper two curves bracket the allowed range for the gravitational-mass profile, while the lower two curves give the upper and lower limit to the gas mass profile.

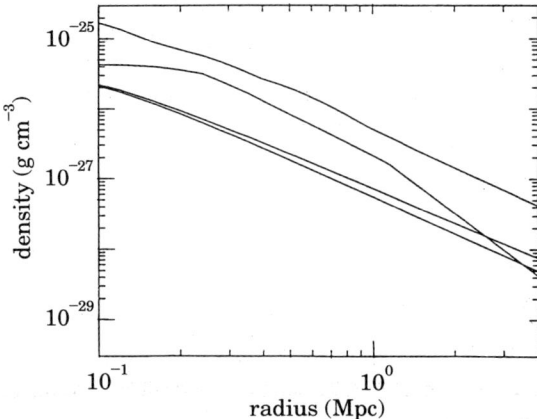

Fig. 8. – Radial density profiles for the X-ray-emitting gas and the gravitating matter in the Perseus cluster. The upper two curves give the upper and lower limit for the gravitational-mass density, while the lower two curves give the upper and lower limit to the gas density.

case of Perseus gravitational-mass models with a total matter density distribution that falls off steeper than r^{-3} are excluded by the values for the gas density deduced for large radii.

A qualitatively very similar result is obtained for the Coma cluster as observed in the ROSAT All Sky Survey [30]. Here the X-ray emission can be followed out to a radius of $(3-4)\,\mathrm{Mpc}\,h_{50}^{-1}$. The total mass determined for a radius of 4 Mpc is $(1.1-2.8)\cdot 10^{15}\,h_{50}^{-1}\,M_\odot$ and the gas mass fraction is $(14-44)\%\,h_{50}^{-1.5}$.

For the observations of Virgo in the ROSAT survey a major part of the cluster can be modelled by subtracting a spherically symmetric X-ray halo around M 87 from the total cluster image. The X-ray emission in this halo component extends to about $(1.5-1.8)\,\mathrm{Mpc}$. Modelling the mass distribution in this halo yields results of $(1.3-6.6)\cdot 10^{14}\,h_{50}^{-1}\,M_\odot$ and a gas mass fraction of $(5-31)\%\,h_{50}^{-1.5}$ for a radius of 1.8 Mpc. Compared to Perseus and Coma the halo of M 87 is a less massive matter complex. Another example for a not so massive system is the Centaurus cluster. The temperature information on this system is more sparse than for the above-mentioned systems and thus only less detailed modelling is possible. From polytropic models alone we obtain the following mass constraints: $(1.9-3.8)\cdot 10^{14}\,h_{50}^{-1}\,M_\odot$ and gas mass fractions of $(13-27)\%\,h_{50}^{-1.5}$.

Two effects have been neglected in the above analysis. First, the hydrostatic equation may have to be modified to account for pressure from turbulence, magnetic fields and cosmic rays. While these contributions may in fact be large in active regions of the cluster like cooling flows, they are probably not so important in the outer regions of the cluster. Second, if the gas is clumped on a scale smaller than the resolution of the telescope, the gass mass is overestimated by a factor $f = \langle n_e^2 \rangle / \langle n_e \rangle^2$. Information on f can be obtained from an analysis of the inhomogeneous mass deposition in cooling flows (e.g., [40]) which shows that the overestimate is probably smaller than a 25%. Taking these corrections into account one may conclude that the gas fraction in richer clusters is about $(10-30)\%$.

5. – Cosmological implications of the gas content in galaxy clusters.

The gas mass fractions observed in clusters of galaxies do have very interesting implications if we take the values found as typical for the rest of the Universe. To justify this extrapolation we may consider the formation process of a cluster in the standard bottom-up hierarchical clustering scenario in the frame of a dark-matter cosmological model. At recombination the protocluster consists of a density fluctuation primarily in the dark-matter distribution. But baryons soon follow these fluctuations and over the rest of the linear-growth phase of the fluctuations the ratio of the dark matter to baryon mass prevailing in the Universe is preserved in the overdense protocluster region. If we now assume that no matter is expelled from the cluster during virialization and thermalization in the collapse phase, one should still observe the universal matter ratio if one takes account of the whole cluster entity.

The present ROSAT observations offer for the first time the possibility of studying the mass distribution in a cluster out to a radius of about 3 Mpc (one Abell radius). Assuming that the Abell radius is close to the edge of the virialized cluster, one can in fact take the baryon-to-dark-matter ratio found in the cluster as approximately representative for the whole Universe.

These results and assumptions would have several important consequences. Most interesting is a comparison with the current understanding of the primordial nucleosynthesis. In a recent paper by WALKER et al. [41], for example, a constraint on the baryon density in the Universe was derived from nucleosynthesis models and a comparison with observations for the primordial abundance of D, ^{3}H, ^{4}H and ^{7}Li. Compared to a critical-density Universe these results give a constraint for the baryon density of $\Omega_B = 0.05 \pm 0.01$. Taking these numbers at face value together with the above result of $\Omega_B/\Omega_0 \geq 0.1$, one would conclude that $\Omega_0 \leq 0.6$ and that we are living in an undercritical-density Universe. It may be too preliminary to make such a strict use of our error margins. But this line of thought emphasizes how important the precise determination of cluster masses is and that we need even better data to decide on this problem.

Another interesting consequence is the fact that also outside clusters the integrated galaxy mass cannot account for all the baryon mass. If one takes a critical-density Universe with a mass-to-light ratio of $M/L = (500-1000) h_{50} M_\odot / L_\odot$ (e.g., [42]) and a typical galaxy mass-to-light ratio of $M/L = 5 M_\odot / L_\odot$, one finds that galaxies account only for a fraction of $\Omega_{gal} =$

$= (0.005\text{--}0.01) h_{50}^{-1}$ of the Universe. Making the extreme assumption that galaxies are exclusively made up by baryons, we find an efficiency factor for the formation of galaxies out of the gas in the Universe of $\varepsilon = 0.02\text{--}0.1$ for an $\Omega_0 = 1$ Universe or $\varepsilon = 0.1\text{--}0.5$ for an $\Omega_0 = 0.2$ Universe. This efficiency factor is especially low in the critical-density Universe. Recalling the scenario of galaxy formation one would expect that a large fraction of the mass in the Universe is contained in overdense fluctuations on galaxy scale at the time of galaxy formation (that is when the r.m.s. of the fluctuations on galaxy scale is about 1 at a redshift around 2) and that these fluctuations rapidly collapse and cool to form galaxies. If the galaxy formation efficiency is in fact lower than 0.1, a mechanism has to be at work that prevents a larger fraction of the potential galaxies from forming. This offers a very interesting clue to the concept of biased galaxy formation which was introduced mostly to reconcile the observed distribution of galaxy peculiar velocities with the observed galaxy density variations [43, 44]. If the above-inferred mechanism that inhibits galaxy formation is designed in a proper way, it could naturally lead to biasing.

At last one can wonder how such a high baryon density outside galaxies could have escaped detection so far. The most simple assumption is that they may be in the form of a tenuous, homogeneous plasma. Recently BARCONS, FABIAN and REES [45] have investigated the constraints on such a plasma set by the Gunn-Peterson effect and the upper limits on the soft-X-ray background and the comptonization of the microwave background radiation. The constraints would allow a plasma density in the range of $\Omega_B = 0.1\text{--}0.3$ if the plasma temperature is around 10^5 K. The cooling time of this plasma would slightly be larger than the Hubble time and thus there is no problem for its existence.

6. – Conclusions.

One of the surprising results of the ROSAT observations on nearby clusters presented in this lecture is the rich degree of substructure revealed by the high-sensitivity, low-background X-ray telescope. The substructure features are not only detected in spatial images, but, as in the case of Perseus, an unrelaxed dynamical state of the cluster may also be inferred from a peculiar temperature distribution. The occurrence of substructure is interesting because it is a signature that clusters are still growing at present by the merging of subcomponents. This is only possible when the mean density in the Universe is not too low.

The second interesting result is the large amount of gas that is observed in the clusters. The gas mass (together with the integrated galaxy mass) is clearly still not large enough to account for all the gravitational mass needed to bind the cluster. The gas component is a factor of 4 to 6 times more massive than the galaxies in the rich clusters (see also [6]) and has a broader distribution than the galaxies. One may conclude that most of this gas has, therefore, to be primordial.

* * *

I like to thank the ROSAT team for the help in the data preparation and data reduction and for making these unique ROSAT observations possible.

REFERENCES

[1] I. R. KING: *Astron. J.*, **71**, 64 (1966).
[2] A. DRESSLER: *Annu. Rev. Astron. Astrophys.*, **22**, 185 (1984).
[3] F. MATTEUCCI and P. VETTOLANI: *Astron. Astrophys.*, **202**, 21 (1988).
[4] L. P. DAVID, W. FORMAN and C. JONES: *Astrophys. J.*, **380**, 39 (1991).
[5] R. E. WHITE III: *Astrophys. J.*, **367**, 69 (1991).
[6] M. ARNAUD, R. ROTHENFLUG, O. BOULADE, L. VIGROUX and E. VANGIONI-FLAM: *Astron. Astrophys.*, **254**, 49 (1992).
[7] J. TRÜMPER: *Science*, **260**, 1769 (1993).
[8] G. O. ABELL: *Astrophys. J. Suppl.*, **3**, 211 (1958).
[9] G. O. ABELL, H. G. CORWIN and R. P. OLOWIN: *Astrophys. J. Suppl.*, **70**, 1 (1989).
[10] J. P. HUCHRA, J. P. HENRY, M. POSTMAN and M. J. GELLER: *Astrophys. J.*, **365**, 66 (1990).
[11] W. J. SUTHERLAND: *Mon. Not. R. Astron. Soc.*, **234**, 159 (1988).
[12] C. S. FRENK, S. D. M. WHITE, G. EFSTATHIOU and M. DAVIS: *Astrophys. J.*, **351**, 10 (1990).
[13] J. TRÜMPER: *Adv. Space Res.*, **2**, 241 (1983).
[14] B. ASCHENBACH: *Appl. Opt.*, **27**, 1404 (1988).
[15] E. PFEFFERMANN, U. G. BRIEL, H. HIPPMANN, G. KETTENRING, G. METZGER, P. PREDEHL, G. REGER and K. H. STEPHAN: *Proc. SPIE*, **733**, 519 (1986).
[16] U. G. BRIEL, E. PFEFFERMANN, G. HARTNER and G. HASINGER: *Proc. SPIE*, **982**, 401 (1988).
[17] W. VOGES: in *Optical Digital Sky Surveys*, edited by H. T. MACGILLIVRAY and E. B. THOMSON (Kluwer Academic Publishers, Amsterdam, 1992), p. 453.
[18] W. FORMAN, C. JONES and M. DEFACCIO: *The Virgo Cluster*, edited by O.-G. RICHTER and B. BINGGELI, ESO Workshop Proceedings No. 20 (1985), p. 323.
[19] S. TAKANO, H. AWAKI, K. KOYAMA, H. KUNIEDA, Y. TAWARA, S. YAMAUCHI, K. MAKISHIMA and T. OHASHI: *Nature (London)*, **340**, 289 (1989).
[20] B. BINGGELI, G. A. TAMMANN and A. SANDAGE: *Astron. J.*, **94**, 251 (1987).
[21] J. C. RAYMOND and B. W. SMITH: *Astrophys. J. Suppl.*, **35**, 419 (1977).
[22] G. C. STEWART, C. R. CANIZARES, A. C. FABIAN and P. E. J. NULSEN: *Astrophys. J.*, **278**, 536 (1984).
[23] K. KOYAMA, S. TAKANO and Y. TAWARA: *Nature (London)*, **350**, 135 (1991).
[24] A. C. FABIAN, E. M. HU, L. L. COWIE and J. GRINDLEY: *Astrophys. J.*, **248**, 47 (1981).

[25] R. A. SCHWARZ, A. C. EDGE, W. VOGES, H. BÖHRINGER, H. EBELING and U. G. BRIEL: *Astron. Astrophys.*, **256**, L11 (1992).
[26] G. BRANDUARDI-RAYMONT, D. FABRICANT, E. FEIGELSON, P. GORENSTEIN, I. GRINDLEY, A. SOLTAN and G. ZAMORANI: *Astrophys. J.*, **248**, 55 (1981).
[27] W. A. SNYDER, M. P. KOWALSKI, R. G. CRUDDACE and G. G. FRITZ: *Astrophys. J.*, **365**, 460 (1990).
[28] S. M. KENT and W. L. W. SARGENT: *Astron. J.*, **88**, 697 (1983).
[29] C. J. EYLES, M. P. WATT, D. BERTRAM, M. J. CHURCH, T. J. PONMANN, G. K. SKINNER and A. P. WILLMORE: *Astrophys. J.*, **376**, 23 (1991).
[30] U. G. BRIEL, J. P. HENRY and H. BÖHRINGER: *Astron. Astrophys.*, **259**, L31 (1992).
[31] Y. MELLIER, G. MATHEZ, A. MAZURE, B. CHAUVINOEAU and D. PROUST: *Astron. Astrophys.*, **199**, 67 (1988).
[32] G. A. KRISS, D. S. CIOFFI and C. R. CANIZARES: *Astrophys. J.*, **272**, 439 (1983).
[33] A. C. EDGE and G. C. STEWART: *Mon. Not. R. Astron. Soc.*, **252**, 428 (1991).
[34] U. G. BRIEL, J. P. HENRY, R. A. SCHWARZ, H. BÖHRINGER, H. EBELING, A. C. EDGE, G. D. HARTNER, S. SCHINDLER and W. VOGES: *Astron. Astrophys.*, **246**, L10 (1991).
[35] A. C. FABIAN and S. J. DAINES: *Mon. Not. R. Astron. Soc.*, **252**, 17p (1991).
[36] H. BÖHRINGER, R. A. SCHWARZ, U. G. BRIEL, W. VOGES, H. EBELING, G. HARTNER and R. G. CRUDDACE: in *Clusters and Superclusters of Galaxies*, edited by A. C. FABIAN (Kluwer Academic Publishers, Amsterdam, 1992), p. 71.
[37] D. RICHSTONE, A. LOEB and E. L. TURNER: *Astrophys. J.*, **393**, 477 (1992).
[38] C. JONES and W. FORMAN: *Astrophys. J.*, **276**, 38 (1984).
[39] J. P. HUGHES: *Astrophys. J.*, **337**, 21 (1989).
[40] P. A. THOMAS: *Mon. Not. R. Astron. Soc.*, **220**, 949 (1986).
[41] T. P. WALKER, G. STEIGMAN, D. N. SCHRAMM, K. A. OLIVE and H.-S. KANG: *Astrophys. J.*, **376**, 51 (1991).
[42] V. TRIMBLE: *Annu. Rev. Astron. Astrophys.*, **25**, 423 (1987).
[43] S. D. M. WHITE, C. S. FRENK, M. DAVIS and G. EFSTATHIOU: *Astrophys. J.*, **313**, 505 (1987).
[44] N. KAISER: *Astrophys. J.*, **284**, L9 (1984).
[45] X. BARCONS, A. C. FABIAN and M. REES: *Nature (London)*, **350**, 685 (1991).

Clusters of Galaxies as Probes for the Large-Scale Structure.

H. BÖHRINGER and S. SCHINDLER

Max-Planck-Institut für Extraterrestrische Physik - D-8046 Garching, B.R.D.

1. – Introduction.

The formation of galaxies is closely coupled to the evolution of the large-scale structure of the matter distribution in the Universe as a whole. In the currently popular models of cosmic evolution galaxies form from density fluctuations on a scale of the order of 1 Mpc (in comoving units) in the post-recombination Universe. The large-scale density distribution is thus controlling the local conditions for galaxy formation; while as a consequence the distribution of the galaxies provides information on the large-scale structure.

In fact our present knowledge on the large-scale structure of the Universe has almost exclusively been derived from the investigation of the spatial distribution of galaxies (*e.g.*, [1-5]). Hereby it is still a matter of debate how closely the galaxy distribution actually follows the density distribution of all matter [6, 7]. In this lecture we will discuss the study of the large-scale structure not by means of galaxies but by using the next larger astronomical objects, clusters of galaxies, as probes for the cosmic structure. In the hierarchical sequence of structures of increasing order of integration clusters of galaxies are probably the last clearly defined step. All structures of larger size, like superclusters, sheets, filaments and voids, are far from being describable by a proper physical state. The reason why one has not made much use so far of clusters of galaxies as tracers of the large-scale structure is the sparcity and imprecision of the currently available data.

This situation is now rapidly changing. The use of automatic scanning machines for astronomical photographic plates, like AMP and COSMOS [5, 8-10], to produce large galaxy and cluster catalogues, the increased velocity with which redshifts can be obtained, and the possibility of detecting many clusters independently in X-ray surveys (as described in the second lecture) will offer the possibility of using clusters of galaxies to study the mass distribution in the Universe in a similar fashion as with galaxies but at an even larger scale.

In this lecture we, therefore, like to investigate how clusters of galaxies can be used to explore the large-scale structure of the Universe and to obtain information on the cosmological initial conditions. We will concentrate here mainly on the study of the cluster mass function as cosmological diagnostics. It will be discussed in the following how the primordial density fluctuation field is characterized (sect. 2) and how clusters form (sect. 3). The cluster mass function is derived and compared to observations in sect. 5 and the intracluster medium and the X-ray appearance of clusters is discussed in sect. 6. The last section provides a conclusion.

We will refer to the most popular cosmological models, that is Friedmann-Lemaître cosmologies with primordial Gaussian density fluctuations and weakly interacting particles as the major mass component of the Universe. For numerical examples we will prefer a value for the Hubble constant of $H_0 = 50 h_{50}$ km s^{-1} Mpc^{-1} with $h_{50} = 1$ which is most commonly used in observational X-ray astronomy, and we will use h_{50} as a scaling parameter where it is useful.

About the formation of galaxies and the large-scale structure of the Universe a number of excellent reviews were written as, for example, by PEEBLES[11], EFSTATHIOU and SILK[12], SZALAY[13], LONGAIR[14], WHITE[15], EFSTATHIOU[16]. A similar, more extended version of this lecture can be found in [17].

2. – Cosmological framework.

In the following we briefly recall some relations concerning the cosmological evolution of the Universe and of density fluctuations. A derivation of these relations can be found in standard textbooks (*e.g.*, [18-21]).

The expansion of the Universe is governed by the following equations (assuming that the cosmological constant is zero):

$$\dot{a}^2 = H_0^2 \left(1 - \Omega_0 + \frac{\Omega_0}{a^\varepsilon}\right), \tag{1}$$

where a is the scaling parameter of the Universe defined such that $a(t = 0) = 1$ (for the present epoch). The Hubble constant, H, and the density parameter, Ω, are defined as $H = \dot{a}/a$ and $H^2 \Omega = (8\pi/3) G\rho$. H_0 and Ω_0 are the values at the present time. In eq. (1) $\varepsilon = 1$ for a matter-dominated Universe and $\varepsilon = 2$ for the radiation-dominated Universe. Thus the expansion is characterized by a scaling parameter $a(t)$ which for a flat Universe is $\propto t^{2/3}$ for matter and $\propto t^{1/2}$ for radiation while for an open Universe with $\Omega \ll 1$ free expansion ($a \propto t$) prevails (for $\Omega_0 = 1$ the transition from the radiation-dominated dynamics to the matter-dominated dynamics of the Universe occurs at $z \sim 10^4$ when $a \sim 9.5 \cdot 10^{-5}$).

The evolution of the density fluctuations with small amplitude, $\Delta = \delta\varrho/\varrho$, is characterized by a growing and a decaying mode. If pressure and damping effects are neglected for a moment, the time dependence of the growing mode is given by $\Delta \propto t^{2/3}$ for matter and $\Omega = 1$, $\Delta \propto t$ for radiation and $\Omega = 1$, and $\Delta =$ = const for $\Omega \ll 1$.

In the radiation-dominated era two further effects modify the growth of density perturbations in a wavelength-dependent way. The radiation pressure prevents the growth of fluctuations on a scale of the size of the horizon of the Universe. And, if most of the matter density is made up by dark matter, density fluctuations are washed out due to diffusion of the dark-matter particles over a scale length comparable to the horizon radius as long as the particles are relativistic. The epoch when the various proposed dark-matter particles become nonrelativistic leads to a distinction between «cold», «warm» and «hot» dark matter (CDM, WDM, HDM). The redshifts for the transitions are $z \sim 4 \cdot 10^{12}$, $z \sim 4 \cdot 10^6$ and $z \sim 4 \cdot 10^4$ and the corresponding masses contained within the horizon are $M < M_\odot$, $M \sim 10^{10} M_\odot$ and $M \sim 10^{15} M_\odot$, respectively.

For a quantitative analysis of the evolution of the density fluctuation spectrum describable by a Gaussian field one usually chooses a Fourier representation. The fluctuation field is then statistically completely described by a power spectrum, $P(k)$ (e.g., [22]).

Of interest is now the variation of the mass or mean density in regions of the size of protoclusters. This can be investigated mathematically by means of a filtering of the original Gaussian field with a filter of the size of protoclusters. Popular filters for this purpose are Gaussian filters

$$(2) \qquad \Delta(\boldsymbol{x})_{R_f} = \int \frac{1}{(2\pi R_f^2)^{3/2}} \exp\left[-\frac{|\boldsymbol{x}-\boldsymbol{x}'|^2}{2R_f^2}\right] \Delta(\boldsymbol{x}') \, d\boldsymbol{x}'^3$$

with the corresponding modification of the power spectrum

$$(3) \qquad P(k)_{R_f} = P(k) W^2(k, R_f) \quad \text{with } W(k, R_f) = \exp\left[-\frac{k^2 R_f^2}{2}\right]$$

and top-hat filters

$$(4) \qquad \Delta(\boldsymbol{x})_{R_t} = \frac{3}{4\pi R_t^3} \int \Theta\left(1 - \frac{|\boldsymbol{x}-\boldsymbol{x}'|}{R_t}\right) \Delta(\boldsymbol{x}') \, d\boldsymbol{x}'^3$$

with the effective power spectrum

$$(5) \qquad P(k)_{R_t} = P(k) W^2(k, R_t) \quad \text{with } W(k, R_t) = \frac{3(\sin kR_t - kR_t \cos kR_t)}{(kR_t)^3} \, .$$

The variance of the density in the filtered field is then given by

(6) $$\langle \Delta(R)^2 \rangle = \sigma(R)^2 = \frac{1}{(2\pi)^3} \int W(k, R)^2 P(k) 4\pi k^2 \, dk \,.$$

If $P(k)$ is approximated by a power law with exponent n one gets

(7) $$\langle \Delta(R)^2 \rangle \propto R^{-(n+3)}$$

independent of the sort of filter used.

Let us now turn to the origin and evolution of the density fluctuation spectrum. The most popular assumption about the power spectrum of the density fluctuations that prevails in the very early Universe just after inflation is

(8) $$P(k)_{\text{init}} \propto k$$

which is often termed Zel'dovich spectrum (after ZEL'DOVICH [23]). From the early epoch to the post-recombination era the fluctuation field grows linearly, independent of the wavelength, but it is modified due to the radiation pressure and due to diffusion as mentioned above. This modification is in practice usually expressed in the form of a transfer function, $T(k)$, that modifies the initial power spectrum: $P(k) = T(k)^2 P(k)_{\text{init}}$ (see, e.g., [24-27]), where $P(k)$ is the power spectrum after recombination. Figure 1 shows, for example, the post-recombination power spectra for CDM, WDM and HDM for a Zel'dovich initial spectrum and transfer functions taken from [22] (BBKS in the following). The asymptotic behaviour of the CDM power spectrum is characterized by $P(k) \propto k$ at small k and $P(k) \propto k^{-3}$ at large k.

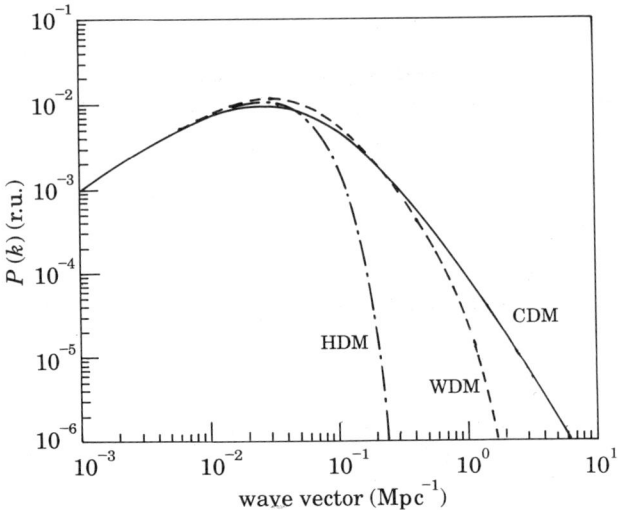

Fig. 1. – Post-recombination power spectra of the density fluctuation field for a Zel'dovich initial spectrum and hot (HDM), warm (WDM) and cold dark matter (CDM) (for $h_{50} = 1$).

3. – The formation of clusters of galaxies.

An overdense region with a mean density larger than the critical density of the Universe, $\rho > \rho_{\text{crit}}$, is bound to collapse eventually. The collapse time for an overdensity depends only on its amplitude:

$$\text{(9)} \qquad t_{\text{coll}} \propto \frac{R}{\langle v \rangle} \propto \frac{R}{\langle \dot{v} \rangle} \propto \frac{R^3}{\Delta M(r)} \propto \Delta.$$

Thus objects form simultaneously on all scales for a power spectrum with $P(k) \propto k^{-3}$. For flatter power spectra small objects form first. This is the case for all the regions of the power spectra shown in fig. 1 except for the steep cut-off regions in the WDM and HDM spectra. We will in the following only consider purely hierarchical clustering models like CDM.

If we approximate the overdense region (protocluster) by a sphere with homogeneous density, the dynamics of the protocluster can be calculated analytically. If we define the parameter Ω_i as the ratio of the mean density of the protocluster to the critical density at an initial time, t_i, we can express the time when the overdensity decouples from the Hubble flow and turns around by

$$\text{(10)} \qquad t_{\text{turn}} = \frac{\pi}{2} \left(\frac{b_t^3}{\Omega_i H_i^2} \right)^{1/2} = \frac{\pi}{2} \frac{\Omega_i}{H_i} \left(\frac{1}{\Omega_i - 1} \right)^{3/2}.$$

The recollapse of the protocluster takes again as long as the expansion and leads to an object with a finite radius in virial equilibrium. The collapse is characterized by a «thermalization» process termed «violent relaxation» described by LYNDEN-BELL [28].

For the case of $\Omega_0 = 1$ the density at turn around is higher by a factor of $(3\pi/4)^2 \sim 5.55$ than that of the background medium, and assuming that the mean radius decreases by a factor of 2 during violent relaxation one finds an overdensity of the collapsed region of ~ 177 at $t = t_{\text{rel}} = 2 t_{\text{turn}}$.

If one takes a look at one of the best-studied fairly relaxed clusters of galaxies, the Coma cluster, one finds for a characteristic radius of 3 Mpc and a deduced mass of the order of $10^{15} M_\odot$ an overdensity of about 200 (for $h_{50} = 1$). This implies that the Coma cluster collapsed very recently. Galaxies for comparison have characteristic overdensities that are several orders of magnitude larger due to an earlier formation time as well as dissipative processes that occur during their formation.

We want in the following to apply a statistical analysis to a linearly growing density field (assuming that small-amplitude approximation holds). Therefore, we need to know the linearly extrapolated values for the overdensity at t_{turn} and t_{rel} to identify those overdense regions that will have collapsed at a certain time. We use the result that the overdensities are growing proportional to the scaling factor a (for an $\Omega_0 = 1$ Universe). Conventionally one starts with overdensities that have no internal particular velocities (which is, of course, somewhat arbitrary). For this case only 3/5 of the initial protocluster contributes to the linearly growing mode (as shown in [11], chapt. 15). This leads to the following linearly extrapolated overdensities:

$$(11) \quad \begin{cases} \Delta_{\text{turn}} = \dfrac{3}{5}\Delta_i \dfrac{a_t}{a_i} = \dfrac{3}{5}\left(\dfrac{3\pi}{4}\right)^{2/3} \sim 1.06 \\ \text{and} \\ \Delta_{\text{rel}} = \dfrac{3}{5}\left(\dfrac{3\pi}{2}\right)^{2/3} \sim 1.69. \end{cases}$$

Therefore, we follow the custom and identify those peaks for which $\Delta \geqslant 1.69$ as collapsed and virialized clusters.

The effects occurring when the assumption of spherical symmetry in the collapse is relaxed have been studied extensively by N-body simulations (*e.g.*, [29-34]). The models display a wide range of collapse morphologies from protoclusters collapsing almost spherically to protoclusters that first form two or more compact subclusters before these finally merge to form a relaxed unit. It is observed in the simulations that subclustering during the collapse can prolong the collapse phase by up to a factor of 3 or 5 [31, 34]. One should, therefore, expect a spread in the collapse times for peaks with the same size and the same mean overdensity in the filter region. This constitutes a major complication that is not easily accounted for in analytical models of cluster formation.

4. – The mass function of clusters of galaxies.

For the hierarchical clustering models described above a simple approach to derive the mass function of the forming objects has been developed first by PRESS and SCHECHTER [35] and has been applied to the current problem and to the problem of galaxy formation by many authors (*e.g.*, [36-42]). The mass function can in turn be used to derive the distribution functions of other physical cluster parameters that can be observed.

The basic point in this approach is that in a random Gaussian field filtered on the scale R_f the quantity $\Delta(R_f)$ is also Gaussian distributed (due to the central-limit theorem):

$$(12) \quad p(\Delta, R_f) = \dfrac{1}{\sqrt{2\pi}\,\sigma(R_f)} \exp\left[-\dfrac{\Delta^2}{2\sigma(R_f)^2}\right].$$

Now we are only interested in those regions of the filtered fluctuation field which are above a certain overdensity limit, Δ_*, at an initial time t_i. They are bound to collapse and form virialized objects at an epoch when $\Delta_*(t) = \Delta_c \sim 1.69$.

In the following we will also use the normalized threshold parameter, $\nu_\star = \Delta_\star / \sigma(R_f)$. The probability for a given point in the field to be in an area with height $> \Delta_\star$ is given by

$$P(\Delta, R_f) = \int_{\Delta_\star}^{\infty} p(\Delta, R_f) \, d\Delta = \frac{1}{2} \mathrm{erfc}\left(\frac{\nu_\star}{\sqrt{2}}\right), \qquad (13)$$

where $\mathrm{erfc}(x) = 1 - \mathrm{erf}(x)$ is the complementary error function.

Equation (13) gives the probablility of finding an arbitrary point in an overdense region for one specific filter radius or filter mass. To obtain the differential probability distribution of overdensities for regions with given mass M, $P(\Delta, R_f)$ has to be differentiated with respect to $M(R_f)$:

$$Q(\nu_\star, M) = -\frac{dP(\nu_\star, R_f)}{dM} = -\frac{1}{\sqrt{2\pi}} \frac{\nu_\star}{\sigma(R_f)} \exp\left[-\frac{\nu_\star^2}{2}\right] \frac{d\sigma(R_f)}{dM}. \qquad (14)$$

Using $M = 4/3 \pi \rho R_f^3$ (which is correct for a top-hat filter; for other filters one may integrate the area under the filter function to calculate $M(R_f)$) one obtains

$$Q(M, \nu_\star) = -\frac{1}{3\sqrt{2\pi}} \nu_\star \frac{1}{M} \exp\left[-\frac{\nu_\star^2}{2}\right] \frac{d\log \sigma(R_f)}{d\log R_f}. \qquad (15)$$

While this formula gives the volume fraction occupied by the objects in a given mass interval, the number density is obtained by multiplying with ρ/M (where ρ is the mean density of the Universe):

$$n(M, \nu_\star) = -\frac{\sqrt{2}}{3\sqrt{\pi}} \nu_\star \frac{\rho}{M^2} \exp\left[-\frac{\nu_\star^2}{2}\right] \frac{d\log \sigma(R_f)}{d\log R_f}. \qquad (16)$$

An extra factor of two has been added to eq. (16). Omitting that factor of two, the integration of eq. (16) from $\nu_\star = 0$ to infinity would lead to only half of the mass in the Universe that could be incorporated into objects which is an uncomforting result. Therefore, PRESS and SCHECHTER argued in their original paper that the remaining mass is somehow consumed by later accretion onto the compact objects formed. To account for that they added a factor of two to eq. (16) which was shown by BOND et al. [43] to be correct for a sharp-k-space filter. In a more rigorous approach this problem has to be solved in connection with the «cloud-in-cloud problem» (see below).

Unfortunately one cannot compare these theoretical results directly to observations, since a reliable mass function of a statistical sample of clusters of galaxies has not been determined yet. An available data set that comes closest to a mass function is the X-ray temperature function of a statistical sample of clusters [40, 44]. In the paper by HENRY and ARNAUD the temperature function is already compared to the predictions of the Press-Schechter theory and we will follow their derivation to illustrate the application of this method.

For the derivation of the relation between the mass and temperature function one can refer to the virial theorem

$$\sigma_r^2 = \frac{1}{3}\langle v_{\text{gal}}^2 \rangle \propto \frac{GM}{r_{\text{turn}}} \propto \frac{GM}{R_f}(1+z_f), \tag{17}$$

where $M(R_f) = 4\pi/3 \rho_b R_f^3$ is the cluster mass, ρ_b is the mean background density, z_f the formation epoch of the cluster, and σ_r is the line-of-sight velocity dispersion of the cluster galaxies. σ_r is usually related to the intracluster plasma X-ray temperature by an empirical parameter β through

$$\beta = \frac{\sigma_r^2 \mu m_p}{kT}, \tag{18}$$

where μm_p is the mean particle mass. To determine the proportionality constant in these relations HENRY and ARNAUD refer to the results of N-body simulations including the hydrodynamics of the intracluster medium by EVRARD [45], who finds

$$M_{15} = \left(\frac{kT}{4.03\,\text{keV}}(1+z_f)^{-1}\right)^{3/2} h_{50}^{-1} \tag{19}$$

(where M_{15} is the cluster mass in units of $10^{15} M_\odot$) which with a proper evaluation of the constants in the above relations corresponds to $\beta \sim 1.2$.

If the post-recombination power spectrum of the density fluctuation field is approximated by a power law in the wavelength region relevant for clusters

$$P(k) = k_0^{-(n+3)} k^n \tag{20}$$

(where k_0 is a constant in units of the wave number), the mass function and temperature function of clusters of galaxies can be calculated analytically. The mass function can be obtained by means of eqs. (5) and (16). HENRY and ARNAUD chose a top-hat filter—eq. (4)—to smooth the fluctuation field on cluster scales. And using the relation $n(T) = n(M(T))\,dM/dT$ together with eq. (19) yields the temperature function.

Figure 2 shows the theoretically determined temperature function where the best-fitting parameters for the primordial power spectrum were $n = -1.7[^{+0.65}_{-0.35}]$, $k_0 = 0.0225[^{+0.0075}_{-0.0125}]\,h_{50}\,\text{Mpc}^{-1}$ (see [40]). The observed cluster sample contains 25 objects with known temperatures and the data were binned such that each bin contains five clusters. The cluster temperatures for the sample span a range from about 3 to more than 8 keV. It is interesting to consider the statistical properties of the corresponding overdensities. The 3 keV clusters have a mass of about $6 \cdot 10^{14} M_\odot$ and correspond to 2.5σ overdensities,

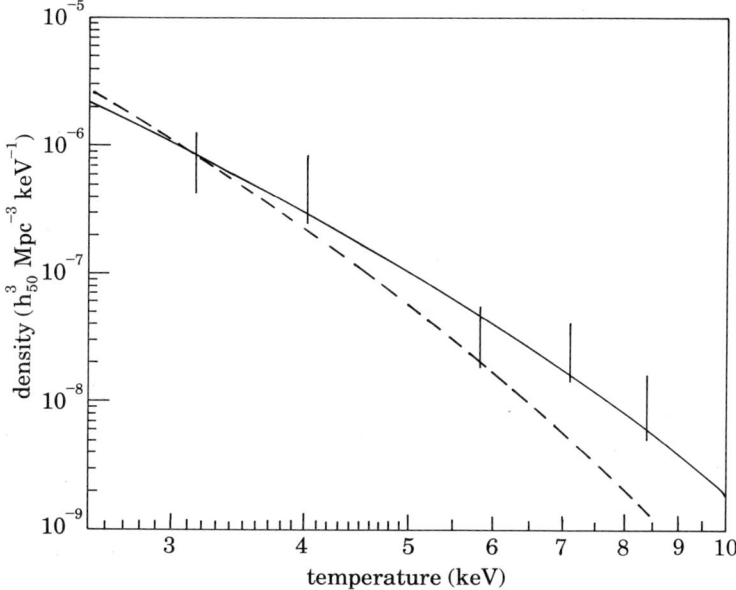

Fig. 2. – Temperature function of clusters of galaxies with experimental data points adapted from [40] and the fit of the Press-Schechter model result (solid line). Also shown is the Press-Schechter result obtained for the same power spectrum but Gaussian filtering of the fluctuation field (dashed line).

while the 8 keV clusters have masses around $3 \cdot 10^{15} M_\odot$ and can be identified with rare objects of an amplitude higher than 3σ. These high σ amplitudes were the justification for the linear approximation for the dynamics and statistics of overdensities in the fluctuation field. When the overdensities become nonlinear the main field is still close to linear evolution. Also shown in fig. 2 is the theoretical result for the temperature function for the same parameters n and k_0 if a Gaussian filter—eq. (2)—is chosen as filter function. It shows how sensitive the results are to the choice of the filter function.

5. – The intracluster medium and its evolution.

In the last section we used the temperature of the intracluster plasma as derived from X-ray observations as a means to estimate the mass of the cluster. To understand how reliably one can use observable parameters of the intracluster plasma some knowledge about the constitution and origin of the plasma in clusters is required. Detailed investigations of some nearby clusters in X-rays show that the plasma content in rich clusters is around 10% and may even reach up to 30% of the estimated virial mass of clusters. The uncertainty of this ratio is caused mainly by the possible errors in the virial-mass estimate. These results were described in detail in the firts section (see also [46, 47]).

There is good evidence that the ratio of the gas mass to the stellar mass increases with the size of the cluster [46, 47]. Figure 3 shows the gas-to-stellar-mass ratio as a function of the temperature deduced by DAVID et al. [46] for a sample of clusters observed with EINSTEIN where the temperatures have been determined by use of the EINSTEIN MPC data. The stellar mass was calculated from the observed luminosity using a galactic-mass-to-light ratio of $8 M_\odot / L_\odot$. If the gas temperature is taken as a measure of the clusters mass, then the ratio of the gas mass to total mass increases from smaller to larger systems from a factor of 1 to about 5 or 6. A similar result with better data for a sample of clusters with temperatures from GINGA observation was found by ARNAUD et al. [47]. They also find an indication that the relative iron abundance decreases with the gas mass probably implying that rich clusters contain a larger fraction of primordial gas.

If spectroscopically determined gas temperatures are compared with galaxy velocity dispersions by means of the β parameter defined in eq. (18), one usually finds that β scatters around 1 (e.g., [44]). Thus the specific heat of the gas is comparable to the specific kinetic energy of the galaxies. Therefore, the gas can have acquired its presently observed temperature from the release of the potential energy during the formation of the cluster and the small scatter in β is a sign that the temperature is a good measure for the cluster mass.

How the gas acquires its temperature during cluster formation and thermalization is nicely pictured in recent N-body/hydrodynamic calculations by EVRARD [45] and SCHINDLER and MÜLLER [48]. Figure 4 shows the evolution of

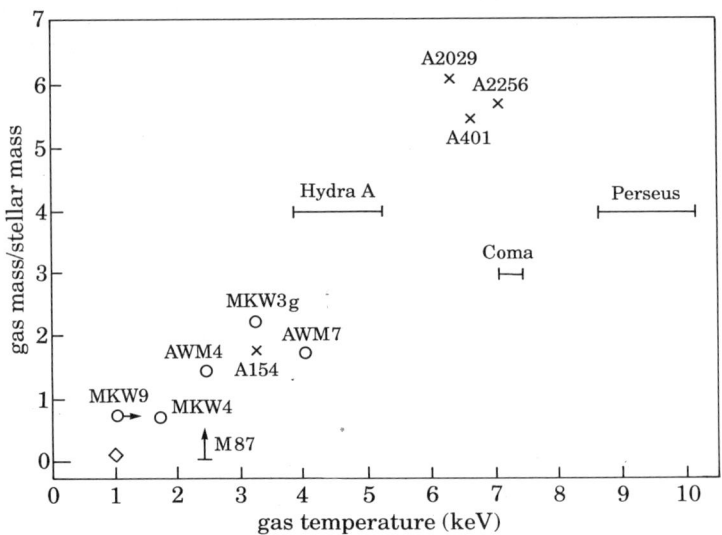

Fig. 3. – Correlation of the gas-to-stellar-mass ratio in clusters of galaxies with the observed gas temperatures from [46]. The assumed galactic-mass-to-light ratio is $8 M_\odot / L_\odot$.

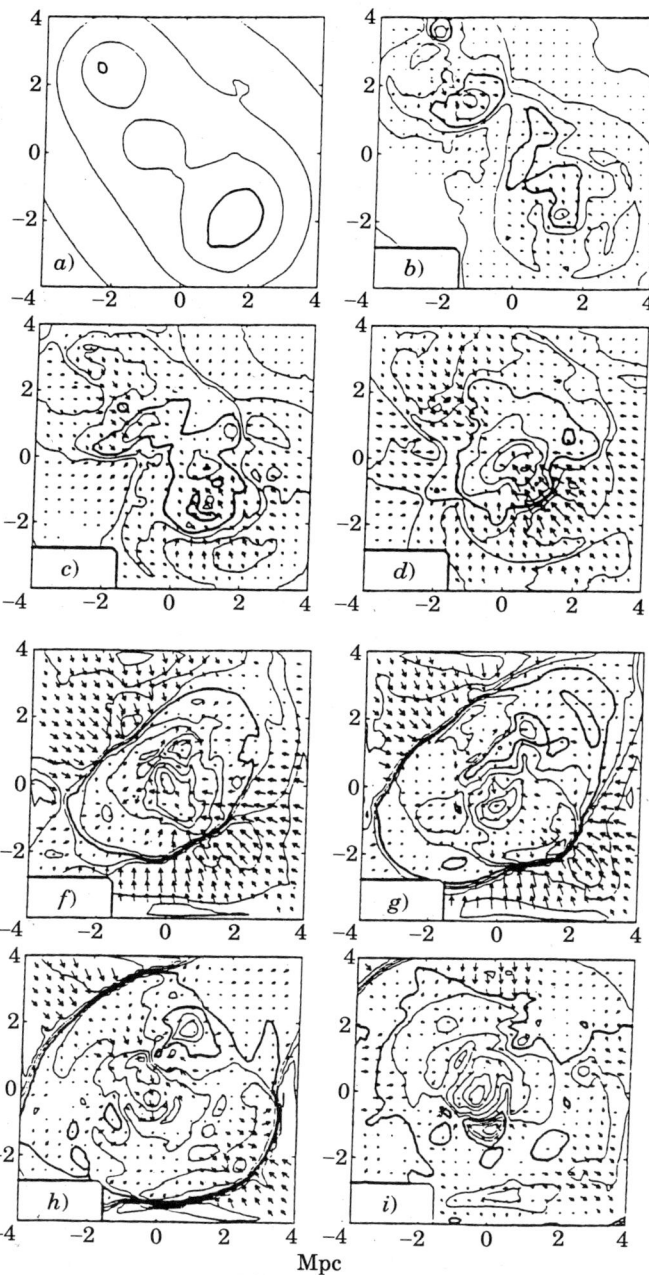

Fig. 4. – Evolution of the temperature distribution during the merger of two subclusters which constitutes the final stage of this model of cluster formation (from [49]). A major shock wave that displays as an easily noticeable discontinuity in the temperature distribution is seen to move radially outward and heats the gas in the outer regions producing a very slowly decreasing radial temperature profile.

the temperature distribution in the final stage of cluster formation: the merger of two major subclumps [48]. Several shock waves pass through the cluster during this process, but the major event that is easily seen in the picture is an almost spherical shock wave set off when the two densest gas regions finally merge. These shock waves are an efficient way of transporting energy throughout the intracluster gas and they produce an almost isothermal gas distribution with a slightly negative radial temperature gradient. The cooling time is larger than the Hubble time except for the central regions ($\sim$ 100 kpc) in a large part of the clusters which consequently have cooling flows [49]. Most of the cluster gas retains the temperature acquired during the cluster formation, however. This reasoning is supported by the relatively good correlation between the X-ray temperature and the galaxy velocity dispersion which is shown in fig. 5 with data from [44].

There is also a good correlation of the X-ray luminosity with the temperature and velocity dispersion. HENRY and ARNAUD [40] find the following relation for the bolometric luminosity:

$$kT = 2.9(h_{50}^2 L_{44}[\text{bol}])^{0.265 \pm 0.035} , \tag{21}$$

where $L_{44}[\text{bol}]$ is the bolometric luminosity in units of $10^{44}\,\text{erg s}^{-1}$. The good correlation gives some hope that L_x can also be used to obtain a rough mass estimate for a cluster. This is quite advantageous since the X-ray luminosity can be

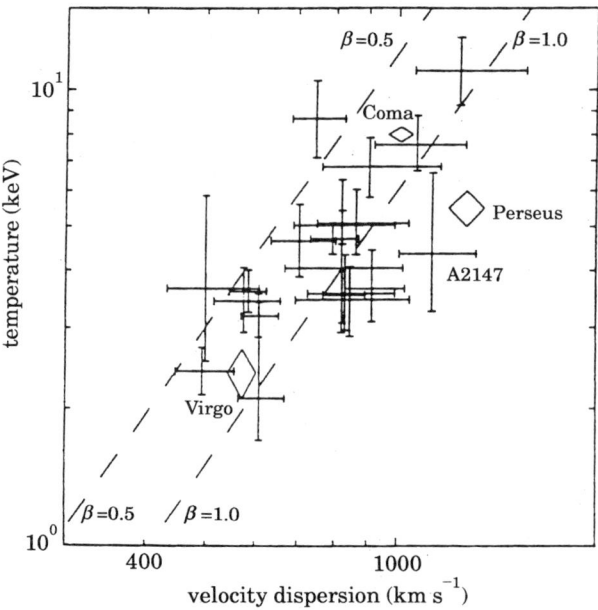

Fig. 5. – Correlation of the observed gas temperature and the galactic velocity dispersion for the sample of clusters observed with EXOSAT [44].

determined for a much larger sample of clusters than the X-ray temperature.

The high amount of iron in the intracluster gas requires that a large fraction of the gas comes from galactic winds and the cluster galaxies must have lost up to 40% of their previous mass in early starburst and wind phases [50-53]. The model calculations show that the supernova explosions responsible for the iron production and the galactic winds can also contribute to the heating of the intracluster medium. The good correlation of the gas temperature with the galaxy velocity dispersion shows that this energy input is not so significant since it would not show a correlation with the cluster mass.

6. – The X-ray luminosity function of clusters.

If we assume that the dark matter is distributed as in a King model and the gas follows the dark matter, and if we further assume that the gas has reached its virial temperature, we can calculate the expected X-ray luminosity of a cluster. That is we assume

$$(22) \qquad L_x(\text{bol}) \propto n_e^2 R^3 T^{1/2} \propto g^2 M R^3 \left(\frac{M}{R}\right)^{1/2},$$

where n_e is the electron density, R the characteristic radius, M the total cluster mass, and g the gas-to-total-mass ratio. To evaluate the constants involved we also have to make an assumption about the ratio of the turn around radius to the present core radius of the cluster. We have chosen a value of $r_{\text{turn}}/r_c \sim 20$ (at $z = z_f$) suggested by results of N-body simulations and obtain the following result:

$$(23) \qquad L_{44} \sim 35 g^2 M_{15}^{4/3} (1 + z_f)^{3.5},$$

where L_{44} is the bolometric luminosity in 10^{44} erg s^{-1} and M_{15} is the cluster mass in units of $10^{15} M_\odot$. Equation (23) now allows one to construct the X-ray luminosity function of clusters of galaxies in the frame of the Press-Schechter theory.

The best statistical sample of clusters of galaxies with known distances and X-ray luminosities comes from the EINSTEIN Medium Sensititivity Survey involving 93 clusters (EMSS, [54]). HENRY et al. [42] have analysed the EMMS results in the light of the Press-Schechter formalism. They have fitted the theoretically predicted X-ray luminosity functions for different redshifts to their data and they could very well reproduce luminosity functions for different redshift intervals. The best fit for a primordial spectrum with power law form was obtained for values of $n = -2.1[^{+0.27}_{-0.15}]$, $k_0 = 0.0145[^{+0.004}_{-0.0065}]$ h_{50} Mpc^{-1} which is well within the error limits of the previous determination of these parameters by HENRY and ARNAUD [40] using the observed temperature function.

We can now compare these results with the form of the primordial power spectrum predicted by the standard CDM model as shown in fig. 1. While the observations suggest a slope of $P(k) \propto k^{-2}$, the theory predicts a slope closer to $P(k) \propto k^{-1}$ in the interesting range around 10 Mpc. One suggestion to rectify the discrepancy is to introduce a cosmological constant that corresponds to a flat Universe with $\Omega \sim 0.2$. This was suggested by EFSTATHIOU *et al.* [55] to explain the new results in the galaxy-galaxy correlation function in the data from the APM survey and the IRAS survey. The same modification would help here.

7. – Conclusions.

It was shown that the cluster mass function is an interesting tool for the study of the primordial density fluctuation spectrum. X-ray observations are a good means to obtain information on the cluster mass spectrum via the X-ray temperature and X-ray luminosity function. In an approximate way the Press-Schechter approach provides the link between the initial conditions and the mass function. For detailed cosmological tests the theory has to be refined and complemented by N-body calculations.

Other statistical measures of the cluster population in the Universe are also interesting tracers of the large-scale structure and the cosmic initial conditions which could not be described in the limited time. This includes in particular the N-point correlation functions of the spatial distribution of clusters and the statistics of substructure appearing in clusters of galaxies. The several thousand clusters detected in X-rays in the ROSAT All Sky Survey are an interesting sample for these types of statistical studies. Redshifts are needed for these clusters to derive the interesting physical parameters, and thus several redshift surveys for this cluster sample have started that should provide interesting new cosmological data within the next four of five years.

* * *

We like to thank J. P. HENRY for discussions.

REFERENCES

[1] M. J. GELLER and J. P. HUCHRA: *Science*, **246**, 897 (1990).
[2] V. DE LAPPARENT, M. J. GELLER and J. P. HUCHRA: *Astrophys. J.*, **332**, 44 (1988).
[3] V. DE LAPPARENT, M. J. GELLER and J. P. HUCHRA: *Astrophys. J.*, **369**, 273 (1988).
[4] W. SAUNDERS, C. FRENK, M. ROWAN-ROBINSON, G. EFSTATHIOU, A. LAWRENCE,

N. Kaiser, R. Ellis, J. Crawford, X.-Y. Xia and I. Parry: *Nature (London)*, **349**, 32 (1991).

[5] L. Guzzo, A. Iovino, G. Chincarini, R. Giovanelli and P. M. Haynes: *Astrophys. J.*, **382**, L5 (1991).

[6] A. Yahil, M. A. Strauss, M. Davis and J. P. Huchra: *Astrophys. J.*, **372**, 380 (1991).

[7] M. Davis, M. A. Strauss and A. Yahil: *Astrophys. J.*, **372**, 394 (1991).

[8] S. J. Maddox, G. Efstathiou, W. J. Sutherland and J. Loveday: *Mon. Not. R. Astron. Soc.*, **242**, 43 (1990).

[9] N. H. Heydon-Dumbleton, C. A. Collins and H. T. MacGillivray: *Mon. Not. R. Astron. Soc.*, **238**, 379 (1989).

[10] G. B. Dalton, G. Efstathiou, S. J. Maddox and W. J. Sutherland: *Astrophys. J.*, **390**, L1 (1992).

[11] P. J. E. Peebles: *The Large Scale Structure of the Universe* (Princeton University Press, Princeton, N.J., 1980).

[12] G. Efstathiou and J. Silk: *Fundam. Cosmic Phys.*, **9**, 1 (1983).

[13] A. Szalay: in *Large Scale Structure of the Universe, 17th Advanced Course of the Swiss Society of Astronomy and Astrophysics*, edited by L. Martinet and M. Mayer (Observatory, Geneva, 1988), p. 175.

[14] M. S. Longair: in *Lecture Notes in Physics, 333, Evolution of Galaxies*, edited by I. Appenzeller, H. J. Habing and P. Léna (Springer-Verlag, Berlin, 1989), p. 1.

[15] S. D. White: in *Physics of the Early Universe*, edited by J. A. Peacock, A. E. Heavens and A. T. Davies, *Proceedings of the 36th Scottish Universities Summer School in Physics 1989* (a NATO Advanced Study Institute, 1989), p. 1.

[16] G. Efstathiou: in *Physics of the Early Universe*, edited by J. A. Peacock, A. E. Heavens and A. T. Davies, *Proceedings of the 36th Scottish Universities Summer School in Physics 1989* (a NATO Advanced Study Institute, 1989), p. 361.

[17] H. Böhringer and G. Wiedenmann: in *New Insights into the Universe*, edited by V. Martinez, V. Portilla and D. Saez (Springer, Heidelberg, 1992), p. 127.

[18] G. C. McVittie: *General Relativity and Cosmology* (Chapman and Hall, London, 1964).

[19] S. Weinberg: *Gravitation and Cosmology* (John Wiley and Sons, New York, N.Y., 1972).

[20] P. J. E. Peebles: *Physical Cosmology* (Princeton University Press, Princeton, N.J., 1971).

[21] E. W. Kolb and M. S. Turner: *The Early Universe* (Addison-Wesley Publishing Co., Reading, Mass., 1990).

[22] J. M. Bardeen, J. R. Bond, N. Kaiser and A. S. Szalay: *Astrophys. J.*, **304**, 15 (1986).

[23] Ya. B. Zel'dovich: *Mon. Not. R. Astron. Soc.*, **160**, 1P (1972).

[24] J. R. Bond and G. Efstathiou: *Astrophys. J.*, **285**, L45 (1984).

[25] P. J. E. Peebles: *Astrophys. J.*, **258**, 415 (1982).

[26] J. R. Bond and A. S. Szalay: *Astrophys. J.*, **274**, 443 (1983).

[27] J. R. Bond, S. Cole, G. Efstathiou and N. Kaiser: *Astrophys. J.*, **379**, 440 (1991).

[28] D. Lynden-Bell: *Mon. Not. R. Astron. Soc.*, **136**, 101 (1967).

[29] S. D. White: *Mon. Not. R. Astron. Soc.*, **177**, 717 (1976).

[30] S. D. White: *Mon. Not. R. Astron. Soc.*, **179**, 33 (1977).

[31] A. Cavaliere, P. Santangelo, G. Tarquini and N. Vittorio: *Astrophys. J.*, **305**, 651 (1986).

[32] A. E. Evrard: *Astrophys. J.*, **310**, 1 (1986).

[33] A. E. EVRARD: *Astrophys. J.*, **316**, 36 (1987).
[34] S. SCHINDLER and H. BÖHRINGER: *Astron. Astrophys.*, **269**, 83 (1993).
[35] W. H. PRESS and P. SCHECHTER: *Astrophys. J.*, **187**, 425 (1974).
[36] S. C. PERRENOD: *Astrophys. J.*, **236**, 373 (1980).
[37] N. KAISER: *Mon. Not. R. Astron. Soc.*, **222**, 323 (1986).
[38] A. E. EVRARD: *Astrophys. J. Lett.*, **341**, L71 (1989).
[39] A. J. PEACOCK and A. F. HEAVENS: *Mon. Not. R. Astron. Soc.*, **243**, 133 (1990).
[40] J. P. HENRY and K. A. ARNAUD: *Astrophys. J.*, **372**, 410 (1991).
[41] R. G. BOWER: *Mon. Not. R. Astron. Soc.*, **248**, 332 (1991).
[42] J. P. HENRY, I. M. GIOIA, T. MACCACARO, S. L. MORRIS, J. T. STOCKE and A. WOLTER: *Astrophys. J*, **386**, 408 (1992).
[43] J. R. BOND, A. S. SZALAY and M. S. TURNER: *Phys. Rev. Lett.*, **48**, 1636 (1982).
[44] A. C. EDGE and G. C. STEWART: *Mon. Not. R. Astron. Soc.*, **252**, 414, 428 (1991).
[45] A. E. EVRARD: in *Clusters of Galaxies*, edited by G. FITCHETT and W. OEGERLE (Cambridge University Press, Cambridge, 1990), p. 287.
[46] L. P. DAVID, K. A. ARNAUD, W. FORMAN and C. JONES: *Astrophys. J.*, **356**, 32 (1990).
[47] M. ARNAUD, R. ROTHENFLUG, V. BOULADE, L. VIGROUX and E. VANGIONI-FLAM: *Astron. Astrophys.*, **254**, 49 (1992).
[48] S. SCHINDLER and E. MÜLLER: *Astron. Astrophys.*, **272**, 137 (1993).
[49] A. C. EDGE, G. C. STEWART and A. C. FABIAN: *Mon. Not. R. Astron. Soc.*, **258**, 177 (1992).
[50] L. P. DAVID, W. FORMAN and C. JONES: *Astrophys. J.*, **359**, 29 (1990).
[51] L. P. DAVID, W. FORMAN and C. JONES: *Astrophys. J.*, **369**, 121 (1990).
[52] R. E. WHITE III: *Astrophys. J.*, **367**, 69 (1991).
[53] A. RENZINI: this volume, p. 303.
[54] I. M. GIOIA, T. MACCACARO, R. E. SCHILD, A. WOLTER, J. T. STOCKE, S. L. MORRIS and J. P. HENRY: *Astrophys. J. Suppl.*, **72**, 567 (1990).
[55] G. EFSTATHIOU, W. J. SUTHERLAND and S. J. MADDOX: *Nature (London)*, **348**, 705 (1990).

ROSAT Deep Surveys.

G. HASINGER

Max-Planck-Institut für Extraterrestrische Physik - D-8046 Garching, B.R.D.

1. – Introduction.

Almost thirty years after the discovery by GIACCONI et al. [1] the study of the X-ray background (XRB) has been revitalized by new observations from recent space astronomy missions. Two conferences have been dedicated to this topic in 1991, the proceedings of which [2,3] provide a comprehensive overview of the subject, while this lecture will concentrate on recent imaging observations with the X-ray satellite ROSAT [4]. This work is done in collaboration with G. HARTNER, J. TRÜMPER (MPE), R. BURG, R. GIACCONI (STScI), M. SCHMIDT (Caltech) and G. ZAMORANI (Osservatorio Bologna).

The XRB, originally observed in the (2–10) keV range, has now been measured with high precision over an energy range of at least five decades, from about 0.1 keV up to several MeV as shown in fig. 1. At energies above 3 keV, where the Galaxy is completely transparent, the (*hard*) XRB is highly isotropic and almost certainly of extragalactic origin. The (3–40) keV spectrum, where the bulk of the energy density of the XRB resides, can be very well fitted by a thermal bremsstrahlung model with a temperature of 40 keV [5], a fact which originally led to the interpretation that a significant fraction of the XRB originates from a diffuse, optically thin plasma with this temperature. COBE measurements [6] have ruled out this possibility, because the expected distortion of the microwave background spectrum due to Comptonization of the 3 K photons on the hot electrons has not been observed.

The alternative interpretation of the XRB as due to the sum of discrete extragalactic sources [7]—supposedly active galactic nuclei (AGN)—is now the only feasible one. Unfortunately, until imaging hard-X-ray satellites will be flown in the near future, the instrumental capabilities do not allow us to resolve a significant fraction of the hard XRB into discrete sources. The only information on the AGN number counts in the hard band is from the HEAO-1 survey at extremely bright fluxes [8] and from fluctuation analyses of HEAO-1 and

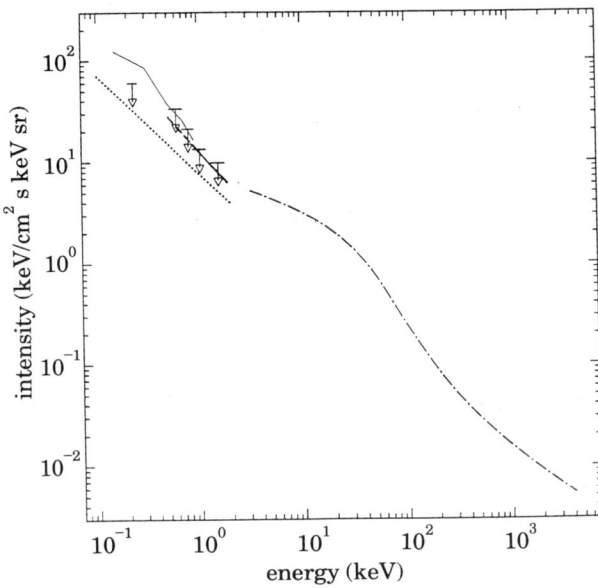

Fig. 1. – Spectrum of the extragalactic X-ray background from the soft-X-ray to the gamma-ray waveband. The dot-dashed curve is a comprehensive fit to a large body of data by Gruber (see [9] and references therein). The thick solid line refers to the ROSAT estimate of the soft-X-ray extragalactic background, while the thin solid line gives the total background, including galactic components (this work). The arrows are upper limits to the extragalactic XRB from the Wisconsin rocket flights [10]. The dotted line gives the contribution from resolved sources in the deepest ROSAT field [4].

GINGA data (see [11] and references therein). Moreover, the canonical AGN spectra do not match the spectrum of the background [12]. This so-called «spectral paradox» can be solved by invoking AGN spectra which are more complicated than a single power law, and in particular flatten at higher energies [13]. GINGA observations of AGNs have revealed that simple power law models are indeed not adequate to describe the hard-X-ray emission for most of these objects [14] and that a Compton reflection model [15] can reproduce the spectral shape. These data have led to new attempts to fit the overall XRB spectrum as a composition of AGN spectra, folded with an appropriate luminosity and redshift distribution [16-19].

The situation at photon energies below 2 keV is diametrically opposite: the extragalactic-background spectrum is very poorly known there because of the interstellar absorption and the «contamination» by galactic foreground emission [4] like, *e.g.*, the hot local bubble. On the other hand, instrumentation is well advanced so that imaging of very faint AGN is possible. Deep imaging X-ray studies were first performed with the EINSTEIN IPC and HRI [20-23]. These observations were able to resolve on the order of 20% of the background at 2 keV (extrapolated downward from the HEAO-1 measurements) into directly observed discrete sources. First results on medium-deep ROSAT pointed observations have already shown that a significant fraction of the soft XRB is resolved into discrete sources [24-27]. Here I will describe shortly the Deep Survey results obtained with the Position Sensitive Proportional Counter (PSPC [28]) aboard ROSAT [29] in the *Lockman Hole*, a region with the absolutely lowest neutral-hydrogen column density [30], as well as observations in a total of 26 shallower fields comprising the heart of the *ROSAT Medium Sensitivity Survey*, RMSS. A much more detailed description of the data analysis and results is given in [4].

2. – X-ray observations.

The PSPC pointed observations of the *Lockman Hole* (fig. 2) were carried out in the time periods April 27-May 20, 1991, October 10-November 2, 1991, and April 15-25, 1992. A total of 152 ks observing time has been accumulated. The average cosmic-background count rate in the direction of the *Lockman Hole*, measured by the PSPC in the energy band (0.1–2) keV, is ~ 10 counts/s

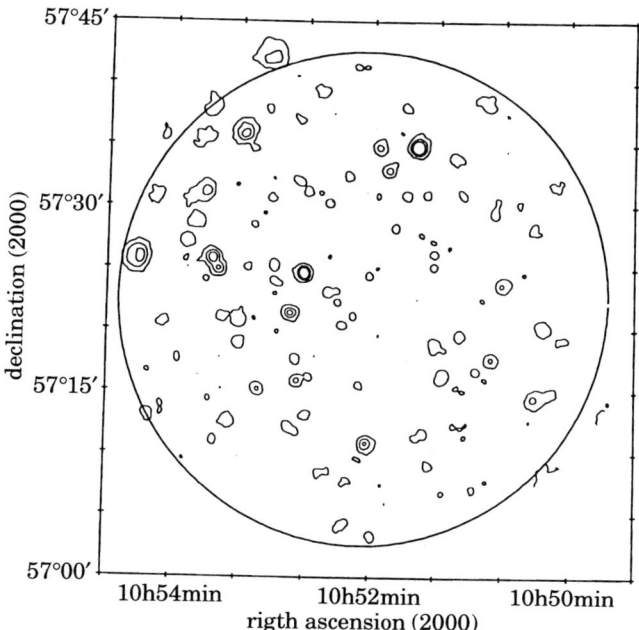

Fig. 2. – PSPC image of the Lockman Hole (inner 20 arcmin radius) in the (0.4–2.4) keV band. The contours are drawn at 0.6, 1.9 and $3.9 \cdot 10^{-3}$ counts s^{-1} arcmin^{-2}. The lowest contour corresponds approximately to the $L = 10$ detection threshold. The circle with radius 20' marks the approximate boundary of the ring in the PSPC support structure (from [4]).

TABLE I. – *Observation summary.*

Field name	Galactic longitude (degrees)	Galactic latitude (degrees)	Column density (**)	Exposure time (s)	Night time (s)	No. of sources total sample	No. of sources clean sample	S_{min} ((0.5–2) keV) (***)
α Boo(*)	15	69	2.2	17100	—	40	29	0.97
B21215(*)	189	82	1.8	22400	—	55	28	0.78
CY Uma(*)	159	59	1.2	9400	—	27	19	0.99
δ Leo	224	67	1.2	20600	—	54	28	0.80
Dra 4(*)	126	48	1.7	15350	—	43	32	0.81
EF Eri(*)	214	−58	1.6	15840	—	27	18	0.78
EX Hya(*)	303	34	6.1	15620	—	19	9	1.68
F 1557(*)	279	−42	4.8	21300	—	25	16	0.98
GRB790406(*)	326	−40	2.2	19500	—	40	25	0.78
HR 857(*)	192	−58	3.9	10680	—	10	7	1.14
HR 8905(*)	99	−35	5.0	26300	—	40	22	0.77
HZ 43	54	84	1.0	15400	—	36	26	0.72
LHS 2924(*)	55	68	1.1	18250	—	34	16	0.72
Lockman	149	53	0.6	143700	2540	166	75	0.21
Meaty	111	53	4.1	38450	14980	44	26	0.62
NEPNNN	96	30	3.9	80880	2810	69	39	0.39
NEPN	96	30	3.9	41760	—	57	29	0.67
NEP Raster	96	30	3.9	70000	3490	—	—	—
Nower 1	202	55	2.5	38910	—	46	28	0.49
Nower 2	136	68	1.2	30590	—	56	24	0.63
ON231(*)	202	83	2.0	10400	—	21	9	1.34
Pavo(*)	319	−77	6.1	25080	—	40	22	1.08
QSO0000(*)	36	−79	1.6	39600	18630	59	37	0.46
QSO1116(*)	223	68	1.2	24950	—	67	38	0.70
QSO1202(*)	206	80	1.6	17400	—	41	17	1.10
SEP	276	−30	4.2	20220	—	23	19	0.80
UMA(*)	178	56	1.1	9200	—	23	13	1.00
VW Hyi(*)	285	−38	8.1	8760	—	14	10	1.58

(*) Courtesy ROSAT PIs.
(**) In units of 10^{20} cm^{-2} [31].
(***) In units of 10^{-14} erg cm^{-1} s^{-1}.

over the entire detector. A small fraction of the data with unfavourable viewing conditions and contaminated by solar scattering, geocoronal or auroral backgrounds had to be removed yielding a net observation time of 143.7 ks.

The observation in the *Lockman Hole* represents the most sensitive soft-X-ray exposure ever taken and, therefore, will constrain the source counts at extremely faint fluxes ($\sim 2 \cdot 10^{-15}$ erg cm^{-2} s^{-1}), but in a relatively small solid angle. For a better definition of the complete log N-log S function over a substantial flux range it is, however, important to obtain reliable statistical accuracy at brighter fluxes, where the number density of objects is much smaller. This is only possible by considerably increasing the survey area. A series of shallower fields, the RMSS, have, therefore, been collected in parallel to the Deep Survey observations. Table I gives a summary of the selected fields, with galactic coordinates, interstellar column densities, exposure times, number of sources detected and limiting sensitivities.

3. – Data analysis.

The detection algorithm is described in detail in [4]. In order to minimize source confusion effects, the effects of interstellar absorption as well as the «contamination» by galactic stars with soft spectra, and to allow a straightforward calculation of the detection thresholds a maximum off-axis angle of 15.5' has been chosen, and the sample was restricted to detections in the H band. A total of 661 objects with fluxes ((0.5–2.0) keV) between $2.6 \cdot 10^{-15}$ and $4.8 \cdot 10^{-13}$ erg cm^{-2} s^{-1} were detected in a solid angle of ~ 5.5 degrees^2. Of those, 75 sources comprise a statistically complete sample in the Lockman Hole.

For each source detected in the inner field of view (radius 20') of the *Lockman Hole* a hardness ratio and its appropriate 1σ error have been calculated from the counts in the S and H band:

$$\text{HR1} = \frac{H - S}{H + S}.$$

Figure 3 shows this ratio as a function of the source flux derived from the total (T) band for all sources detected significantly in T (*i.e.* not in H in order to avoid selection biases). At bright fluxes, where the statistical errors are small, most of the sources have their HR1 value in the relatively narrow range [-0.1, -0.7]. The only exception at these bright fluxes is an AGN with an extremely hard spectrum (HR1 = 0.85 ± 0.04). The hard spectrum of this source can be explained in terms of an intrinsic absorption column ($N_\text{H} \sim 3 \cdot 10^{21}$ cm^{-2} for a redshift of zero). The four objects optically identified as foreground stars (asterisks in fig. 3) have relatively hard spectra—their optically thin thermal spectra have considerable Fe-L and O-K line emission in the range (0.7–1) keV which in the case of the two brighter objects is clearly visible in the PSPC spectra. There is an indication that the average source spectrum becomes harder with decreasing

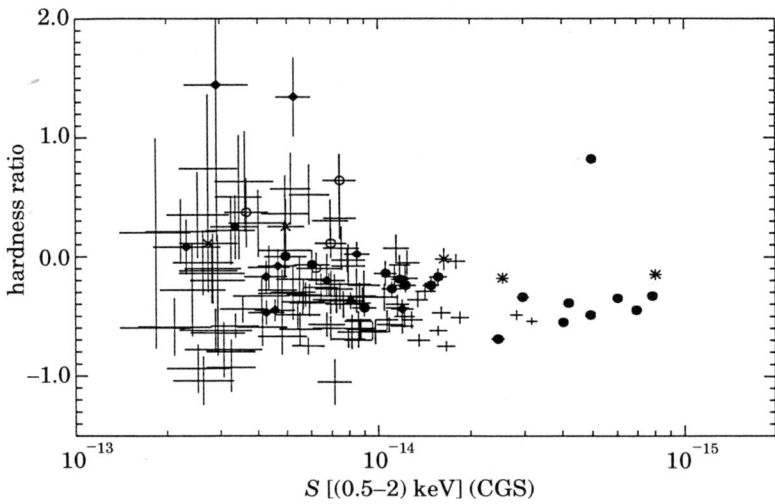

Fig. 3. – Hardness ratios (HR1) as a function of source flux in the Lockman Hole. The (0.5–2) keV fluxes have been determined from the total (T) band, and are in units of erg cm^{-2} s^{-1}. Asterisks, open circles and solid circles refer to objects optically identified [32] as foreground stars, galaxies and emission line AGN, respectively. Diamonds mark counterparts in deep VLA radio maps [33].

flux: at low fluxes there is a substantial fraction of objects with positive values of HR1. Dividing the sample into sources brighter and fainter than $0.8 \cdot 10^{-14}$ erg cm^{-2} s^{-1} yields average hardness ratios which are different at the 3.3σ level.

4. – Average source spectrum.

An average spectrum of all sources fainter than $4 \cdot 10^{-14}$ erg cm^{-2} s^{-1} in the clean *Lockman* sample (*i.e.* H-band detections only within an off-axis angle $\leqslant 15.5'$) was computed. Counts were summed in rings around all sources and the average background, determined from the remaining adjacent pixels, was subtracted. The average source spectrum was corrected for vignetting, detector dead time and the contribution from sources brighter than the upper flux threshold (a factor 1.41). The latter correction factor was derived from an extrapolation to infinity of the source counts discussed below. A single power law model with cold-gas absorption could be fitted to the summed source spectrum over the entire (0.1–2) keV band. The best-fit parameters are $N_\mathrm{H} = (8.7 \pm 2.4) \cdot 10^{19}$ cm^{-2} for the average absorbing column density, $\alpha = 0.96 \pm 0.11$ for the average energy spectral index and (7.8 ± 0.3) keV cm^{-2} s^{-1} sr^{-1} keV^{-1} for the normalization of the power law [4]. This spectrum is shown as a dotted line in fig. 1.

5. – Preliminary cosmic X-ray background spectrum.

An average total X-ray background spectrum has been constructed from five RMSS observations which were taken at low proportional counter gain, yielding a formal energy range up to 3.75 keV. To minimize the contamination by solar scattered and geocoronal light only night-time observation intervals with Sun-Earth-satellite angles larger than 120 have been accepted, leaving a net observing time of 45250 s (see table I for individual night-sky low-gain observing time). To improve on the statistical quality of the data, the spectrum has been accumulated over the outer part of the PSPC field of view ((27–50)' radius, the inner FOV has been omitted to avoid any influence of the observation target). The particle background model given in [34] was subtracted, normalized to the energy channels above 3.0 keV, where the sky contribution can be neglected.

Although the extragalactic X-ray background spectrum dominates at energies above ~ 1 keV [27] and is the main component of interest in the current context, a three-component model had to be fitted to the data because of the relatively coarse energy resolution of the instrument. Two thermal line spectra at one million and (2–3) million degrees, respectively, were added to the extragalactic power law (the absorption value for the power law component has been fixed at $2.5 \cdot 10^{20}$ cm^{-2}, which is roughly the average absorption of the five different fields and an uncritical parameter). This fit yielded an acceptable χ^2 value only for energies below ~ 1.8 keV. At higher energies the currently available best detector response matrix still has some known deficiencies as judged from a recent fit to the low-gain spectrum of the Crab nebula. Therefore, the current result and all previous fits to the ROSAT X-ray background spectrum are only preliminary and have to be treated with some caution. The best-fit normalization and power law energy index are (13.0 ± 0.2) keV cm^{-2}s^{-1}sr^{-1}keV^{-1} (at 1 keV) and $\alpha_E = 1.12 \pm 0.12$, respectively. The latter error is purely the statistical uncertainty, the systematic error on the slope is ~ 0.2. The best-fit power law model is shown in fig. 1 in context with other spectral measurements of the XRB. There is good agreement between the present fit and the Wisconsin data above 1 keV; an extrapolation of the simple power law model down to 0.2 keV is still compatible with the C-band SMC shadow experiment [10]. The total and the estimated extragalactic background spectrum is plotted in fig. 1.

6. – Systematic effects of source detection.

Before a final $\log N$-$\log S$ relation can be derived, the data have to be corrected for the biases and selection effects present in the source detection procedure. The most famous of those is the Eddington bias, which produces a net gain of the number of sources detected above a given flux limit as a consequence of statistical errors in the measured flux (Eddington, 1940). For simple assumptions about the log N-log S relation this bias can be corrected for analytically [35]. Another selection effect, most important in the deep fields considered here, is source confusion. The net effect of source confusion is difficult to quantify analytically, because it can affect the derived log N-log S relation in different ways:

1) Two subthreshold sources could be present in the same resolution element and thus produce a single detected source. This corresponds to a net gain in the number of sources, similar to the Eddington bias.

2) Two sources above the threshold could merge into a single brighter source. In this case one source is lost and one is detected at a considerably higher flux. Whether the total flux is conserved or not depends on the distance between the two sources and on the details of the source detection algorithm.

3) The specific detection algorithm used in the RMSS is incomplete for sources closer than $32''$, which results in a net loss of fainter sources.

4) Finally, if the PSF radius is significantly larger than the above match radius (*i.e.* at large off-axis angles and at low energies), spurious faint sources can appear in the PSF wings of brighter sources, resulting in a net gain of fainter sources.

Because of the complexity of these possible errors, a full account of the systematic effects can only be obtained through detailed Monte Carlo simulations of a large number of realistic fields. Therefore, 140 fields with the same distribution of exposure times, N_H values and background count rates and thus of limiting fluxes as the 27 fields in the RMSS sample have been simulated. In each field X-ray sources have been drawn at random positions, with input fluxes taken from a variety of differential log N-log S relations. At faint fluxes the slope of the log N-log S relation has been varied between 1.5 and 2.2. Details of the simulations are described in [4].

The artificial fields created this way were run through exactly the same source detection and flux estimate algorithms as the real fields thus producing a simulated sample in the H band similar by construction to the clean data sample. Comparing the input and output fluxes no systematic error could be detected in the flux determination for brighter sources. At the faintest fluxes, however, confusion effects can be clearly seen. On the basis of these data, comparing the input and output source count relations, an analytic «efficiency function» could be derived, which is roughly unity for fluxes above 10^{-14} erg cm^{-2} s^{-1} and decreases to ~ 0.75 at $3 \cdot 10^{-15}$ erg cm^{-2} s^{-1}. Although individual sources change their flux, lose their identity or can be lost completely in the detection process, the efficiency function can be used *post facto* for an analytic correction of the derived log N-log S function.

7. – The $\log N$-$\log S$ relation.

For the construction of the final $\log N$-$\log S$ curve the data have been corrected for systematic effects by giving every source a weight corresponding to the inverse of the above efficiency function. Figure 4 shows the corrected integral $\log N$-$\log S$ relation in comparison with EINSTEIN data and the fluctuation analysis (see below).

We have fitted the corrected flux distribution of the 661 sources in the clean sample with a model in which the differential counts $(\mathrm{d}N/\mathrm{d}S \equiv N(S))$ are represented by two power laws:

$$N(S) = N_1 S^{-\beta_1} \qquad \text{for } S > S_b,$$
$$N(S) = N_2 S^{-\beta_2} \qquad \text{for } S < S_b.$$

By requiring continuity in the differential counts, the following relation holds:

$$N_2 = N_1 S_b^{\beta_2 - \beta_1}.$$

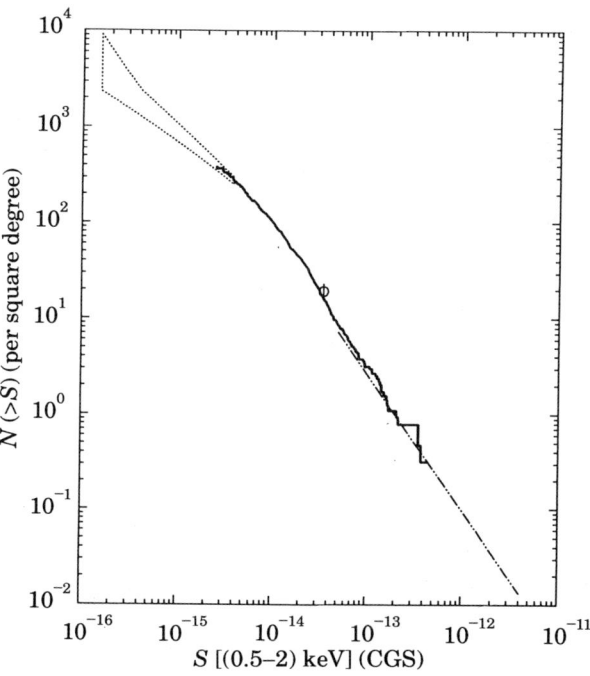

Fig. 4. – Integral source counts from the clean RMSS sample, corrected for systematic effects (histogram). The dash-dotted lines represent the EMSS source counts (see text). The open circle corresponds to the Einstein Extended Deep Survey (EDS) point [22] scaled by a factor of 1.32 to account for stars in the sample. The dotted area at faint fluxes shows the 90% confidence regions from the fluctuation analysis of the deepest ROSAT field in the Lockman Hole (from [4], but corrected for a $\sim 10\%$ offset in [4]).

For convenience the flux S is in units of 10^{-14} erg cm^{-2} s^{-1}. The best-fit parameters are $\beta_1 = 2.72 \pm 0.27$, $\beta_2 = 1.94 \pm 0.19$, $S_b = 2.66 \pm 0.66$ and normalization $N_2 = 111 \pm 10$ (90% errors taking into account the interdependence of the fit parameters). The reduced χ^2 is 1.1 for 18 degrees of freedom. A single power law model (i.e. $\beta_1 = \beta_2$) is excluded at 99.9% confidence (reduced χ^2 of 2.3 for 20 d.o.f.), thus showing the reality of the detected flattening at faint fluxes. The possible presence of a flattening was already inferred from the fluctuation analysis of deep EINSTEIN fields [36,37] and also suggested by shallower ROSAT surveys [24,25].

Our data appear to be in good agreement with the EMSS: the two dotted lines at high fluxes represent the best fit to the EMSS extragalactic counts [38], scaled by a factor 1.37 to account for the stars in the EMSS. To convert from the EMSS energy band ((0.3–3.5) keV) to the RMSS ((0.5–2.0) keV) a power law spectrum with energy index 1 has been assumed (conversion factor 0.56). There is also good agreement with the EINSTEIN deep-survey counts [22,23].

Integrating the $\log N$-$\log S$ function above a flux of $2.5 \cdot 10^{-15}$ erg cm^{-2} s^{-1}, the «corrected» integral surface density is ~ 413 degrees^{-2} and the total flux in resolved sources amounts to $1.48 \cdot 10^{-8}$ erg cm^{-2} s^{-1} sr^{-1} or about 59% of the (1–2) keV X-ray background.

8. – Fluctuation analysis.

In order to obtain constraints on the shape of the $\log N$-$\log S$ relation below the discrete source detection threshold a fluctuation analysis of the intensity distribution (called $P(D)$ function in radioastronomy) was performed in the inner region (15.5 arcmin radius) of the *Lockman* field. The analysis has been done in the J band ((0.9–2) keV) because this band is essentially free from diffuse galactic and geocoronal X-ray emission which starts to dominate below 0.9 keV. The particle background contamination is about 10% in this band. The image was binned to a pixel size $48'' \times 48''$, well matched to the average point spread function in the central area.

Figure 5 shows the intensity distribution of the *Lockman* J-band image, calculated from the inner FOV with a radius of 15.5 arcmin. The average surface brightness is $1.22 \cdot 10^{-4}$ counts s^{-1} arcmin^{-2}.

We chose an analysis method very similar to that of Hamilton and Helfand [36], which does not calculate the expected $P(D)$ functions analytically [39,40], but tries to obtain a best fit to the measured intensity distribution through a large number of detailed simulations. This has the advantage that systematic instrumental effects, like, e.g., the radial dependence of the point spread function, can be incorporated quite naturally in the simulation process. Source and background photons were simulated according to the procedure described in [4]. We assumed a total extragalactic background flux of

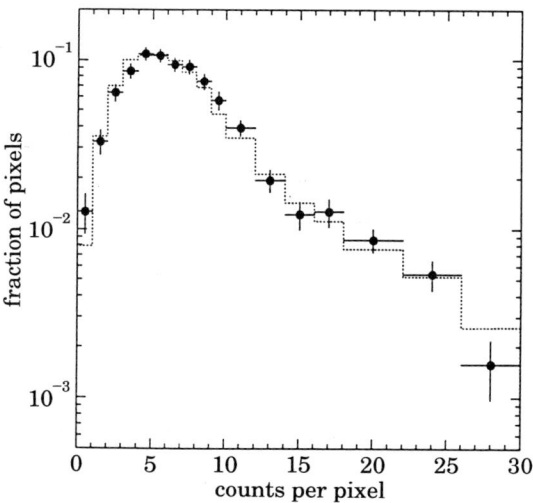

Fig. 5. – Intensity distribution of the inner 15.5′ of the Lockman Hole image in the J band ((0.9–2.4) keV) in pixels of 48″ × 48″ (filled circles). The dotted histogram shows the best-fit simulation, based on 10 fields with a sensitivity corresponding to the Lockman data. The best-fit slope of $\beta_2 = 1.8$ has been used for the faint end of the source counts, while the bright end is consistent with the direct determination from the RMSS sample (from [4]).

$1.25 \cdot 10^{-8}$ erg cm^{-2} s^{-1} sr^{-1} in the (1–2) keV band, consistent with the Wisconsin data [10], with the preliminary spectral fits to the data from the RMSS fields (see above and ref. [27]) and with the actually measured surface brightness in the *Lockman* field. For each trial $\log N$-$\log S$ function 10 realizations of PSPC fields with exposure times corresponding to the *Lockman* data were drawn, in order to reduce the noise in the models. For each simulated image the pixel intensity distribution was calculated in exactly the same way as for the real data, the 10 realizations were averaged. The full simulated distribution was compared to the data in fig. 5 with a χ^2 statistics, taking into account the statistical errors of both the data and the simulations. Note that with this choice of pixel size a source detection threshold of $ML = 10$ used for the analysis of the $\log N$-$\log S$ relation corresponds to about 11 counts per pixel in the J band. Instrumental effects on the $P(D)$ distribution are negligible [4].

The best-fit power law slope and normalization for the faint end of the $\log N$-$\log S$ relation are $\beta_2 = 1.80 \pm 0.08$ and $N_2 = 116 \pm 10$ with a reduced χ^2 of 1.17 for 12 d.o.f., using the above broken power law description of $N(S)$. The best-fit $N(S)$, integrated to fainter fluxes, resolves 100% of the (1–2) keV background at $S = 5 \cdot 10^{-18}$ erg cm^{-2} s^{-1}. The flattest power law extrapolation allowed by the $P(D)$ analysis resolves 85% of the background, while the steepest allowed slope resolves all of the background already at a flux of 10^{-16} erg cm^{-2} s^{-1}, *i.e.* only a factor of ~ 20 below the flux limit of the resolved sample. The upper limit for the contribution of a truly diffuse background component is 25% (90% confidence), *i.e.* at least 75% of the (1–2) keV background originates from discrete sources. The dotted areas in fig. 4 indicate the 90% confidence regions for $N(S)$ allowed by the fluctuation analysis.

9. – A new population?

Models for the AGN X-ray luminosity function (XLF) have been constructed from the EMSS by MACCACARO *et al.* [41] and from a combination of the EMSS AGN with 42 new QSOs from ROSAT medium-deep pointings by BOYLE *et al.* [42]. These data are consistent with pure luminosity evolution, terminated at a redshift of about 2. Both XLF models predict a steep slope in the $\log N$-$\log S$ at bright fluxes, with a significant flattening at fainter fluxes. In particular, integration up to $z_{max} = 3.0$ of the XLF of Boyle *et al.* produces a $\log N$-$\log S$ with differential slopes ~ 2.65 and ~ 1.7 with a change of slope at $\sim 10^{-14}$ erg cm^{-2} s^{-1}. The values for the two slopes are in reasonably good agreement with the RMSS best-fit parameters (see above). However, a significant difference is found in the number counts at the faintest flux limit: at $\sim 2.5 \cdot 10^{-15}$ erg cm^{-2} s^{-1} about a factor of 1.5–1.9 more sources are observed than predicted by these XLF models (see fig. 2 in ref. [42]). One way to remedy this discrepancy might simply be to adopt different parameters or more complicated evolutionary models for the QSO XLF. Future optical identifications of the faint sources in the Lockman field will show whether this possibility is viable or not.

Another possible explanation for the discrepancy in the total number of faint sources would be the existence of a new population of faint sources with a rather steep $\log N$-$\log S$ function: such a population could be practically absent at fluxes greater than a few 10^{-14} erg cm^{-2} s^{-1}, but could contribute a significant fraction of sources at our faint flux levels. The relatively large fraction of faint sources with hard-X-ray spectra might be very interesting in this context. If these are really a population of intrinsically absorbed AGN, their soft-X-ray $\log N$-$\log S$ relation might be steeper than implied by their cosmological evolution, simply because at greater distances a larger fraction of their unabsorbed flux might be redshifted into the ROSAT band. The number of these objects would be sufficient to explain the discrepancy of about a factor of 3 between the EMSS/ROSAT soft-X-ray source counts and the GINGA-HEAO-1 fluctuation analysis in the (2–10) keV range (see ref. [11] and references therein). The intrinsic absorption columns in these objects would have to be less than $\sim (3\text{--}10) \cdot 10^{21}$ cm^{-2}, depending on their redshift, *i.e.* much less than those typically observed in obscured Seyfert-2 AGN [43] in order to still observe residual soft-X-ray emission. The summed X-ray spectrum of these objects may, therefore, well be compatible with the spectrum of the GINGA fluctuations which requires an average N_H value of $\leq 3 \cdot 10^{21}$ cm^{-2} [11]. Such a population of sources might have profound consequences also for the composition of the hard ((2–10) keV)-X-ray background.

REFERENCES

[1] R. GIACCONI, H. GURSKY, F. R. PAOLINI and B. B. ROSSI: *Phys. Rev. Lett.*, **9**, 439 (1962).
[2] X. BARCONS and A. C. FABIAN, Editors: *The X-Ray Background* (Cambridge University Press, Cambridge, 1992).
[3] W. BRINKMANN and J. TRÜMPER, Editors: X-*Ray Emission from Active Galactic Nuclei and the Cosmic X-ray Background*, MPE report 235 (1992).
[4] G. HASINGER, R. BURG, R. GIACCONI, R. HARTNER, M. SCHMIDT and G. ZAMORANI: *Astron. Astrophys.*, **275**, 1 (1993).
[5] F. E. MARSHALL, E. A. BOLDT, S. S. HOLT, R. B. MILLER, R. F. MUSHOTZKY, R. E. ROSE, R. E. ROTHSCHILD and P. SERLEMITSOS: *Astrophys. J.*, **235**, 4 (1980).
[6] J. MATHER, E. S. CHENG, R. E. EPLEE, R. B. ISAACMAN and S. S. MEYER: *Astrophys. J. Lett.*, **354**, L37 (1990).
[7] G. SETTI and L. WOLTJER: *Astrophys. Space Sci.*, **9**, 185 (1970).
[8] G. PICCINOTTI, R. F. MUSHOTZKY, E. A. BOLDT, S. S. HOLT and F. E. MARSHALL: *Astrophys. J.*, **253**, 485 (1982).
[9] D. E. GRUBER: in *The X-Ray Background*, edited by X. BARCONS and A. C. FABIAN (Cambridge University Press, Cambridge, 1992), p. 45.
[10] D. MCCAMMON and W. T. SANDERS: *Annu. Rev. Astron. Astrophys.*, **28**, 657 (1990).
[11] G. STEWART: in *The X-Ray Background*, edited by X. BARCONS and A. C. FABIAN (Cambridge University Press, Cambridge, 1992), p. 259.
[12] E. BOLDT and D. LEITER: *Astrophys. J.*, **276**, 427 (1984).
[13] D. A. SCHWARTZ and W. H. TUCKER: *Astrophys. J.*, **332**, 157 (1988).
[14] K. A. POUNDS, K. NANDRA, G. C. STEWART, I. M. GEORGE and A. C. FABIAN: *Nature (London)*, **344**, 132 (1990).
[15] A. P. LIGHTMAN and T. R. WHITE: *Astrophys. J.*, **335**, 92 (1988).
[16] A. C. FABIAN, I. M. GEORGE, S. MIYOSHI and M. J. REES: *Mon. Not. R. Astron. Soc.*, **242**, 14 (1990).
[17] N. TERESAWA: *Astrophys. J. Lett.*, **378**, L11 (1991).
[18] R. D. ROGERS and G. B. FIELD: *Astrophys. J. Lett.*, **378**, L17 (1991).
[19] A. ZDZIARSKI, P. T. ZYCKI, R. SVENSSON and E. BOLDT: *Astrophys. J.*, **405**, 125 (1993).
[20] R. GIACCONI, J. BECHTOLD, G. BRANDUARDI, W. FORMAN, J. P. HENRY, C. JONES, E. KELLOGG, H. VAN DER LAAN, W. LILLER, H. MARSHALL, S. S. MURRAY, J. PYE, E. SCHREIER, W. L. W. SARGENT, F. SEWARD and H. TANANBAUM: *Astrophys. J. Lett.*, **234**, L1 (1979).
[21] R. E. GRIFFITHS, S. S. MURRAY, R. GIACCONI, J. BECHTOLD and P. MURDIN: *Astrophys. J.*, **273**, 458 (1983).
[22] F. A. PRIMINI, S. S. MURRAY, J. HUCHRA, P. SCHILD and R. BURG: *Astrophys. J.*, **374**, 440 (1991).
[23] T. T. HAMILTON, D. J. HELFAND and X. WU: *Astrophys. J.*, **379**, 576 (1991).
[24] G. HASINGER, M. SCHMIDT and J. TRÜMPER: *Astron. Astrophys.*, **246**, L2 (1991)

[25] T. SHANKS, I. GEORGANTOPOULOS, G. C. STEWART, K. A. POUNDS and B. J. BOYLE: *Nature (London)*, **353**, 315 (1991).
[26] S. F. ANDERSON, R. A. WINDHORST, T. MACCACARO, D. BURSTEIN, B. E. FRANKLIN, R. E. GRIFFITHS, D. C. KOO, D. F. MATHIS, W. A. MORGAN and L. W. NEUSCHAEFER: in X-*Ray Emission from Active Galactic Nuclei and the Cosmic X-ray Background*, edited by W. BRINKMANN and J. TRÜMPER, MPE report 235 (1992), p. 227.
[27] G. HASINGER: in *The X-Ray Background*, edited by X. BARCONS and A. C. FABIAN (Cambridge University Press, Cambridge, 1992), p. 229.
[28] E. PFEFFERMANN, U. G. BRIEL, H. HIPPMANN, G. KETTENRING, G. METZNER, P. PREDEHL, G. REGER, K. H. STEPHAN, M. V. ZOMBECK, J. CHAPPELL and S. S. MURRAY: *Proc. SPIE*, **733**, 519 (1986).
[29] J. TRÜMPER: *Phys. Blätter*, **46**, 137 (1990).
[30] F. J. LOCKMAN, K. JAHODA and D. MCCAMMON: *Astrophys. J.*, **302**, 432 (1986).
[31] J. M. DICKEY and F. J. LOCKMAN: *Annu. Rev. Astron. Astrophys.*, **28**, 215 (1990).
[32] M. SCHMIDT, R. BURG, D. SCHNEIDER and J. GUNN: private communication (1992).
[33] H. DE RUITER *et al.*: private communication (1992).
[34] P. P. PLUCINSKY, S. L. SNOWDEN, U. G. BRIEL, G. HASINGER and E. PFEFFERMANN: *Astrophys. J.*, **418**, 519 (1993).
[35] J. H. M. M. SCHMITT and T. MACCACARO: *Astrophys. J.*, **310**, 334 (1986).
[36] T. T. HAMILTON and D. J. HELFAND: *Astrophys. J.*, **318**, 93 (1987).
[37] X. BARCONS and A. C. FABIAN: *Mon. Not. R. Astron. Soc.*, **243**, 366 (1990).
[38] I. M. GIOIA, T. MACCACARO, R. E. SCHILD and A. WOLTER: *Astrophys. J. Suppl.*, **72**, 567 (1990).
[39] P. A. G. SCHEUER: *Mon. Not. R. Astron. Soc.*, **166**, 329 (1974).
[40] J. J. CONDON: *Astrophys. J.*, **188**, 279 (1974).
[41] T. MACCACARO, R. DELLA CECA, I. M. GIOIA, S. L. MORRIS, J. T. STOCKE and A. WOLTER: *Astrophys. J.*, **374**, 117 (1991).
[42] B. J. BOYLE, R. E. GRIFFITHS, T. SHANKS, G. C. STEWART and I. GEORGANTOPOULOS: *Mon. Not. R. Astron. Soc.*, **260**, 49 (1993).
[43] H. AWAKI, K. KOYAMA, H. KUNIEDA and Y. TAWARA: *Nature (London)*, **346**, 544 (1990).

QSO Absorption Line Systems.

D. G. York

Department of Astronomy and Astrophysics and The Enrico Fermi Institute
University of Chicago - Chicago, Ill.

PART I
Interstellar Absorption Lines in the Galaxy.

1. – Introduction.

If galaxies form at redshifts lower, on average, than the epoch of QSO's, some background QSO's randomly sample sight lines through forming galaxies. Absorption lines in spectra of QSO's, at redshifts lower than that of each QSO itself (referred to as QSO absorption line systems, or QSOALS), may help us probe galaxy formation. To explore this possibility, I divide this lecture into three parts. First, I deal with what is known about absorption lines in our own Galaxy (part I). In part II, I deal with the observational facts of QSO absorption lines, with little regard for their interpretation. How frequently do they occur? What species are present? What is the likely ionization source? Finally, I deal in part III with the evidence that absorption lines arise in forming galaxies—principally through arguments concerning clustering of components, ionization, abundances and association of absorbers with emission line objects.

I selectively refer to my own work below. This lecture is meant to be educational on certain basic points, not to be a general review of the literature, and it is naturally easier to draw on the pedagogical tools at my disposal. Where comprehensive reviews exist, I have referred to them, and the reader is urged to consult them for the original references. For general reviews, the reader is referred to ref.[1-3], on the interstellar medium; ref.[4], on the halo of our Galaxy; and ref.[5] on QSOALS.

2. – General comments on the interstellar medium.

The interstellar medium consists of the gas and various energy-containing fields between the stars. The gas exists in neutral and ionized components, with reference to whether neutral (H I) or ionized hydrogen (H II) is dominant. The primary ionization stages of the elements are determined by the diluted blackbody spectrum of hot stars and the recombination properties of the various ions with electrons. For times $\leq (10^4-10^6)\,y$, equilibrium exists, so an ionization equilibrium can be assumed: $\alpha n_e n(X^{q+1}) = \Gamma n(X^q)$. X denotes an element; q, the charge on the ion or atom; n, the number of particles/cm^3; α, the recombination coefficients, and Γ, the ionization rate, an integral over the ionizing spectrum of starlight or other sources and the wavelength-dependent ionization cross-section of the species being ionized.

Densities are so low that local thermodynamic equilibrium cannot be assumed in deriving population levels of different species: Virtually all atoms and ions are in the ground state.

This part is divided into three sections. First, I deal with the observations of absorption lines. Then diagnostics for learning the physical state of gas in our Galaxy are reviewed, followed by comments on how the gas is distributed. Finally, the results needed for the next two parts on QSOALS are summarized.

3. – The absorption line observations.

3`1. *Equivalent widths and velocity distributions.* – Interstellar lines show up as narrow, steep-edged absorption in spectra of background galactic stars. Even with the highest-resolution spectra, one must be careful to use stars with $v \sin i > 30$ km/s, to avoid confusion with stellar absorption lines. The earliest type stars can, further, have various forms of circumstellar matter that may confuse observations of truly interstellar lines. Figure 1 shows the interstellar lines near 960 Å in the B1 III star, β Cen, against the rather bumpy stellar continuum.

The attenuation caused by interstellar atoms leaves a residual intensity $I = I_0 \exp[-\tau]$, where I_0 is the intensity that would be observed in the absence of interstellar atoms and τ is the optical depth. τ is directly related to the column

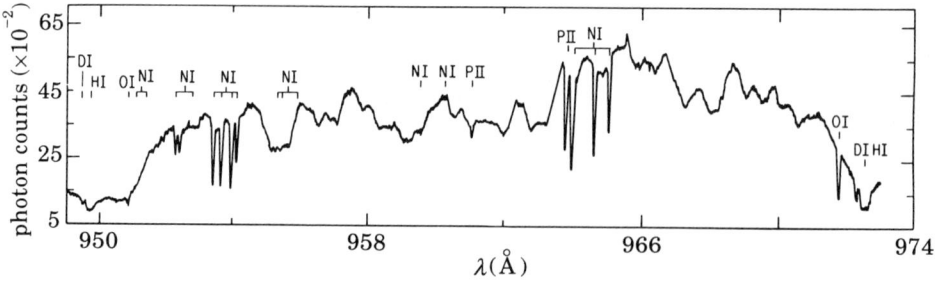

Fig. 1. – Interstellar absorption lines near 960 Å, in the star β Cen. Most of the lines are N I, with some O I, D I and H I visible at 948 and 972 Å, and lines of P II near 962 Å. The true zero can be seen at the bottom of the H I lines, at $\lambda\lambda 948, 972$.

density of atoms (N, cm^{-2}) at each velocity and to the oscillator strength for each transition. The velocity distribution is related to the natural line width, the temperature of the gas, the separate velocities of the clouds and any systematic flows, as in shocks, in the gas. Evaporation could, for example, also contribute such systematic flows. In practice, the velocity distribution is seldom derivable, and $N(v)$ is not known. Rather, it is necessary to represent the data as an integral over $1 - \exp[-\tau]$, called the equivalent width (W). The most common practice is to refer to obvious minima as clouds and to associate with each cloud a value $W_i(v)$, where i is an index referring to one of the many components, or clouds. Further, it is common to assume that the velocity distribution within each cloud is Gaussian, parameterized by a quantity $b = \sqrt{2}\,\sigma$, where σ is the standard deviation in the case of a Maxwellian velocity distribution. However, this parameterization shields the observer from a large amount of physics. A complete discussion of profiles of absorption lines for the interstellar case is given in ref.[6].

Detailed component models can be used to derive curves of growth for each sight line. These plots of $\log W_\lambda/\lambda$ vs. $\log Nf\lambda$ have a linear portion at low N (W/λ is proportional to $Nf\lambda$), a flat portion (W/λ is insensitive to large changes in N or f) and a square-root portion (W_λ proportional to $N^{1/2}$). Such models are helpful in understanding the errors in analysis of N from absorption line profiles.

3'2. *Absorption by atoms and ions.* – Generally speaking, all directions through the interstellar medium, outside dense clouds, are transparent at $\lambda > 912$ Å. Shorter-wavelength radiation is shielded by neutral hydrogen. When $N(\text{H I}) \leq 10^{17}$ cm^{-2}, H I is ionized at interstellar densities, and the interstellar medium is transparent at all wavelengths. When $N(\text{H I}) > 4 \cdot 10^{21}$ cm^{-2}, the continuous dust opacity blocks over 95% of the radiation at $\lambda < 3000$ Å.

In the ultraviolet region of the spectrum, some 30 elements, a few in multiple-ionization stages, can be observed[7]. Typical results from the *Copernicus* satellite[8] are shown in fig. 2[9]. The 21 cm results are in the center of the figure, inverted to show predicted absorption line components. (The damped Lyman α line contains no information on velocity structure.) O VI and Si III are shown at the bottom of the figure. Ar I and Fe II are shown at the top. Note the differences in profiles between Ar I (incipient separation of components by $\sim (6\text{-}10)$ km/s), Si III (saturated, broad, with steep edges) and O VI (shallow but very broad, no obvious structure). These differences will be shown to derive from quite different properties of the regions containing the three species.

The broad O VI lines reflect the general trend that O VI follows the velocity distribution of other interstellar lines. Observations[10] of line widths indicate that the O VI gas is hot, $T \sim 2 \cdot 10^5$ K.

Among the stars within 500 pc of the Sun, the Orion stars have the strongest

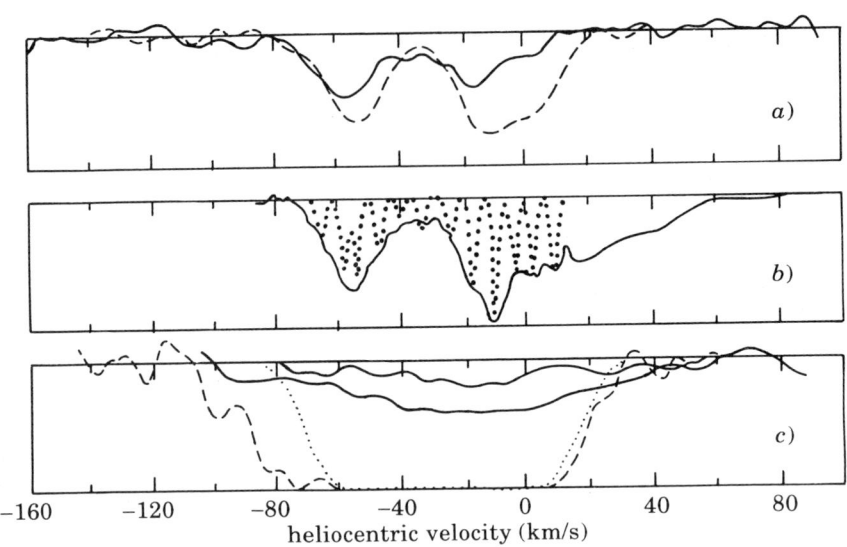

Fig. 2. – Profiles for several lines of different ionization stages in interstellar gas, in the star HD 93521, a high-latitude star about 1 kpc distant: a) —— Ar I 1048 Å, --- Fe II 1143 Å; b) —— H I, 21 cm; ··· velocity model fit, 21 cm; c) —— O VI, --- Si III, ····· N I, N II. See ref.[9].

interstellar absorption lines. Figure 3 gives a full set of absorption line profiles for several ions in ζ Ori and ι Ori. Note that Si IV is barely present, Si III is ⩾ 100 km/s wide, and Si II is broad, but hardly visible from (50–100) km/s. Gas at the highest velocities must have thermal electrons with $T_e > 30\,000$ K, so all Si^+ is ionized, or must be affected uniquely by radiation, $E \geqslant h\nu = 16$ eV, the ionization energy of Si^+. Yet Si IV is not present. Therefore, a much higher temperature, or a radiation field much bluer than a 30 000 K black-body are not present. The global behavior of Si II interstellar absorption in Orion stars (Orion's cloak [11]) indicates the effects are thermal, possibly due to shocks from supernovae (SNe) in Orion. Note the presence of high-velocity neutral hydrogen (bottom frames of ι Ori spectra).

Soft-X-ray maps in C band (centered near 44 Å) show that much of the sky emits X-rays from a yet hotter gas that is local to our Galaxy [12]. At least some of this radiation comes from local gas at $T \geqslant 10^6$ K, which fills much of interstellar space, but contains little mass. X-ray spectrometers operating between 20 Å and 70 Å are required to see it directly in absorption but sensitive instruments are not yet available [2].

3˙3. *Absorption by molecules.* – The cool phase of the interstellar medium contains most of the mass. It is easily recognized by large H I column densities ($\geqslant 10^{20}$ cm^{-3}); the presence of the molecules, H_2 and CO; and the presence of extinction, ever increasing blueward but of little effect by $\lambda \geqslant 2$ μm.

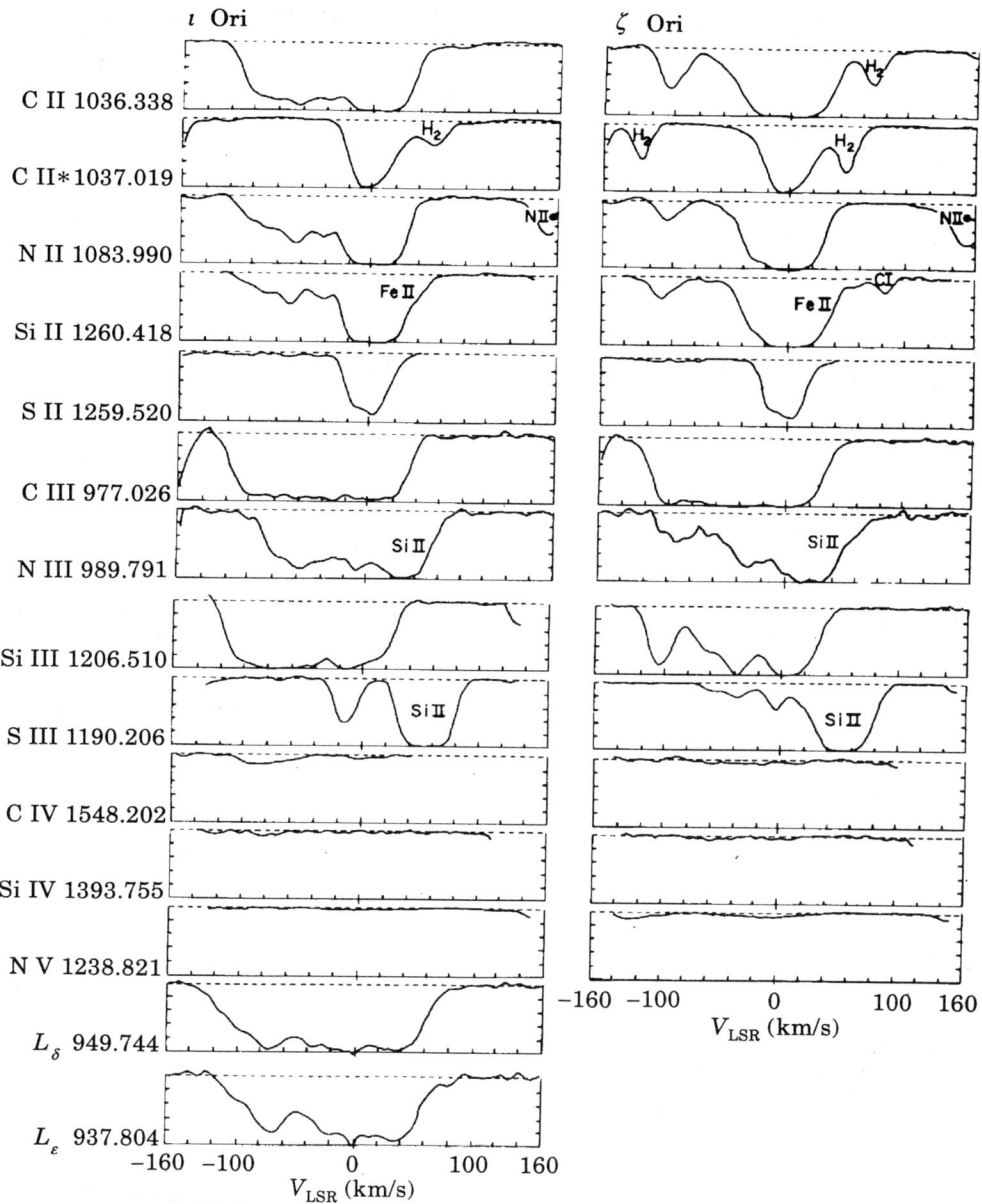

Fig. 3. – Velocity profiles of interstellar lines in two stars in the constellation Orion, ι Ori and ζ Ori, showing high-velocity gas in intermediate ionization ions (i.e. Si^{++}, N^+, N^{++}, C^+, C^{++}) accompanied by H I. See ref. [11]. The species plotted is noted at the right of each frame. Lines from other species are noted with the appropriate symbol on the plots. Note that the lines of D I occur at -80 km s^{-1} in the H I frames for ι Ori.

Molecular hydrogen has an easily recognizable, periodic spectrum for each vibration band [13]. The lowest rotation levels give the relative population of ortho- and para-hydrogen, which yields a temperature for the main components of the ~ 100 K, comparable to the spin temperature of H I [14, 15]. H_2 is thought to be formed on interstellar grains and to be dissociated by UV radiation near 1000 Å [14]. The population of high rotational levels shows that H_2 often occurs in shocks or near an intense UV radiation field. Traces of H_2 occur in unreddened stars (see interloping lines labeled in fig. 3), but strong lines only occur in more reddened stars where self-shielding to the dissociating radiation field occurs.

CO is the next most dominant molecule, and is seen only in the highest-column-density diffuse clouds. UV absorption lines occur near 1400 Å [16]. Radiodetection of CO is just barely possible at the values of N encountered in diffuse absorbing clouds, so strong CO detections in the radiospectrum correspond to extremely massive cloud systems. Profiles showing the rotational structure of CO in three stars are given in ref. [16]. Other molecules that may be studied using UV spectra include H_2O, HCl and N_2. Diatomic molecules (OH^+, CH, CH^+, CN) are detected towards diffuse clouds in the visible part of the spectrum [17]. The material giving rise to these molecules, locally, is so dense and opaque that it is unlikely that detectable QSO's would be seen through such clouds in other galaxies.

3˙4. *Dust.* – To complete this overview of types of material seen in absorption, in our own interstellar medium, we mention the continuous extinction. The particles or large molecules that form the dust produce continuous absorption, increasing at shorter wavelength [18], a pronounced bump (extinction increase) at (2200 ± 250) Å and broad absorption lines called diffuse interstellar bands [19]. The continuous extinction amounts to 3 magnitudes of absorption at 5500 Å when $N(H) \sim 6 \cdot 10^{21}$ cm^{-2} and the slope of the visible part of the spectrum is steepened to give a red excess $E(B - V) = 1.0$. (By $N(H)$, we refer to the total number of H nuclei, in H^0, H^+ and H_2.) The extinction at the center of the 2200 Å feature and at 1100 Å can entail more than seven astronomical magnitudes of additional extinction for this column density of H I. Small amounts of polarization of starlight are observed generally and are attributed to magnetically aligned grains ($P < 2\%$ [20]).

4. – Diagnostic techniques.

4˙1. *Selected abundances.*

4˙1.1. The heavy elements. Using the available interstellar lines and the techniques outlined above, quantitative abundance information can be obtained for two nucleosynthesis groups: O, Mg, Al, Si, S and Ca from type-II SNe; and

Mn, Fe, Zn, Ni and Cu from type-I SNe. The abundance [X/H] for species X is defined as $N(X)/N(H)$ divided by the corresponding ratio in the Sun [21]. Some abundances (*e.g.*, [Fe/H]) are so variable from cloud to cloud, and differ so much from the values expected from the theory of nucleosynthesis, that it is generally presumed that the abundance is modulated by their condensation into interstellar grains. By careful intercomparison of abundances of a number of species, one may hope to decipher the integrated effects of nucleosynthesis in the interstellar medium, as well as to learn the physical makeup of the grains that lead to the extinction of starlight noted above.

4˙1.2. Be, B and Li. The ratio [Be/H] (Be II, $\lambda 3130$) is thought to be formed by spallation of heavy elements by high-energy cosmic rays [22]. As such, it is a barometer for cosmic-ray activity, at any epoch [23].

Similar comments apply to boron (B II, $\lambda 1362$) [24]. Boron is also a control that must follow any nonprimordial source of deuterium [25]. Lithium is only detectable as a trace element Li I, 6707 Å. If ionization equilibrium (Li I is less than 1% of all Li in the interstellar medium) and dust depletion are properly accounted for, it provides a rough indicator of Li/H today and an indicator of ^{6}Li/Li [26].

4˙1.3. Deuterium. Deuterium measurements at any epoch provide an important cosmological indicator of the baryon density in the Universe, as detailed in lectures in this volume and in ref. [27]. Normally, in order to derive the cosmological value of [D/H], direct measurements of [D/H] [28] must be coupled with an estimate of ^{3}He and with some theory to estimate the effect of deuterium destruction due to processing of gas through stars, called astration.

4˙1.4. Neutral fluorine. Neutral fluorine is the dominant stage of fluorine in the interstellar medium (F I, $\lambda 954$). The abundance of fluorine is of interest because it may be especially sensitive to neutron processes, independent of the formation of the main elements [29].

4˙1.5. Chlorine and H_2. Neutral chlorine is formed by dissociation of HCl, which in turn is formed in the charge exchange reaction $H_2 + Cl^+ \to$ $\to HCl^+ + H, HCl^+ + e^- \to HCl$ [30]. This reaction and the dissociation to $H + Cl^0$ are so fast that most chlorine is in Cl^0, even though Cl^+ is the expected dominant ionization stage based on the earlier discussion. Thus Cl I ($\lambda 1347$) provides a way to know of the presence of H_2, even when it is not directly observable (*e.g.*, with *Hubble Space Telescope*, or in high-z QSOALS, when the lines of H_2 will be badly blended in the Lyman α forest).

4˙1.6. CN and the cosmological background. CN is a molecule of cosmological interest because the rotational splittings of the ground state are

~ 10^{-3} eV and the upper levels are populated predominantly by the cosmic 2.74 K background radiation [31]. This radiation, at a redshift above 2, populates fine-structure upper levels of C I and C II [32], as well as rotational splittings of other molecules. (The temperature of the background radiation at redshift z is $T_z = 2.74[1 + z]$.)

4˙2. *Temperatures*. – Temperatures can be determined by using the dependence of thermal motion on atomic mass, by using ratios of neutral species that have different dependences of their recombination coefficients on temperature, or by comparing ratios of column densities in different ionization stages of the same element with collisional and photoionization models.

4˙2.1. Thermal b-values. For a gas at 10 000 K, the thermal Doppler widths for H I, D I, N I and Ar I are 12.8, 9.1, 3.4 and 2.0 km s^{-1}, respectively. The values scale as $T^{0.5} m^{-0.5}$, where m is the nuclear mass of the element in question. For gas toward two early-type stars, derived values of b for these species are, respectively, 12, 7, 3.5 and 2 km s^{-1}, implying a temperature of several thousand degrees and turbulence of about 2 km s^{-1} [33, 34].

Of course, ions may be used in this way as well, providing ions common to H I and H II regions can be resolved. In any case, a turbulent b-value, independent of mass, must be assumed. Turbulent and thermal b-values should add quadratically.

4˙2.2. Recombination. Neutral Mg and Ca are formed especially rapidly, near $(3\text{--}10) \cdot 10^3$ K, compared to C I, Na I and Fe I. The latter three atoms form from the first ions mainly by radiative recombination, whereas dielectronic recombination is important for Mg and Ca first ions in the noted temperature range. Thus, if the rough abundances are known, the ratios $N(\text{Mg I})/N(\text{C I})$, or $N(\text{Ca I})/N(\text{Fe I})$ can be used as temperature indicators [33]. The former ratio is the one most likely to be well determined in low-density warm gas such as B-star H II regions or QSO absorption lines.

4˙2.3. Ionization. In collisional ionization equilibrium in the steady state, calculations [35] predict ratios of different ions for different temperatures. Ratios of ions for C, N, Al, Si, S and Fe can be used to determine temperatures by comparison with models. The ratios are substantially modified by the presence of ionizing photons, and by the inclusion of time-dependent photon or mechanical-energy input [36-39]. In practice, the data show all nearby volumes of space (within 500 pc) are dominated by neutrals and first ions [33], so the early models do not seem to apply. However, variants of these models may be relevant in the QSOALS, where ubiquitous C IV must be explained, and exotic ionization mechanisms may exist.

The profiles of C IV, N V and O VI for α Vir are given in ref. [10]. O VI is clearly visible, but the other thermally broadened lines of C IV and N V are shallow and only ambiguously detected. Models for solar abundance and collisional ionization explain the observations, if $T = 2 \cdot 10^5$ K, in agreement with the earlier cited results from line widths. There may be a hotter component of O VI, which becomes weaker and broader at higher T [40].

Our knowledge of ionization-related physics will be much better when observations of S I, S II, S III, S IV and S VI become available, and when the relevant atomic-physics calculations can be improved, through modern calculations for stellar atmospheres.

4˙3. *Densities.* – The final diagnostic to discuss is density. While most interstellar atoms are in the ground state, fine-structure levels ≲ 0.01 eV above the ground state are populated by collisions and depopulated by radiative processes. Thus the density is directly related to the number of atoms or ions in upper states. C II, N II and Si II are sensitive to the electron density when $0.1 \text{ cm}^{-3} < n_e < 100 \text{ cm}^{-3}$ [33, 41]. C I is sensitive to neutral hydrogen or H_2 collisions [42] in H I regions and to collisions with H^+ in H II regions. See ref. [41] for the best-developed excited lines of Si II, which are, in this case, in a dense H II region around an X-ray source, V Pup.

5. – What have we learned?

In this section, I give a brief overview of the distribution and state of the interstellar gas that will be of help in contemplating the QSOALS.

5˙1. *Gas distribution near the Sun.* – Except for a large cloud nestled up against the Sun on the galactic-center side, the volume of space within 100 pc is virtually empty [43]. On scales of 150 pc, there are expected H II regions from *B*-stars, which were detected, along with some very diffuse clouds on these sight lines, by the *Copernicus* satellite. These only fill about (1–5)% of the volume. When the volume out to 500 pc is examined in H I absorption, three major dust cloud complexes exist: one toward Perseus, one toward Ophiucus and one in Orion, just at the edge of the volume noted. There is an extended, low-density, 400 pc hole in galactic longitude 270°, toward the star β CMa.

Within a 1 kpc^3 volume, the interstellar components spread over a full range of 80 km s^{-1} [44], in apparent random motion not related to galactic rotation. Exceptions (larger spreads) are the Si III lines in Orion and C II in some halo stars [45].

Beyond 500 pc, in the disk of the Galaxy, are found the large molecular gas cloud complexes that harbor continuing star formation in our Galaxy. Extinction precludes absorption line studies in such regions.

On lines of sight through the halo, one encounters the nearby galactic fountain[46], as well as the 21 cm high-velocity clouds[47]. A subset of the latter may be related to the former. The high-velocity clouds are (2-40) degrees in extent, some within 2 kpc and others as far away as the Magellanic clouds. The distances are only poorly known. They cover about 10% of the sky when all velocities, $|v| > 50$ km s^{-1}, are considered.

At the edge of the disk of gas is now thought to be a zone called the galactic fountain[46] that probably develops when supernovae in low-density regions blow holes in the disk, injecting energy and material into the halo. The material eventually recycles to the disk. The C IV and Si IV, so obviously missing in spectra of O and B stars near the Sun[48], in fact show up in the halo. (See examples in fig. 1 of part III.)

To complete this survey of the distribution of absorbing material, we point out the extensive absorption line profiles towards several O and B stars in the LMC. There are numerous studies of UV absorption lines in the LMC and SMC[49-51], showing widely dispersed gas covering 450 km/s, much stronger than any possible galactic absorption along the sight line. In particular, quite strong Si IV and C IV lines occur, empirically, despite the unlikelihood of such lines occurring near O and B stars[48].

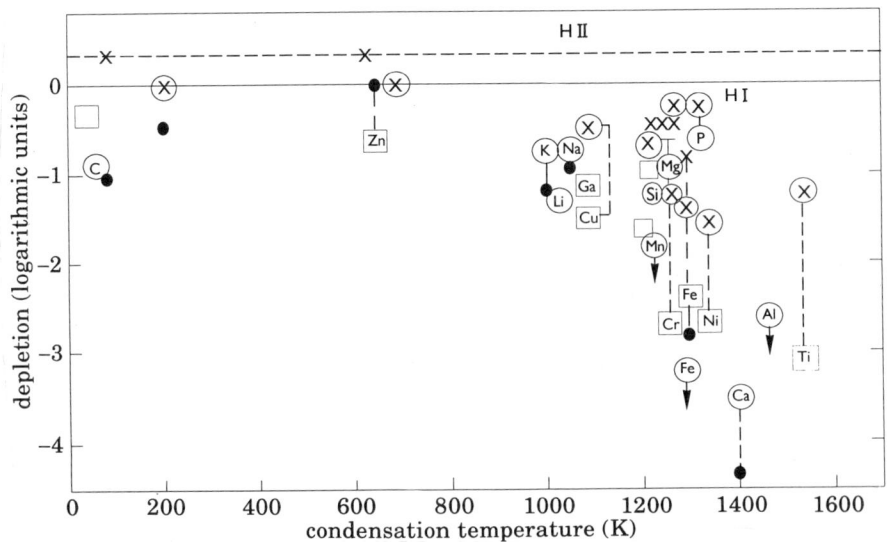

Fig. 4. – Depletions of interstellar species, referenced to the solar abundances. By symbols, ○ ζ Ophiuchi (*Copernicus*); ● ο Persei (*Copernicus*); × unreddened stars; ⊗ ζ Ophiuchi, HST, − 15 km/s component; □ ζ Ophiuchi, ξ Persei, main clouds, HST, this figure shows several effects elaborated on in the text: the new results from HST differ little from the general conclusions drawn from *Copernicus* data: gas towards reddened stars show significantly more depletion than gas on unreddened sight lines; gas in intermediate- and high-velocity clouds show reduced depletions. Data are taken from ref.[33, 34, 52, 53] and references therein, and [54].

5'2. *Dust.*

5'2.1. Depletions. Numerous studies of interstellar-gas phase abundances for the first 30 elements of the periodic table have been done using *Copernicus* (see references above, plus [55, 56]). These studies show that there is little if any depletion of C, N, O, Mg, S, P, Cl and Zn, while Si, Mn, Fe are depleted by factors of 5–100 (fig. 4). Ground-based studies of Ca and Ti confirm detectable depletion to low values of $E(B - V)$ [57]. Recent improvements in instruments used for these measurements reduce the early fears about the accuracy of abundances in reddened stars. (Compare, for instance, ref. [52, 53] from 1974 with ref. [54] from 1992.) Depletions are selectively reduced in high-velocity gas [58], presumably because of grain destruction in shocks [59].

It is generally thought, but not proven, that extinction is caused by a mixture of small grains, with radii, r, < 100 μm, large grains, r < 1000 μm, and large molecules. The small grains explain the rise in the far-UV curve [18]; the large grains account for the visible extinction, in numbers consistent with the atomic depletion [60, 61]; and the large molecules are required to account for the diffuse interstellar bands [62-64].

6. – Conclusion.

Our goal in this part has been to lay the basis for understanding QSOALS as interstellar media of intervening galaxies. Evidently, we can make several inferences concerning the galaxies, about the latter providing the full range of absorption line diagnostics discussed above are available for QSOALS.

1) Gas at $T < 5 \cdot 10^5$ K can be directly detected. The column densities detectable correspond to cloud masses ≤ 100 solar masses.

2) The presence of dust in intervening galaxies can be discerned or constrained by observations of various features of the extinction curve (the 2200 Å bump, diffuse interstellar bands, steeply falling fluxes at the shortest wavelengths of background QSO's) or by detection of a depletion pattern in gas phase abundances.

3) Nucleosynthesis products in early galaxies can be specifically searched for: fluorine for neutrino processes, sulphur for type-II SNe products, zinc for SNe type-I products, deuterium and lithium as tracers of big-bang nucleosynthesis and boron and beryllium as tracers of cosmic-ray effects.

4) Numerous physical parameters related to the global properties of the intervening galaxies can be derived:

a) Dominant nucleosynthesis sources should be identifiable.

b) Gas temperatures follow from several techniques.

c) Densities of neutral hydrogen, molecular hydrogen, electrons and protons can be directly discussed.

d) Items *b*) and *c*) yield global pressures in the ISM of any galaxy sampled.

e) The velocity dispersion of the aggregate of clouds reflects the rate of SNe at the epoch of observation.

f) The stellar radiation field from 100 Å to 2500 Å should be derivable from observed distributions of ions, especially for C, Si and S.

5) Using our discussion of the ISM on different scales, the projected covering factors, in projection, can be estimated. Hence, the statistical frequency of occurrence can be predicted for the hypothetical experiment of a QSO shining through our Galaxy. The converse is also true. For our Galaxy, the chances of a background QSO illuminating different physical domains, as observed by a distant observer of our Galaxy in absorption, can be listed. The chance

a) to go through diffuse clouds with significant H_2, given a sight line through the disk, is ~ 0.07;

b) to go through the disk and an HVC (to get high-velocity spread) is 0.2;

c) to go through an H II region in the disk is 0.1;

d) to go through an H II region including strong C IV and Si IV is ≤ 0.0025;

e) to go through the disk of a gas-rich dwarf in the Local Group (*e.g.*, LMC or SMC) is ~ 0.0005;

f) to go through the disk and see Si IV or C IV from the galactic fountain is < 1.0.

The cross-section of our Galaxy is $\sim \pi(10 \text{ kpc})^2$ to a Holmberg radius, and the cross-sections for the coincidences *a*)-*f*) are cited probabilities multiplied by $\pi(10 \text{ kpc})^2$. As we shall see later, the chance of intercepting an intervening disk of the modern type in a random background QSO is about $1/10$ that required of the entity that explains the absorbers. The chances that any of the above galactic entities, in the form with which we are acquainted in low-z normal galaxies, produce the QSOALS are very small. This fact highlights a final property of high-z galaxies that can be inferred from observations of QSO absorption lines:

6) evidently, the size of the objects that evolved to produce modern galaxies should be derivable, once we understand the segment of interstellar media that leads to the QSOALS.

REFERENCES

[1] L. L. COWIE and A. SONGAILA: *Annu. Rev. Astron. Astrophys.*, **24**, 499 (1986).
[2] L. SPITZER: *Annu. Rev. Astron. Astrophys.*, **28**, 71 (1990).
[3] D. G. YORK: in *Encyclopedia of Physical Science and Technology* (Academic Press, Inc., San Diego, Cal., 1992), p. 8285.
[4] D. G. YORK: *Annu. Rev. Astron. Astrophys.*, **20**, 221 (1982).
[5] R. J. WEYMANN, R. G. CARSWELL and N. G. SMITH: *Annu. Rev. Astron. Astrophys.*, **19**, 41 (1981).
[6] L. SPITZER: *Physical Processes in the Interstellar Medium* (Wiley, New York, N.Y., 1978).
[7] L. M. HOBBS, D. E. WELTY, D. C. MORTON, L. SPITZER and D. G. YORK: *Astrophys. J.*, **411**, 750 (1993).
[8] J. B. ROGERSON, L. SPITZER, J. F. DRAKE, K. DRESSLER, E. B. JENKINS, D. C. MORTON and D. G. YORK: *Astrophys. J.*, **181**, L110 (1973).
[9] J. A. R. CALDWELL: Thesis, Princeton University Observatory (1979).
[10] D. G. YORK: *Astrophys. J.*, **213**, 43 (1977).
[11] L. L. COWIE, A. SONGAILA and D. G. YORK: *Astrophys. J.*, **230**, 469 (1979).
[12] D. MCCAMMON and W. T. SANDERS: *Annu. Rev. Astron. Astrophys.*, **28**, 657 (1990).
[13] D. C. MORTON: *Astrophys. J.*, **204**, 1 (1976).
[14] J. M. SHULL and S. BECKWITH: *Annu. Rev. Astron. Astrophys.*, **20**, 163 (1982).
[15] J. M. DICKEY and F. J. LOCKMAN: *Annu. Rev. Astron. Astrophys.*, **28**, 215 (1990).
[16] E. B. JENKINS, J. F. DRAKE, D. C. MORTON, J. B. ROGERSON, L. SPITZER and D. G. YORK: *Astrophys. J.*, **181**, L122 (1973).
[17] E. F. VAN DISHOECK and J. H. BLACK: *Astrophys. J. Suppl.*, **62**, 109 (1986).
[18] J. MATHIS: *Annu. Rev. Astron. Astrophys.*, **28**, 37 (1990).
[19] G. HERBIG and D. R. SODERBLOM: *Astrophys. J.*, **252**, 610 (1985).
[20] R. H. HILDEBRAND: *Q. J. R. Astron. Soc.*, **29**, 327 (1988).
[21] N. GREVESSE and E. ANDERS: in *Cosmic Abundances of Matter*, edited by C. J. WADDINGTON (American Institute of Physics, New York, N.Y., 1989), p. 1.
[22] M. MENEGUZZI and H. REEVES: *Astron. Astrophys.*, **40**, 99 (1975).
[23] B. FIELDS, D. N. SCHRAMM and J. W. TRURAN: *Astrophys. J.*, **405**, 559 (1993).
[24] M. MENEGUZZI and D. G. YORK: *Astrophys. J.*, **235**, L111 (1980).
[25] S. COLGATE: *Astrophys. J.*, **181**, L53 (1973).
[26] D. MEYER, E. HAWKINS and E. WRIGHT: *Astrophys. J.*, **409**, L61 (1993).
[27] A. M. BOESGAARD and G. STEIGMAN: *Annu. Rev. Astron. Astrophys.*, **23**, 319 (1985).
[28] D. G. YORK and J. R. ROGERSON: *Astrophys. J.*, **203**, 378 (1976).
[29] T. P. SNOW and D. G. YORK: *Astrophys. J.*, **247**, L39 (1987).
[30] M. JURA and D. G. YORK: *Astrophys. J.*, **219**, 86 (1978).
[31] D. MEYER and K. C. ROTH: *Astrophys. J.*, **363**, 57 (1990).
[32] D. M. MEYER, J. H. BLACK, F. CHAFFEE, C. B. FOLTZ and D. G. YORK: *Astrophys. J.*, **308**, L37 (1987).
[33] D. G. YORK and B. KINAHAN: *Astrophys. J.*, **228**, 127 (1979).
[34] D. G. YORK: *Astrophys. J.*, **264**, 172 (1983).
[35] P. R. SHAPIRO and R. T. MOORE: *Astrophys. J.*, **207**, 460 (1976).
[36] J. WEISHEIT: *Astrophys. J.*, **190**, 735 (1974).
[37] P. MESZAROS: *Astrophys. J.*, **191**, 79 (1974).

[38] J. SCHWARZ: *Astrophys. J.*, **182**, 449 (1973).
[39] S. M. LEA and J. SILK: *Astrophys. J.*, **182**, 731 (1973).
[40] L. COWIE, E. B. JENKINS, A. SONGAILA and D. G. YORK: *Astrophys. J.*, **232**, 467 (1979).
[41] D. G. YORK, B. FLANNERY and J. N. BAHCALL: *Astrophys. J.*, **210**, 143 (1976).
[42] K. S. DE BOER and D. MORTON: *Astron. Astrophys.*, **71**, 141 (1979).
[43] P. C. FRISCH and D. G. YORK: *Astrophys. J.*, **271**, L59 (1983).
[44] L. COWIE and D. G. YORK: *Astrophys. J.*, **220**, 129 (1978).
[45] L. DANLY, F. J. LOCKMAN, M. R. MEADE and B. D. SAVAGE: *Astrophys. J. Suppl.*, 81, 125 (1992).
[46] P. R. SHAPIRO and R. A. BENJAMIN: *Publ. Astron. Soc. Pac.*, **103**, 923 (1991).
[47] D. G. YORK, G. BURKS and T. GIBNEY: *Astron. J.*, **91**, 354 (1986).
[48] L. L. COWIE, W. TAYLOR and D. G. YORK: *Astrophys. J.*, **248**, 528 (1981).
[49] B. D. SAVAGE and K. S. DE BOER: *Astrophys. J.*, **243**, 460 (1982).
[50] E. L. FITZPATRICK and B. D. SAVAGE: *Astrophys. J.*, **267**, 93 (1983).
[51] J. C. BLADES, J. M. WHEATLEY, N. PANAGIA, M. GREWING, M. PETTINI and W. WAMSTEKER: *Astrophys. J.*, **334**, 308 (1988).
[52] D. C. MORTON: *Astrophys. J.*, **197**, 85 (1975).
[53] T. P. SNOW: *Astrophys. J.*, **204**, 759 (1976).
[54] B. D. SAVAGE, J. CARDELLI and U. J. SOFIA: *Astrophys. J.*, **401**, 706 (1992).
[55] E. B. JENKINS, B. D. SAVAGE and L. SPITZER: *Astrophys. J.*, **301**, 355 (1986).
[56] E. B. JENKINS and L. SPITZER: *Annu. Rev. Astron. Astrophys.*, **13**, 133 (1975).
[57] G. STOKES: *Astrophys. J. Suppl.*, **36**, 115 (1978).
[58] R. GIOVANELLI, M. HAYNES, D. G. YORK and J. M. SHULL: *Astrophys. J.*, **219**, 60 (1978).
[59] L. SPITZER: *Comments Astrophys.*, **6**, 177 (1977).
[60] E. M. PURCELL: *Astrophys. J.*, **158**, 433 (1969).
[61] L. M. HOBBS, D. G. YORK and W. OEGERLE: *Astrophys. J.*, **252**, L21 (1982).
[62] J. PLATT: *Astrophys. J.*, **123**, 486 (1956).
[63] W. SMITH, T. P. SNOW and D. G. YORK: *Astrophys. J.*, **218**, 124 (1977).
[64] J. L. PUGET and A. LEGER: *Annu. Rev. Astron. Astrophys.*, **27**, 161 (1989).

PART II
Properties Observed for QSOALS.

1. – The observed phenomena.

Absorption lines with $z_{abs} < z_{em}$ are commonly observed in spectra of QSO's. The number is higher by a factor of more than 10–12 than expected from the simple picture that galaxies like those we observe at $z < 0.05$ lie randomly in front of QSO's at all epochs.

Sample spectra of QSO's are given in ref. [1]. For reference, note that strong Ly α emission from QSO's is almost always observed. The absorption phenomena observed are general, and may be noted as follows.

a) Narrow absorption lines are seen longward of Lyman α emission that correspond to patterns of heavy-element lines at high redshift. However, the absorption redshifts, z_{abs}, are often well below the QSO emission line redshifts, z_{em}. The lines at wavelengths longward of the QSO Ly α emission are almost always identifiable as multiple heavy-element lines from gas in a few intervening systems, called QSO absorption line systems (QSOALS). These systems appear to have very low densities, similar to those normally found in the interstellar medium (ISM), as described in part I. Virtually all atoms are in the ground state, $n_e \lesssim n_H \lesssim 10$ to 100 cm^{-3}. (Unless otherwise noted, the notation in this and the third part is as in part I.)

b) Profiles similar to mass loss profiles in hot stars (redshifted emission, blueshifted absorption) are often seen around the QSO emission lines. QSO's in which this effect is seen are often called broad absorption line QSO's, or BALs. It is unclear how they are related to the narrow heavy-element systems at $z_{abs} < z_{em}$. We shall not concern ourselves further with the BAL phenomenon.

c) There is a general attenuation below Ly α owing to a «forest» of strong lines. These lines are normally assumed to be Ly α lines from a multitude of systems with $z_{abs} < z_{em}$, because few of the lines fit patterns expected of heavy-element lines from well-established systems known from lines outside the forest.

d) At wavelengths below 912 Å in the QSO rest frame, it is not uncommon for the QSO signal to drop to zero, leaving only residual signal from instrumental scattering. This phenomenon arises from Lyman continuum absorption, and an absorption line system that displays this feature is referred to as a Lyman limit system. Convergence of absorption lines of the Lyman series of H I with high quantum number causes a gradual roll-off on the long-wavelength side of this feature. For $N(\text{H I}) \sim 3 \cdot 10^{17} \text{ cm}^{-2}$, the discontinuity corresponds to $\tau = 1$ at 912 Å in the rest frame; then τ decreases as ν^{-3} at shorter wavelengths. When $\tau_{912} \gg 1$, the wavelength at which τ decreases to 1, so that light from the background QSO can be seen again, may be at very short wavelengths, often below the short-wavelength cut-off of the observing system.

The Lyman α forest lines and the heavy-element lines arise in gas clouds that lie between the QSO and the observer. The multitude of systems thus seen, at redshifts much lower than the QSO's in which the lines are seen, are generally thought to be intergalactic gas clouds that are related in some manner to galaxy formation. The systems are observed serendipitously as QSO spectra are obtained.

The heavy-element systems have been catalogued [2, 3]. Some of these systems have very strong, damped Lyman α absorption lines. These are

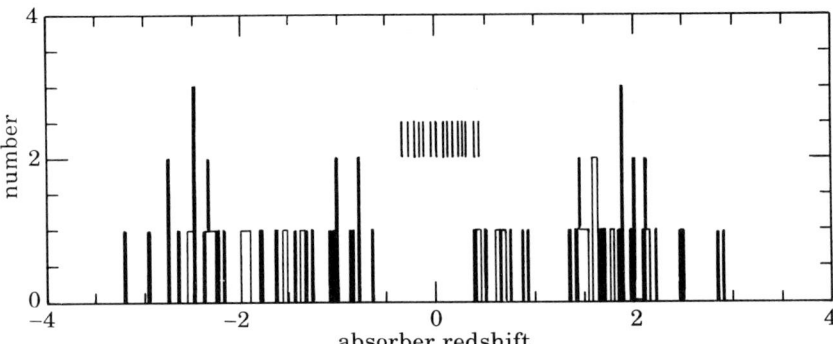

Fig. 1. – Number of absorbers at each redshift (89 total) in a 22° cone centered (for the North) at right ascension $12^h 40^{min}$, declination $+27° 24'$ [2]. Tick marks at low z represent the peaks in the galaxy distribution with redshift [4]. (Positive redshift denotes the direction of the Galactic North Pole, negative redshift denotes the direction of the Galactic South Pole.)

generally searched for separately [5], as they require less sensitivity and more spectral coverage, because they are rare.

The importance of the absorbers as cosmological probes is shown in fig. 1, which is a histogram of absorbers in a 22° cone extending toward both galactic poles [2]. Tick marks show the redshifts of peak counts in the corresponding histogram of faint galaxies [4] in a smaller cone along the same directions. Clearly, galaxies can be probed to higher redshifts using absorbers than using the deepest galaxy samples obtainable today, or in the near future.

2. – A stroll through the forest.

The absorbers are divided into two groups—Ly α absorbers and heavy-element absorbers. Figures 2, 3 and 4 show portions of the spectrum of the QSO 1331 + 170 [6, 7]. Spectra with KP4m in the bottom left corner were recorded at 18 km s^{-1} resolution [6], while those with MMTF in the bottom left corner are from the Multi-Mirror Telescope 1 Å (~ 70 km s^{-1} resolution) resolution spectrograph [7].

Lines deemed to be real are numbered at the bottom. Labels as to the redshift system and element identification are just above the spectra. Labels with solid horizontal marks are firm identifications. Labels with dashed horizontal marks correspond to predicted lines in definite systems, but cannot be confirmed independently because of the possibility of blends with the Ly α forest. Labels with no horizontal line are predicted positions of lines, which, however, are not consistent with being detected.

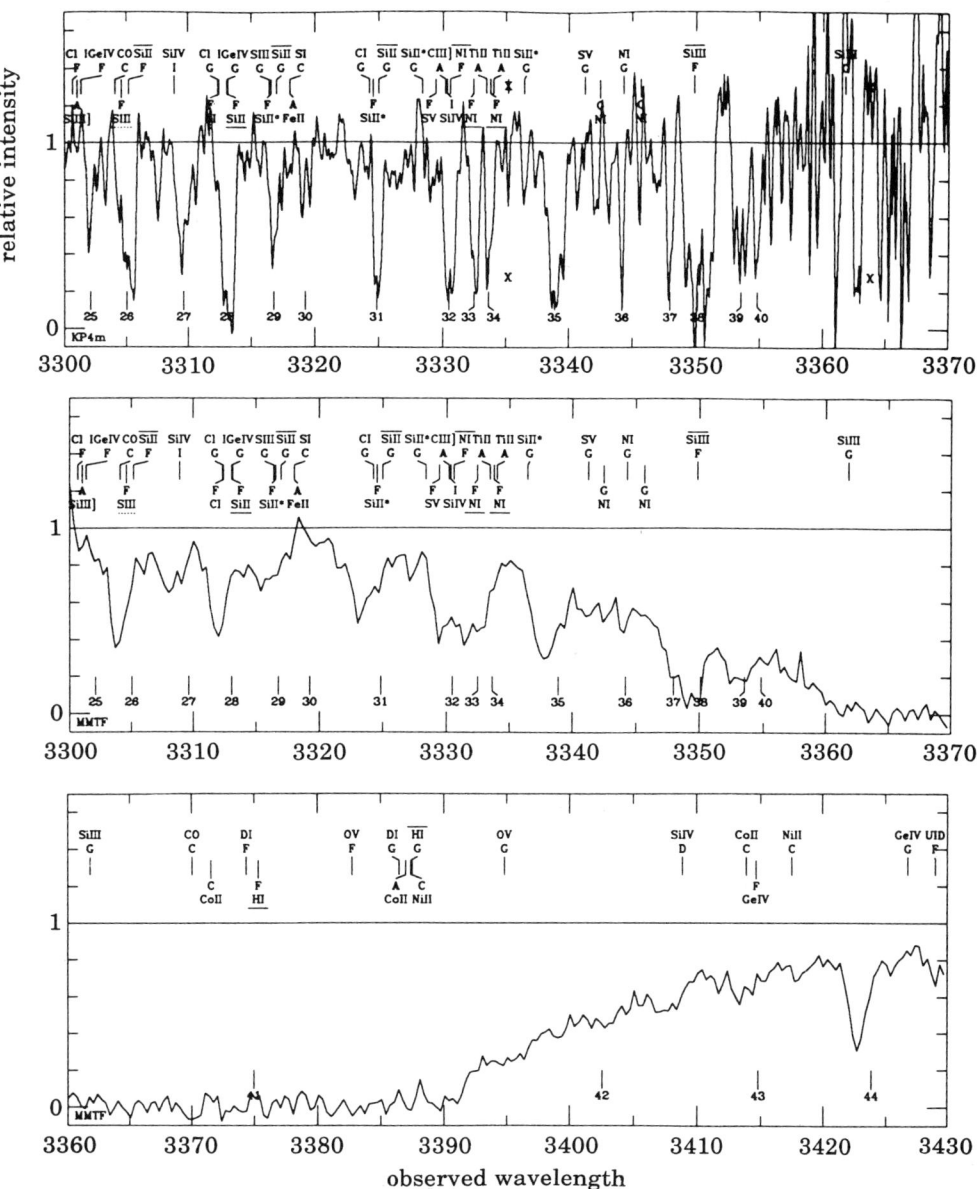

Fig. 2. – The bottom two panels show data from [7], covering the damped Lyman α absorption line from the $z = 1.77$ system in Q1331 + 170. The resolution is about 1 Å. The top panel shows data from [6], at a resolution higher by a factor of five (0.2 Å, 18 km s^{-1}). The wavelength scales are misaligned by about 0.7 Å. Note the narrower lines evident in the high-resolution data and the separation of the parts of the broad blend at 3330 Å (top panel) into at least three components. Labels across the top show the positions of heavy-element lines that might be present. When these labels represent positive identifications, the species labels have solid horizontal bars above or below them. Dotted horizontal bars are questionable identifications. The letters A through J denote the redshift system to which the labels apply.

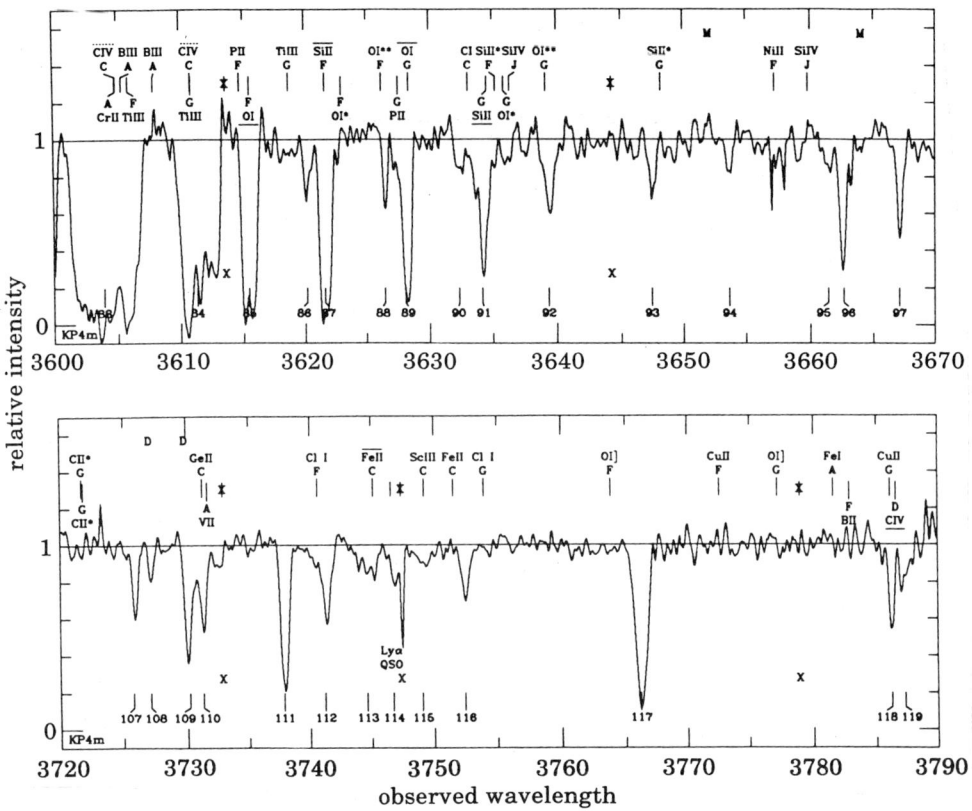

Fig. 3. – The regions (3600–3670) Å and (3720–3790) Å in Q1331 + 170 [6]. Notation for labelling is as for fig. 2. The first region is in the Lyman α forest, the second includes the position of Lyman α emission from the QSO (3747 Å) but includes unidentified lines longward of that point that may be Lyman α lines as well.

The panels in fig. 2, 3 and 4 contain small segments of the spectrum of QSO 1331 + 170. Several points are noteworthy.

a) From fig. 2, the strong absorption feature, split between the bottom two frames, is a damped Lyα line in a system at $z = 1.77$, $N(\text{H I}) \sim$ $\sim 3 \cdot 10^{21}$ cm^{-2}.

b) The top panel of fig. 2, showing the same spectral region as the middle panel, but at 5 times higher resolution, is from an echelle spectrograph. The Lyman α line covers more than one echelle order, and it is hard to calibrate the continuum, so the strong line has been normalized out. Note the difference in appearance of the lines in the leftmost 2/3 of the top two frames. While lines such as those at 3339 Å and 3310 Å have no suggested identification from known heavy-element line systems, others, such as those near 3313 Å and 3332 Å, coin-

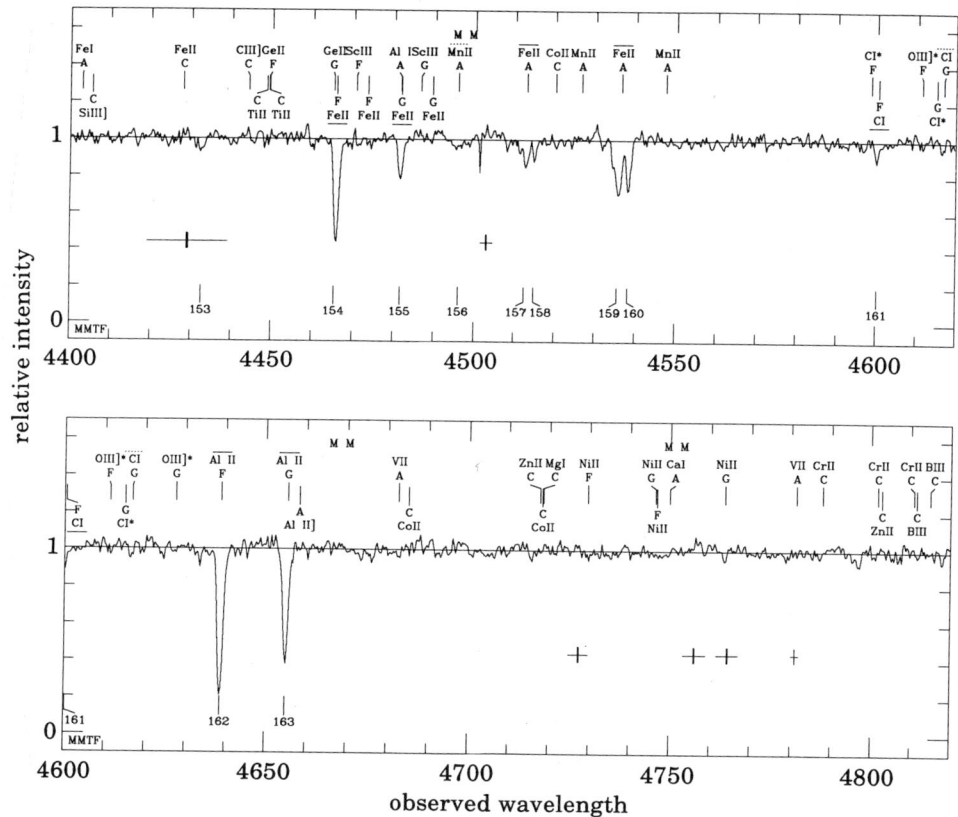

Fig. 4. – The region (4400-4800) Å in Q1331 + 170 [7], outside the Lyman α forest, including only lines of heavy elements and free of lines that are unidentified. Labelling is as for fig. 3, except that the label «M» at the very top of the frame denotes the locations of manmade emission lines in the night sky spectrum. Heavy vertical lines, at the center of the frames vertically, and crossed by long thin lines, are the positions and widths, respectively, of diffuse interstellar bands from our own Galaxy that could affect the drawing of the continuum.

cide with lines of heavy-element systems. In fact, in this short stretch of spectrum (3300-3355) Å, there are positive identifications of one-half of the lines as heavy-element lines from two systems, historically called system F and system G (1000 km s^{-1} longward of F). There are two strong points illustrated here. First, without high resolution it is difficult to be sure that lines in the forest are single, not multiple. Second, blends of heavy-element lines make some apparent Ly α line identifications ambiguous without further information.

c) Figure 3 shows two more regions of the forest. In the top panel, note again the large number of lines from systems F and G, including strong O I and Si II lines, along with lines with no identification (supposed Lyman α lines).

But the main new point to notice here is the pair of lines at 3604 Å and 3611 Å. These are identified as possible C IV from a real system, C, $z = 1.33$. But the two lines are obviously not matched pairs of a doublet. They have different shapes and centers, relative to the predicted position (vertical tick marks, labelled «C IV»). Most of both lines must be (multiple) Lyman α lines, accidentally overlapping the region where possible C IV lines would appear. It is not even clear there is any C IV from system C. It may, like system G, have mainly low ions (see fig. 5, below).

d) The second panel of fig. 3 shows the transition from the forest to the region of the spectrum where redshifted Lyman α should not exist if the absorbers are in front of the QSO's and have cosmological redshifts. Yet there is a very strong line at 3766 Å that is unidentified. Perhaps it comes from an absorbing object that is moving towards the QSO at 1500 km s^{-1}.

e) Several interesting weak lines may be noted in the second panel of fig. 3. At 3740 Å, there is a possible detection of Cl I, a feature that may signal the presence of H_2 (see part I). The spin-forbidden line of O I, rest wavelength 1355 Å, should be detectable at 3764 Å, given the large H I equivalent width. The absence of this line implies that the oxygen abundance in system F is well below solar. (The strong line of O I from the same system in fig. 2 is saturated, and does not yield a column density without further assumptions being made).

f) Longward of Lyman α emission, QSO spectra are relatively simpler. Sample plots are shown in fig. 4. All the features are identified with well-established systems, with the exception of the feature at 4432 Å, which may be a residual of the galactic diffuse band at the same wavelength, the wings of which may have been lost in our normalization procedure.

3. – The Ly α forest.

Now that the vagueries and various problems of line identification have been pointed out in specific cases, we discuss in turn the Ly α forest lines and the heavy-element systems, from the point of view of large samples of data.

Various statistical studies of line strengths, line frequencies and line clustering have been done in the Ly α forest. No correlations have been consistently reported by different authors. There has been considerable discussion of voids in the Ly α forest [8, 9] and of clustering on small scales [10, 11]. Larger samples are needed to make general statements. It is likely that the absence of high resolution has led to blends of clustered components, masking clustering effects on scales of 100 km/s.

There is a well-defined equivalent width distribution for Ly α lines, and a histogram in column density [12] shows an approximate power law behavior of all Ly α lines, Lyman limit systems and damped Ly α lines. However, the conversion to column density from equivalent width is highly uncertain. Moreover, unresolved blends of closely spaced Ly α components and blends with heavy-element lines (such as those pointed out above) will obviously contribute to making a smooth distribution out of what could be a bimodal or more complicated distribution of components.

Evidently, Ly α studies must come to terms with the problem of blends, using high resolution and detailed long-wavelength studies to identify all heavy-element systems so heavy-element lines can be removed from the Ly α forest.

Three particularly widely accepted results concerning the Ly α forest are that the systems are metal free, hence «intergalactic»; that the number of systems rises very rapidly compared to expectations for an unevolving population of constant co-moving density; and that QSO's ionize a volume adjacent to them, removing Ly α clouds (the proximity effect).

The first point is an extreme oversimplification. To give an abundance requires observations or model calculations providing column densities in all ionization stages, or limits, and an accurate H I + H II column density. Most Ly α lines are saturated, yet not damped. As the velocity distribution is not known, the column densities are ambiguous. That is, the lines often lie on the flat portion of the curve of growth (part I). In ref. [13, 14], it is shown that what appear to be isolated Ly α lines in data of modest signal-to-noise ratio in fact show C IV doublets at the same redshifts when data with a high signal-to-noise ratio are available. It has been shown, further, that by accepting Ly α identifications, isolating the expected location of C IV, then co-adding the regions of the C IV lines in the rest frame of many Ly α systems, C IV absorption can be detected [15]. Many Ly α lines are so weak that C IV from regions with a carbon abundance, [C/H], of 1/1000th solar value (termed 10^{-3} solar metallicity) could not possibly produce a detectable signal. Such regions cannot be said to be free of heavy elements, yet they give rise to Lyman α lines that are unaccompanied by any other detectable features.

In general, the very-model-dependent result is that there are not systems with [C/H] less than 10^{-4} of the solar value known for sure. There are numerous systems in the Ly α forest that could have higher abundance.

The rapid rise with redshift in Ly α line density is documented by several authors, see, e.g., ref. [16]. A population of objects with a constant, co-moving space density that are not changing will show a number per unit redshift $(\mathrm{d}M/\mathrm{d}z)$ proportional to $(1+z)^\alpha$, where α is 0.5 to 1.0. The much steeper rise in $\mathrm{d}M/\mathrm{d}z$ for Lyman α lines ($\alpha = 1.5$–2.5) can be taken as a sign that the population that gives rise to these lines is changing with redshift. At least the weaker lines, for which the best statistics are available, seem to be disappearing at low z, that is, as the Universe evolves.

The proximity effect refers to a reported diminution of Lyman α lines near the redshift of the background QSO. The proximity effect [17] is important because the larger the redshift space near the QSO that is diminished in Ly α

lines, the weaker must the extragalactic field of integrated QSO light be. That is, if the Ly α lines are intergalactic and affected by an intergalactic radiation field, the ionizing radiation of an individual QSO will be more important, at some distance, than the background. The observed effect implies that the intergalactic flux at $z \sim 2$ is 10^{-21} erg cm^{-2} s^{-1} sr^{-1} Hz^{-1}. The local value of this flux is $< 5 \cdot 10^{-23}$ erg cm^{-2} s^{-1} sr^{-1} Hz^{-1}, a value derived from the number of H_α photons emitted by a Oort high-velocity clouds in our own halo (one H_α photon per ionizing photon) [18, 19]. These values are roughly consistent with the known drop in the number density of QSO's with decreasing z between $z = 2.5$ and $z = 0$, since the background is thought to be due to the integrated light of the QSO's themselves.

The abundance arguments given above are robust. The rest of the arguments as to the intergalactic nature of the Lyman α lines depend on careful pruning of the Ly α forest to get proper line lists. In particular:

a) There may be a population of low-metallicity «Ly α clouds», which, if they have C IV doublets, probably have Si III 1206 Å detectable, and these must be thrown out.

b) Whether to remove the Ly α lines with C IV doublets before doing statistical studies is a matter of context, depending on the problem.

c) Blends with heavy elements must be removed. If they are ignored, more or stronger Lyman α lines will be recorded than are actually real. If they are not removed by profile fitting, but contaminated Ly α lines are simply thrown out, the number of true Ly α lines is reduced. Subtle effects such as clustering will obviously not be clarified until all of these effects are properly dealt with.

In summary, it is not proven that the Ly α lines are intergalactic. Detailed clustering studies of properly pruned data samples may help discern several populations of clouds now all categorized as Ly α clouds. There could be strong correlations on small scales (< 100 km/s), correlations with galaxy peaks [4], correlations or anticorrelations with individual galaxies, strength-dependent correlations, and no Lyman α lines that are truly free of heavy elements. The Ly α clouds could be detritus of galaxy formation, with high speeds of ejecta from supernova shocks because of the low mass of the clouds that would remove obvious velocity correlations. If several uncorrelated populations exist in the Ly α line sample, treating the entire sample as one could smear out correlation effects.

4. – Heavy-element systems.

4'1. *Low-resolution results.* – Because of the velocity widths of the heavy-element systems (see the lines of system F in fig. 2, where the widths are (1–2) Å, or (80–160) km s^{-1}), a resolution of 1 Å is particularly effective in surveys for the absorbers. Numerous surveys at or near this resolution have been carried out. Additionally, high signal-to-noise ratios can be obtained on the bright objects, so that even rare elements, or elements with only low f-value lines, such as S, Cr and Zn[20,21], can be detected.

While spectra of this latter quality are not yet common, hundreds of systems are less well observed and tabulated[2,3]. The key statistical results may be noted as follows.

a) The number of systems per unit z, dM/dz, is ~ 1 for Mg II from $z = 0.5$ to 2, and ~ 1 for Si II from 1.5 to 2.5. Mg II ($\lambda\lambda 2795, 2802$) and Si II (1526, 1304 Å) are observable using ground-based spectrographs, but at different redshifts. On the other hand, data for Mg and Si at different redshifts are complementary because Mg and Si have the same nucleosynthetic origin, similar ionization potentials and similar f-values. Therefore, it is safe to assume as a working hypothesis that Mg II and Si II have $dM/dz \sim 1$ from $z = 0.5$ to $z = 2.5$. This, and several other results cited here, can be seen graphically in the figures of ref.[2].

b) For C IV, dM/dz is near 0.5 at $z = 1$, is about 1.8 at $z = 1.5$–2.2, and drops to 0.4 at $z = 3$. The data at $z \sim 1$ are suspect because most QSO's have higher z's. The low value may result from the inability of observers to sort C IV out of the Ly α forest. The range $z = 0$ to $z = 1$ should be well covered by the HST key project[22]. The drop at high z is demonstrated in [23], and interpreted as an abundance effect, namely: the value [C/H] decreases, on average, as z increases at $z > 2$.

c) For Si IV, $dM/dz \sim 1.4$ from $z = 1.5$ to 2.5, but drops to 0.3 at $z = 3.5$. The same trend occurs in Si II. Thus both Si IV and Si II show the same behavior as C IV. The fact that Si II and Si IV both drop at high z indicates that the drop in high ions is not caused by a dropping radiation field, with Si IV going to Si II. The Lyman limit systems evolve as a population of constant co-moving density[24]. These systems represent the high-column-density end of the absorption line systems. If detections of Si IV were dropping because of higher ionization at high z (favoring Si V), some drop in dM/dz for H I might be expected to occur, but evidently does not. Knowing the dependence of N V and O VI on redshift would prove useful in this regard.

d) C IV is usually used as the key search doublet, and seems to virtually always be present. On the other hand, the difficulty of separating the neutral species in interstellar media from the Lyman α forest, as in fig. 2 (N I) and 3 (O I) above, and the presence of some systems with little or no Si IV or C IV (see below), may mean that several systems have been missed because of the bias of using C IV as a key search parameter.

Identifications made using coincidences in the Lyman α forest are often misleading (the C IV system in fig. 3, top frame). Ideally, very-long-wavelength spectra, beyond the Lyman α forest, should be used to identify systems, using the plethora of Fe II lines, and possibly Al III at $\lambda\lambda 1854, 1862$.

e) The equivalent-width histogram for C IV and Mg II shows the typical width to be about 80 km s^{-1}, though for Mg II the widths can approach 1000 km s^{-1}[2]. As discussed later, the widths of the absorption line systems represent the velocity spread of multiple components. Since these are often unsaturated, the equivalent width in velocity units is a lower limit to the actual velocity spread. Thus over half the systems are wider than the 80 km/s noted in the last part as applying to our Galaxy, at least near the Sun.

4'2. *High-resolution studies.* – As noted in part I, and in the discussion in this part of fig. 2, high-resolution data on the absorbers are essential. All the high-resolution data available (for example, ref.[25-28]) show that the QSO absorbers split into narrow components. The edges of saturated lines are very sharp, indicating unresolved Gaussian profiles for at least the outer velocity components of the blends, as in the top panel of fig. 2.

Figure 5 shows the profiles of Si IV and Si II from system F in the QSO $1331 + 170$[11]. The plots in velocity space clearly show the component structure in the two ions and the mismatch of the obvious components in each. Material at -180 and $+100 \text{ km s}^{-1}$ in Si IV has no counterpart in Si II. Conversely, the velocity extremes of the Si II profile have only minimal Si IV present. A detailed comparison of profiles of C II, C IV and of Al II and Al III leads to a similar conclusion in this system, and the conclusion carries throughout the small sample of systems for which high-resolution data are available. It is noteworthy, given the dominance of Si IV in system F, that the Si IV in system G is barely detectable, if at all. Thus, whether we discuss net detection of C IV and Si IV (compare system F to G), or component-to-component structure in Si IV

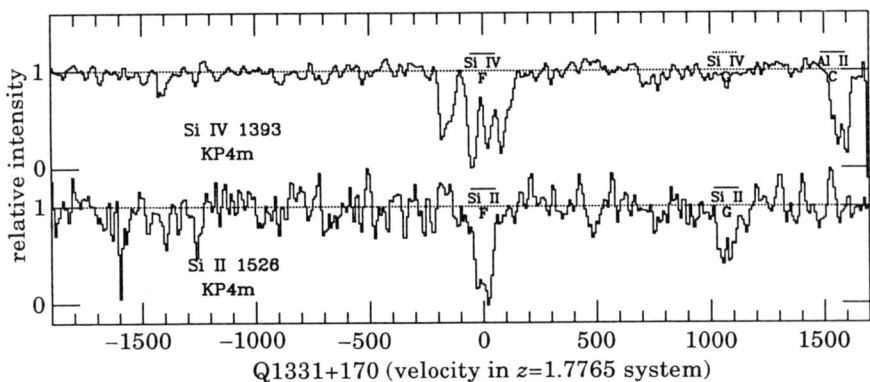

Fig. 5. – Velocity profiles for Si IV and Si II for the system at $z = 1.77$ in Q1331 + 170. Note the dramatic variations in the ratio of the line strengths of the two ions from component to component. The additional system (G) at 1000 km s^{-1} shows Si II, but little if any Si IV[11].

and C IV (within system F), it is clear that the properties vary dramatically from one system to another and from one component to another.

4˙3. *Ionization models.* – In principle, the radiation field giving rise to the observed ionization, together with collisional ionization effects, can be modeled. In practice, there are various problems. For carbon, the lines of C II, C III and C IV are sometimes observable. However, both detectable lines of C II ($\lambda\lambda 1036$, 1334) and the one line of C III ($\lambda 977$) are usually, at least ambiguously, saturated. All three lines are at such short rest frame wavelengths that they are blended with the Ly α forest. Thus blending and line saturation make the ratio of ions hard to derive. For Si, Si II (multiple lines, several of which appear in the figures of this part) and Si IV are useful, but the single line of Si III is blended with the Ly α forest and is often saturated.

Ionization calculations have been done for numerous systems [29-32]. There are two major problems with such calculations. First, the atomic physics is still poorly known. Using new cross-sections for ionization and new recombination coefficients from the Opacity Project [33], ion ratios can differ by as much as a factor of 10 from the older models [32]. The Opacity Project has not done all the relevant elements yet. Second, the calculations apply to systems observed at low resolution, whereas systems studied at high resolution show, not only multiple components with $b \leq 5$ km/s, but also widely varying ion ratios from component to component. The standard assumptions made concerning a uniform radiation field and uniform density over hundreds of kilometers per second cannot both be correct. Ultimately, the ionization calculations will need to be done on a component-by-component basis.

Once the atomic physics is done for sulphur, it should prove a useful ionization discriminant. All ions S I, S II, S III, S IV and S VI have unsaturated lines because of the low f-values and the lower cosmic abundance of S compared to C and Si.

Models are normally parameterized by the ratio $n_\gamma/n_H = U$ (called the ionization parameter), where n_γ is the density of H ionizing photons and n_H is the total hydrogen density. For a given abundance and shape of the ionizing spectrum the column density of each ion can be calculated in terms of $N(\text{H I})$.

Typical calculations to date assume an ionizing flux of 10^{-21} to 10^{-22} erg cm^{-2} s^{-1} sr^{-1} Hz^{-1} at the Lyman limit, with a power law spectrum ($\alpha \nu^{-1.5}$), and require an ionization parameter 10^{-2} to 10^{-3}. It is controversial whether a typical value for the ionizing flux of $3 \cdot 10^{-22}$ erg cm^{-2} s^{-1} sr^{-1} Hz^{-1} at the Lyman limit is too high, or too low by as much as a factor of three, on grounds of the uncertainty in the existence of UV-creating sources [34] and on grounds of the opacity of intervening absorbing (Lyman limit) systems that reduces the source flux reaching distant intergalactic clouds [35]. Models now exist for H II regions in absorption, showing the effect of local stellar ionization expected in star-forming regions of galaxies [36].

It is apparent that the ionization and density of the interstellar components of QSOALS vary on small and large velocity scales. There may be an extragalactic ionizing component of the radiation field that affects regions at the edges of intergalactic clouds, where neutral zones as thin as $N(\text{H I}) \sim 10^{17}\,\text{cm}^{-2}$ may exist, and for which penetration of radiation below 912 Å could explain the high ions seen in QSOALS. However, many systems have higher H I column densities that preclude penetration of the UV photons, and in such cases an interior, local source of photons is suggested. As demonstrated for system F in Q1331 + 170, the high ions still exist when the total H I column density gets very large, and in fact the strongest total system strengths of C IV and Si IV are encountered in just these cases. It follows that the internal sources of radiation may be of such a nature as to produce ionizing photons, at least over short length scales. As noted in part I, the conditions around many O and B stars in our own Galaxy do not permit strong lines of C IV and Si IV to exist. They do, however, appear in globally extended, low-density regions affected by O and B star radiation fields and dynamical processes, in the galactic halo. Considering systems F and G in Q1331 + 170, they must both be impinged upon by the same intergalactic field, and they both have $N(\text{H I}) \gg 10^{17}\,\text{cm}^{-2}$. The Si IV column densities might be expected then to be quite similar, unless the outer zones have quite different densities. On the other hand, there is no reason to suppose the internal sources of ionization are the same. The existence of internal sources thus may be a more viable explanation of the properties of these and similar systems than an intense extragalactic radiation field. Observations of more ions in the two systems could provide conclusive evidence, one way or the other.

Collisional ionization may be important, but, in at least some cases, the components in C IV are too narrow to arise in a gas hot enough to yield adequate collisional ionization (80 000 K) [27].

5. – Conclusions.

a) Observationally, most QSOALS contain Ly α lines with no observable lines of heavy elements.

b) The heavy-element systems contain lines of H I plus a variety of heavy-element lines. They may represent a continuous population, in terms of rising column density and rising abundance, with the Ly α forest.

c) The ionization source for the systems with heavy elements is ambiguous because of model dependences. Candidate sources include the extragalactic background radiation from QSO's or early galaxies, ionizing photons from stellar sources local to the absorbers and mechanical energy from stellar activity local to the absorbers.

d) Statistically there are three populations in dM/dz: The number of Ly α systems rises from $z = 0$ to $z = 5$, though there is an excess at $z < 0.3$ [37]; Mg II is flat in dM/dz from 0 to probably 2.5; C IV is the most common system at $z = 2$, but drops gradually in dM/dz to high z. From various arguments, the drop in number of detected C IV systems with redshift is probably an abundance effect: [C/H] decreases to high z.

e) While systems of mixed ionization are common, there are systems with little if any C IV or Si IV, and many systems with C IV and/or Si IV but no lower ions detected. The existence of higher ionization stages is uncertain, because O VI, N V and S VI are always in the Ly α forest. More careful study of these ions at high resolution is essential.

f) The overall value of dM/dz for normal intervening galaxies is 0.1/unit z, compared to values for QSOALS of 1 for Mg II, 1.5 for Si IV and 1.8 for C IV, near $z = 2$. Thus it is unlikely that normal galaxies can produce the absorbers, given the very small cross-sections of the types of regions in our own Galaxy that could actually produce systems similar to the QSOALS, as summarized at the end of part I.

REFERENCES

[1] W. L. W. SARGENT, C. C. STEIDEL and A. BOKSENBERG: *Astrophys. J. Suppl.*, **69**, 703 (1988).
[2] D. G. YORK, B. YANNY, A. CROTTS, C. CARILLI, E. GARRISON and L. MATHESON: *Mon. Not. R. Astron. Soc.*, **250**, 24 (1991).
[3] V. JUNKKARINEN, A. HEWITT and G. BURBIDGE: *Astrophys. J. Suppl.*, **77**, 203 (1991).
[4] T. J. BROADHURST, R. S. ELLIS, D. C. KOO and A. S. SZALAY: *Nature (London)*, **343**, 726 (1990).
[5] K. M. LANZETTA, A. M. WOLFE, D. A. TURNSHEK, L. LU, R. G. MCMAHON and C. HAZARD: *Astrophys. J. Suppl.*, **77**, 1 (1991).
[6] V. KULKARNI, K.-L. HUANG, R. F. GREEN, J. BECHTOLD, D. E. WELTY and D. G. YORK: submitted (1994).
[7] F. CHAFFEE, J. BLACK and C. FOLTZ: *Astrophys. J.*, **335**, 584 (1988).
[8] A. CROTTS: *Astrophys. J.*, **336**, 550 (1989).
[9] R. C. DUNCAN, J. P. OSTRIKER and S. BAJTLIK: *Astrophys. J.*, **345**, 39 (1989).
[10] M. RAUCH, R. F. CARSWELL, F. H. CHAFFEE, C. B. FOLTZ, J. K. WEBB, R. J. WEYMANN, J. BECHTOLD and R. F. GREEN: *Astrophys. J.*, **390**, 387 (1992).
[11] D. G. YORK, D. E. WELTY, R. GREEN, J. BECHTOLD, V. KULKARNI, P. A. KHARE and M. CARLSON: submitted (1994).
[12] D. TYTLER: *Astrophys. J.*, **321**, 49 (1987).
[13] D. MEYER and D. G. YORK: *Astrophys. J.*, **315**, L5 (1987).
[14] W. L. W. SARGENT, C. C. STEIDEL and A. BOKSENBERG: *Astrophys. J.*, **351**, 364 (1990).

[15] L. LU: *Astrophys. J.*, **379**, 99 (1991).
[16] R. W. HUNSTEAD, H. S. MURDOCH, M. PETTINI and J. C. BLADES: *Astrophys. J.*, **329**, 527 (1988).
[17] S. BAJTLIK, R. C. DUNCAN and J. P. OSTRIKER: *Astrophys. J.*, **327**, 570 (1988).
[18] A. S. KUTYREV and R. J. REYNOLDS: *Astrophys. J.*, **344**, L9 (1989).
[19] A. SONGAILA, W. BRYANT and L. L. COWIE: *Astrophys. J.*, **345**, L71 (1989).
[20] J. S. CHEN and D. C. MORTON: *Mon. Not. R. Astron. Soc.*, **208**, 167 (1984).
[21] D. M. MEYER, D. E. WELTY and D. G. YORK: *Astrophys. J.*, **343**, L37 (1989).
[22] J. N. BAHCALL, J. BERGERON, A. BOKSENBERG, G. F. HARTIG, B. T. JANNUZ, S. KIRHAKOS, W. L. W. SARGENT, B. D. SAVAGE, D. P. SCHNEIDER, D. P. TURNSHEK, R. J. WEYMANN and A. M. WOLFE: *Astrophys. J. Suppl.*, **87**, 1 (1993).
[23] C. C. STEIDEL: *Astrophys. J. Suppl.*, **72**, 1 (1990).
[24] W. L. W. SARGENT, C. C. STEIDEL and A. BOKSENBERG: *Astrophys. J. Suppl.*, **69**, 703 (1989).
[25] J. C. BLADES, R. W. HUNSTEAD, H. S. MURDOCH and M. PETTINI: *Astrophys. J.*, **288**, 580 (1985).
[26] M. PETTINI, R. W. HUNSTEAD, L. J. SMITH and D. P. MAR: *Mon. Not. R. Astron. Soc.*, **246**, 545 (1990).
[27] J. BECHTOLD, R. F. GREEN and D. G. YORK: *Astrophys. J.*, **312**, 50 (1987).
[28] J. E. WAMPLER, J. BERGERON and P. PETITJEAN: *Astron. Astrophys.*, **273**, 15 (1993).
[29] C. F. MCKEE, B. TARTER and J. WEISHEIT: *Astrophys. Lett.*, **13**, 13 (1972).
[30] W. L. W. SARGENT, P. J. YOUNG, A. BOKSENBERG, R. F. CARSWELL and J. A. J. WHELAN: *Astrophys. J.*, **230**, 49 (1979).
[31] C. C. STEIDEL: *Astrophys. J. Suppl.*, **74**, 37 (1990).
[32] M. DONAHUE and M. J. SHULL: *Astrophys. J.*, **383**, 511 (1991).
[33] M. J. SEATON: *J. Phys. B*, **20**, 6363 (1987).
[34] C. C. STEIDEL and W. L. W. SARGENT: *Astrophys. J.*, **343**, L33 (1989).
[35] P. MADAU: *Astrophys. J.*, **389**, L1 (1992).
[36] R. B. GRUENWALD and S. M. VIEGAS: *Astrophys. J. Suppl.*, **78**, 153 (1992).
[37] J. N. BAHCALL, B. T. JANNUZI, D. P. SCHNEIDER, G. F. HARTIG, R. BOHLIN and V. JUNKKARINEN: *Astrophys. J.*, **377**, L5 (1991).

PART III
Where Do the QSOALS Arise?

1. – Introduction.

We can now turn to matters of interpretation. First, the empirical facts about the QSOALS (part II), in comparison with our knowledge of the local interstellar medium (LISM) (part I), are used to derive the possible small-scale nature of the absorbing regions seen in background spectra of QSO's. The QSOALS most probably arise in conjunction with formation of massive stars. Next, observations are described that relate the absorbers to galaxies near them. Some of the absorbers may arise in H II regions that must be extremely numerous. Other origins are discussed. Finally, the consequences of the abundance trends in absorbers are discussed, relating these to the morphological scenarios and theories of the early Universe.

2. – Analogies between QSOALS and LISM.

The key facts about the QSOALS that we attempt to relate to properties of the LISM (here referring to the nearest 500 pc) are listed here.

a) All observable ionization stages below O VI (C IV, C III, C II, C I) are detected in the QSOALS and the LISM. While there are some QSOALS that are typical of interstellar sight lines near the Sun (system G in Q1331 + + 170, part II), showing little C I and no C IV, most absorbers show a dominance of C IV over C II: The latter situation is virtually unknown in the Galaxy or the LISM.

b) In QSOALS, system widths are (10–1000) km s^{-1}, each system being made of multiple narrow components[1,2]. Narrow components exist in the LISM as well, but the spread of components is small, < 80 km s^{-1}.

c) Strong absorption lines of H$_2$ are not observed in the QSOALS. There is little or no evidence for dust. Both are ubiquitous in the LISM.

d) Ratios of ions of a given element change from component to component. This is true on scales < 10 km s^{-1} in both the QSOALS and the LISM, and may be true on even smaller velocity scales, in both cases. Observations at a resolution of 5 km s^{-1} (say) are not available in either case yet.

e) There are few obvious trends in redshift for the QSOALS, using only spectroscopy as a criterion, with the exception of abundance trends noted below. The C IV, Si IV and Si II lines become less frequent at high z[3]; the lines that do appear are weaker[4], and there seem to be more systems with Si IV, but no C IV[5].

f) The data on QSOALS suggest a rate of occurrence of one or two intervening multiple-component systems per unit z per QSO[3,6]. The conditions giving rise to the absorbers must, therefore, have a total cross-section about 10 times greater than that of a typical galaxy (referred to as an L_* galaxy, from a particular characterization of galaxy statistics[7]).

Selection effects may exist that over-emphasize the trend of item *a)*. For example, lines O I, C II and N I in QSOALS at $z_{\text{abs}} < (0.9 \text{ to } 1.0) \times z_{\text{em}}$ may have been selectively misidentified as Lyman α forest lines, owing to overlap with and/or blending with lines in the forest. True forest lines may have been accidentally identified as C IV doublets (fig. 3, part II, system C). The contrast between ISM and QSOALS with regard to C IV may also be distorted because the latter appear on full sight lines through intervening objects, whereas ISM observations most often do not penetrate the outer halo, and always penetrate ≤ 0.5 of the halo of the Milky Way.

Similarly, item b) could be partly a selection effect. Those absorbers with the widest spread of velocity components produce the strongest absorption lines, the easiest to detect in surveys. More sensitive surveys have in fact increased the average number of absorbers per unit z, but the effect is not dramatic.

One possible selection effect that could affect item c) was mentioned in part II: Dust and H_2 should selectively be seen on reddened sight lines, just the sight lines where QSO's would be dimmer, and thus not included in magnitude-limited surveys.

Figure 1 summarizes the comparison of QSOALS and observations of the LISM. The first three frames show Si II, Si III and Si IV toward ζ Ori (see part I). The bottom two frames show Si II and Si III in system F at $z = 1.77$ (part II) in Q1331 + 170. The middle frames show Fe II (similar to Si II) and Si IV toward the QSO 3C273, as well as Si II and Si IV in the spectrum of the LMC SNe 1987A.

The obvious major difference between the ISM toward ζ Ori and system F toward Q1331 + 170 is that ζ Ori shows no Si IV at all, whereas Si IV dominates in system F. There are also similarities: Both spectra have a spread of (150–200) km s^{-1} in obvious components, and both have regions of high ions (Si III or Si IV) with little or no Si II in the same components, with the ratios of line strengths for the relevant lines changing over velocity scales ≤ 50 km s^{-1}. Note that the particular Si II line used for ζ Ori is intrinsically about three times stronger than that shown for the QSOALS. The latter might show traces of Si II at other velocities if data were available for the strong $\lambda 1260$ line of Si II (but it is obliterated by Ly α forest lines in the spectrum of this QSO).

The spectra of 3C273 (frames 4 and 5) do not show the wide velocity spread or the large, component-to-component ionization variations seen toward ζ Ori or Q1331 + 170. Note that the 3C273 spectra were recorded at a resolution of 30 km s^{-1}, while the resolution for ζ Ori is 15 km s^{-1}, and for Q1331 + 170 is 20 km s^{-1}. 3C273 shines through a galactic radio continuum loop (loop I) and the Si IV seen here may be a feature of the galactic fountain [8].

The remaining spectra are of SN 1987A in the H II region 30 Doradus in the Large Magellanic Cloud [9-11]. The component structure is obvious; the combination of galactic and LMC gas gives a spread of 350 km s^{-1} for the lines shown. However, there are few resolved components in Si IV. On the other hand, ab-

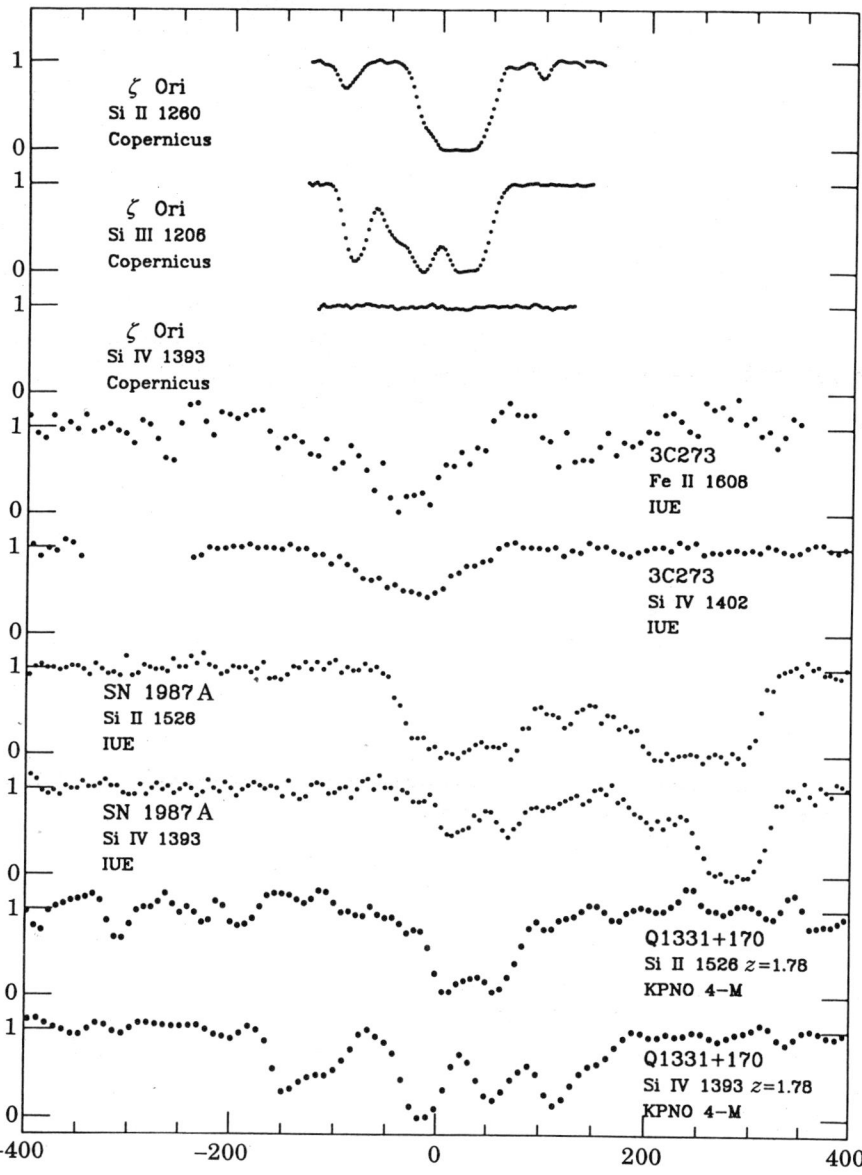

Fig. 1. – Interstellar absorption lines with similarities to QSOALS. References and discussion are in the text. Relative intensity is plotted against velocity. Note the absence of Si IV in the spectrum of ζ Ori, but its strong presence in the QSO 1331 + 170 (system F, bottom frame) and at the LMC velocity ($+230\,\mathrm{km\,s^{-1}}$) in the spectrum of SN1987A. Si IV is less conspicuously present, near $0\,\mathrm{km\,s^{-1}}$, from the Milky Way halo, in spectra of 3C273 and SN1987A. The top three frames are from [11]. The 3C273 data are from [8]. The spectra of SN1987A are from the archives of the *International Explorer Satellite*. Data for Q1331 + 170 (the two bottom panels) are from ref. [11] of the previous part II.

sorption lines of Si IV that have great strength occur in the LMC, compared to the absence of Si IV in ζ Ori.

What would it take to make gas like that toward ζ Ori, if located on a QSO sight line through an intervening galaxy, accountable for QSOALS? An increase in the SN rate (or, possibly, a decrease in ambient interstellar density) over that in Orion would increase the velocity spread. An increase in the volume of space with the Orion O-star formation rate and supernova rate could increase the frequency of occurrence of such lines in QSO's. More O-stars and supernovae imply more stellar remnants which can produce ionization of the gas to Si IV and C IV through the agency of stellar remnants, which may produce hard UV photons, through direct radiation from very hot remnant cores, by accretion of interstellar gas onto neutron stars, or through mechanical interaction with the interstellar medium[12].

However, aggregations of O-stars greater in space density than are seen locally in our Galaxy would produce average blue luminosities of the intervening galaxies that are higher than seen in typical galaxies today. There is no evidence that typical galaxies were more luminous in the past. (See ref.[13].)

Another approach is to imagine the postulated high-mass stars to be distributed in dwarf galaxies. This scenario is suggested by the fact that at least one local gas-rich small galaxy, the LMC, produces absorption line spectra with similarities to the QSOALS (see fig. 1). Many dwarf galaxies could pre-date a galaxy of typical mass, perhaps forming a cloud of gas-rich dwarfs that evolved by mergers to form a modern galaxy[14]. Because the high O-star luminosity is diffused over a large volume in this picture, the precursors of modern galaxies would be very extended (> 100 kpc), mottled, low-surface-brightness objects, easy to miss in magnitude-limited surveys for objects of (10–20) kpc extent.

These alternative pictures are arrived at by using analogies in the spectra of H II regions and absorbers, and asking what galaxies would have to look like to explain the observations. Another approach is to assume that modern galaxies have had their present morphologies for much of the age of the Universe (*i.e.* since before the look-back time corresponding to $z < 5$) and have been for most of that time surrounded by gaseous halos[15]. This picture requires an energy source that extends 3–4 times further from the disk of a spiral galaxy than we know of today, to stir the clouds, as well as a strong extragalactic radiation field to produce the observed ions. Both velocity flows and high ionization might be created by galactic fountains more active than our own[16]. To avoid producing too many galaxies with $L > L_*$, the star-forming region that drives the fountains would have to fill a larger volume than is taken up by modern disks, diffusing the total light and masking such objects from detection, at least from ground-based telescopes. Then, on average, early galaxies, to deep enough limiting magnitudes, should appear larger than the sizes predicted for modern galaxies placed at high z after applying the cosmological angle-length relationship.

S. M. FALL and D. LYNDEN-BELL (Cambridge University Press, Cambridge, 1981), p. 111.
[4] J. BARNES and L. HERNQUIST: *Astrophys. J.*, **370**, L65 (1991).
[5] M. DOPITA: *Astrophys. J.*, **295**, L5 (1985).
[6] J. M. SCALO and C. STRUCK-MARCELL: *Astrophys. J.*, **301**, 77 (1986).
[7] C. A. NORMAN and J. SILK: *Astrophys. J.*, **238**, 158 (1980).
[8] J. FRANCO and D. COX: *Astrophys. J.*, **273**, 243 (1983).
[9] J. SILK: *Austr. J. Phys.*, **45**, 437 (1992).
[10] B. WANG and J. SILK: *Astrophys. J.*, in press (1994).
[11] P. GOLDREICH and D. LYNDEN-BELL: *Mon. Not. R. Astron. Soc.*, **130**, 97 (1965).
[12] A. TOOMRE: *Astrophys. J.*, **139**, 1217 (1964).
[13] T. MOUSCHOVIAS and L. SPITZER: *Astrophys. J.*, **210**, 326 (1976).
[14] T. NAKANO: *Fund. Cosmic Phys.*, **9**, 139 (1984).
[15] C. MCKEE: *Astrophys. J.*, **345**, 782 (1989).
[16] R. PUDRITZ and J. SILK: *Astrophys. J.*, **312**, 213 (1989).
[17] B. G. ELMEGREEN: in *Proceedings of 3rd Canary Islands Winter School*, edited by G. TENORIO-TAGLE, M. PRIETO and F. SANCHEZ (Cambridge University Press, Cambridge, in press).
[18] R. F. G. WYSE: *Astrophys. J.*, **311**, L41 (1986).
[19] R. F. G. WYSE and J. SILK: *Astrophys. J.*, **339**, 700 (1989).
[20] W. W. ROBERTS and M. A. HAUSMAN: *Astrophys. J.*, **277**, 788 (1984).
[21] K. M. OLSON and J. KWAN: *Astrophys. J.*, **361**, 426 (1990).
[22] E. C. VAZQUEZ and J. M. SCALO: *Astrophys. J.*, **343**, 644 (1989).
[23] F. CASOLI and F. COMBES: *Astron. Astrophys.*, **110**, 287 (1982).
[24] R. KENNICUTT: *Astrophys. J.*, **344**, 685 (1989).
[25] A. WHITWORTH: *Mon. Not. R. Astron. Soc.*, **186**, 59 (1979).
[26] L. COWIE: *Astrophys. J.*, **245**, 66 (1981).
[27] E. LADA: *Astrophys. J.*, **393**, L25 (1992).
[28] G. BERTIN and A. G. ROMEO: *Astron. Astrophys.*, **195**, 105 (1988).
[29] D. CIOFFI, C. F. MCKEE and E. BERTSCHINGER: *Astrophys. J.*, **334**, 252 (1989).
[30] I. SHLOSMAN and M. NOGUCHI: *Astrophys. J.*, **414**, 474 (1993).
[31] M. JURA: *Astrophys. J.*, **301**, 624 (1986).
[32] M. G. WOLFIRE and J. P. CASSINELLI: *Astrophys. J.*, **319**, 850 (1987).
[33] C. MCKEE and J. P. OSTRIKER: *Astrophys. J.*, **218**, 168 (1977).
[34] P. M. SOLOMON, A. R. REVOLO, J. BARRET and A. YAHIL: *Astrophys. J.*, **319**, 730 (1987).
[35] W. KEGEL: *Astron. Astrophys.*, **225**, 517 (1990).
[36] M. PERAULT, E. FALGARONE and J.-L. PUGET: *Astron. Astrophys.*, **157**, 139 (1986).
[37] C. A. NORMAN and S. IKEUCHI: *Astrophys. J.*, **345**, 372 (1989).
[38] M.-M. MACLOW, R. MCCRAY and M. NORMAN: *Astrophys. J.*, **337**, 141 (1989).
[39] G. TENORIO-TAGLE, M. ROZYCZKA and P. BODENHEIMER: *Astron. Astrophys.*, **237**, 807 (1990).
[40] K. TOMISAKA: *Astrophys. J.*, **361**, L5 (1990).
[41] C. MCKEE: in *Evolution of the Interstellar Medium*, edited by L. BLITZ (A.S.P., San Francisco, Cal., 1991), p. 3.
[42] S. D. M. FALL and G. EFSTATHIOU: *Mon. Not. R. Astron. Soc.*, **193**, 189 (1980).
[43] K. C. FREEMAN: *Astrophys. J.*, **160**, 811 (1970).
[44] T. BOROSON: *Astrophys. J. Suppl.*, **46**, 177 (1981).

[45] J. S. YOUNG and P. M. KNEZEK: *Astrophys. J.*, **347**, L55 (1989).
[46] P. J. E. PEEBLES: *Astrophys. J.*, **365**, 27 (1990).
[47] G. BYRD and M. VALTONEN: *Astrophys. J.*, **350**, 89 (1990).
[48] A. DRESSLER and J. E. GUNN: in *Evolution of the Universe of Galaxies*, edited by R. G. KRON (A.S.P., San Francisco, Cal., 1990), p. 200.
[49] A. E. EVRARD: *Mon. Not. R. Astron. Soc.*, **248**, 8 (1991).
[50] M. A. STRAUSS, M. DAVIS, A. YAHIL and J. P. HUCHRA: *Astrophys. J.*, **385**, 421 (1992).
[51] M. D. GREGG: *Astrophys. J.*, **384**, 43 (1992).
[52] F. SCHWEIZER, P. SEITZER, S. FABER, D. BURSTEIN, C. M. DALLE ORE and J. J. GONZALES: *Astrophys. J.*, **364**, L33 (1991).
[53] J. SILK: *Astrophys. J.*, **345**, L1 (1989).
[54] G. EFSTATHIOU, N. KAISER, S. W. SAUNDERS, A. LAWRENCE, M. ROWAN-ROBINSON, R. S. ELLIS and C. S. FRENK: *Mon. Not. R. Astron. Soc.*, **247**, 10 (1990).
[55] R. GIOVANELLI, M. P. HAYNES and G. L. CHINCARINI: *Astrophys. J.*, **300**, 77 (1986).
[56] A. DRESSLER: *Astrophys. J.*, **236**, 351 (1980).
[57] M. POSTMAN and M. GELLER: *Astrophys. J.*, **281**, 95 (1984).
[58] S. J. MADDOX, G. EFSTATHIOU, W. J. SUTHERLAND and J. LOVEDAY: *Mon. Not. R. Astron. Soc.*, **242**, 43 (1990).
[59] A. J. S. HAMILTON: *Astrophys. J.*, **331**, L59 (1988).
[60] J. BERGERON and P. BOISSÉ: *Astron. Astrophys.*, **243**, 344 (1991).
[61] C. C. STEIDEL: in *Evolution of Galaxies and their Environment*, edited by J. M. SHULL and H. THRONSON (in press).
[62] T. J. BROADHURST, R. S. ELLIS and T. SHANKS: *Mon. Not. R. Astron. Soc.*, **235**, 827 (1988).
[63] M. COLLESS, R. S. ELLIS, K. TAYLOR and R. N. HOOK: *Mon. Not. R. Astron. Soc.*, **244**, 408 (1990).
[64] W. SAUNDERS, M. ROWAN-ROBINSON, A. LAWRENCE, G. EFSTATHIOU, N. KAISER, R. S. ELLIS and C. S. FRENK: *Mon. Not. R. Astron. Soc.*, **242**, 318 (1990).
[65] A. MEIKSIN, I. SZAPUDI and A. S. SZALAY: *Astrophys. J.*, **394**, 87 (1992).
[66] T. BROADHURST, R. ELLIS, D. KOO and A. SZALAY: *Nature (London)*, **343**, 726 (1990).
[67] C. D. IMPEY, G. D. BOTHUM and D. F. MALIN: *Astrophys. J.*, **330**, 634 (1988).
[68] G. D. BOTHUN, C. D. IMPEY and D. F. MALIN: *Astrophys. J.*, **376**, 404 (1991).
[69] C. J. PRITCHET and L. INFANTE: *Astrophys. J.*, **39**, L35 (1992).
[70] G. R. BLUMENTHAL, S. M. FABER, J. R. PRIMACK and M. J. REES: *Nature (London)*, **311**, 517 (1984).
[71] S. D. M. WHITE, M. DAVIS, G. EFSTATHIOU and C. S. FRENK: *Nature (London)*, **330**, 451 (1987).
[72] J. M. BARDEEN, J. R. BOND, N. KAISER and A. S. SZALAY: *Astrophys. J.*, **304**, 15 (1986).
[73] W. H. ZUREK, P. J. QUINN and J. K. SALMON: *Astrophys. J.*, **330**, 519 (1988).
[74] A. DEKEL and J. SILK: *Astrophys. J.*, **303**, 39 (1986).
[75] J. BREGMAN: *Astrophys. J.*, **224**, 768 (1978).

3. – Directing searches for high-z galaxies using known absorption line.

For the heavy-element line systems, the presence of C, N, O and Fe signals that nucleosynthesis has occurred. Atoms cannot move more than 1 Mpc in a Hubble time, so the nucleosynthesis site must still be within a few arcminutes (≤ 4) of the absorbers.

Unless the absorber has burned up all its gas, leaving only a gas remnant, it should be observable as a continuum or emission line object. Even if it has burned out, sight lines to other QSO's with $z > z_{abs}$ and ≤ 4 arcminutes from the absorber sight line should show absorption at the same redshift, and active emission patches should be seen at least on a statistical basis.

Numerous searches for objects at the redshift of absorbers are under way. One technique is to use narrow-band filters to look for [O II] at $z \sim 0.3$ to 1.0[17], or for Ly α at $z = 2.0$–5.0[18]. Another approach is to identify all galaxies in an annulus around the QSO in question, then obtain redshifts from spectra of these[19,20]. Numerous absorbers have yielded galaxies at the same redshifts within 250 kpc (in redshift and in projection on the sky). Sometimes the objects are nearby galaxies, near L_* luminosity; sometimes there are only objects below L_* that are many arcseconds away from the background QSO. In these latter cases, there may be an H II region directly on top of the QSO[21]. There may be some evidence for a trend that low-z searches tend to yield L_* galaxies near the absorbers ($z < 0.5$), while higher-z absorbers tend to yield multiple small galaxies far from the background QSO, in projection[19,22].

4. – QSOALS and the early Universe.

How can we distinguish these three pictures:

a) Each galaxy once had an extended active star-forming volume, 30 times larger than the present volumes of galaxies.

b) Each galaxy we see today had, at an earlier time, a halo 30 times larger in volume than the present volume, ionized by extragalactic UV radiation.

c) Each galaxy we see today is a merger, the consequence of a cloud of gas-rich dwarfs galaxies, one of which appears to a distant observer as a QSOALS when a QSO is lined up with it.

Obviously, continuing studies of the morphology of high-z galaxies, particularly galaxy profiles and luminosities, are important. Another tool at our disposal is abundance studies.

Elements that are products of both type-I and type-II SNe are detectable in QSOALS. Providing that the ratio $N(\text{Al III})/N(\text{Al II})$ is monitored and demonstrated to be < 1, and that a damped Ly α line is available to obtain $N(\text{H I})$, abundances free of ionization corrections can be obtained for, among other species, S, Zn; Si, Cr; Al, Mn. These pairs list the species that come from type-II SNe first and type-I SNe second[23]. The first pair consists of elements of even nuclear charge that do not show depletion into interstellar dust in our Galaxy; the second consists of elements with even-numbered nuclear charge, but which are strongly affected by dust depletion, in so far as apparent gas phase abundance goes; the third consists of elements affected by dust depletion but of odd-numbered nuclear charge. In halo stars for our own Galaxy, the generally low total abundances are mirrored in the excess of [Si/Fe] and [Zn/Mn]. In QSOALS, the integrated ratio of numbers of type-II SNe to numbers of type-I SNe, prior to the epoch of observation, should be reflected in [S/Zn]; the age of either element should be traceable by the ratios [Al/S] and [Mn/Zn] (odd elements should be lower in abundance when the material from either type SNe is fresh); and the effect of dust presence on Al and Mn should be derivable from the ratios [Si/S] and [Cr/Zn]. [Cr/Zn] is below solar in the absorbers, and [Zn/H] is below solar by a factor of 10–20, in the 10–15 absorbers studied in these species to date [24,25]. From $z = 0.6$ to $z = 3.0$, no trend is detected with redshift [25]. Evidently, the total element production in type-I SN, to the epoch observed in a given case, is well below the total production in gas near the Sun today, based on the net observed production of Zn, which is not depleted on dust grains. The low value of [Cr/Zn] means that Cr is depleted on grains, since known production mechanisms produce solar [Cr/Zn].

These remarks should be put in the light of comments in part II implying that the abundances of Si and C drop as z increases above $z = 2$, reflected in the decreasing frequency with which these species are detected.

Whatever the morphology of early sites of nucleosynthesis, it would appear from the absorbers that the appropriate morphology must satisfy several conditions.

a) Averaged over random sight lines, abundances of C, Si and Fe increase from $\leq 10^{-3}$ at $z = 3.5$ to $\sim 10^{-1}$ at $z = 2.0$.

b) Solar-abundance, disklike gas cannot be more extensive in volume-filling factor than we see today: No solar-abundance QSOALS have been found among the intervening QSOALS.

c) Much of the star formation at $z > 1$ must not lead to abundances much above 10^{-1}. No strong trend in Zn/H has been detected.

d) Independent of redshift, large gas motions are commonly found, but narrow systems are also sometimes seen.

We may postulate from these remarks that most element formation has occurred at $z < 2$. The absence of a trend with redshift in [Zn/H] or [Cr/H], if confirmed in the future, would imply that low-metallicity regions fill most of the space characterized by nonprimordial abundances from the highest observable redshifts down to possibly as low as 0.5. The QSOALS naturally select the populations of objects with the largest cross-section. This implication argues in favor of the dwarf-galaxy scenario, because the alternate explanations of halos or extended star-forming disks would provide enriched gas in most sight lines, the abundances building up to solar values gradually at lower redshift or reaching solar levels at high z. In the dwarf-galaxy picture, the mass of each dwarf allows a natural mechanism that controls star formation and element formation (disruption by supernovae and stellar winds). Mergers then might lead to aggregation of stars from the dwarf galaxies into the modern halos of stars with low metallicity. The associated low-metallicity gas would end up as a gaseous disk from which the last 90% of the stars form, in small enough pockets to escape frequent sampling by background QSO's.

The possibility of detritus from the dwarf galaxies, in that scenario, implies that there could be high-velocity H I gas (unbound) contaminated with heavy elements from the first supernovae and loosely clustered with galaxies. Thus any attempt to find an intergalactic population of primordial Ly α clouds in the Ly α forest must contend with such contaminated systems.

5. – The future.

The QSOALS provide a rich field of research for their own sake. At the same time, they promise to provide new and independent information on galaxy evolution.

In terms of the absorbers themselves, it is important to press the distinction between the three scenarios cited: halos; extended, contiguous star formation; and clouds of gas-rich dwarfs. Presumably there are examples of all three lurking in the QSOALS sample. On the spectroscopic front, a large enough sample is needed to begin to classify absorbers, independent of any preconceptions. Hopefully, natural groups will emerge that will show distinctive properties which can ultimately be related to physical parameters. It may be expected that some parameters of such a classification will be the velocity spread of component systems, the degree of ionization and the degree of element depletion, but other characteristics may emerge as high-resolution data become available.

Imaging of emission and continuum objects near QSO's, at the same redshifts as any QSOALS observed, is the area aside from spectroscopy that holds out promise for revealing the relation of the absorbers to galaxies. It will be particularly important to obtain high-angular-resolution images of QSO's with absorbers to see if lower-z objects lie on top of the QSO images. In this regard, I have not discussed clustering. Obviously, statistical relationships among the absorbers themselves, QSO's and galaxies will shed light on the relationship between absorbers and galaxy formation. However, much larger samples will be needed than are now available.

* * *

D. WELTY produced the figures for this lecture. Discussions with G. BURKS, P. FRISCH, L. KAO, J. LAUROESCH, D. MEYER, J. TRURAN, D. VANDEN BERK, D. WELTY and B. YANNY have been useful in preparing these notes.

REFERENCES

[1] J. BECHTOLD, R. F. GREEN and D. G. YORK: *Astrophys. J.*, **312**, 50 (1987).
[2] D. M. MEYER and D. G. YORK: *Astrophys. J.*, **399**, L121 (1992).
[3] D. G. YORK, B. YANNY, A. CROTTS, C. CARILLI, E. GARRISON and L. MATHESON: *Mon. Not. R. Astron. Soc.*, **250**, 24 (1991).
[4] C. C. STEIDEL: *Astrophys. J.*, **72**, 1 (1990).
[5] P. KHARE, D. G. YORK and R. GREEN: *Astrophys. J.*, **347**, 627 (1989).
[6] G. R. BURBIDGE, S. L. O'DELL, D. H. ROBERTS and H. E. SMITH: *Astrophys. J.*, **218**, 33 (1977).
[7] P. SCHECTER: *Astrophys. J.*, **203**, 297 (1976).
[8] G. BURKS, D. G. YORK, J. C. BLADES, R. C. BOHLIN and W. WAMSTEKER: *Astrophys. J.*, **381**, 55 (1991).
[9] R. P. KIRSCHNER, G. SONNEBORN, M. CRENSHAW and G. E. NASSIOPOULOS: *Astrophys. J.*, **320**, 602 (1987).
[10] J. C. BLADES, J. M. WHEATLEY, N. PANAGIA, M. GREWING, M. PETTINI and W. WAMSTEKER: *Astrophys. J.*, **334**, 308 (1988).
[11] B. D. SAVAGE, E. B. JENKINS, C. L. JOSEPH and K. S. DEBOER: *Astrophys. J.*, **345**, 393 (1989).
[12] Q. D. WANG, Z. Y. LI and M. C. BEGELMAN: *Nature (London)*, **364**, 127 (1993).
[13] D. KOO and R. KRON: *Annu. Rev. Astron. Astrophys.*, **30**, 613 (1992).
[14] D. G. YORK, M. DOPITA, R. GREEN and J. BECHTOLD: *Astrophys. J.*, **311**, 610 (1986).
[15] J. N. BAHCALL and L. SPITZER: *Astrophys. J.*, **156**, L63 (1969).
[16] J. N. BREGMAN: *Astrophys. J.*, **236**, 577 (1980).
[17] B. YANNY, D. G. YORK and T. B. WILLIAMS: *Astrophys. J.*, **351**, 377 (1989).
[18] J. D. LOWENTHAL, C. J. HOGAN, R. F. GREEN, A. CAULET, B. E. WOODGATE, L. BROWN and C. B. FOLTZ: *Astrophys. J.*, **377**, L73 (1991).
[19] J. BERGERON and P. BOISSE: *Astron. Astrophys.*, **243**, 344 (1991).
[20] B. O. NELSON and M. A. MALKAN: *Astrophys. J. Suppl.*, **82**, 447 (1992).
[21] B. YANNY and D. G. YORK: *Astrophys. J.*, **391**, 569 (1992).
[22] B. YANNY: *Astrophys. J.*, **351**, 396 (1990).
[23] J. C. WHEELER, C. SNEDEN and J. W. TRURAN: *Annu. Rev. Astron. Astrophys.*, **27**, 279 (1989).
[24] D. M. MEYER, D. WELTY and D. G. YORK: *Astrophys. J.*, **343**, L37 (1989).
[25] M. PETTINI and R. W. HUNSTEAD: *Aust. J. Phys.*, **43**, 227 (1989).

From Starburst Galaxies to Protogalaxies.

C. A. NORMAN

Department of Physics and Astronomy, Johns Hopkins University - Baltimore, MD
Space Telescope Science Institute

1. – Introduction.

Starburst galaxies are remarkable objects. The most powerful ultraluminous infrared starburst galaxies have luminosities of $10^{12} L_\odot$. They are all merging and interacting systems. The inferred star formation rates are 100 $M_\odot \text{y}^{-1}$. In the H_2 line at 2.2 μm alone they can emit $10^9 L_\odot$. They are frequently associated with megamasers. They have huge outflows with a bipolar morphology emanating from the central regions of the starburst with outflow rates of up to (10–100) $M_\odot \text{y}^{-1}$ and with outflow velocities of 1000 kms^{-1}. It has often been argued that, as we shall discuss later, there is strong circumstantial evidence to associate these starbursts with active galactic nuclei and the fuelling of massive central black-holes. The gas masses associated with the central regions are of order $(10^9-10^{10}) M_\odot$ within a radius of approximately 100 pc.

In this lecture it is not possible to cover more than a small part of this vast and rapidly growing field. I have, therefore, selected topics associated with the starburst phenomenon that have particularly interested me in the last few years. The text of this lecture leans heavily on the papers I have written recently with my collaborators in these areas. The subsequent sections will cover the dynamics of starburst galaxies, the interstellar medium of starburst galaxies and disk galaxies in general including superwinds and outflows. Then we move on to the relation of starburst galaxies to active galaxies.

We examine the high-redshift counterparts to rapidly star-forming galaxies and the relation to protogalaxies. In particular we discuss feedback effects that occur during the galaxy building process which results in a massive injection of energy from supernovae and from any central active nucleus that may be present.

From the beginning, it is important to note that a more detailed knowledge of star formation than we have at present is probably necessary to really understand starburst galaxies. The current theories divide into models of low-mass star formation and high-mass star formation. The low-mass star formation mode is thought to be due to a slow accretion onto a central core[1] where the magnetic field is removed from the gas by ambipolar diffusion. A rough estimate for the time scale for ambipolar diffusion is $\tau_d \sim 10^{14} x$ years, where x is the ionization in the cloud. The star formation process for these low-mass stars can be self-regulated by processes caused by other low-mass stars such as winds[2] or the ionization due to the X-ray emission from the low-mass T-Tauri component[3,4]. The high-mass stars through their ionization effect could self-regulate the ambipolar-diffusion process in the clouds including the formation of the low-mass stars. This latter point may be important for starburst galaxies in that a burst of high-mass stars may be able to completely inhibit the slow ambipolar-diffusion-driven process of low-mass star formation and, therefore, give a plausible basis for a mass function that is truncated at the low-mass end. Both WOLFIRE *et al.* [5] and STACEY *et al.* [6] have shown that the central region of M 82 is very similar to the Orion massive-star-forming region in our own Galaxy.

The high-mass star formation mode seems to be trigerred in associations correlated with spiral arms in disk galaxies and is generally correlated with the giant-molecular-cloud complexes. The high-mass star formation mode is thought to be associated with the magnetic Jeans mass. High masses can occur if the temperature is large and can occur in a self-consistent way in regions of high stellar luminosity due to already existing massive stars. If the central turbulent-velocity dispersion is large as is the case in a medium stirred up by the energy and momentum injection from massive stars, the magnetized Jeans mass can also be very large. Such large turbulent velocities are observed in starburst regions and also in galactic regions of massive-star formation. Therefore, it seems plausible that starbursts might operate in a way that suppresses the low-mass star formation mode. In addition, the presence of existing massive stars and their associated winds and supernovae can act to keep the star formation at the high-mass end.

2. – Dynamics of starbursts.

The general dynamical ingredients to set up the physics of initiating and fuelling a starburst are a merger or interaction event or a strong dynamical asymmetry such as a central and possibly transient bar (see, however, [7]). The ISM in starbursts is obviously very gas rich and, therefore, realistic models of the dynamics of starbursts must include the gas component and in particular the molecular phase. This has been achieved by HERNQUIST and BARNES in their excellent simulations of merging gas-rich disks to form a residual elliptical system. HERNQUIST[8] showed how a companion falling into a cold gas-rich disk galaxy can trigger global dynamical instabilities in the gas that can rapidly drive a large fraction of the gas into the central regions of the disk galaxy. The angular-momentum loss from the gas to the background stars is strong, continuous and monotonic. The tidal torques formed by the asymmetric infall of the companion are initially the most important component of the angular-momentum loss. After the gas has become sufficiently concentrated to be self-gravitating in massive blobs, the angular-momentum loss becomes driven by the dynamical friction against the background halo of stars.

Following the pioneering work of Negroponte and White[9], HERNQUIST and BARNES[10] have used an SPH code to simulate major mergers between equal-mass gas-rich galaxies. Of great interest is the effect in the central regions of the merging galaxies that exhibit strong shocks and strong bars, and which initially fuel gas into the centers of both the galaxies. The final nucleus is formed by an eventual merger of the cores. Very significant fractions of the entire gas mass end up in the central region. This result is extremely interesting in the context of the starburst problem since the models seem consistent with both the large observed central gas masses and the rapid process of gas accumulation necessary to account for the observed frequency of starbursts.

A qualitative way to understand the likely dynamical evolution of starbursts has been indicated earlier[11-13]. There are essentially three phases:

I) *Normal starburst phase.* Interaction and merging generate bars and waves that act to give angular-momentum transfer outward and consequently mass transfer to the center. The central surface density rises and the star formation rate increases. The time scale is $(10^8 - 10^9)$ y depending on the strength of the nonaxisymmetric perturbation.

II) *Very luminous starburst phase.* The gas mass accumulates sufficiently in the center so that the gas itself develops appreciable self-gravity and then subsequently becomes bar unstable. The angular-momentum loss is consequently very rapid.

III) *Ultra-luminous starburst phase.* The gas is now so centrally concentrated that it becomes completely self-gravitating. A massive central star cluster and central object such as a black-hole may result.

Depending on the strength of the interaction and the gas masses involved the starburst systems may find themselves only able to achieve stage I in the case of weak bursts with low gas masses or to complete phase III in the case of strong bursts with high gas masses.

When phase II is reached, we are at the stage where the gas mass is a significant fraction of the total interior mass. This leads us to the almost inevitable conclusion that a bar instability will form in the gas. A robust criterion to assess this effect is the Ostriker-Peebles condition $T/W \sim 0.14$ [14], where T and W are the rotational and potential energy, respectively. It follows that the observed central gas masses are inferred to be bar unstable if $M_{gas}/M_{dynamical} \geq$ $\geq (10\text{--}30)\%$.

After a time delay of order $\sim 10^7$ y the gas will bave become concentrated by a factor of 3–10. Then a significant fraction of the central gas mass will become totally self-gravitating. This can lead to the formation of a central massive black-hole and starburst and should be associated with the phase III, the ultraluminous phase.

PFENNIGER and NORMAN[15] incorporated the essential ingredients of a massive central starburst, namely a barred system, and the gas modelled by dissipative molecular clouds orbiting in the potential of the bar and central mass concentration. The orbits of the clouds were modelled as test particles with a drag component that effectively incorporates the dissipative effects of shocks on gas streamlines or of dissipative collisions. Some very interesting effects were found. In barred systems with sufficiently concentrated central masses strong radial and vertical resonances develop. Any dissipative particle close to a radial resonance will rapidly lose its radial action, and consequently its angular momentum. It is clear from the results of this work that rapid angular-momentum loss occurs in jumps as the dissipative clouds move through the resonance regions. The net effect of this is to very significantly reduce the time scale on which gas clouds can flow into the center.

An additional effect that is clear in the results of many test particle runs is that the vertical-resonance structure allows significant vertical heating from the disk to the bulge or halo. Stars inside the vertical-resonance region enter the chaotic regime where all the integrals of motion, except the Jacobi integral, vanish. The shape of the Jacobi integral inside a rotating bar bounds the motion of stars within concentric, nearly spheroidal shells. Bulge formation and disk thickening may be significantly affected by this process. In fact barred Sc galaxies with central mass concentrations should be forming bulges at present.

This process of inflow is clearly self-enhancing. The inflowing gas increases the central mass concentration which in turn increases the resonances which has the effect of enhancing the rate of angular-momentum loss, thus feeding gas into the nucleus at an increasing rate. The limit of this process is discussed by HASAN and NORMAN[16], who showed that the fundamental family of orbits that support the bar will be dissolved if the central mass concentration grows to between 10% and 20% of the total dynamical mass.

A crude but self-consistent approach exhibiting this limiting process has been calculated by PFENNIGER and FRIEDLI[17], who ran an N-body calculation where each of the test particles had some dissipative drag, so that an initially barred system slowly grew a central mass concentration which then dissolved the bar and turned the system into a hot triaxial bulge or lens.

Further progress in modelling these important secular processes requires more detailed self-consistent N-body simulations and we discuss such simulations in sect. 4.

3. – Interstellar medium.

Active star formation processes are now reported in many galaxies such as M 51, M 82, M 83 and NGC 253, where the prodigious energy input by newly born stars and many supernovae are detected. Ultraluminous IRAS galaxies like Arp 220 show such especially violent star formation activity that the energy output from the galaxy in the infrared band amounts to $\sim 10^{45}$ erg s^{-1}. The star formation activity is generally seen in the central regions of the galaxy, and the huge energy output can produce a bipolar flow from the nucleus[18,19]. These are extreme examples of gigantic chimneys that emit hot gas from both sides of the disk. Such megachimneys may be the cause of formation of enormous halos in distant galaxies at earlier epochs during their formation phase or as a consequence of galaxy interactions, starbursts and the like[11-13].

The interstellar medium in these galaxies has an energy balance dominated by the energy input from many spatially and temporally contiguous supernovae. Even in normal disk galaxies the manifestation of the collective energy input from type-II supernovae is to create supperbubbles, chimneys and winds (cf.[20]). These extreme starbursting environments, vast chimneys and superwinds are formed in the disk and emanate from the central regions of the disk along the polar axis. The energy and momentum input into the circumgalactic medium in these winds is significant[21]. The photoionization balance in these central regions with the very high internal UV flux from the massive-star population can result in a fraction, of order unity, of the gas mass being in the form of C II. The C II is produced in photodissociation regions at the edges of molecular clouds that are subject to an intense UV radiation field. FIR observations with the Fabry-Perot spectrometer on board the KAO indicate that this is indeed the case[6].

The starburst galaxies have a most interesting interstellar medium. For the powerful systems the medium most resembles that of the Orion star-forming region[6] where much of the gas is photodissociated molecular gas in the strong radiation field from the massive stars making up the starburst. These systems are at the extreme end of the chimney model of the interstellar medium as the massive central starburst seems to act coherently to blow a huge chimney or superwind out of the central regions[19,21-23]. The inferred mass outflow rates are $(1-100) M_\odot$ y^{-1} at speeds of 100 kms^{-1}. In milder, but closer, versions of this process chimneylike structures emanating from systems such as NGC 3079 are clearly visible in narrow-band filter images originating from both the central regions and the disk. These structures can be modelled in the context of the chimney models where a relatively normal disk galaxy is put into the starburst mode by increasing the star formation rate in the disk[20]. With sufficient energy release the entire ISM could be driven out in a galactic wind, but this is not observed in general and the energy release is more via localized chimneys. An examination of fig. 3 and 4 in [20] can show the likely states of the halo of a starburst (confined halo, chimney or wind) as a function of two global parameters, the energy input and the associated mass input. A major motivation for developing the chimney model for disk galaxies in general came from the work of Brinks and Bajaja[24], who undertook a detailed H I survey of M 31. They found many H I holes with size ~ 100 pc -1 kpc, projected onto the plane of the disk. If we were to observe the superbubbles from outside the Galaxy, they would be seen as H I holes. Of course, the inclination of the plane of the disk to the line of sight and the inclination of the superbubbles to the disk should lead to various elliptical figures of H I holes. It will be interesting to compare the distributions of H II regions, radio continuum contours and X-ray sources in the same galaxy and between galaxies.

The above picture is described by using the following metaphor. The correlated supernovae with chimneys are lined up along the spiral arms, and the smoke of hot gas is puffed up from chimneys. In this gas, the metals would be abundant because many supernovae contribute. Such a polluted gas is injected into the halo, and it infalls into the disk. Along the Hubble sequence of disk galaxies, the energy and mass input can vary, and the disk-halo connection can change from a wind, to chimneys to two-phase cooling halos. These sequences of halo structures can also be considered as evolutionary sequences. In the early explosive era, a huge extended halo can form and the halo gas can escape as a wind. The halos in this burst phase may be detected as absorption systems with highly ionized gas. The present structure in our Galaxy and Sb type galaxies appears to be in the chimney mode when the halo has shrunk to several kiloparsec. The circulation of gas between the disk and halo occurs. With decreasing star formation activity the blow-out condition for superbubbles cannot be satisfied and the gaseous halo disappears.

During the galaxy formation epoch extended wind-driven halos can form. The subsequent evolution as galaxies settle down and evolve towards their current state may pass through the stages of wind-driven halo, to chimney-driven halo, to cooling halo. Quasar absorption lines of highly ionized species like Fe XV will be seen at high redshift. In the chimney phase a mixed ionization state is expected. At later evolutionary stages typical low-ionization species such as Mg II should be associated with the cooler halos. As the evolution proceeds, the halo size will decrease.

The absorption line systems in quasars probably originate in the intervening material. The narrow metallic-line systems with redshifts smaller than the emission redshifts of quasars are thought to be due to the gas in the halo around distant galaxies. This hypothesis has been strengthened by Bergeron's [25] discovery of a galaxy with the emission lines at the same redshift as the Mg II absorption lines in quasars. In this case, the average extension of the halo gas containing Mg II ions is about 60 kpc because the line of sight to the quasar runs at this distance from the disk. General statistics of observed frequencies of Mg II systems indicates that the average radius of intervening galaxy halos at redshifts $z \sim 0.5$ should be ~ 50 kpc. The high-ionization C IV systems have not been directly confirmed as occurring in distant galactic halos, in contrast to the lower-ionization Mg II systems, but the two-point correlation of C IV systems is at least consistent with that of galaxies [26]. We assume here the C IV systems are also associated with galactic halos. The observed frequency of absorbers indicates that the average radius of distant galaxies at $z \sim 2.0$ is ~ 90 kpc. We expect that the size of big H I disks have shrunk considerably by the present epoch (cf. [27]). It is interesting that the results of Steidel and Sargent [28] indicate that the sizes of the metal absorbers associated with isolated galaxy halos do not evolve significantly.

For normal disk galaxies, a large-scale mass circulation between disk and halo is set up with a commensurate energy and momentum input into the halo from superbubbles. As the gas rains back down onto the disk, there will be a further exchange of mass, energy and momentum—this time from the halo to the disk. Chimneys should be observed as independent entities from hard-X-ray measurements in our own Galaxy and from absorption line studies, H I observations, etc. of external galaxies. The state of the interstellar media in other disk galaxies may be qualitatively described as follows. We expect a trend from three-phase homogeneous to chimney to two phase as we move from early to late type. There are two important points to note. The type-II supernovae rate and the mean ambient density can vary as a function of galactocentric distance, so that even in the same galaxy different phases may occur at different radii. Secondly, these parameters, and particularly the type-II supernova rate, can vary as a function of time particularly for galaxies that undergo bursts of star formation. Thus the state of a galaxy interstellar medium may vary between the phases as a function of both time and galactocentric distance.

4. – Active galaxies and starbursts.

We have analyzed the fate of a massive central star cluster of $\sim (10^8 - 10^9) \, M_\odot$ within a galactocentric radius of 10 pc [29, 30]. We have assumed efficient star formation and analyzed the evolution of the star cluster for a range of initial mass functions and upper and lower mass cut-offs. Specific calculations involve the production of red giants, the mass loss from the stars, the fate of the lost mass most of which goes to feed a central black-hole and the role of stellar collisions. Interesting effects result from the influence of the strong central continuum radiation source on the envelopes of the red giants giving a strong externally induced mass loss. The ionization structure, density and size for these stellar envelopes is consistent with those for the broad-line clouds. At later times the covering factor of the red giants is of order 10%. A black-hole mass of $(10^8-10^9)\,M_\odot$ has now grown and the orbital velocity of the clouds is ~ 5000 kms^{-1}. Much early work is related to this[31-36], but here at last we have good initial conditions for the active-galactic-nuclei and black-hole scenarios. In this version dense central star clusters should be associated with AGNs and in the immediately post-active phase dense stellar cores with A-type spectra should be observable. The existence of these massive central components leads us to some interesting simulations as initially discussed in sect. 2.

We study the evolution of a rotating barred galaxy in the presence of a central mass concentration using high-dynamic-range N-body simulations. The two-dimensional N-body code of Sellwood was used with 50 000 particles. The potential consisted of 75% of the mass in a Kuzmin-Toomre disk accounting for the disk and bar component and 25% of the mass in two unresponsive concentric Plummer spheres accounting for the bulge and core components.

The system was evolved self-consistently until a steadily rotating bar formed. The central mass was formed in a plausible manner by contracting the radius of the Plummer sphere containing a set fraction of the mass over about 4 bar tumbling periods. The system was then followed for a further period of time. The simulations were repeated for different mass concentrations.

We find that a central mass $\geq 5\%$ of the galactic mass will dissolve the bar. This is in agreement with the earlier single-particle orbit studies of Hasan and Norman[16] and Hasan, Pfenniger and Norman[37]. The physical reason is that the inner Lindblad resonance moves to the edge of the minor axis of the bar. Orbits become stochastic as they switch from orbits predominantly along the bar (called x_1 or A-type) to predominantly perpendicular to it (called x_2 or B-type). Similar results have been obtained recently by NISHIDA and WAKAMATSU[38]. Bar dissolution by build-up of central mass has also been demonstrated in N-body simulations by FRIEDLI and PFENNIGER[17] and FRIEDLI and BENZ[39].

In the single-orbit analysis of Pfenniger and Norman[15] and Hasan, Pfenniger and Norman[37] instabilities were found that occurred during the bar dissolution process owing to the action of a central mass that resonantly scattered orbits above the plane. These unstable orbits suffer a complex instability caused by interaction with resonances[15]. The phase space that they fill is bulgelike with a reasonable size and scale length. The zero-velocity surface can trace out not only a bulge but even a thick disk component.

A powerful and nonresonant instability of the firehose type has been shown to exist in barred galaxies by RAHA et al.[40]. This instability and others of a similar ilk discussed by SELLWOOD and MERRITT[41] give substantial heating above the plane and produce peanut-shaped bulges[42-44].

Both these points are fairly well established. Full confirmation would lie with very-high-resolution 3-D simulations of the central-mass growth, the bar dissolution and the bulge-building process. We hope to undertake this challenging task in the near future.

Very complete reviews of the subject of secular evolution, bulge building and bar dissolution have been given by KORMENDY[45, 46]. Comprehensive references to the current state of our knowledge can be found in the proceedings of the IAU Symposium No. 153[47] on galactic bulges.

However, on the basis of what we already know, we are led to speculate that the bar dissolution process discovered here leads to bulge building. The scenario is as follows. In a disk galaxy the ubiquitous bar instability occurs and a relatively stable bar forms in the central regions out to the corotation point. Gas is swept rapidly into the central regions and the central mass builds up (cf.[15] for some interesting slightly dissipative dynamical effects associated with this process). This phase can be most dramatic in starburst on AGN-type galaxies, but will still occur slowly and inextricably in normal barred spirals. The central mass concentration builds up and, when it reaches approximately 5% of the mass interior to the bar, the bar will dissolve and bulge building will occur. Since the bar-driven mass inflow will now cease, the galaxies may return to a less active phase.

Eventually the bar instability will be triggered once more and the cycle will begin again. In the pure disk scenario with bulges built in this fashion the two classic parameters of the Hubble sequence barred and unbarred and bulge-to-disk ratio are inextricably related. The bulge-to-disk ratio would then be merely a function of how many times this bulge-building process has occurred. The fact that about half the galaxies we see are barred indicates that the duty-cycle for this process is about 50%. An important unresolved issue raised by FREEMAN[48] is why there are so many Sc galaxies at the curent epoch.

The most fundamental prediction of this model is that barred Sc galaxies with at least 5% relative central mass concentrations are forming bulges now. KORMENDY[46] points out that there are some small bulge Sc galaxies that are blue and rapidly rotating.

5. – Starbursts and protogalaxies: feedback.

The relation between the starbursting systems that we can observe directly and the process of galaxy formation may be fairly close. If the starburst scenario is correct, then an appreciable fraction of the central regions of a starburst galaxy are being freshly minted in a burst. This is at least partial galaxy formation! Those high-redshift objects that are associated with radiosources may be associated with forming galaxies. CHAMBERS et al. [49] have inferred star formation rates similar to those of ultra-luminous starbursts. Assuming this is so, some interesting progress can be made by using what we know about the physical conditions in starbursts to extrapolate to physical conditions appropriate to galaxy formation models. Feedback considerations from the star formation process in galaxy formation and the effect on the galaxy collapse have been studied by IKEUCHI and NORMAN [50], who used a simple star formation model based on forming massive stars in cloud-cloud collisions. The energy input is taken to be due to supernovae. Characteristic masses, radii and binding energies of the galaxies were calculated in terms of fundamental constants. In addition the basic cloud parameters were inferred in terms of fundamental constants as were the properties of the hot component of the interstellar medium of the protogalactic cloud.

In pursuing the feedback idea further it is possible that the bursts in the high-redshift objects are being triggered by the high internal pressure of the radio lobe. In general triggering of starbursts can occur due to the action of a central source such as an AGN, radiosource or quasar. In analyzing this process NORMAN and IKEUCHI [51] found that such triggering was likely, as a central energy source turned on and began to interact with the multiphase protogalactic medium. One very interesting result that came out of this work is that the binding energy of galaxies and the amount of injected energy from the central source are roughly equal from stability considerations of the collapsing gaseous protogalaxy. Furthermore, for canonical parameters for central luminosities and the speed of the energy input, the binding energy of galaxies turns out to be of order $(10^{60}-10^{61})$ erg which is roughly correct for L_* galaxies.

The binding energy of galaxies is roughly of the same order of magnitude as the energy released by their central AGNs, radiosources and quasars over their individual lifetime. We model the energy input from quasars, BALs and starbursts in a unified manner as a function of the speed at which the energy is injected. We argue that active protogalaxies can have a *self-regulated* collapse phase where the energy and momentum input from the central source acts to slow the collapse. We then show that for the typical values of the various central energy luminosities for these objects the combination of the global energy and momentum balance condition leads to a characteristic binding energy for galaxies of $\approx 10^{60-62}$ erg.

In recent years, many candidates for protogalaxies have been discovered by identifying spectroscopically steep-spectrum radiosources as high-redshift galaxies [49, 52, 53]. The principal observational features are summarized as follows: i) The radio lobes extend to ~ 100 kpc and the line-emitting regions elongate for (30–50) kpc aligning with the radio axes; ii) highly ionized ions of heavy elements are detected, and Lyman-α-emitting regions surround them; iii) spectra from U-band to H-band show relatively flat continua and iv) some of them show strong velocity gradients of the order ~ 1000 kms^{-1}. Concerning feature i), either star formation in radio lobes [54-56] or light scattering by intracluster gas [57] has been proposed. Coupled with feature ii), they show plausibly that an efficient star formation and intense UV flux from young massive stars and active nuclei may be occurring. For feature iii), two competitive models have been proposed. One is that the star formation has just begun at $z = 2$–3 with an initial mass function different from present nearby observed IMFs [49, 58]. In this model the age of such high-redshift galaxies is less than several times 10^8 y. The other proposal is that the H-band intensity must come from relatively old stars with ages $\sim 10^9$ y and the U-band intensity from young stars [53, 59]. Here, the galaxy formation has to occur at $z \geq 7$, and the triggering of star formation in radio lobes does not occur. Feature iv) denotes that these galaxies have systematic motions like rotation as well as chaotic motions of the same order. These velocity fields seem to imply violent gas motions with scales of order ~ 100 kpc.

We now describe a scenario for the evolution of multiphase protogalaxies with central energy sources such as radiosources and investigate the relations among observables which may be checked by radio, optical and X-ray observations. Our model is based on a previous study of isolated protogalaxies without external energy input [50]. However, in protogalaxies with central energy sources such as radio jets, quasars and AGNs the external energy input can become the principal influence on the pressure of the interstellar medium. The increase in ambient pressure can affect the mass spectrum of clouds and can trigger star formation at large galactic radii. We present a simplified model of the type of central energy input based on the speed at which the energy is transferred to the overall protogalactic structure. Many of the specific details here are relevant to radiosources for which we have substantial and interesting data to guide us. However, where possible we incorporate more general considerations of other types of central energy sources such as AGNs, quasars and starbursts. It is a remarkable coincidence that the binding energy of galaxies and the total energy input from radiosources, quasars, BAL quasars and starburst galaxies integrated over the specific lifetime are roughly equal. This could be just a coincidence, but our analysis is motivated by a search for an underlying physical reason to explain it using considerations of energy and momentum balance in the structure of active protogalaxies.

The evolution of such protogalaxies is analyzed from the stage of formation of a central AGN or radiosource through the stage of formation of large-scale optical line-emitting regions. We assume that the evolutionary stages are divided into three major parts: I) The *virial equilibrium* stage with hot diffuse gas and cool cloud components in a pressure equilibrium. Star formation occurs in colliding clouds. II) The *efficient star formation* stage is triggered by the high external pressure from the central source such as that due to the radio lobes acting on the cool clouds lowering their internal Jeans mass. Alignment of radio axes with optical axes is expected for newly born stars formed in this stage. III) The *equipartition* stage amongst the various components of the protogalactic medium such as hot gas, magnetic fields and relativistic electrons with high-pressure extended line-emitting regions. This stage is similar in nature (but is far more powerful) to nearby active galaxies. We outline the influence of BAL quasars, quasars, AGNs and starbursts on protogalaxies. We note also the relationship between binding energy of the host protogalaxy and the luminosity of the central energy source with its associated speed of energy transfer to the larger protogalactic structure.

High-redshift galaxies with steep-spectrum radiosources have active central engines which are ejecting extended radio jets on the scale ~ 100 kpc. We assume here that the central AGN possibly associated with a black-hole is born rapidly and shortly after the galaxy begins forming. We regard quasars to be also of common origin, and the difference of black-hole masses may reflect the difference of activity. The formation of central engines at $z = 5$–10 is an important assumption in the present scenario. We concentrate here on the physical state of the multiphase interstellar medium of the protogalaxy hosting the central AGN.

The protogalactic cloud has a spectrum of fluctuations which results directly in a spectrum of bound clouds. Further fragmentation will be a result of thermal instability [60], which produces unbound pressure-confined clouds. As a result, the gas will undergo a change from a one-component to a two-component medium consisting of clouds and a hot ambient medium. These two components of the protogalactic medium are in a pressure equilibrium, and the virial equilibrium will be attained as a whole. The physical quantities obtained under general assumptions are summarized in the below. The equality of binding energy in galactic-scale objects is naturally explained.

For radiosources the central engine emits prodigious energy in the form of gas flows and relativistic particles associated with collimated jets and radio lobes. The shock waves associated with the jets, and their associated lobes and cocoons, propagate into the surrounding medium composed of clouds and hot gas. The clouds are compressed and hot gas is heated by the propagation of shock waves. BEGELMAN and CIOFFI [56] investigated the effects of shock compression parallel and perpendicular to the jet. The difference of compression causes the difference of star formation efficiencies in compressed clouds, and this can explain the alignment of radio and optical axes.

After the jets and lobes propagate out into the intergalactic space, the physical state inside the galaxies will be in a relatively steady state. In such a state and equipartition of energy among the various components of hot gas, magnetic fields and high-energy particles can, in principle, be realized. The energy supply from stars as UV flux and kinetic energy by stellar explosions determines the ionization states, and the luminosities in radio, optical and X-ray bands which are predicted from the minimum-energy principle. They are summarized below.

A protogalactic cloud is composed of cool clouds and hot ambient gas which are in a pressure equilibrium. The physical quantities in a virial equilibrium stage in a such a multiphase interstellar medium are calculated and discussed by IKEUCHI and NORMAN [50]. The adopted assumptions are as follows:

i) Virial equilibrium

$$\sigma_h^2 = \frac{GM_h}{R_h} = \frac{kT_h}{m_p}.$$

Here, R_h, M_h and T_h are the size of hot gas, total mass of hot gas and its temperature, respectively. We assume the mass of hot gas exceeds the cloud mass for simplicity, in this lecture.

ii) Equality of the cooling time and free-fall time

$$t_{cool} = t_{ff}.$$

By using these two equations we can obtain the equilibrium mass as a function of temperature as follows:

$$M_h = G^{-2} m_p^{-5/2} k^{1/2} T_h^{1/2} \Lambda(T_h),$$

$$R_h = G^{-1} m_p^{-3/2} k^{-1/2} T_h^{-1/2} \Lambda(T_h),$$

where $\Lambda(T_h)$ is the cooling function.

As REES and OSTRIKER[61] and SILK[62] indicated, the mass is nearly constant in the region of bound-free cooling in the temperature range $\sim (10^6 - 10^8)$ K and the radius is nearly constant in the free-free cooling range above $\sim 10^8$ K.

iii) Pressure equilibrium between clouds and hot ambient gas,

$$\tilde{p} = n_c T_c = n_h T_h.$$

Here, we introduce a characteristic process of the two-phase protogalactic medium.

iv) Energy balance condition,

$$L_r = \Lambda(T_h) n_h^2 R_h^3 = L_{cc} = \eta N_c M_c c^2 (N_c R_c^2 \sigma_h)/R_h^3,$$

where σ_h is the random velocity of clouds. This condition means that the energy loss from hot gas, $L_r = R_h^3 n_h^2 \Lambda(T_h)$, is equal to the energy supply triggered by cloud-cloud collisions, $L_{cc} = \eta M_c c^2 (N_c R_c^2 \sigma_h / R_h^3)$. We assume the energy liberated by cloud-cloud collisions is a fraction η of the rest mass energy which is the order of $\sim 10^{-5} - 10^{-6}$. The subsidiary parameters which characterize the physical state of a protogalactic cloud are as follows:

v) Covering factor of clouds,

$$f_s = N_c R_c^2 / R_h^2.$$

vi) Cloud temperature, T_c.

From the above we can obtain the equilibrium radius and mass of clouds,

$$R_c = G^{-1} m_p^{-5/2} c^{-2} k^{1/2} \eta^{-1} f_s^{-2} T_c T_h^{-1/2} \Lambda(T_h),$$

$$M_c = G^{-2} m_p^{-11/2} c^{-6} k^{7/2} \eta^{-3} f_s^{-6} T_c^2 T_h^{3/2} \Lambda(T_h).$$

vii) The cloud column density is greater than the critical column density for the cloud to cool during a collision and to consequently survive,

$$R_{c,\,\mathrm{crt}} = G^{-1} m_p^{-3/2} k^{-1/2} T_c T_h^{-3/2} \Lambda(T_h),$$

$$M_{c,\,\mathrm{crt}} = G^{-2} m_p^{-5/2} k^{1/2} T_c^2 T_h^{-3/2} \Lambda(T_h).$$

The curves described by these two equations cross at a temperature given by $kT_h/m_p c^2 = \eta f_s^2$; above this temperature the model given here is valid. Note that this temperature is associated with a critical binding energy for galaxies given by $\sigma_h^2 = c^2 \eta f_s^2$.

The Jeans mass of a pressure-confined cloud is given by [32]

$$M_J = \frac{1.9 c_c^4}{G^{3/2} P_h^{1/2}},$$

where c_c is the sound speed in a cool cloud. This can be rewritten as an equation for pressure giving

$$P_h = 1.9^2 c_c^8 M_J^{-2} G^{-3}.$$

We have found a range for the masses and radii of clouds and for the ambient gas pressure. Typical values for these physical parameters can be expressed by establishing various intersection points of the critical curves. In the thermal bremsstrahlung limit, the typical cloud mass is given by

$$M_{\mathrm{cross}} = \beta \alpha^3 \alpha_G^{-2} \left(\frac{kT_c}{m_p c^2}\right)^2 \left(\frac{m_p}{m_e}\right)^{3/2} (\eta f_s^2)^{-1} m_p,$$

where α and α_G are the fine-structure constants of electromagnetic and gravitational interactions.

The corresponding cloud radius is given by

$$R_{\mathrm{cross}} = \beta \frac{kT_c}{m_p c^2} \left(\frac{m_p}{m_e}\right)^{1/2} \alpha^2 \alpha_G^{-1} (\eta f_s^2)^{-1} r_e,$$

where r_e is the classical electron radius.

These can both the expressed in terms of fundamental constants, except for the cloud temperature, and the parameter ηf_s^2. The characteristic ambient pressure is determined from the intersection of the curves defined by equations describing the energy balance and the cloud survivability condition and is given by

$$P_{\text{cross}} = \beta^{-2} 1.9^2 \alpha_G \alpha^{-3} (\eta f_s^2)^2 \frac{m_p c^2}{r_e^3},$$

which is independent of T_c and is a typical pressure for protogalaxies.

The external pressure can increase due to an external energy input such as that from a radiosource, a starburst via supernova energy input, a quasar via radiation pressure, or mechanical-energy input from a BAL quasar. For radiogalaxies typical pressures are of order $(10^{-11}$–$10^{-12})$ dyn cm^{-2} for luminosities of order 10^{45} erg^{-1}. For quasars the radiation pressure is of order $10^{-8}(L/10^{48}$ erg$^{-1})(1$ kpc$/r)^2$ dyn cm^{-2}, for starbursts it is

$$10^{-8}(\dot{M}_*/100\ M_\odot\ \text{y}^{-1})(t_{\text{burst}}/3 \cdot 10^7\ \text{y})(10\ \text{kpc}/r)^3\ \text{dyn cm}^{-2}$$

and for BAL systems it is

$$2 \cdot 10^{-9}(\dot{M}/10\ M_\odot\ \text{y}^{-1})(V/10^4\ \text{kms}^{-1})(10\ \text{kpc}/r)^2\ \text{dyn cm}^{-2}.$$

Depending on the type of energy input and the luminosity, the effect on the star formation in the protogalaxy can be as follows. At pressures of order 10^{-8} dyn cm^{-2} these actual pressures acting on clouds can be sufficiently high that they can trigger all the clouds into star formation in the region of influence. The duration of the active star formation phase could then be very short with the remaining intersteller medium very hot and expanded. This energy input could easily give the observed chaotic motions of the gas with velocities of order $\sim 10^3$ kms^{-1}. At the lower end with pressures of order 10^{-12} dyn cm^{-2} the effect on the evolution of the protogalaxy is still important. At least two modes of star formation are occurring owing to external effects. One star formation mode is triggered by cloud collisions for all clouds and the other is triggered for more massive clouds by the lowering of the Jeans mass due to the action of the AGN or radiosource.

We now study the relationship between total energy of the system, binding energy and the energy input from active galactic nuclei, radiosources and quasars. The total binding energy or thermal energy in the virial equilibrium state is given by

$$E = G^{-2} m_p^{-7/2} k^{3/2} T_h^{3/2} \Lambda(T_h).$$

The thermal bremsstrahlung limit is given by

$$E = \beta \left(\frac{m_p}{m_e}\right)^{3/2} \alpha^3 \alpha_G^{-2} m_p c^2 \left(\frac{kT_h}{m_p c^2}\right)^2.$$

Note that for velocity dispersions of the order of ~ 300 kms^{-1} the binding energy is of order 10^{60} erg.

Now let us consider the momentum balance equation for the protogalaxy as a whole. First let us characterize the pressure acting on the protogalaxy as

$$P = \frac{L}{4\pi R^2 V},$$

where L is the luminosity of central source and V its characteristic speed of propagation. For quasars V is the speed of light, for BAL QSOs, for example, $V \sim 10^4$ kms^{-1}, and for starbursts we take the age of the burst to be of order the crossing time of the system. We further assume that the pressure acts globally on the system in a well-coupled manner. In particular for considerations of radiation pressure we note that the covering factor of clouds is of order unity and the clouds are optically thick. Then for the protogalaxy as a whole the force balance condition is

$$\frac{GM^2}{R^2} = \frac{L(4\pi R^2)}{4\pi R^2 V}$$

giving

$$\frac{GM^2}{R} = \frac{LR}{V}.$$

The binding energy of the protogalaxies in approximate momentum balance with the central source is calculated as a function of luminosity for a wide range of velocities encompassing starbursts, quasars, BAL QSOs and AGNs. Similar results are easily found for radiogalaxies.

We argue that the process of protogalaxy contraction with a central energy source or starburst (which is a natural consequence of the galaxy formation process) is *self-regulated*. If the value of L is too high, the system will expand and the central source will be cut off from fuel. If the value of L is too low, the system will contract and can enhance the fuelling of the central source or starburst. (Note that, with the time delays built into the system, instabilities and bursts can occur.) Since, for the lower-velocity systems in particular, the factor R/V is comparable to the lifetime of the starburst of BAL QSO, we see that the total energy put out in the AGN and/or starburst phase LR/V is comparable to the binding energy of galaxies as a natural consequence of the overall momentum balance equation.

It is interesting to note that for the typical range of luminosities considered here the binding energy of surviving protogalaxies is $\sim (10^{60}$–$10^{62})$ erg.

6. – Summary.

In this lecture we have discussed some of the important aspects of starburst galaxies, but, of course, vast areas of this field could not be treated. A good overview of the field can be found in the book *Massive Stars and Star Formation* by LEITHERER, WALBORN, HECKMAN and NORMAN. For the classic i.r. luminous starburst galaxies we have concentrated on dynamical studies, the nature of the interstellar medium and the outflows associated with superwinds. To make the connection with galaxy formation, we discussed the high-redshift counterparts of starbursts including radiosources and quasars. In most protogalaxy models the star formation is extremely rapid and exceeds even the most luminous nearby starbursts. The energy and momentum injected into the interstellar medium during this phase is prodigious. The major sources of energy injection are from supernovae and active galactic nuclei. We have examined how the structure of the protogalaxies may be determined internally by the feedback processes due to input of energy and momentum.

Since this field is so rapidly moving and it is such a rich field for further studies, this lecture could only be a sketch of some of the points that I personally think may be amongst the most productive to examine further. I encourage the students at this school to do so!

* * *

It is a pleasure to acknowledge the stimulating collaborations with Drs. H. HASAN, S. IKEUCHI, D. PFENNIGER, N. Z. SCOVILLE and J. A. SELLWOOD. The recent papers with these co-authors formed the basis of the text of this lecture. I also wish to acknowledge very useful conversations on this topic with Drs. K. C. FREEMAN, T. M. HECKMAN, C. LEITHERER, N. PANAGIA, B. WANG and R. WYSE.

I also thank Drs. J. SILK and N. VITTORIO for organizing such a stimulating course and the students for their very active and enthusiastic interest and participation.

REFERENCES

[1] F. SHU, F. H. ADAMS and S. LISANO: *Annu. Rev. Astron. Astrophys.*, **25**, 23 (1987).
[2] C. A. NORMAN and J. SILK: *Astrophys. J.*, **238**, 158 (1980).
[3] C. A. NORMAN and J. SILK: *Astrophys. J. Lett.*, **272**, L49 (1983).
[4] C. F. MCKEE: *Astrophys. J.*, **345**, 782 (1989).
[5] M. G. WOLFIRE, D. HOLLENBACH and A. TIELENS: *Astrophys. J.*, **344**, 770 (1989).
[6] G. J. STACEY, H. GEIS, R. GENZEL, J. B. LUGTEN, A. POGLITSCH, A. STERNBERG and C. H. TOWNES: *Astrophys. J.*, **373**, 423 (1990).
[7] S. POMPEA and G. H. RIEKE: *Astrophys. J.*, **356**, 391 (1990).
[8] L. HERNQUIST: *Nature (London)*, **340**, 687 (1989).
[9] J. NEGROPONTE and S. D. M. WHITE: *Mon. Not. R. Astron. Soc.*, **205**, 1009 (1983).

[10] J. E. BARNES: in *Dynamics and Interactions of Galaxies*, edited by R. WIELEN (Springer-Verlag, Berlin, 1990), p. 186.
[11] C. A. NORMAN: in *Starbursts and Galaxy Evolution*, edited by T. X. THUAN and T. MONTMERLE (Editions Frontieres, Gif-sur-Yvette, 1988), p. 183.
[12] C. A. NORMAN: in *Galactic and Extragalactic Star Formation*, edited by R. PUDRITZ and M. FICH (Reidel, Amsterdam, 1988), p. 495.
[13] C. A. NORMAN: in *Comets to Cosmology*, edited by A. LAWRENCE (Springer-Verlag, Berlin, 1988), p. 177.
[14] J. P. OSTRIKER and P. J. E. PEEBLES: *Astrophys. J.*, **186**, 467 (1973).
[15] D. PFENNIGER and C. NORMAN: *Astrophys. J.*, **363**, 391 (1990).
[16] H. HASAN and C. NORMAN: *Astrophys. J.*, **361**, 69 (1990).
[17] D. FRIEDLI and D. PFENNIGER: in *Dynamics of Galaxies and their Molecular Cloud Distributions, IAU Symposium* No. 146, edited by F. COMBES and F. CASOLI (Reidel, Dordrecht, 1991), p. 362.
[18] T. HECKMAN: in *Proceedings of IRAS Caltech Conference on Star Formation*, edited by C. PERSSON (NASA, 1986), p. 451.
[19] K. TOMISAKA and S. IKEUCHI: *Astrophys. J.*, **330**, 695 (1988).
[20] C. A. NORMAN and S. IKEUCHI: *Astrophys. J.*, **395**, 372 (1989).
[21] T. M. HECKMAN, L. ARMUS and G. K. MILEY: *Astrophys. J. Suppl.*, **74**, 833 (1990).
[22] R. CHEVALIER and A. W. CLEGG: *Astrophys. J. Lett.*, **300**, L107 (1985).
[23] W. H. WALLER: *Astrophys. J.*, **370**, 144 (1991).
[24] E. BRINKS and E. BAJAJA: *Astron. Astrophys.*, **169**, 14 (1986).
[25] J. BERGERON: in *Quasar Absorption Lines*, edited by C. BLADES, D. TURNSHEH and C. A. NORMAN (Cambridge University Press, Cambridge, 1988), p. 127.
[26] A. P. YOUNG, W. L. W. SARGENT and A. BOHRENBERG: *Astrophys. J.*, **252**, 10 (1982).
[27] A. WOLFE: in *Quasar Absorption Lines: Probing the Universe*, edited by J. C. BLADES, C. A. NORMAN and D. A. TURNSHEK (Cambridge University Press, Cambridge, 1988), p. 297.
[28] C. STEIDEL and W. L. W. SARGENT: *Astrophys. J. Suppl.*, **80**, 1 (1992).
[29] N. Z. SCOVILLE and C. A. NORMAN: *Astrophys. J.*, **332**, 163 (1988).
[30] C. A. NORMAN and N. Z. SCOVILLE: *Astrophys. J.*, **238**, 158 (1988).
[31] YA B. ZEL'DOVICH and I. D. NOVIKOV: *Dokl. Acad. Nauk SSSR*, **344**, 770 (1964).
[32] L. SPITZER: *Diffuse Matter in Space* (Wiley Interscience, New York, N.Y., 1966).
[33] M. BEGELMAN and M. J. REES: *Mon. Not. R. Astron. Soc.*, **185**, 847 (1978).
[34] M. E. BAILEY: *Mon. Not. R. Astron. Soc.*, **191**, 195 (1980).
[35] J. M. SHULL: *Astrophys. J.*, **264**, 446 (1983).
[36] W. C. MATTHEWS: *Astrophys. J.*, **305**, 187 (1986).
[37] H. HASAN, D. PFENNIGER and C. NORMAN: *Astrophys. J.*, **409**, 91 (1993).
[38] M. T. NISHIDA and K. WAKAMATSU: preprint (1994).
[39] D. FRIEDLI and W. BENZ: *Astron. Astrophys.*, **268**, 65 (1993).
[40] N. RAHA, J. A. SELLWOOD, R. A. JAMES and F. D. KAHN: *Nature (London)*, **352**, 411 (1991).
[41] J. A. SELLWOOD and D. MERRITT: *Astrophys. J.*, in press (1994).
[42] F. COMBES, F. DEBASCH, D. FRIEDLI and D. PFENNIGER: *Astron. Astrophys.*, **233**, 82 (1990).
[43] J. A. SELLWOOD: in *Galactic Bulges, IAU Symposium* No. 153, edited by H. DEJONGHE and H. HABING (Kluwer, Dordrecht, 1993), p. 391.

[44] D. PFENNIGER: in *Galactic Bulges, IAU Symposium* No. 153, edited by H. DEJONGHE and H. HABING (Kluwer, Dordrecht, 1993), p. 387.
[45] J. KORMENDY: in *Morphology and Dynamics of Galaxies, Saas Fee Course* No. 12, edited by L. MARTINET and M. MAYOR (Geneva Observatory, Suaverny, 1982), p. 113.
[46] J. KORMENDY: in *Galactic Bulges, IAU Symposium* No. 153, edited by H. DEJONGHE and H. HABING (Kluwer, Dordrecht, 1993), p. 209.
[47] H. DEJONGHE and H. HABING, Editors: *Galactic Bulges, IAU Symposium* No. 153 (Kluwer, Dordrecht, 1993).
[48] K. C. FREEMAN: in *Galactic Bulges, IAU Symposium*, No. 153, edited by H. DEJONGHE and H. HABING (Kluwer, Dordrecht, 1993), p. 263.
[49] K. C. CHAMBERS, G. K. MILEY and W. J. M. VAN BRUEGEL: *Astrophys. J.* (1990).
[50] S. IKEUCHI and C. A. NORMAN: *Astrophys. J.*, **375**, 479 (1991).
[51] C. A. NORMAN and S. IKEUCHI: *Astrophys. J.*, in preparation (1995).
[52] H. SPINRAD: in *Proceedings of the Durham NATO Workshop on Epoch of Galaxy Formation, ASI Series* (Kluwer, Dordrecht, 1989), p. 39.
[53] S. J. LILLY: *Astrophys. J.*, **340**, 77 (1989).
[54] M. J. REES: *Mon. Not. R. Astron. Soc.*, **239**, 1p (1989).
[55] D. S. DE YOUNG: *Astrophys. J.*, **342**, L59 (1989).
[56] M. C. BEGELMAN and D. F. CIOFFI: *Astrophys. J.*, **345**, L21 (1989).
[57] A. C. FABIAN: *Mon. Not. R. Astron. Soc.*, **238**, 41p (1989).
[58] M. BITHELL and M. J. REES: *Mon. Not. R. Astron. Soc.*, **242**, 570 (1989).
[59] B. ROCCA-VOLMERANGE and B. GUIDERDONI: *Astron. Astrophys.*, **252**, 435 (1991).
[60] S. M. FALL and M. J. REES: *Astrophys. J.*, **298**, 18 (1985).
[61] M. J. REES and J. P. OSTRIKER: *Mon. Not. R. Astron. Soc.*, **179**, 541 (1977).
[62] J. SILK: *Astrophys. J.*, **211**, 638 (1977).

Stellar-Population Tools and Clues to Galaxy Formation.

Three Lectures with Exercises.

A. Renzini

Dipartimento di Astronomia dell'Università - Bologna, Italia

The stellar content of galaxies represents a record of their star formation history, and as such it tells about the formation of galaxies themselves. Stellar-population studies have indeed a variety of applications that are relevant for our understanding of galaxy formation processes, as they offer the unique opportunity to date galaxies (or galaxy components), and, therefore, answer questions such as «when did the galaxies form?», or «how long did it take to form the bulk of stars in elliptical galaxies?», etc. Moreover, coupling stellar-population studies to stellar nucleosynthesis one can gather additional insight into the time scale and energetics of galaxy formation. Several other lectures at this summer school deal with the formation of galaxies and structures. Many of them do so in a *deductive* approach in which the formation and evolution of structures is followed starting from postulated initial conditions. The study of stellar populations, instead, tends to move in the opposite direction, inferring about the galaxy formation process from its observable output, *i.e.* it is more an *inductive* approach. The two approaches are clearly complementary to each other, and we hope that they will eventually meet at some point—as miners do when they escavate the tunnel starting from the opposite sides of a mountain.

Another aspect worth making explicit is that stellar populations are made up of baryons, which may well be little more than a trace constituent of the Universe. As such, they are not explicitly treated by some theory for the formation of large-scale structure and we should bear in mind that by *galaxy formation* different people occasionally mean rather different things. For example, in the classical CDM theory galaxy formation is the process of assembling CDM particles, while for a baryon chauvinist the epoch of galaxy formation is the time when the first major burst took place forming the bulk of stars in ellipticals and bulges [1]. These differences reflect in the different physical ingredients which enter the simulations, and correspondingly in the different tools which are used. Thus in CDM simulations the only physical ingredient is gravity, and the tool is represented by N-body calculations. Once the primordial perturbations have grown enough, and many dips in the gravitational potential have grown, we would like to know how baryons have been collected in such dips, and what kinds of observable galaxies these baryons are going to build up. Besides gravity, this latter process includes atomic physics (which describes the interaction of matter and radiation) and hydrodynamics (which in particular should describe star formation). Given its intimately hydrodynamical nature this is a very messy physical problem, and we do not have a first-principle theory of star formation, *i.e.* we do not know how to relate the star formation rate and IMF to the local conditions. Even parametrizing our ignorance of star formation, the process of (baryonic) galaxy formation remains highly conjectural. But once stars are formed they shine, and from their light we may learn about the messy beginnings. Now the relevant physics includes gravity, atomic and nuclear physics, and the main tool is represented by the output of stellar-evolution codes coupled to observations. Curiously enough, the inclusion of yet more physics (*i.e.* nuclear reactions) makes things somewhat simpler, as now matter tends to organize itself in hydrostatic equilibrium structures: stars.

In the first of these three sections I will introduce the basic properties of stellar populations, *i.e.* a few fundamental concepts which—like the pieces of a kit—can be of fairly ubiquitous use in astrophysics, and which often allow one to get the relevant numbers out in the quickest possible way. The second section deals with the stellar dating methods for globular clusters and low-redshift elliptical galaxies, while the third section gives some examples on how stellar nucleosynthesis can provide information on the energetics and time scale of the galaxy formation process.

1. – The basic properties of stellar populations.

By simple stellar population (SSP) we mean an assembly of coeval, chemically homogeneous, single stars. Besides age and composition, only the initial mass function (IMF) remains to be specified in order to completely characterize the global properties of a SSP, and this is all. The theory of stellar evolution—by describing the detailed behaviour of the individual constituent stars—can then be used to predict several global properties of SSPs (cf. [2, 3] and [4], hereafter RB).

Galaxies are complex stellar populations, characterized by age and metallicity distributions yet to be determined, and it is then worth becoming familiar with simple populations before attacking higher complexities. In the following of this section the main global properties of SSPs are briefly recalled. For some of them a more detailed discussion may be found in RB.

1`1. *The stellar-evolution clock.* – The stellar lifetime increases with decreasing initial mass, and for roughly 90% of their lifetime stars burn hydrogen at the centre, and lie near the main sequence (MS) in the H-R diagram. The MS turn-off (TO) is—at any age t of the population—the pont of maximum effective temperature along the corresponding isochrone, which in turn is the locus in the H-R diagram occupied by stars (models) of different mass but the same age. The TO nearly coincides with the point where stars exhaust hydrogen at the centre, and, therefore, below the isochrone turn-off one finds stars which are still burning hydrogen at the centre, and beyond it stars burn hydrogen in a shell or are in even more advanced evolutionary stages.

Stellar-evolution calculations naturally provide the luminosity and effective temperature of the TO, together with the mass M_{TO} of stars at TO, all as a function of time and composition. Analytical approximations to tabular values of these quantities are particularly useful in a number of astrophysical applications. For example, for solar composition $(Y, Z) = (0.28, 0.02)$ the TO mass can be approximated by the relation [4]

$$(1.1) \qquad \log M_{TO}(t) = 0.0558 \log^2 t - 1.338 \log t + 7.764,$$

where M_{TO} is in $M_\odot$ units and t is in years. Past TO, stars spend a post-MS time t_{PMS} in a sequence of various nuclear burning stages, before dying as either supernovae or white dwarfs. A fair approximation to t_{PMS} is given by

$$(1.2) \qquad t_{PMS} \simeq 1.66 \cdot 10^9 M_i^{-2.72} \text{ (y)},$$

where M_i is the initial mass. Correspondingly, the initial mass of stars in the verge of completing their thermonuclear evolution is given by $M_D \simeq M_{TO}(t - t_{PMS})$. Figure 1.1 shows M_D, M_{TO} and its time derivative $\dot M_{TO}$ as a function of the age of the population. So, at any given time stars with $M_i < M_{TO}$ are still burning hydrogen at the centre and lie close to the MS, stars with $M_i > M_D$ are already dead remnants, while stars in the rather narrow range $M_{TO} < M_i < M_D$ are venturing in their post-MS evolutionary stages. Correspondingly, we do not commit a large error if we approximate all evolved stars as having the same initial mass $M_i \simeq M_{TO} \simeq M_D$.

Analytical approximations are also available for the turn-off luminosity and temperature, or for their observable counterparts: the TO magnitude and colour. For example, for old populations ($t \gtrsim 1$ Gy) one can use

$$(1.3) \qquad \log t_9 \simeq -0.41 + 0.37 M_V^{TO} - 0.43 Y - 0.13 \text{ [Fe/H]},$$

where t_9 is the age in Gy units, M_V^{TO} the absolute visual magnitude of the main-sequence turn-off, Y the helium abundance, and [Fe/H] the iron abundance in standard notations (*). For $Y = 0.23$ and ages in excess of several Gy the TO

(*) This relation has been derived by BUONANNO et al. [5] from the isochrones of VANDEN-BERG and BELL [6].

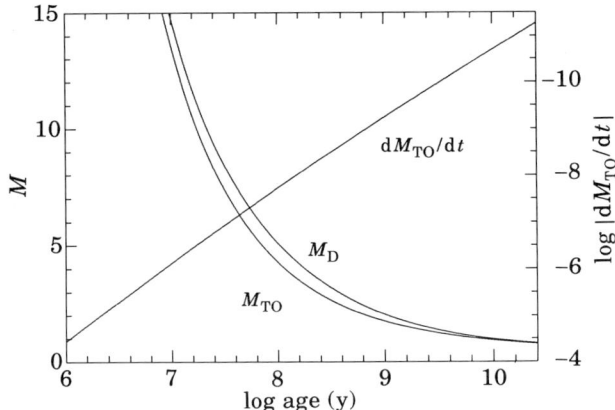

Fig. 1.1. – The turn-off mass M_{TO}, the mass of dying stars M_D and the time derivative $|\dot{M}_{TO}|$ of the turn-off mass, all as a function of the age of a population.

$(B - V)$ colour can be approximated by the relation [7]

(1.4) $\quad (B - V)_{TO} = 0.3140 + 0.3092 \log t + 0.2713\,[M/H] + 0.0543\,[M/H]^2$,

where [M/H] is the overall metallicity in standard notations(*). Equations (1.1), (1.2) and (1.4) are optimized for old populations, but similar relations can be obtained for any desired age range, depending on the needs (see, for instance, [8, 9]).

1˙2. *Evolutionary flux.* – In the aging process of a SSP stars continuously leave the main sequence, venture through the various post-MS stages, and eventually become dead remnants. The rate at which this process occurs is the *evolutionary flux* of the population:

(1.5) $\quad b(t) = \psi(M_{TO})|\dot{M}_{TO}| = A M_{TO}^{-(1+x)}|\dot{M}_{TO}|$,

and is measured in stars per year. Here ψ is the IMF, and for $M_{TO}(t)$ one can use eq. (1.1). The two factors in (1.5) have a different rôle in determining the flux. The IMF $\psi = A M^{-(1+x)}$ represents the reservoir of stars which is progressively depleted at a rate which is established by the evolutionary *clock* of the population, *i.e.* $|\dot{M}_{TO}|$. Initially, for large M_{TO} values, the clock is very fast, but its rate drops dramatically as M_{TO} decreases. For example, $|\dot{M}_{TO}|$ drops by almost five orders of magnitude as M_{TO} decreases from $\sim 10 M_\odot$ ($t \simeq 10^7$ y) to $\sim 1 M_\odot$ ($t \simeq 10^{10}$ y). However, in eq. (1.5) this drop of $|\dot{M}_{TO}|$ can be partly balanced by the modulation of the IMF, which increases for decreasing mass. All this is

---

(*) Note that [Fe/H] = [M/H] if elemental proportions are solar.

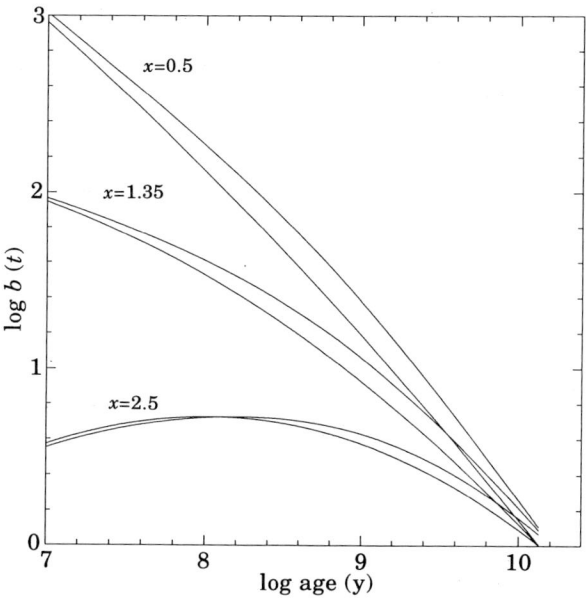

Fig. 1.2. – The evolutionary flux $b(t)$ from eq. (1.5) for three choices of the IMF slope as indicated, and normalized at $b = 1$ for $t = 15$ Gy. For each value of the IMF slope the upper line represents the death rate $b(t - t_{\rm PMS})$, thus making fully apparent the error which is introduced when the evolutionary flux is assumed to be the same for all post-MS stages at a given epoch.

shown in fig. 1.2, along with the death rate of the population, i.e. $b(t - t_{\rm PMS})$: since at any time there is little difference between the rate at which stars leave the MS and that at which they die, one can state that the evolutionary flux of a population is nearly constant along the post-MS evolutionary path. In other words, to a fair approximation the flux $b(t)$ also gives the rate at which stars enter or leave any specific post-MS evolutionary stage.

If t_j is the duration of the generic post-MS evolutionary stage j, then we have $b(t) t_j$ stars entering phase j during the time interval t_j, which is the number of stars N_j that are present in phase j, i.e.

(1.6) $$N_j = b(t) t_j ,$$

which says that N_j is proportional to the stage duration t_j and the coefficient of proportionality is nothing else but the evolutionary flux $b(t)$.

1˙3. *The total luminosity.* – One of the most conspicuous properties of a stellar population is certainly its total (bolometric) luminosity $L_{\rm T}$. Following RB, we can split $L_{\rm T}$ into two terms, one for the contribution of core hydrogen-burning stars (the MS stars) and one for the post-MS stars. After some little algebra, such terms can be written as

(1.7) $\qquad L_T(t) = L_{MS} + L_{PMS} = KAM_{TO}^{a-x} + 9.75 \cdot 10^{10} b(t) \sum_j F_j(M_{TO}),$

where the numerical factor K is a simple function of several other quantities such as the slope of the IMF $(1 + x)$ and the local slope of the ZAMS mass-luminosity relation (cf. RB for details). F_j is the *fuel consumption*—i.e. the amount of nuclear fuel burned by stars with $M = M_{TO}$ in their j-th evolutionary stage (*). The two terms at the r.h.s. of eq. (1.7) are of comparable size, and for order-of-magnitude estimates it is convenient to take

(1.8) $\qquad\qquad\qquad L_T(t) \simeq 10^{11} k' b(t) F_T(M_{TO}),$

where the numerical coefficient $k' = L_T/L_{PMS}$ is the inverse of the relative contribution of post-MS stars, and $F_T = \sum_j F_j(M_{TO})$. The quantity k' is a very slow function of age, which for a Salpeter's IMF decreases from ~ 3 in a 10^7 y old population, down to ~ 1.3 in a 10^{10} y old population (see subsect. 1˙6). Note that $k' = 1 + L_{MS}/L_{PMS}$, i.e. it can be expressed in terms of the *dwarf-to-giant* ratio. It is also worth noting that the luminosity evolution between, say, 1 billion and 15 billion years is entirely determined by the behaviour of the IMF between $\sim 2 M_\odot$ and $\sim 0.85 M_\odot$, i.e. between the corresponding M_{TO} values. From these relations one can rather easily realize that, for $t \gtrsim 10^8$ y, $L_T(t) \propto \sim t^{-(4-x)/3}$, i.e. the luminosity evolution of a SSP is controlled by the slope of the IMF (see also fig. 4 in RB).

1˙4. *The specific evolutionary flux*. – Dividing the evolutionary flux (1.5) by the total luminosity (1.8) one gets the *specific evolutionary flux* of the SSP, i.e. the flux per unit luminosity of the parent population, which is measured in stars $y^{-1} L_\odot^{-1}$:

(1.9) $\qquad\qquad\qquad B(t) = \dfrac{b(t)}{L_T(t)} \simeq \dfrac{10^{-11}}{k' F_T},$

where we have also used eq. (1.4). Note the remarkable cancellation of the $b(t)$'s in the approximate relation, which makes the specific evolutionary flux virtually independent of the IMF. Moreover, since the total post-MS fuel consumption F_T does not change much with stellar mass (**), $B(t)$ turns out to be almost inde-

---

(*) The fuel consumption F_j is given by $m_j^H + 0.1 m_j^{He}$, with m_j^H and m_j^{He} being, respectively, the amount of hydrogen and helium burned during stage j, both measured in $M_\odot$ units. The 0.1 coefficient comes from the helium fuel giving $\sim 1/10$ the energy released by burning hydrogen.
(**) The total post-MS fuel consumption decreases from $\sim 0.75 M_\odot$ for $M_{TO} = 10 M_\odot$ (age about 10^7 y) to $\sim 0.35 M_\odot$ for $M_{TO} = 0.85 M_\odot$ (age about 10^{10} y, cf. RB fig. 3). The change is only a factor of 2 while age increases by 3 orders of magnitude.

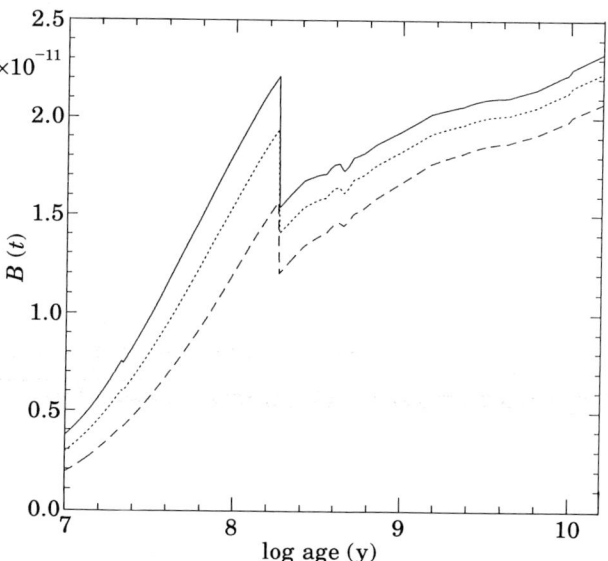

Fig. 1.3. – The specific evolutionary flux $B(t)$ as a function of age for three IMF slopes $(1+x)$: —— $x = 0.35$, ···· $x = 1.35$, --- $x = 2.35$.

pendent of age, increasing from $\sim 0.5 \cdot 10^{-11}$ to $\sim 2 \cdot 10^{-11}$ (stars y^{-1} $L_\odot^{-1}$) as the population age increases from 10^7 to 10^{10} y, as shown in fig. 1.3.

By substitution of eq. (1.9) into eq. (1.6), one obtains the notable relation

(1.10) $$N_j = B(t) L_T t_j \, ,$$

which is extremely useful in a number of astrophysical applications. Also, the death rate of a SSP is given by $b(t) = B(t) L_T$: even without making any guess about the age of a population (or the stellar-age distribution within a galaxy), by adopting $b(t) = 10^{-11} L_T$ the death rate of the population is obtained within a factor of 2. If we know the age within a factor of 2 or so, then we can use a better guess for $B(t)$ from fig. 1.3, and the resulting death rate can be obtained with an accuracy which is as good as (10-20)%. To my understanding of these matters, this is one of the most robust predictions of stellar-evolution theory applied to stellar populations.

1˙5. *The IMF scale factor.* – The scale factor A appearing in eq. (1.5) is clearly a measure of the size of the stellar population one is dealing with. It is useful to express this scale factor in terms of some directly observable quantity which is also proportional to the size of the population, such as its integrated luminosity L_T. The equations so far obtained give us the handle. We can solve for A either eq. (1.7) or eq. (1.9). In this latter case one gets

(1.11) $$A = B(t) L_T(t) M_{TO}^{1+x} |\dot{M}_{TO}|^{-1}.$$

The r.h.s. of this equation can be evaluated for an arbitrary age, A being a constant. Thus for $t = 15$ Gy one has $B(t) \simeq 2 \cdot 10^{-11}$ (see fig. 1.3), while M_{TO} and its derivative can be evaluated from eq. (1.1). In this way one finally gets (for $x = 1.35$)

(1.12) $$A \simeq 1.2 L_T(t = 15 \text{ Gy}),$$

where, as usual, L_T is expressed in $L_\odot$ units.

Unfortunately, the bolometric luminosity of a stellar population is not so directly reachable by observations. The luminosity in a given band is much more directly accessible, and it too is proportional to either L_T or A. In this respect, we can write

(1.13) $$L_T = B_c(Z, x, t) L_B,$$

where L_B is the blue luminosity of the population and $B_c(Z, x, t)$ is the bolometric correction factor to the blue luminosity, which is a function of metallicity, IMF slope and age. The evaluation of this function requires the construction of appropriate synthetic stellar populations, such as those of Buzzoni[10], and fig. 1.4 is drawn from his tabulations. One sees that values between 2 and 3 are appropriate in quite a range of ages, metallicities and IMF slopes. For this lecture I conservatively adopt $B_c(15 \text{ Gy}) \simeq 2.4$, and correspondingly get $A \simeq 2.9 L_B$. This is to say that if we know (or assume) the age of the bulk stellar population of, e.g., an elliptical galaxy to be 15 Gy, then the scale factor to use in its IMF is $\sim 2.9 L_B$, with an accuracy of $(10\text{--}20)\%$, and fairly independent of the assumed IMF slope. This last aspect follows from the fact that most of the light of

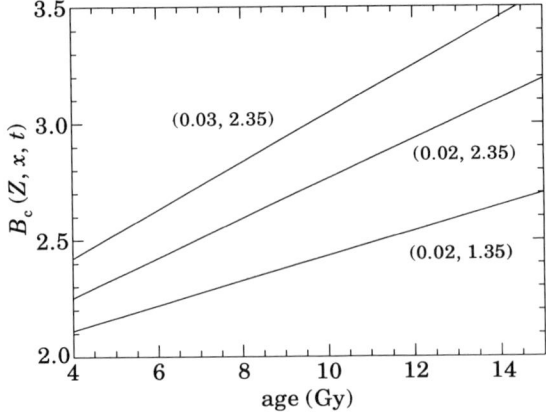

Fig. 1.4. – The bolometric correction factor $B_c(Z, x, t)$ as a function of age for three combinations of Z and x.

the population is produced by stars in a very narrow mass range ($\sim M_{\text{TO}} - M_{\text{D}}$), and, therefore, the present ($t = 15$ Gy) luminosity primarily depends on the number of stars in this mass range. This is true in so far as the IMF is not too steep for masses below $\sim 0.5 M_\odot$, i.e. if $1 + x \geqslant 2.5$, when the population would turn from giant to dwarf dominated (cf. RB, for a detailed discussion of this aspect).

1'6. *The fuel consumption theorem.* – The construction of the synthetic spectra of stellar populations is a rather complex affair, and before embarking in such a tiresome work it may be worth having an idea of what is important and should be taken into account, and what is less important and could be neglected so as to make our life simpler. The integrated light of a population results from the overlapping contributions of stars in the various evolutionary stages. Giants and dwarfs, horizontal-branch and asymptotic-giant-branch stars, all have their rôle to play. At some point we may like to know such relative contributions, i.e. to know the relative importance of the various evolutionary stages. As far as the bolometric luminosity is concerned, this is established by the *fuel consumption theorem*, which is just a little more than the principle of energy conservation: *The contribution of stars in any post-MS stage is proportional to the amount of fuel burned during that stage.* The F_j's appearing in, e.g., eq. (1.7) may not be familiar to nonpractitioners, but are among the most natural products of stellar evolutionary calculations (although they are often left unrecorded). This *theorem* was implicitly used in deriving eq. (1.7), as the total luminosity (in $L_\odot$) provided by stars in the j-th stage is given by

(1.14) $$L_j(t) = 9.75 \cdot 10^{10} \, b(t) \, F_j(M_{\text{TO}}),$$

where in turn $M_{\text{TO}} = M_{\text{TO}}(t)$ as given, e.g., by eq. (1.1)(*). The relative contributions are then obtained by dividing eq. (1.14) by the total luminosity, and

---

(*) CHARLOT and BRUZUAL [11] have repeatedly criticised this approximation in which one assumes a star with $M_i = M_{\text{TO}}$ to be representative of the evolution of all post-MS stars at a given epoch. They claim that «*By assigning the same IMF weight and post-main-sequence lifetime to all stars above the turn-off, the fuel consumption theorem model overestimates* the contribution of luminous evolved stars to the integrated light of the population». In fact the opposite is true; approximation (1.14) slightly *underestimates* (!) the contribution of post-MS stars. CHARLOT and BRUZUAL correctly notice that the IMF weight for evolved stars is smaller than $M_{\text{TO}}^{-(1+x)}$, but forget the effect of the other two terms $|\dot{M}_{\text{TO}}|$ and F_j, that are actually larger than assumed in the simple approximation. Note that what matters is the flux $b(t)$, which *decreases* with time (see fig. 1.2). When properly computing the contributions L_j's, one sees that approximation (1.14) actually *underestimates* by (10–20)% the contribution of luminous evolved stars. This means, e.g., that the models of Bruzual [12]—which do not include some advanced stages—have omitted from the synthesis some 60% of the light of an intermediate-age population, rather than 50% as indicated by eqs. (1.7) and (1.14).

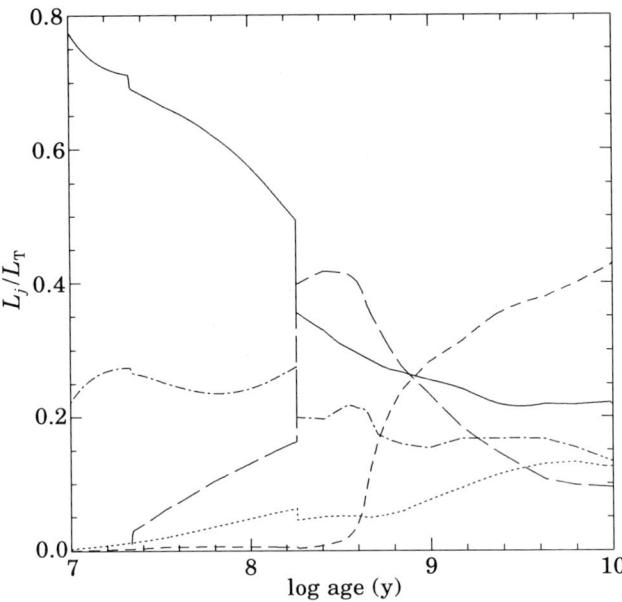

Fig. 1.5. – The relative contributions of the various evolutionary stages to the integrated light of a stellar population are shown as a function of age [15]: ——— MS, ···· SGB, ––– RGB, –·–·– HB, – – – AGB; $Y = 0.30$, $Z = 0.01$, $\eta = 2/3$, $x = 1.35$.

using eq. (1.9) we finally get

$$\frac{L_j}{L_T} = 9.75 \cdot 10^{10} B(t) F_j \, . \tag{1.15}$$

This relation gives us another opportunity for appreciating the virtues of the specific evolutionary flux.

The whole history of the various contributions is displayed in fig. 1.5, which supersedes the corresponding figure in RB and [13]. The input models are the same as those used in RB, apart from the use of newer models [8] which nicely delineate the RGB phase transition at $t \simeq 5 \cdot 10^8$ y. As pointed out in RB, the early generation of synthetic AGB models largely overproduce bright AGB stars compared to what is observed in Magellanic Cloud globular clusters younger than a few 10^8 y. The fuel consumption during the thermally pulsing AGB phase of stars more massive than $3 M_\odot$ has been correspondingly decreased by a factor of 10, so as to roughly take into account the effects of the envelope burning process as found by BLÖCKER and SCHÖNBERNER [14], whose result provides a natural solution to this long-lasting discrepancy [15]. Note from fig. 1.5 that for ages in excess of $\sim 3 \cdot 10^8$ y the relative contribution of the main-sequence stars is nearly constant, which justifies the approximation $k' \simeq$ const used in subsect. 1˙3.

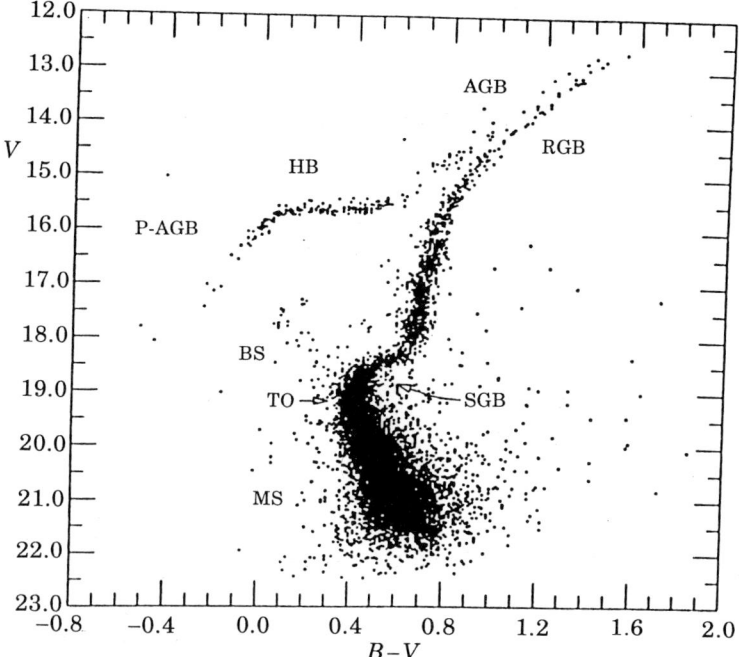

Fig. 1.6. – The CMD of the cluster M3 [13] with the main evolutionary stages indicated by their achronims: MS = Main Sequence, BS = Blue Stragglers, SGB = Subgiant Branch, RGB = Red-Giant Branch, HB = Horizontal Branch, AGB = Asymptotic Giant Branch, P-AGB = Post-AGB.

An experimental check of these complex theoretical predictions is rather laborious but feasible, and at least in part it has been already positively undertaken. The ideal check is to obtain complete colour-magnitude diagrams of populous star clusters—such as galactic and Magellanic globulars—, identify all the stars in the various evolutionary stages, and obtain empirical L_j/L_T ratios to be compared to the predictions of eq. (1.15). So far this has been done only for the galactic globular M3 [13], whose colour-magnitude diagram is shown in fig. 1.6, and where the main evolutionary phases are identified. The state of the art in stellar model testing is discussed by RENZINI and FUSI PECCI [13] and ROOD [16].

1˙7. *The rate of mass return.* – The mass return to the interstellar medium is one of the major metabolic functions of a stellar population. This results from each star losing a significant fraction of its mass before concluding its evolution as a dead remnant. In a SSP the rate of this process is given by the product of the death rate (*i.e.* the evolutionary flux) times the total mass loss suffered by one evolving star:

(1.16) $$\dot{\mathscr{M}} = b(t)\Delta M = \psi(M_{TO})|\dot{M}_{TO}|\Delta M,$$

where $\Delta M = M_{TO} - M_f(M_{TO})$, M_f being the final mass, or mass of the remnant. It may be worth calling attention to the obvious, albeit remarkable similarity between eq. (1.8) for the total luminosity and eq. (1.6) for the rate of mass return. In both equations the evolutionary flux $b(t)$ provides most of the time dependence, and it is multiplied either by the mass of burned fuel, or by the mass lost by one evolving star. This says that the initial mass M_{TO} is partly burned (the remnant is composed of thermonuclear *ashes*) and partly lost. The former activity inevitably produces light (L_T), the latter produces exhaust ($\dot{\mathcal{M}}$). One cannot reduce the exhaust without increasing the ashes, and, therefore, without getting a brighter population. Note also that some ashes can be contained in the exhaust, as a result of either dredge-up processes or explosions, the rest remaining locked into the dead remnants (white dwarf or neutron star). Once properly quantified, these considerations may find numerous applications for a better understanding of galaxy formation and evolution, and also of AGN absorption line systems.

Equation (1.16) can be generalized to describe the rate of mass return for each individual, element, isotope, or kind of dust grain. For this purpose it is sufficient to substitute to ΔM the mass of the particular species which is lost by $M = M_{TO}$ stars. For example, in the case of dust grains, a SSP initially returns just silicates until M_{TO} drops below $\sim 2.5 M_\odot$ (at $t \simeq 4 \cdot 10^8$ y), when carbon stars first appear on the AGB and the particulate exhaust turns to graphite and other carbon-rich grains.

Dividing the rate of mass return by the total luminosity, one gets the rate of mass return per unit luminosity of the parent population, *i.e.* the specific rate of mass return. From eqs. (1.16), (1.7) and (1.9) one obtains

$$(1.17) \qquad \dot{m} = \frac{\dot{\mathcal{M}}}{L_T} = B(t) \Delta M \simeq \frac{\Delta M}{k' 10^{11} \sum_j F_j}.$$

This rate is measured in $M_\odot \, y^{-1} L_\odot^{-1}$. Note that, apart from the nearly constant coefficient, the specific rate of mass return is given by the *exhaust-to-ash* ratio for $M = M_{TO}$, and is, therefore, almost independent of the IMF, and fairly independent of age as well. This follows from the very basic fact that most of the light *and* most of the mass emitted by an evolving SSP come from evolved stars, which at any time span a fairly narrow range of initial masses [2, 4].

Equation (1.17) allows us to derive a very useful and practical expression for the rate of mass return, which is valid for every age of the population. One can in fact write

$$(1.18) \qquad \dot{\mathcal{M}} = B(t) \Delta M L_T,$$

which relates $\dot{\mathcal{M}}$ to the luminosity of the population. For example, for a 15 billion year old population $B(t) \simeq 2 \cdot 10^{-11}$, $M_{TO} \simeq 0.85 M_\odot$, $M_f \simeq 0.55 M_\odot$ ($\Delta M = 0.3 M_\odot$), and, therefore, $\dot{\mathcal{M}} \simeq 6 \cdot 10^{-12} L_T$, and, *e.g.*, the metabolism of a

$2 \cdot 10^{11} L_\odot$ passive galaxy generates an exhaust of $1.2 M_\odot \, y^{-1}$. I believe that these estimates are accurate to better than the 10% level. Neither the specific evolutionary flux, nor the total mass lost by an old star can in fact be off by more than that (see [17] for a further discussion on the rate of mass return).

The fact that most of the light is produced by stars in a rather narrow mass interval makes several quantities rather insensitive to the slope of the IMF. In summary, with the limitations discussed in RB, the list of such IMF-insensitive quantities includes the integrated spectrum of a population, the specific evolutionary flux, the relative contributions of the various evolutionary stages, the specific rate of mass return, and to some extent also the IMF scale factor. Very dependent on the IMF is instead the time evolution of the flux $b(t)$ and, therefore, of all the quantities (nearly) proportional to it, such as the total luminosity and the rate of mass return. Some applications of conceptual tools so far introduced are presented in the next two sections.

REFERENCES

[1] D. C. Koo: in *Spectral Evolution of Galaxies*, edited by C. Chiosi and A. Renzini (Reidel, Dordrecht, 1986), p. 419.
[2] B. M. Tinsley: *Fund. Cosmic Phys.*, **5**, 387 (1980).
[3] A. Renzini: *Ann. Phys. (Paris)*, **6**, 87 (1981).
[4] A. Renzini and A. Buzzoni: in *Spectral Evolution of Galaxies*, edited by C. Chiosi and A. Renzini (Reidel, Dordrecht, 1986), p. 195 (RB).
[5] R. Buonanno, C. E. Corsi and F. Fusi Pecci: *Astron. Astrophys.*, **216**, 80 (1989).
[6] D. A. VandenBerg and R. A. Bell: *Astrophys. J. Suppl.*, **58**, 561 (1985).
[7] O. Straniero and A. Chieffi: *Astrophys J. Suppl.*, **76**, 525 (1991).
[8] A. V. Sweigart, L. Greggio and A. Renzini: *Astrophys J. Suppl.*, **364**, 527 (1990).
[9] L. Greggio and A. Renzini: *Astrophys. J.*, **364**, 35 (1990).
[10] A. Buzzoni: *Astrophys. J. Suppl.*, **71**, 817 (1989).
[11] S. Charlot and G. Bruzual: *Astrophys. J.*, **367**, 126 (1991).
[12] G. Bruzual: *Astrophys. J.*, **273**, 105 (1983).
[13] A. Renzini and F. Fusi Pecci: *Annu. Rev. Astron. Astrophys.*, **26**, 199 (1988).
[14] T. Blöcker and D. Schönberner: *Astron. Astrophys.*, **244**, L43 (1991).
[15] A. Renzini: in *The Stellar Populations of Galaxies*, edited by B. Barbuy and A. Renzini (Kluwer, Dordrecht, 1992), p. 325.
[16] R. T. Rood: in *Astrophysical Ages and Dating Methods*, edited by E. Vangioni-Flan, M. Cassé, J. Audouze and J. Tran Thanh Van (Editions Frontières, Gif-sur-Yvette, 1990), p. 313.
[17] L. Ciotti, A. D'Ercole, S. Pellegrini and A. Renzini: *Astrophys. J.*, **376**, 380 (1991).

Exercises.

Exercise 1.1: How many planetary nebulae are sampled by a CCD camera with a $2' \times 3'$ field of view looking at an elliptical galaxy with average blue surface brightness of 21 mag per square arcsec and distance modulus 25 mag? Assume $5 \cdot 10^3$ y for the PN lifetime.

Exercise 1.2: HOLLAN FORD (*Astrophys. J.*, **219**, 595 (1978)) estimates that in the bulge of M31 there are 775 PN's in the top 3 magnitudes of the PN luminosity function. Knowing that this bulge has a luminosity $L_B = 1.14 \cdot 10^{10} L_\odot$, determine the average duration of the PN stage (within 3 mag from the tip of the luminosity function).

Exercise 1.3: What fraction of a globular-cluster light is provided by red-giant branch (RGB) stars?

Exercise 1.4: How many supernovae of type II have exploded in an elliptical galaxy whose present luminosity is $L_B = 10^{11} L_\odot$? Suppose the galaxy can be approximated with a 15 Gy old single-burst population, and adopt the Salpeter IMF slope $x = 1.35$.

Exercise 1.5: Write L_{PMS} in eq. (1.7) without making the simplifying assumption of a constant evolutionary flux through the post-MS evolutionary phases.

The answers to all the exercises can be found at the end of the lecture.

2. – The age of galaxies.

Some of the tools sketched in the first section are now put at work to get estimates of the age of galaxies (or parts of them), which is certainly a strong bound for any galaxy formation scenario. The items to be covered include the age of galactic globular clusters, the population synthesis approach to galaxy ages and the age of low-redshift elliptical galaxies. Some qualitative conjectures about galaxy formation will conclude this second section.

2`1. *The age of galactic globular clusters.* – The age of galactic globular clusters can be best estimated by using the relation between age and turn-off luminosity, as approximated, *e.g.*, by eq. (1.3). To get the cluster age we need to measure the input quantities M_V^{TO}, Y and [Fe/H]. In turn, $M_V^{\text{TO}} = V^{\text{TO}} - \text{mod}$, where V^{TO} (the TO apparent magnitude) is the directly *observable* quantity, and mod is the cluster distance modulus.

Determining the age of globular clusters has been a very lively field of research in the last 20 years or so, with many active groups especially in Canada, Italy and the USA. Virtually all studies are consistent with

$$\text{age} = (13 - 15 \pm 3) \text{ Gy};$$

for recent reviews on the subject see, *e.g.*, [1-3]. The 2 Gy range represents an attempt at estimating the possible systematic uncertainty of stellar models whose use is unavoidable in the dating procedure [4], *i.e.* in determining the numerical coefficients in eq. (1.3). The sources of ±3 Gy *error* are discussed below. Many attempts at substantially reducing the age of globulars have failed, and this result should now be regarded as rather robust and taken seriously in a cosmological context. Yet, it is unavoidable to ask how accurate and reliable this determination is.

Equation (1.3) allows one to estimate the relative importance of the uncertainty in each of the four input quantities in establishing the total uncertainty in the age determination. Clearly the error in the distance of the clusters dominates. The current distances are typically affected by a 1/4 magnitude error in the modulus—$\sigma(\text{mod}) \simeq 0^m.25$—which immediately translates into a ~ 22% error in the derived cluster age (~ 3 Gy for an age of 15 Gy). All other input quantities convey substantially smaller errors. The high photometric accuracy of CCDs now allows one to determine a cluster's V^{TO} with an accuracy perhaps better than $0^m.1$, which translates into a ~ 9% error in age. The helium abundance is very well known, from either the R method, primordial nucleosynthesis, or empirical determinations of the *pregalactic* abundance, which all indicate $Y = 0.23$–0.24 [5]. Anyway, a ± 0.02 uncertainty in Y gives a negligible 2% error in age. I consider the metal content of the best-studied clusters to be uncertain by perhaps 0.3 dex, which translates into a ~ 9% uncertainty in age. There is a problem with the *composition* of metallicity (*e.g.*, enhanced [O/Fe], see [6]), but the possible effect on the estimated age is marginal. Clearly the *great villain* is the error in the distance of the cluster, and the question «*How good are globular-cluster ages?*» immediately becomes: «*How good are globular-cluster distances?*». There are three main standard candles that have been used (or are being considered) for the determination of accurate GC distances, namely, RR Lyrae, subdwarfs and white dwarfs. Pros and cons of each method are discussed elsewhere in some detail [2], suffice here a few schematic considerations.

RR Lyrae are rather bright, but rare objects, and even the nearest one is too distant for a trigonometric parallax to be obtained with an interesting accuracy. The calibration of this standard candle then relies on indirect physical or astrophysical methods, each making extensive use of pulsation, evolution and stellar-atmosphere models in some combination. Unfortunately, the more theory is used, the less we are able to quantify errors. Ultimately, even if we reach self-consistent distances (and then ages), we are left with the doubt that the models may have been slightly inaccurate in some way, thus introducing an unknown bias. After early enthusiasm, I later came to the rather skeptical conclusion that these methods may not pay the effort.

Having adopted a linear relation between the absolute magnitude of RR Lyrae and their metallicity ($M_V^{RR} = a[\text{Fe/H}] + b$), then the question becomes: «*how can we directly measure the slope (a) and the zero point (b)?*». A calibration of the zero point with $0^m.1$ accuracy would need trigonometric parallaxes with ~ 0.1 m.a.s. accuracy for a suitable sample of RR Lyrae stars, something we will not reach soon. The perspective looks more favourable for the slope a, which also plays an important rôle in the dating process. Some of the indirect methods give $a \simeq 0.35$, which would imply virtually coeval (to within $\lesssim 1$ Gy) globular clusters in our Galaxy, while other methods suggest $a \simeq 0.2$ which may imply a several Gy trend of the cluster age with metallicity if not compensated by a suitable trend in [O/Fe] (e.g., [3]). Therefore, disentangling among much different scenarios for the formation of the Galaxy ultimately relies on measuring the slope with better than $\sim 0^m.05$ accuracy per dex in [Fe/H]. Obtaining CMDs for the globular clusters in M31 would provide a direct determination of the slope, as such clusters are virtually all at the same distance, and span the full range of metallicity from [Fe/H] $\simeq -2$ to ~ 0. HST was supposed to provide the necessary data, but now we will have to wait for the recovery of its full imaging capabilities.

Subdwarf distances are currently based on five such stars with reasonably well-known trigonometric parallax [7]. The average error in the modulus of the five calibrators is $\langle \sigma(\text{mod}) \rangle = 0^m.15$, which is rather good. However, the MS location is sensitive to [Fe/H], and distance determinations by this method need the metallicity of the subdwarfs and of the cluster to be specified, and both are subject to errors. Unfortunately the five subdwarfs span a very narrow range in [Fe/H], and to extend the calibration to other metallicities one has to rely on theoretical ZAMS models (or on the main sequence of the Hyades cluster) and proceed to interpolations and extrapolations. When paying attention to the propagation of the errors, one finds that $\sigma(\text{mod}) \simeq \sigma[\text{Fe/H}]$, and, therefore, the relative error in age is nearly equal to the error in [Fe/H]: i.e. a 0.3 dex error in either the cluster or the subdwarf [Fe/H] propagates into an error of $\sim 30\%$ in age. All in all, while future efforts (and specially the Hipparcos mission) can improve the trigonometric-parallax determinations for very many subdwarfs, the problem with the uncertainty in the metallicity scale may remain.

The basic idea of using white dwarfs as standard candles is very simple: to fit the WD cooling sequence of a globular cluster to either the appropriate theoretical or empirical WD cooling sequence [2]. The procedure is analogous to the classical main-sequence fitting to the local subdwarfs, but with some nontrivial advantages: the method does not involve metallicity determinations which inevitably bring along their uncertainties, and there are no complications with convection. In fact, WDs have virtually metal-free atmospheres, coming either in the DA or non-DA varieties (nearly pure hydrogen or pure helium, respectively), and their radius is insensitive to the adopted mixing length. Moreover, WDs are locally much more abundant than subdwarfs, and, therefore, an accurate trigonometric parallax can be obtained for a much larger sample of calibrators. All in all—with HST working at nominal performance—the distance modulus of a cluster should be obtained with an accuracy better than $\sim 0^m.1$, which translates into a better than 10% accuracy in age. Unfortunately, WDs are very faint, and the usual ones are fainter than $V \simeq 24$ even in the closest globular cluster. The spherical aberration affecting HST makes presently impossible to reach such faint objects in the crowded field of the clusters, and again we may have to wait for the deployment of WFPC II, unless adaptive optics on large ground-based telescopes comes first. But I believe that, ultimately, white dwarfs will prove to be the best standard candles, able in fact to calibrate all other calibrators (RR Lyrae, subdwarfs, etc.).

In conclusion, unaberrated HST observations are required to reduce the age uncertainty from the present 25% down to $\sim 10\%$. If in the meantime one will get H_0 with the same accuracy, a fair fraction of the present cosmological ambiguities may be removed. In particular, it should be indicated whether a cosmological constant $\Lambda \neq 0$ is required. For the time being, it is worth emphasizing that our Milky Way is a spiral galaxy in a very loose group located rather away from major density peaks in the distribution of galaxies. Yet, she looks to be about one Hubble time old. Common sense would argue that ellipticals within high-density peaks should be even older, in any case.

2˙2. *Dating galaxies by evolutionary population synthesis.* – The first step in evolutionary population synthesis (EPS) is to construct the integrated spectrum of a SSP as a function of its age and composition:

$$(2.1) \qquad F_\lambda(t, Z) = F_\lambda^{ik},$$

where the indices i and k label age and metallicity, respectively. Having constructed such a spectral *library* for a suitable number of ik combinations, given a real galaxy spectrum F_λ^{GAL} one often attempts to find the matrix elements a^{ik} which minimize the discrepancies δ_λ:

$$(2.2) \qquad \delta_\lambda = \left| F_\lambda^{\text{GAL}} - \sum_{ik} a^{ik} F_\lambda^{ik} \right|,$$

and then pretends that the matrix a^{ik} represents the history of the galaxy, *i.e.* the time evolution of the star formation rate and the resulting chemical enrichment.

In this play there are three main characters. There is the observational aspect, where the whole point is to get a good spectrum of the target galaxy. There is the optimization procedure, which is also fairly trivial: commercial codes such as those aimed at optimizing the profit of supermarkets are perfectly adequate. Finally, there is the library, and the real thing is indeed to get adequate F_λ^{ik}'s. This is where the whole astrophysics is contained, after all, and, therefore, special attention must be paid to the construction of the population library. The most common limitations affecting existing libraries include:

1) Use of uncalibrated or erroneously calibrated isochrones. Stellar effective temperatures (and correspondingly the synthetic F_λ^{ik}) depend on our current, still very rough way of parametrizing the efficiency of the convective en-

ergy transfer, which is perhaps the most rudimentary of all ingredients entering in the construction of stellar models. Thus, when using uncalibrated theoretical isochrones to construct synthetic SED's and get ages, one hard-wires into the EPS clock the most obviously inadequate result of theoretical models: certainly not a clever way of proceeding. The calibration of the convection parameter is nothing else than the synchronization of the EPS clock [8]. Widely used isochrones such as the «Yale» and «revised Yale» ones are not adequately calibrated, and their use can generate errors in age by a factor of two or more [1, 8].

2) Missing evolutionary stages. In some existing EPS models advanced evolutionary stages such as, *e.g.*, the HB, AGB and P-AGB are not included at all. Yet these advanced stages contribute from $\sim 20\%$ to over 50% of the total light of a stellar population (see fig. 1.5).

3) Use of isochrones based on erroneous stellar models (see, *e.g.*, the case pointed out by BERTELLI, BRESSAN and CHIOSI [9]).

4) Incomplete spectral libraries. In particular missing the super-metal-rich component that exists in the bulge of our Galaxy and presumably exists in elliptical galaxies.

5) K giants replacing M giants in $Z \geqslant Z_\odot$ populations.

6) Use of solar elemental proportions ([M/Fe] = 0) assuming they are universal.

7) No attempt at including interacting binary stars, whose fuel consumptions are very different from those of single stars with the same mass.

None of the existing EPS's is completely immune from these diseases. Actually, all are affected by deficiencies 4)-7), and some are seriously undermined by 1)-3). I do not want to give here a too explicit *consumer guide* to existing EPS's, but the potential EPS user is advised to check items 1)-3) above before applying literature EPS spectral libraries to his/her own astrophysical problem, and to consider whether 4)-7) may have an impact on it.

Dating galaxies via EPS is an exercise one should undertake not with one, but two grains of salt. The first problem is that optimization methods are too smart, and want to give an answer anyway, no matter how bad the library is. At a first sight an optimization procedure looks something robust and impersonally objective, but we must be aware that in fact it is terribly effective in transmitting any of the 1)-7) genetic loads of the library F_λ^{ik} into the inferred galaxy history a^{ik}. The concerned astrophysicist should really beware of this.

The second problem arises even in the ideal case we would dispose of a perfect and complete library, since it is inherent in the intimate nature of the EPS clock itself. One should be aware that the core of the EPS dating machine is represented by the relation between the main-sequence turn-off temperature (colour) and age, such as, *e.g.*, eq. (1.4). This is so because the spectral energy distribution (SED) of a stellar population is primarily sensitive to the temperature distribution of the constituent stars, and for red giants such distribution is fairly insensitive to age, in particular in the case of old populations. I have already mentioned the difficulties introduced by convection and the need of a careful calibration, but convection is not the only concern when using the TO colour as an age indicator. As frequently emphasized, metallicity is another serious problem (see, *e.g.*, [8, 10]). Suppose in fact we have been able to determine the TO colour of a population with extreme precision, from either its CMD or its SED; to get the age we then need an independent estimate of the metallicity of the population. With little algebra, from eq. (1.4) one gets

(2.3) $$\left(\frac{\partial \log t}{\partial [\text{Fe/H}]}\right)_{(B-V)^{\text{TO}}} = -0.88 - 0.35 [\text{Fe/H}],$$

i.e. $\delta \log t \simeq \delta [\text{Fe/H}]$ for near solar metallicity, and a 0.3 dex error in the estimated metallicity translates into a factor-of-two error in age. A comparison with eq. (1.3) immediately reveals how much more vulnerable to metallicity errors ages estimated from TO colours are, compared to ages derived from the TO luminosity. In the case of colours the error is up to ten (!) times larger than in the latter case. This is a most disturbing intrinsic weakness of the EPS dating method, a drawback that simply cannot be eliminated: the interesting quantity to determine is age, but the SED of a population is more sensitive to metallicity (and to its detailed distribution) than it is to age.

Determining the age (distribution) of stars in nearby elliptical galaxies has been a widely entertained exercise in recent years, and many have claimed evidence for the bulk of stars being (5–8) Gy old, thus favouring a late formation of these galaxies, for example by merging spirals (see, *e.g.*, [10]). Others (*e.g.*, [8]) have argued that having neglected the metallicity dispersion that must exist in ellipticals(*) the optimization method will ask intermediate-age stars to provide the bluish light that in real galaxies is radiated by stars in the metal-poor half of the metallicity distribution. I find it hard to believe that the EPS method applied to individual galaxies can at once determine both the metallicity *and* age distributions with such an accuracy to really put interesting constraints on galaxy formation. I also believe that EPS is in real trouble even distinguishing between a solar-metallicity, 8 Gy old simple stellar population and a half-solar, 16 Gy one, not to say if the task is to find internal metallicity/age distributions. My impression is that, given the internal degeneracy of age and metallicity in the EPS clock (and given its other current deficiencies), the debate on whether

---

(*) In a closed-box model for chemical evolution one expects ellipticals to exhibit a ~ 2 dex metallicity dispersion [11]. Direct observations show that this is indeed the case for the stars in the galactic bulge [12].

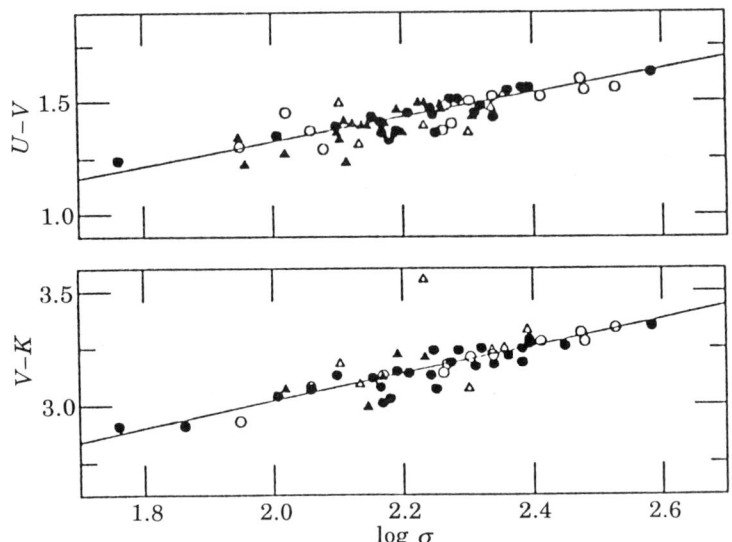

Fig. 2.1. – The $U-V$ and $V-K$ vs. velocity dispersion $\sigma_\star$, for ellipticals (circles) and S0 galaxies (triangles) in the Virgo and Coma clusters (open and filled symbols, respectively). Reproduced from [13].

ellipticals are definitely old or just middle aged may last forever, as frequently happens to ill-posed questions. In this case the problem arises from a very inefficient use of the EPS clock.

A much more clever use of the EPS clock is one in which one renounces to the absolute dating of each individual galaxy, but rather sets a useful limit to the age of groups of galaxies such as those in clusters, i.e. «We need to ask simpler questions and aim for *statistical* conclusions» [13]. In this mood BOWER, LUCEY and ELLIS [14] notice the remarkable homogeneity of the Virgo and Coma ellipticals, with their colour scatter for a given central-velocity dispersion $\sigma_\star$ being fully consistent with mere observational error (typically (0.03-0.04) mag). Figure 2.1 shows examples of such colour-$\sigma_\star$ relations. Were the formation of ellipticals a continuous process—such as in a late merging scenario—one would expect the scatter in formation epoch to translate into a scatter in colour for a given $\sigma_\star$. Clearly, the tightness of the colour-$\sigma_\star$ relation sets a limit on the age scatter of ellipticals. BOWER et al. put this argument on a quantitative basis, and write

$$\delta(U-V) = \frac{\partial(U-V)}{\partial t}(t_H - t_F),\tag{2.4}$$

where $\delta(U-V)$ is the scatter in colour resulting from the scatter in age ($t_H - t_F$), as it is assumed that galaxy formation started one Hubble time t_H ago, and was completed a time t_F ago. The partial derivative of the colour can be obtained from EPS models, which give $\partial(U-V)/\partial t \simeq (0.02\text{--}0.03)$ mag/Gy for $t \gtrsim 10$ Gy. Thus from the observed scatter in $U-V$ one gets $t_H - t_F \lesssim 0.04/0.02 = 2$ Gy, and for a Hubble time of 15 Gy one concludes that Virgo and Coma ellipticals ought to be older than $\sim 15 - 2 = 13$ Gy. It would be possible to have younger ellipticals only if their delayed formation were well *coordinated*, i.e. if there were a *golden age* for forming such galaxies. Thus, assuming that ellipticals formed only in the last fraction β of the $t_H - t_F$ time interval, BOWER *et al.* argue that the epoch of elliptical formation is constrained by

$$(2.5) \qquad t_H - t_F \lesssim \frac{0^m.04}{\beta}\left(\frac{\partial(U-V)}{\partial t}\right)^{-1},$$

which even with $\beta = 0.3$ still requires $t_F \gtrsim 10$ Gy. With the rather high synchronization of galaxy formation in this illustrative example, ellipticals would have formed in a ~ 2 Gy time interval centred ~ 5 Gy after the big bang. The more one attempts to move close to us this $\lesssim 2$ Gy lasting golden age of galaxy formation, the less plausible such a short period of activity becomes. Correspondingly, BOWER *et al.* conclude that «the main body of elliptical galaxies must have been formed beyond $z = 2$». Later additions should not provide more than 10% of the present luminosity.

It is really worth emphasizing that only a minimal use of EPS models has been made in order to reach such an important conclusion, which provides compelling evidence for an early formation of ellipticals. In practice, only the time derivative of the integrated $U-V$ colour has been used. This quantity is fairly insensitive to metallicity, and even EPS models which differ dramatically in other respects (such as those of Bruzual [15] and Buzzoni [16]) actually closely agree on this and other colour derivatives. This makes the conclusion about the age of cluster ellipticals very robust, just the opposite of the inconclusive case of a galaxy-by-galaxy optimization approach. One can really say that the optimization method is very poor in dating the trees, while the colour derivative approach is very effective in dating the forest.

Similar conclusions about the age of ellipticals could have been reached by considering other observational evidences, such as the tightness of the $Mg_2\text{-}\sigma_\star$ relation [17], or the narrowness of the distribution of ellipticals about the so-called *fundamental plane* (for the latest developments see [18,19]). For example, judging by eye from Bender's *et al.* fig. 1 the dispersion about the fundamental plane is $\delta \log L \lesssim 0.1$ (for given galactic mass). If we assume the IMF slope of ellipticals to be $x = 1.35$, then we get $\delta \log t \lesssim 1.1\, \delta \log L \simeq 0.11$, which implies an age dispersion less than ~ 3 Gy for a typical age $\gtrsim 10$ Gy. Again, this limit becomes narrower and narrower the more one wants to move forwards the epoch of elliptical formation.

All the considerations above must be only marginally relaxed when considering field galaxies. Indeed, in the field the correlations are not so tight as they are in clusters [20], which argues for some field elliptical being a genuine latecomer. Still, the majority of field ellipticals are indistinguishable from their cluster counterparts, which should imply they have quite similar ages.

2˙3. *Forming elliptical galaxies.* – As the tightness of the correlations helps constraining the epoch of elliptical-galaxy formation, also the trends of such correlations have profound implications. Indeed, the colour-$\sigma_\star$ and Mg_2-$\sigma_\star$ trends imply that stars formed *in situ*, because their metallicity (which correlates with colours and line indices) knows about the depth of the potential well ($\propto \sigma_\star^2$) in which they are contained. Thus, if merging was important in assembling ellipticals, such merging took place between mostly still gaseous units, and star formation and galaxy build-up proceeded simultaneously. Colour and line radial gradients within individual ellipticals support the same notion [21, 22].

A scenario in which stars form in small irregulars and spirals, these later merge, experience a dissipationless violent relaxation, and thus form larger and larger ellipticals does not really work. It may form roundish galaxies resembling real ellipticals in some (but not all, see [23]) structural detail, but such galaxies would not exhibit the observed colour-magnitude, colour-$\sigma_\star$ and line strength-$\sigma_\star$ relations. They would be dominated by the metal-poor stellar populations ($[M/H] \lesssim 0$) which are typical of today irregulars and spirals. They would also lack the observed radial gradients, unless a strong metallicity gradient was present already in the merging units, which still leaves open the question of how such gradients were established in the first place. A step sequence of star formation in small galaxies → merging of these stellar systems → dissipationless violent relaxation may be easy to simulate for the terrestrial mathematician, as at least starting from step two an N-body code will suffice. But the result is not the elliptical galaxies that we see: their stellar populations would be those of aged populations in dwarf irregulars or spirals, while giant ellipticals are so strongly lined that they must contain stellar populations which are german to them, with no counterparts anywhere in the realm of galaxies apart from the very central regions of galactic bulges.

These arguments do not exclude merging and dynamical violent relaxation as important processes in elliptical-galaxy formation. The implication is rather that star formation, metal enrichment, merging and violent relaxation did not take place *sequentially*; they more likely were all concomitant processes, along with supernova heating and radiative cooling in a multiphase ISM. In a word, elliptical-galaxy formation was a mess. A mess much more difficult to simulate for the terrestrial physicist, where all the uncertainties in the mentioned processes interfere with each other. Predictions become difficult to make under these circumstances, and simulations rather aim at *fitting* some properties of the observed galaxies, something *algebraically* not too difficult to achieve, given the number of fudge parameters. The *old-fashioned* wind models for elliptical-galaxy formation, such as those of Larson [24] and Arimoto and Yoshii [11], certainly oversimplify most of the mentioned physical processes, but have at least one virtue: they produce galaxies which exhibit colour-$\sigma_\star$ and similar correlations in qualitative agreement with the observations, a clear advantage over dissipationless merging models. Some aspects of the energetics of galaxy formation will be discussed in the third section.

In concluding this section, I cannot omit mentioning that several elliptical galaxies exhibit various types of isophotal distortions, ripples, gas disks, counterrotating cores and the like, which have been widely attributed to recent merging events [25]. Like old whales, giant ellipticals may indeed be infested by all sort of parasites, but this does not mean that ellipticals formed piecemeal by eating dwarfs. The mentioned tightness of the correlations above argues that accreting dwarfs had at most a marginal rôle in the formation of cluster ellipticals.

REFERENCES

[1] D. A. VANDENBERG: in *Astrophysical Ages and Dating Methods*, edited by E. VANGIONI-FLAN, M. CASSÉ, J. AUDOUZE and J. TRAN THANH VAN (Editions Frontières, Gif-sur-Yvette, 1990), p. 241.
[2] A. RENZINI: in *Observational Tests of Cosmological Inflation*, edited by T. SHANKS, A. J. BANDAY, R. S. ELLIS, C. S. FRENK and A. W. WOLFENDALE (Kluwer, Dordrecht, 1991), p. 131.
[3] B. CARNEY: *Mem. Soc. Astron. Ital.*, **63**, 409 (1992).
[4] R. T. ROOD: in *Astrophysical Ages and Dating Methods*, edited by E. VANGIONI-FLAN, M. CASSÉ, J. AUDOUZE and J. TRAN THANH VAN (Editions Frontières, Gif-sur-Yvette, 1990), p. 313.
[5] A. M. BOESGAARD and G. STEIGMAN: *Annu. Rev. Astron. Astrophys.*, **23**, 319 (1985).
[6] O. STRANIERO and A. CHIEFFI: *Astrophys. J. Suppl.*, **76**, 525 (1991).
[7] W. F. VAN ALTENA, J. T. LEE, R. B. HANSON and T. E. LUTZ: in *Calibration of Stellar Ages*, edited by A. G. D. PHILIP (L. Davis, Schenectady, N.Y., 1988), p. 175.
[8] A. RENZINI: in *Stellar Populations*, edited by C. A. NORMAN, A. RENZINI and M. TOSI (Cambridge University Press, Cambridge, 1986), p. 213.
[9] G. BERTELLI, A. BRESSAN and C. CHIOSI: *Astrophys. J.*, **392**, 522 (1992).
[10] R. W. O'CONNELL: in *Stellar Populations*, edited by C. A. NORMAN, A. RENZINI and M. TOSI (Cambridge University Press, Cambridge, 1986), p. 167.
[11] N. ARIMOTO and Y. YOSHII: *Astron. Astrophys.*, **173**, 23 (1987).
[12] R. M. RICH: in *The Stellar Populations of Galaxies*, edited by B. BARBUY and A. RENZINI (Kluwer, Dordrecht, 1992), p. 29.
[13] R. S. ELLIS: in *The Stellar Populations of Galaxies*, edited by B. BARBUY and A. RENZINI (Kluwer, Dordrecht, 1992), p. 297.
[14] R. G. BOWER, J. R. LUCEY and R. S. ELLIS: *Mon. Not. R. Astron. Soc.*, **254**, 613 (1992).
[15] G. BRUZUAL: *Astrophys. J.*, **273**, 105 (1983).

[16] A. BUZZONI: *Astrophys. J. Suppl.*, **71**, 817 (1989).
[17] R. BENDER: in *The Stellar Populations of Galaxies*, edited by B. BARBUY and A. RENZINI (Kluwer, Dordrecht, 1992), p. 267.
[18] S. DJORGOVSKI: in *Cosmology and Large Scale Structure in the Universe*, edited by R. R. DE CARVALHO, *ASP Conf. Ser.*, **24**, 19 (1992).
[19] R. BENDER, D. BURSTEIN and S. FABER: *Astrophys. J*, **399**, 462 (1992).
[20] R. R. DE CARVALHO and S. DJORGOVSKI: *Astrophys. J.*, **389**, L49 (1992).
[21] J. GORGAS, G. EFSTATHIOU and A. ARAGÓN SALAMANCA: *Mon. Not. R. Astron. Soc.*, **245**, 217 (1990).
[22] R. L. DAVIES, E. M. SADLER and R. F. PELETIER: *Mon. Not. R. Astron. Soc.*, **262**, 650 (1993).
[23] J. P. OSTRIKER: *Comments Astrophys.*, **8**, 177 (1980).
[24] R. B. LARSON: *Mon. Not. R. Astron. Soc.*, **166**, 586 (1974).
[25] F. SCHWEIZER, P. SEITZER, S. M. FABER, D. BURSTEIN, C. M. DALLE ORE and J. J. GONZALES: *Astrophys. J.*, **364**, L33 (1990).

Exercises.

Exercise 2.1: Show that the narrowness of the fundamental plane $\delta \log(M_\star/L) \lesssim 0.1$ implies a tight limit on the galaxy-to-galaxy variations of the IMF slope above M_{TO}. Assume all galaxies are 15 Gy old and all start at $t = t_0$ with the same $M_\star/L$ ratio.

Exercise 2.2: Suppose we live in a universe with $H_0 = 100$, $\Omega_0 = 1$, and $\Lambda = 0$. What would be the upper limit on the age scatter of ellipticals following the argument of Bower *et al.* [14]? Adopt $\partial(U-V)/\partial t = 0.05$ for $t \simeq 5$ Gy.

The answers to all the exercises can be found at the end of the lecture.

3. – The iron bound on elliptical-galaxy formation.

Supernovae make metals, and in particular they make iron, an element that nonexploding stars are unable to synthetize. Thus the total amount of iron in a portion of the Universe today (*e.g.*, a cluster of galaxies) keeps the record of the overall SN activity in the past, provided it has evolved as a *closed box*, at least in the sense of having not suffered any loss of mass. But SN's do not make only iron. For example, each SN injects a sizable amount of kinetic energy into the ISM, and each SN progenitor burns a definite amount of fuel and radiates the corresponding amount of photons. There is, therefore, a direct link between the total amount of iron and the integrated, past SN heating and stellar-light production. In this section I follow in some detail these connections, but I do so with no recourse to elaborate modelling—thus renouncing to their very detailed predictions—but with the advantage of keeping track in a most direct and transparent way of the implications of the various assumptions, on the one hand, and of the observed amount of iron in clusters, on the other. In doing so much less algebra will be used—but not necessarily less physics—than, *e.g.*, in detailed chemical and dynamical evolutionary models. The aim is to gather direct information and useful constraints on the formation processes of galaxies—especially of elliptical galaxies and bulges—and on their subsequent evolution. The rôle of SN activity in galaxy formation has been widely entertained in recent years, and along with it the notion that some kind of feedback process has contributed shaping galaxies as we see them today [1-7]. This section practically replicates a recent preliminary report of an ongoing investigation [8].

3˙1. *The iron-mass-to-light ratio in clusters of galaxies.* – In clusters of galaxies iron is partly dispersed in the intracluster medium (ICM), partly it is locked into the stellar component of galaxies. The interstellar medium (ISM) of individual galaxies also contains some iron, but its total amount is negligible compared to the two former contributions.

The amount of iron in the various components of a cluster can be expressed in several ways, the most traditional being the abundance, *i.e.* the mass fraction of iron in either the ICM or in stars, sometimes expressed in solar units. For our purposes this is not the most useful way of describing the iron content of clusters. For example, in the case of the ICM iron abundances depend not only on the total amount of iron produced, but also on its dilution with unpredictable amounts of pristine gas, an effect that we know to exist (*e.g.*, [9], [10] hereafter CDPR). ARNAUD *et al.* [11] have used the iron-mass-to-stellar-mass ratio, where the total iron mass in the ICM is divided by the total mass in stars present in the cluster. Operationally, the latter quantity comes from the observed total luminosity of the cluster, and assuming a stellar-mass-to-light ratio for individual galaxies, for which ARNAUD *et al.* adopt $M_\star/L_V = 10$. This is better than the straight iron abundance, but it requires the assumption of a stellar-mass-to-light ratio, and the relation between iron production and stellar mass is rather indirect. For example, two stellar populations differing only in the lower-mass cut-off of the IMF would produce the same amount of iron but would have different mass-to-light ratios. Moreover, the observed stellar-mass-to-light ratio comes from dynamical arguments and from the central velocity dispersion of galaxies, and it may, therefore, be contaminated by unknown amounts of dark matter. To avoid all these assumptions and complications, CDPR have adopted the concept of «*iron-mass-to-ligh ratio*» (IMLR), defined as the ratio of the iron mass (either in a specific cluster component, or total) to the total light of the cluster, *i.e.*

$$(3.1) \qquad \frac{M_{\text{Fe}}}{L_\text{B}} = \frac{M_{\text{Fe}}}{\sum_i L_\text{B}^i} \quad (M_\odot/L_\odot),$$

where the L_B^i's are the (blue) luminosities of the individual galaxies of the cluster. The IMLR is expressed in terms of quantities whose link with real observables is most direct, and presents several other advantages that will become apparent in the following subsections.

3'1.1. The ICM iron-mass-to-light ratio. The presence of vast amounts of iron in the ICM was theoretically anticipated by LARSON and DINERSTEIN [12], shortly before its actual discovery via the iron line emission in the X-ray spectra of galaxy clusters [13]. From literature data, CDPR have estimated $M_{\rm Fe}^{\rm ICM}/L_B$ values in the range from 0.7 to $1.6 \cdot 10^{-2} M_\odot/L_\odot$ for the nearby clusters Virgo, Coma and Perseus. This value is substantially confirmed when using the relevant input data from recent more extended compilations [11, 14], which indicate that the IMLR for the ICM of rich clusters is in the range

$$(3.2) \qquad \frac{M_{\rm Fe}^{\rm ICM}}{L_B} = 0.01\text{--}0.02 \quad (M_\odot/L_\odot),$$

which means that the ICM contains $(0.01\text{--}0.02) M_\odot$ of iron for every $L_\odot$ of the galaxy (star) population of the cluster. For example, a cluster such as Perseus has a total luminosity $L_B \simeq 10^{13} L_\odot$, and its ICM contains $\sim 0.015 \cdot 10^{13} = 1.5 \cdot 10^{11} M_\odot$ of iron.

One caveat is, however, in order: this value follows from the assumption of a chemically homogeneous ICM, an assumption which has been recently questioned as a gradient in the ICM iron abundance may be present, with iron being concentrated in the central regions [15, 16] (but see [17] for a different conclusion).

3'1.2. The stellar iron-mass-to-light ratio. The IMLR of individual galaxies (i.e. of their stellar component) is simply given by

$$(3.3) \qquad \frac{M_{\rm Fe}^\star}{L_B} = \langle Z^{\rm Fe} \rangle \frac{M_\star}{L_B},$$

where $\langle Z^{\rm Fe} \rangle$ is the average iron abundance (by mass) in the stellar component. This average refers to all stars in the cluster, while the iron abundance ranges over a wide interval within galaxies, and from one galaxy to another. Within individual galaxies metallicities span from $\sim 1/100$ solar for the extreme population-II component, to several times solar in the central regions of galactic bulges and in giant elliptical galaxies. Since here we are dealing with rich clusters of galaxies, I make the simplifying assumption that clusters contain only early-type galaxies, which is certainly a reasonable approximation. In practice, this is equivalent to ignoring spiral disks and dwarf irregulars. Early-type galaxies span *average* metallicities (iron abundances) from $\sim 1/3$ solar for dwarf ellipticals such as M32 [18] to a few times solar for giant ellipticals [19]. I assume the general average iron abundance for all the stars in a cluster of galaxies to be nearly *solar*, a reasonable estimate that cannot be off by more than a factor of ~ 2, either way. I correspondingly adopt $\langle Z^{Fe} \rangle \simeq Z_\odot^{Fe} \simeq 0.002$ [20] in eq. (3.3). For the stellar mass-to-light ratio I adopt typical values for early-type galaxies and bulges, *i.e.* $M_\star/L_B = 5\text{--}10$, and from eq. (3.3) one concludes

$$(3.4) \qquad \frac{M_{Fe}^\star}{L_B} = 0.01\text{--}0.02 \quad (M_\odot/L_\odot).$$

Comparing this estimate to eq. (3.2) it appears that *in rich clusters of galaxies there is nearly an equal amount of iron dispersed in the ICM, as there is locked into the stars within individual galaxies*. This is an important constraint for any theory of galaxy formation. Combining eqs. (3.2) and (3.4) one gets the *total* cluster IMLR:

$$(3.5) \qquad \frac{M_{Fe}}{L_B} = \frac{M_{Fe}^\star + M_{Fe}^{ICM}}{L_B} \simeq 0.03 \pm 0.01 \quad (M_\odot/L_\odot),$$

which means that for each $L_\odot$ of cluster blue light there are $\sim 0.03 M_\odot$ of iron, nearly evenly shared between stars and the ICM.

3˙2. *Supernova type Ia vs. type II iron production*. – In this subsection I discuss the amount of iron that SN's of type Ia and type II can produce in clusters of galaxies, and their relative rôle in establishing the observed IMLR. Following the conclusion of sect. 2, I take the view that elliptical galaxies in rich clusters are dominated by an old stellar population of age close to one Hubble time. This allows us to use the single-burst approximation for the whole stellar content of a cluster of galaxies, a welcome simplification.

3˙2.1. Type-Ia supernovae. The total amount of iron produced by SNIa's over one Hubble time (~ 15 Gy) is given by the total number of SNIa's which have exploded times the mass of iron produced by each SNIa. There is general consensus that a typical SNIa produces $\sim 0.7 M_\odot$ of iron (after the ^{56}Ni decay). This follows from the success with which the «W7» carbon deflagration model of Nomoto, Thielemann and Yokoi [21] accounts for both the light curve [22] and the spectrum [23,24] of SNIa's. I correspondingly adopt $M_{Fe}^I = 0.7 M_\odot$ for the amount of iron yield of each SNIa, while the yield for other abundant species (*e.g.*, O, Mg and Si) is negligible.

Much less definite is the situation concerning the total number of SNIa's, which in turn is given by the time integral of the SN rate. Up to now, observations tell only about the *present* rate, while for its time evolution over the preceding Hubble time no direct evidence is available. The present rate in ellipticals is estimated to be $0.88 h^2$ SNU [25,26], where $h = H_0/100$ and 1 SNU $= 10^{-12} L_B$ SN's y^{-1}, and L_B is the blue luminosity of the parent stellar population (*e.g.*, of one galaxy, of a cluster of galaxies, etc.). I adopt $h = 1/2$ and correspondingly the present rate becomes 0.22 SNU, or $2.2 \cdot 10^{-13} L_B$ SN's y^{-1}, a value whose observational uncertainty is estimated to be of the order of a factor of 2.

In the absence of direct observational evidence, on purely theoretical grounds the past evolution of the SNIa rate in ellipticals—and more generally its evolution following a burst of star formation—both remain highly conjectural. This is so for two main reasons: 1) because there is no general consensus on whether the single degenerate (SD) or the double degenerate (DD) model for the SNIa precursors applies [27], and 2) because in either model the evolution of the rate is controlled by an unknown distribution function describing either the mass distribution of the secondary binary components (SD model), or the distribution of the initial separation of the two white dwarfs in the DD model (see CDPR for an extensive discussion). Still, for both models the *qualitative* behaviour of the rate is similar: it departs from zero at least $\sim 3 \cdot 10^7$ y after the burst of star formation, climbs rather quickly to a maximum in $\sim (10^8\text{–}10^9)$ y, and then declines reaching the present value (0.22 SNU) one Hubble time after the burst. The precise run of the SNIa rate is, however, highly model dependent; examples of plausible evolutions of such rate are offered by the models of Greggio and Renzini [28] for the SD model, and Tornambè [29] for the DD model. Given these uncertainties, following CDPR I adopt a convenient parametrization for the rate:

(3.6) $$R_{\text{SNIa}}(t) = 2.2 \cdot 10^{-13} \vartheta_{\text{SN}} L_B t_{15}^{-s} \quad (y^{-1}),$$

for $t_{15} > t_{15,0}$, where t_{15} is time in units of 15 Gy, in such a way that for the parameter $\vartheta_{\text{SN}} = 1$ and $t_{15} = 1$ the standard SNIa rate in ellipticals is recovered, *i.e.* 0.22 SNU. For $t_{15} < t_{15,0}$ I assume the rate to increase linearly with time in such a way as to join the value given by eq. (3.6) at $t_{15} = t_{15,0}$, *i.e.*

(3.7) $$R_{\text{SNIa}}(t) = 2.2 \cdot 10^{-13} \vartheta_{\text{SN}} L_B t_{15,0}^{-(s+1)} t_{15} \quad (y^{-1}).$$

In the frame of the adopted parametrization, the time $t_{15,0}$ is the *rise time* of the SNIa rate, from zero to its maximum, and I shall explore values in the range between $1.5 \cdot 10^8$ and $1.5 \cdot 10^9$ y, or $t_{15,0} = 0.01\text{–}0.1$.

The total number N_{Ia} of SN's that have exploded over a time interval of 15 Gy is given by the integral of the rate, and from eqs. (3.6) and (3.7) one obtains

(3.8) $$N_{\text{Ia}} = \int_0^1 R_{\text{SNIa}}(t_{15}) \, dt_{15} = 3.3 \cdot 10^{-3} \vartheta_{\text{SN}} L_B f(s, t_{15,0}),$$

TABLE 3.1. – *Values of the function* $f(s, t_{15, 0})$.

$t_{15, 0}$	$s = 1.3$	$s = 1.4$	$s = 1.5$
0.01	11.9	16.4	23.0
0.05	6.1	7.4	9.2
0.10	4.3	5.0	5.9

where

$$(3.9) \qquad f(s, t_{15, 0}) = \frac{1}{2} t_{15, 0}^{1-s} + \frac{t_{15, 0}^{1-s} - 1}{s - 1},$$

and where the first term corresponds to the number of SNIa's exploded during the rate rise time, and the second term refers to the subsequent power law decline. The contribution of SNIa's to the IMLR follows naturally from eq. (3.8):

$$(3.10) \qquad \left(\frac{M_{\text{Fe}}}{L_{\text{B}}}\right)_{\text{SNIa}} = \frac{N_{\text{Ia}} M_{\text{Fe}}^{\text{Ia}}}{L_{\text{B}}} = 2.3 \cdot 10^{-3} \vartheta_{\text{SN}} f.$$

For a past SNIa rate constant in time one has $s = 0$, $t_{15, 0} = 0$, and so $f = 1$. With $\vartheta_{\text{SN}} = 1$ the corresponding IMLR is more than a factor of 10 smaller than the observed value: *had their rate been constant in the past, SNIa's would have contributed* ≲ 10% *of the observed iron in clusters of galaxies*. Since it is unlikely that the present rate has been underestimated by such a large factor, it follows that the past rate had to be significantly higher if SNIa's ought to contribute a fair fraction of the observed iron, *i.e.* for $\vartheta_{\text{SN}} = 1$ a value of $f \simeq 10$ is required. Formally, the SNIa contribution to the IMLR in eq. (3.10) depends on three parameters, of which only one (ϑ_{SN}) is directly determined by the observations. The function $f(s, t_{15, 0})$—which contains the other two parameters—is tabulated in table 3.1 for reasonable values of s and $t_{15, 0}$. It appears that values of the exponent s in excess of ~ 1.3 are to be preferred if one wants SNIa's to play a major rôle in iron production.

3˙2.2. Type-II supernovae. In the case of SNIa's we believe to have a fairly precise knowledge of the amount of iron released by each event, while the ambiguities affecting the progenitors make theory unable to predict the corresponding evolution of the SN rate. The case of type-II SN's is quite the opposite. In fact, we believe to have a perfect understanding of what are the progenitors (*i.e.* stars more massive than ~ $8 M_\odot$), while a great uncertainty affects the amount of iron $M_{\text{Fe}}^{\text{II}}(M)$ produced by each SNII event as a function of the initial mass of the progenitor. This follows from the current uncertainty in SN theory, with the iron delivered by a SNII event being crucially dependent on the precise position of the *mass cut* between the neutron star remnant and the

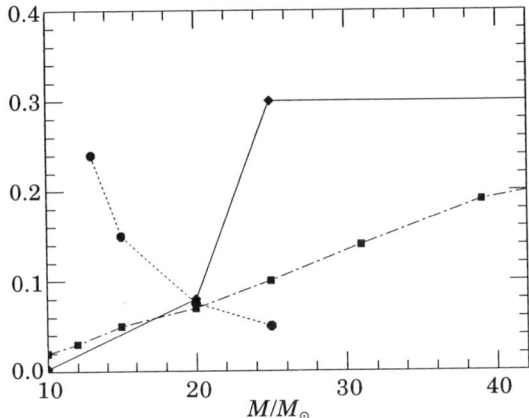

Fig. 3.1. – The iron yield of massive stars in $M_\odot$ units is displayed as a function of the initial mass, from three different sources. Solid line: WOOSLEY (private communication to WHITE [34]); dotted line: THIELEMANN and HASHIMOTO [35]; dot-dashed line: ARNETT [36].

ejecta, an extremely model-dependent quantity [30]. Figure 3.1 shows three options for $M_{\rm Fe}^{\rm II}(M)$. Not by chance do all the three options coincide for $M = 20 M_\odot$, the progenitor mass of SN1987A, for which the amount of iron produced—$0.07 M_\odot$—comes from the very precise determination obtained from fitting the light curve of this SN [31-33].

The total amount of iron produced by SNII's is obtained by integrating the yield per star $M_{\rm Fe}^{\rm II}(M)$ over the stellar IMF, or, equivalently, by the product of the mass-weighted average yield per star, times the total number of SNII's (*). Following the prescription in sect. 1, for the IMF I take

(3.11) $$\psi(M) = AM^{-(1+x)} = 2.9 L_{\rm B} M^{-(1+x)},$$

with $L_{\rm B}$—like everywhere in this lecture—being the luminosity of the stellar population at the conventional age of elliptical galaxies, i.e. 15 Gy. For the average iron production by SNII's one finds $\langle M_{\rm Fe}^{\rm II} \rangle \simeq 0.07 M_\odot$, fairly insensitive to the slope x of the IMF, and for all the options shown in fig. 3.1. Correspondingly, the SNII contribution to the IMLR becomes

(3.12) $$\left(\frac{M_{\rm Fe}}{L_{\rm B}}\right)_{\rm SNII} = \langle M_{\rm Fe}^{\rm II} \rangle 2.9 \int_8^{100} M^{-(1+x)} \, dM \simeq 0.2(8^{-x}/x) \quad (M_\odot/L_\odot).$$

(*) Here I assimilate to SNII producers also those massive stars which may have entirely lost their hydrogen envelope, and for this reason would be spectroscopically classified in SNI subtypes other than SNIa.

Clearly, the flatter the IMF slope the larger the number of massive stars per unit present luminosity, the larger the number of SNII's, and, therefore, the larger the implied IMLR. Thus from eq. (3.12) one gets

$$\left(\frac{M_{\rm Fe}}{L_{\rm B}}\right)_{\rm SNII} \simeq \begin{cases} 0.003 & \text{for } x = 1.7, \\ 0.009 & \text{for } x = 1.35, \\ 0.035 & \text{for } x = 0.9. \end{cases}$$

I conclude that, if the galactic IMF slope ($x = 1.7$ [37]) applies also to ellipticals in rich clusters, then SNII's underproduce iron by about a factor of 10 compared to what demanded by eq. (3.5). Instead, making all the observed iron by SNII's would require a very flat IMF ($x \simeq 0.9$).

3'2.3. *The solar supernova proportion.* – Before proposing a recipe for the production of the observed IMLR one needs to consider the evidence about the relative rôle of the SN types in galactic nucleosynthesis. There is certainly no *a priori* guarantee that such a rôle was the same in elliptical galaxies, but we should not renounce to explore the possibility of accounting for the cluster IMLR without appealing to exotic nucleosynthesis scenarios, which would inevitably introduce extra adjustable parameters.

It is well known that the O/Fe ratio in the population-II stars of the galactic halo is 3–4 times solar [38-40]. The most popular interpretation of the galactic trend of [O/Fe] *vs.* [Fe/H] appeals to the different nucleosynthesis rôles of SNIa's and SNII's: the enrichment of the halo was sufficiently rapid to sample almost exclusively the prompt release of SNII products (α-elements such as oxygen, r-process elements, but little iron), while it would take a longer time for SNIa's to release the bulk of their iron [28, 38, 41-45]. Correspondingly, in this scenario SNII's produce almost 100% of the iron in extreme population-II stars, and $\sim 1/4$ of the iron present in the Sun, while the residual $\sim 3/4$ comes from SNIa's. The same 1/4–3/4 proportion—that hereafter I call the *solar SN proportion*—would be expected for any system old enough to have already experienced the bulk of SNIa iron release, provided the IMF and the frequency of SNIa precursors are the same of the solar neighbourhood. Rich clusters of galaxies are likely to meet the first requirement: the bulk of their stellar content is very old, and, therefore, most of SNIa's should have already exploded by now. The likelihood of the second requirement is further discussed in the next subsections.

3'2.4. *Producing the observed IMLR.* – We are now in a position to attack the question of the cluster IMLR, for which I discuss three possibilities.

1) Making all the iron with SNIa's requires a strong evolution of the SN rate, with $f \simeq 15$ in eq, (3.10) if the canonical Tammann's rate applies to present-day ellipticals (*i.e.* $\vartheta_{SN} = 1$). From table 3.1 we see that large values of the exponent s and short rise times $t_{15,0}$ are favoured. Such values are consistent with those demanded by the evolutionary scenario for gas flows in ellipticals, in which the SNIa rate secularly declines and galactic flows evolve from an early supersonic wind, to a subsonic outflow, to eventually an inflow regime (CDPR). Such early winds are instrumental in establishing the observed even share of iron between stars and the ICM (see subsect. 3.2). However, no doubt some iron is also contributed by SNII's.

2) Making all the iron with SNII's requires star formation in cluster elliptical galaxies to have occurred with a very flat IMF ($x \simeq 0.9$), much flatter than the galactic IMF. Equivalently, one can appeal to a *bimodal* IMF, as argued by ARNAUD *et al.* [11], who dismiss SNIa's as major iron producers. In the bimodal IMF scenario quiescent star formation proceeds with a normal IMF while strong bursts of star formation produce only massive stars (*e.g.*, more massive than $\sim 2 M_\odot$). Quite obviously, in such a scenario one can generate an arbitrarily large IMLR, simply because the burst component produces iron from SNII's, but no light at all after 15 Gy as stars more massive than $2 M_\odot$ stars live less than ~ 1 Gy. Clearly, bimodal star formation can be tuned to produce the observed IMLR in rich clusters.

3) The third scenario allows for a contribution from both SN types, and does so in the *solar SN proportion*, with 1/4 of the iron in clusters coming from SNII's and 3/4 from SNIa's. In this case, indeed, we have that massive stars with a galactic IMF ($1.35 \lesssim x \lesssim 1.7$) contribute for about 1/4 to the cluster IMLR, while with $\vartheta_{SN} f \simeq 10$ SNIa's contribute the residual 3/4. In this way, with the same combination of parameters x for massive stars and $\vartheta_{SN} f$ for SNIa's, one accounts for the galactic chemical evolution as well as for the IMLR in clusters: a clear advantage for this scenario. Moreover, with this choice of parameters one satisfactorily accounts also for the present X-ray properties of elliptical galaxies and one preserves the early winds necessary to replenish the ICM with iron (CDPR). Although very attractive, this scenario requires independent confirmation, as several ingredients are in fact rather uncertain (*e.g.*, the iron yield from SNII's [46]). I shall further explore its implications in the following subsections.

3'3. *Supernova heating vs. binding energy of galaxies.* – Each SN injects into the ISM $\sim 10^{51}$ erg of kinetic energy, irrespective of the SN type. Knowing how much iron is in a cluster, and the average iron yield per SN, we also know how much kinetic energy has been injected into the ISM during the formation of ellipticals or shortly thereafter. From eq. (3.5) we get the number of SN's that have exploded, as the ratio of the total amount of iron in the cluster, over the average iron yield of each SN:

$$(3.13) \quad N_{SN} = \frac{M_{Fe}}{\langle M_{Fe} \rangle} = \frac{0.03 L_B}{\langle M_{Fe} \rangle} = \begin{cases} 0.04 L_B & \text{for all iron from SNIa's}, \\ 0.40 L_B & \text{for all iron from SNII's}, \\ 0.14 L_B & \text{for the \emph{solar SN proportion}}, \end{cases}$$

respectively for the three possibilities discussed in paragraph 3'3.4. Correspondingly, the kinetic energy injected is given by

$$(3.14) \quad E_{SN} = 10^{51} N_{SN} \; (\text{erg}) = \begin{cases} 0.4 \cdot 10^{50} L_B & \text{for all iron from SNIa's}, \\ 4.0 \cdot 10^{50} L_B & \text{for all iron from SNII's}, \\ 1.4 \cdot 10^{50} L_B & \text{for the \emph{solar SN proportion}}. \end{cases}$$

Note that in these relations L_B can be interpreted as the luminosity of a whole cluster, or of an individual galaxy. This latter interpretation is adopted in the rest of this subsection.

Certainly part of this kinetic energy has to be dissipated and radiated away during galaxy formation, while the rest can be used to drive gas out of galaxies, thus interrupting the star formation process, establishing galactic winds, and replenishing the ICM with iron. One way of figuring out how *big* is E_{SN} is to ask how much baryonic mass it would be able to extract from the galactic potential well. For the bright- and dark-matter distributions used in CDPR, the energy required to extract one $M_\odot$ from the potential well of the galaxy is given by

$$(3.15) \quad E^{LIFT}(M_\odot) \simeq 2 \cdot 10^{44} (0.34 + 0.09 R) \sigma_\star^2 \; (\text{erg}),$$

where $R = M_{dark}/M_\star$ is the galactic dark-matter-to-star-matter ratio, and $\sigma_\star$—the galactic central velocity dispersion—is in km s^{-1} units (*). By good fortune, in this class of models E^{LIFT} is fairly insensitive to the actual distribution of dark matter, in so far as it is definitely less centrally concentrated that the shining matter.

Clearly, the ratio of the available energy E_{SN} to the energy required to lift one $M_\odot$ gives the total mass that could be extracted from the potential well of a galaxy, provided all the available energy could be used. Thus from eqs. (3.14) and (3.15) we can derive the *liftable-mass-to-light ratio* (LMLR):

$$(3.16) \quad \frac{M^{LIFT}}{L_B} = \frac{E_{SN}}{L_B E^{LIFT}} \simeq \frac{5 \cdot 10^5 F_{SN}}{(0.34 + 0.09 R) \sigma_\star^2} \quad (M_\odot/L_\odot),$$

(*) This eq. (3.15) comes from CDPR's equation (15) where $\sigma_\star$ was expressed in cm s^{-1} (see also [47]). Here I have further adopted the ratio of the dark- to bright-matter core radii to be 10, and the ratio of the tidal to core radii to be 140.

TABLE 3.2. – SN-*energy-to-galaxy-binding-energy* ratios.

$\sigma_\star$ (km s^{-1})	L_B ($L_\odot$)	$E_{SN}/E_\star^{BIN}$			
		$R = 0$ (^{a})	$R = 9$ (^{a})	$R = 0$ (^{b})	$R = 9$ (^{b})
200	$2.0 \cdot 10^{10}$	29	9	10	3.0
300	$8.5 \cdot 10^{10}$	13	4	5	1.3
400	$2.0 \cdot 10^{11}$	7	2	3	0.8

(a) For all iron from SNII's.
(b) For the solar SN proportion.

where the numerical factor F_{SN} depends on the mix of SNIa's to SNII's, and is equal to 0.4, 4.0 and 1.4, respectively, for the three cases discussed in paragraph 3˙3.4, see also eq. (3.14). Furthermore, the ratio of this LMLR to the usual stellar mass-to-light ratio gives

$$(3.17) \qquad \frac{M^{LIFT}}{L_B} \frac{L_B}{M_\star} = \frac{M^{LIFT}}{M_\star} \simeq \frac{E_{SN}}{E_\star^{BIN}},$$

which is the ratio of the SN energy to the present binding energy of the bright (sometimes called baryonic) component of galaxies. This ratio is given in table 3.2 for $M_\star/L_B = 5$, for two values of the dark-to-bright-matter ratio R, and for the two options 1) all iron from SNII's and 2) the solar SN proportion. Here L_B and $\sigma_\star^2$ are connected by the Faber-Jackson relation.

A glance at table 3.2 reveals several aspects which are relevant for our understanding of galaxy formation.

1) Were most of the iron in the Universe manufactured by massive stars (SNII's) and were ellipticals deprived of dark matter, then the energy released by SN's would exceed the present binding energy of galaxies by more than an order of magnitude.

2) Allowing for 90% of galactic mass to be in a dark-matter halo reduces this figure by nearly a factor of ~ 3, with the SN energy still exceeding the binding energy of the bright matter by a factor between 2 to 9.

3) The SN-to-binding-energy ratio is similar to the previous case if no dark matter is present, but $\sim 3/4$ of the iron in the Universe is manufactured by SNIa's.

4) When SNIa's make $\sim 3/4$ of the iron, *and* $\sim 90\%$ of galactic mass is in a dark-matter halo, then the early SN heating becomes comparable or smaller than the binding energy of the stellar component, and from these facts we can derive the following qualitative inferences.

5) To make the (baryonic component of) galaxies one has to radiate away a fair fraction of the SN energy, which is at least as large as the binding energy of this component.

6) As mentioned in sect. 2, SN-driven wind models for the formation and early chemical evolution of elliptical galaxies [48-50] may be inadequate in many details, but remain the only ones able to establish (at least qualitatively) a trend of average metallicity (as measured, *e.g.*, by the Mg_2 index) with the depth of the galactic potential well (as measured by $\sigma_\star^2$), as is observed. This basically comes from the Faber-Jackson relation, which implies that matter is more tightly bound in more luminous galaxies, with star formation and chemical enrichment proceeding further in them before SN heating succeeds in establishing a global wind.

7) The numbers in table 3.2 enlight that in this class of models the detailed results must be very sensitive to the particular SNIa/SNII mix that is adopted, to the amount of dark matter assumed around protogalaxies, and especially to the treatment of radiative losses, as a major fraction of SN heating has to be radiated away. This has to be a major concern for SN-driven wind models for elliptical-galaxy formation.

8) Given these difficulties, we are just left with the rather obvious consideration that it must be easier to make galaxies the larger the share of iron made by SNIa's, the larger the amount of dark matter around protogalaxies, and the larger the ability to cool of the ISM of forming galaxies. All in all, we are left with an even stronger suspicion than before that some feedback, self-regulating mechanism must be at work during galaxy formation [5].

3˙4. *The iron-light connection.* – Given the rôle played by the actual SN mix in galaxy formation and evolution, and in establishing the IMLR of clusters of galaxies, additional clues that may help to distinguish among the various possibilities are certainly welcome. One such clue comes from considering the total amount of light that SN precursors emit in the whole course of their evolution, before finally delivering their iron yield. There is in fact a dramatically different behaviour in this respect between the two types of SN's.

We assume the typical SNII iron producer to be a $20 M_\odot$ star, which delivers $0.07 M_\odot$ of iron, and exhausts hydrogen in a $\sim 6 M_\odot$ core before exploding. Conversely, SNIa's deliver 10 times more iron, and result from the explosion of a C-O white dwarf which is driven to exceed the Chandrasekhar limit, *i.e.* SNIa precursors need to exhaust hydrogen in only a $\sim 1.4 M_\odot$ core. In conclusion, the iron-yield-to-light-pollution ratio is some $(0.7/0.07) \cdot (6/1.4) \simeq 40$ times higher for SNIa's than it is for SNII's: protogalaxies would be much brighter if most of the iron is made by SNII's, rather than by SNIa's.

We can put this argument on more quantitative grounds. Since the burning of 1 g of material with primordial composition releases $\sim 6 \cdot 10^{18}$ erg, the amount of early (photon) energy release produced by SNII precursors is given by the amount of energy released by burning $6M_\odot$ of fuel times the number of SNII's that have contributed to the observed IMLR. Using eq. (3.5), and adopting the solar SN proportion, we easily get

$$(3.18) \qquad E_{\mathrm{ph}} \simeq 7 \cdot 10^{51} L_{\mathrm{B}} \, (\mathrm{erg}),$$

for the energy radiated by SNII precursors during the formation phase of galaxies. Here, again, L_{B} (in solar units) can be interpreted as either the present luminosity of a whole cluster, or of one individual elliptical. For example, a today $L_\star$ elliptical galaxy with $L_{\mathrm{B}} \simeq 5 \cdot 10^{10} L_\odot$ had to radiate some $4 \cdot 10^{62}$ erg during its formation. In general, for a star formation time scale τ_{SF}, the bolometric luminosity of forming galaxies was

$$(3.19) \qquad L = E_{\mathrm{ph}}/\tau_{\mathrm{SF}} \simeq 2.5 \cdot 10^3 (\tau_{\mathrm{SF}}/10^8 \, \mathrm{y})^{-1} L_{\mathrm{B}},$$

i.e. galaxies were up to a few thousand times brighter than they are today. Clearly, this figure would be a factor of 4 larger (or protogalaxies ~ 1.5 mag brighter) if most of the iron was made by SNII's, rather than in the solar SN proportion.

Here we are concerned with purely energetic arguments. The actual, present spectrum of such bright protogalaxies is clearly affected by several poorly-known factors: reprocessing to longer wavelengths due to gas and dust absorption within the forming galaxies themselves, and redshift by an amount z_{GF}, the redshift of galaxy formation. A discussion of these aspects goes beyond the scope of the present lecture (but see [7], for a recent review on the search for primeval galaxies).

3˙5. *The galaxy-ICM chemical asymmetry*. – The elemental ratios of various nuclear species in the ICM and in the stellar component offer another possibility of checking whether the integrated SN activity in clusters of galaxies has—or has not—been characterized by the solar SN proportion. At the same time, such ratios may offer useful information on the typical time scale of star formation in early elliptical galaxies in clusters.

Ratios such as [O/Fe], [Mg/Fe] and [Si/Fe] may be different in the stellar component of galaxies from what they are in the ICM, and the size of this *galaxy-ICM chemical asymmetry* would have important implications for our understanding of the galaxy formation process. Again, this follows from the fact that elements almost exclusively produced by SNII's (*e.g.*, O, Mg and Si) are released on a time scale $\sim \tau_{\mathrm{SF}}$, since the lifetime of SNII precursors is very short, while most of iron is realeased at a much slower pace following the evolution of the rate of SNIa's. So, the particular «*chemical trajectory*» traced by the stellar component of young ellipticals in, *e.g.*, the [O/Fe] *vs.* [Fe/H] plane depends on the relative size of τ_{SF} and a typical time scale τ_{SNIa} for SNIa's [51]. The shorter τ_{SF} relative to τ_{SNIa}, the higher the stellar [O/Fe], [Mg/Fe] and [Si/Fe] for given [Fe/H]. Yet, neither of the two times scales is *a priori* known, a condition which leaves them as mere adjustable (or hidden) parameters in chemical-evolution models (*).

In this scenario for the formation of elliptical galaxies, star formation ceases after $t \simeq \tau_{SF}$, as gas starts flowing out of galaxies into the ICM. The composition of this gas reflects the average stellar abundances, further enriched *in fly* by SNIa's. It follows that the shorter the time scale of galaxy formation, the less SNIa iron remains locked into stars, and a larger fraction of it flows into the ICM. If the solar SN proportion applies to clusters as a whole, then one predicts [O/Fe], [Mg/Fe] and [Si/Fe] ratios *above solar* in the stellar component, and *below solar* in the ICM, the larger the size of this chemical asymmetry, the shorter τ_{SF} with respect to τ_{SNIa}. In the limit $\tau_{SF} \ll \tau_{SNIa}$, all the SNIa iron would flow in the ICM, and one would have [O/Fe] = [Mg/Fe] = [Si/Fe] $\simeq$ $\simeq +0.6$ within galaxies, while the same ratios would be negative in the ICM. However, the value of [O/Fe], [Mg/Fe] and [Si/Fe] in the ICM cannot be predicted with simple arguments such as those developed so far, as it depends on the fraction of the original baryonic mass of the galaxy which has participated in the nucleosynthesis and which is later ejected from the galaxy. For example, if $1/3 M_\star$ is ejected, this mass will contain 1/3 of the whole Mg, 1/3 of the Fe from SNII's (which is $1/3 \times 1/4 = 1/12$ of the total), plus the whole iron from SNIa's, which is 3/4 of the total. Thus all in all the ICM would contain 5/6 of the total iron but only 1/3 of the total magnesium, or [Mg/Fe] = -0.4. On the other hand, by construction the overall cluster Mg/Fe ratio is solar, where by overall ratio one means the total mass of magnesium in the ICM *and* galaxies, over the corresponding mass of iron.

The observational check of this predicted galaxy-ICM chemical asymmetry raises formidable technical problems, first of due to by the inevitable *asymmetry* in experimental procedures between optical and X-ray-based observations, with the risk that the real effect is masked by differential systematic errors. The determination of the elemental abundances (and masses) in galaxies can only come from pupulation synthesis techniques; the raw data are already available and indeed there is a hint for Mg being enhanced compared to Fe [52]. For the ICM, X-ray line observations are required, but no data for elements other

---

(*) The definition itself of τ_{SNIa} is unavoidably ambiguous, given the particular shape of the SNIa rate past a burst of star formation. For example, following an analogy with radioactive decay, τ_{SNIa} could be defined as the time after the burst of star formation when half of SNIa's have exploded. In the adopted parametrization—eqs. (3.6) and (3.7)—τ_{SNIa} is a function of $t_{15,0}$ and s that can be easily evaluated.

than iron are presently available. The only exception I am aware of is a very uncertain estimate of O/Fe in Virgo and Perseus [53, 54], according to which O/Fe would be 3–5 times solar. Such a large ratio would definitely rule out SNIa's as important iron producers in clusters of galaxies as argued by WHITE [34], who makes an attempt at reproducing it with chemical-evolution models. However, his solution implies an IMLR ~ 0.0003, or ~ 100 times smaller than the observed value. The conclusion is that within our current knowledge of stellar nucleosynthesis there is no way of reconciling a high O/Fe ratio in the ICM with the observed IMLR (and with the solar O/Fe ratio!). Future X-ray observations will tell: though difficult, the observational check of the predicted chemical asymmetry would be of great value for our understanding of galaxy formation and evolution.

REFERENCES

[1] J. P. OSTRIKER and L. COWIE: *Astrophys. J.*, **243**, L127 (1981).
[2] S. IKEUCHI and J. P. OSTRIKER: *Astrophys. J.*, **301**, 502 (1986).
[3] A. DEKEL and J. SILK: *Astrophys. J.*, **303**, 39 (1986).
[4] S. WHITE: in *The Epoch of Galaxy Formation*, edited by C. FRENK, R. S. ELLIS, T. SHANKS, A. F. HEAVENS and J. A. PEACOCK (Kluwer, Dordrecht, 1989), p. 15.
[5] T. M. HECKMAN, L. ARMUS and G. K. MILEY: *Astrophys. J. Suppl.*, **74**, 833 (1990).
[6] S. IKEUCHI and C. NORMAN: *Astrophys. J.*, **375**, 479 (1991).
[7] S. DJORGOVSKI: in *Cosmology and Large Scale Structure in the Universe*, edited by R. R. DE CARVALHO, *ASP Conf. Ser.*, **24**, 73 (1992).
[8] A. RENZINI, L. CIOTTI, A. D'ERCOLE and S. PELLEGRINI: in *Structure, Dynamics, and Chemical Evolution of Early-Type Galaxies*, edited by I. J. DANZIGER, W. W. ZEILINGER and K. KJÄR (ESO, Garching, 1992), p. 443.
[9] F. MATTEUCCI and G. VETTOLANI: *Astron. Astrophys.*, **202**, 21 (1988).
[10] L. CIOTTI, A. D'ERCOLE, S. PELLEGRINI and A. RENZINI: *Astrophys. J.*, **376**, 380 (1991).
[11] M. ARNAUD, R. ROTHENFLUG, O. BOULADE, L. VIGROUX and E. VANGIONI-FLAM: *Astron. Astrophys.*, **254**, 49 (1992).
[12] R. B. LARSON and H. L. DINERSTEIN: *Publ. Astron. Soc. Pac.*, **87**, 511 (1975).
[13] R. J. MITCHELL, J. L. CULHANE, P. J. DAVISON and J. C. IVES: *Mon. Not. R. Astron. Soc.*, **175**, 29 (1976).
[14] T. TSURU, Y. IKEBE, T. OHASHI, K. MAKISHIMA, I. HATSUKABE, K. YAMASHITA, W. FORMAN and M. ARNAUD: in *Frontiers of X-Ray Astronomy*, edited by Y. TANAKA and K. KOYAMA (Universal Academy, Tokyo, 1992), p. 485.
[15] T. J. PONMAN, D. BERTRAM, M. J. CHURCH, C. J. EYLES, M. P. WATT, G. K. SKINNER and A. P. WILLMORE: *Nature (London)*, **347**, 450 (1990).
[16] K. KOYAMA, S. TAKANO and Y. TAWARA: *Nature (London)*, **350**, 35 (1991).
[17] J. P. HUGHES, J. A. BUTCHER, G. C. STEWART and Y. TANAKA: *Astrophys. J.*, **404**, 611 (1993).
[18] W. FREEDMAN: *Astrophys. J.*, **94**, 1285 (1989).

[19] R. W. O'CONNELL: in *Stellar Populations*, edited by C. NORMAN, A. RENZINI and M. TOSI (Cambridge University Press, Cambridge, 1986), p. 167.
[20] E. ANDERS and N. GREVESSE: *Geochim. Cosmochim. Acta*, **53**, 197 (1989).
[21] K. NOMOTO, F.-K. THIELEMANN and K. YOKOI: *Astrophys. J.*, **286**, 644 (1984).
[22] K. NOMOTO and T. SHIGEYAMA: in *Supernovae*, edited by S. E. WOOSLEY (Springer, New York, N.Y., 1991), p. 572.
[23] D. BRANCH, J. B. DOGGETT, K. NOMOTO and F.-K. THIELEMANN: *Astrophys. J.*, **294**, 619 (1985).
[24] R. P. HARKNESS and J. C. WHEELER: in *Supernovae*, edited by A. G. PETSCHEK (Springer, New York, N.Y., 1990), p. 1.
[25] G. TAMMANN: in *Supernovae: A Survey of Current Research*, edited by M. REES and R. STONEHAM (Reidel, Dordrecht, 1982), p. 371.
[26] S. VAN DEN BERGH and G. TAMMANN: *Annu. Rev. Astron. Astrophys.*, **29**, 363 (1991).
[27] U. MUNARI and A. RENZINI: *Astrophys. J.*, **397**, L87 (1992).
[28] L. GREGGIO and A. RENZINI: *Astron. Astrophys.*, **118**, 217 (1983).
[29] A. TORNAMBÈ: *Mon. Not. R. Astron. Soc.*, **239**, 771 (1989).
[30] S. E. WOOSLEY and T. A. WEAVER: *Annu. Rev. Astron. Astrophys.*, **24**, 205 (1986).
[31] S. E. WOOSLEY: *Astrophys. J.*, **330**, 218 (1988).
[32] Y. SHIGEYAMA, K. NOMOTO and M. HASHIMOTO: *Astron. Astrophys.*, **196**, 141 (1988).
[33] D. ARNETT and A. FU: *Astrophys. J.*, **340**, 396 (1989).
[34] R. R. WHITE III: *Astrophys. J.*, **367**, 69 (1991).
[35] F.-K. THIELEMANN, K. NOMOTO, T. SHIGEYAMA, T. TSUJIMOTO and M. HASHIMOTO: in *Elements and the Cosmos*, edited by R. J. TERLEVICH, B. E. J. PAGEL, R. CARSWELL and M. EDMUNDS (Cambridge University Press, Cambridge, 1991), p. 68.
[36] D. ARNETT: in *Frontiers of Stellar Evolution*, edited by D. L. LAMBERT, *ASP Conf. Ser.*, **20**, 389 (1991).
[37] J. M. SCALO: *Fundam. Cosmic Phys.*, **11**, 1 (1986).
[38] C. SNEDEN, D. L. LAMBERT and R. W. WHITAKER: *Astrophys. J.*, **234**, 964 (1979).
[39] R. GRATTON and S. ORTOLANI: *Astron. Astrophys.*, **169**, 201 (1986).
[40] B. BARBUY: in *The Stellar Populations of Galaxies*, edited by B. BARBUY and A. RENZINI (Kluwer, Dordrecht, 1992), p. 143.
[41] L. GREGGIO and A. RENZINI: *Mem. Soc. Astron. Ital.*, **54**, 311 (1983).
[42] F. MATTEUCCI and L. GREGGIO: *Astron. Astrophys.*, **154**, 279 (1986).
[43] C. ABIA, R. CANAL and J. ISERN: *Astrophys. J.*, **366**, 198 (1991).
[44] K. NOMOTO *et al.*: in *Elements and the Cosmos*, edited by R. J. TERLEVICH, B. E. J. PAGEL, R. CARSWELL and M. EDMUNDS (Cambridge University Press, Cambridge, 1991), p. 55.
[45] F. MATTEUCCI and P. FRANÇOIS: *Astron. Astrophys.*, **262**, L1 (1992).
[46] S. E. WOOSLEY and T. A. WEAVER: preprint (1992).
[47] L. CIOTTI and S. PELLEGRINI: *Mon. Not. R. Astron. Soc.*, **255**, 561 (1991).
[48] R. B. LARSON: *Mon. Not. R. Astron. Soc.*, **166**, 586 (1974).
[49] N. ARIMOTO and Y. YOSHII: *Astron. Astrophys.*, **173**, 23 (1987).
[50] F. MATTEUCCI and A. TORNAMBÈ: *Astron. Astrophys.*, **185**, 51 (1987).
[51] A. RENZINI: in *Stellar Populations*, edited by C. NORMAN, A. RENZINI and M. TOSI (Cambridge University Press, Cambridge, 1986), p. 213.
[52] R. L. DAVIES, E. M. SADLER and R. F. PELETIER: *Mon. Not. R. Astron. Soc.*, **262**, 650 (1993).

[53] C. R. CANIZARES, G. W. CLARK, J. G. JERNIGAN and T. H. MARKERT: *Astrophys. J.*, **262**, 33 (1982).
[54] C. R. CANIZARES, T. H. MARKERT and M. E. DONAHUE: in *Cooling Flows in Clusters and Galaxies*, edited by A. FABIAN (Kluwer, Dordrecht, 1988), p. 63.

Exercises.

Exercise 3.1: In this section it was assumed that ellipticals are 15 Gy old. What would become the predicted contribution of SNII's to the IMLR if instead one assumed ellipticals to be only 5 Gy old? Assume $x = 1.35$.

Exercise 3.2: Suppose that Prof. Gustav TAMMANN has overestimated by a factor of 3 the rate of type-Ia supernovae in elliptical galaxies. What should be the exponent s in eq. (3.6) in order for the solar SN proportion to be preserved for clusters of galaxies? Assume a SNIa rate rise time of 150 million years ($t_{15,0} = 0.01$).

Answers to all the exercises.

Exercise 1.1: There are $3600 \times 6 = 21\,600\,\square''$ in the FoV of the CCD camera. Each $\square''$ samples an absolute magnitude $M_B = 21 - 25 = -4$, which corresponds to a luminosity $L_B = 10^{-0.4(-4 - M_{B,\odot})} = 6309 L_\odot$ (having used $M_{B,\odot} = 5.5$). Thus the camera samples a total $L_B = 21\,600 \times 6309 = 1.36 \cdot 10^8 L_\odot$ population. Multiplying by the bolometric correction factor (~ 2.4) we get $L_T = 3.3 \cdot 10^8 L_\odot$. One now enters eq. (1.10):

$$N_{PN} = B(t) L_T t_j = 2 \cdot 10^{-11} \times 3.3 \cdot 10^8 \times 5 \cdot 10^3 = 33,$$

i.e. there are 33 PN's in the FoV.

Exercise 1.2: From eq. (1.10) one has $t_{PN} = N_{PN}/B(t) L_T = 775/(2 \cdot 10^{-11} \times 2.7 \cdot 10^{10}) \simeq 1500$ y, having adopted $L_T = 2.4 L_B$ (see L. GREGGIO and A. RENZINI: *Astrophys. J.*, **364**, 35 (1990)).

Exercise 1.3: The globular is $\sim 10^{10}$ y old, so we have $B(t) \simeq 2 \cdot 10^{-11}$, and from eq. (1.15) we get $L_{RGB}/L_T \simeq 2 F_{RGB}$. The RGB fuel consumption can be obtained from tables of RGB models such as those of Sweigart and Gross (*Astrophys. J.*, **36**, 405 (1978)) from which one gets $F_{RGB} = 0.22 M_\odot$ of hydrogen, and, therefore, RGB stars contribute $\sim 2 \times 0.22 \times 100 = 44\%$ of the total cluster light (in bolometric).

Exercise 1.4: The IMF scale factor is $2.9 \cdot 10^{11}$, and one integrates the IMF from $8 M_\odot$ (minimum mass to get a SNII) to, say, $100 M_\odot$:

$$N_{\rm SNII} = 2.9 \cdot 10^{11} \int_8^{100} M^{-2.35} \, dM = 1.3 \cdot 10^{10} \text{ SNII's},$$

which is a rather respectable number, as discussed in sect. 3.

Exercise 1.5: Let $\tilde{t}_j$ be the time taken by a star with $M = M_{\rm TO}$ to evolve from the turn-off to the beginning of phase j. At time t, the mass of stars just entering phase j is, therefore, $M_j \simeq M_{\rm TO}(t - \tilde{t}_j)$, and the evolutionary flux through phase j is $b_j(t) \simeq \psi(M_j)\dot{M}_j = b(t - \tilde{t}_j)$. The luminosity contribution of post-MS stars is, therefore,

$$L_{\rm PMS} \simeq 9.75 \cdot 10^{10} \sum_j b_j(t) F_j(M_j).$$

Exercise 2.1: As shown in subsect. 1˙3, the galaxy luminosity (passively) evolves as

$$L = L_0(M_\star)(t/t_0)^{-(4-x)/3}.$$

Since the present $M_\star/L$ ratio is still the same to within $\delta \log(M_\star/L) \lesssim \pm 0.1$,

$$\delta \log(M_\star/L) \simeq \delta(x/3) \log(t/t_0) \lesssim 0.1$$

or $\delta x \lesssim 0.3/\log(t/t_0) \simeq 0.2$, having assumed $t_0 = 0.5$ Gy, which is a remarkable result. As a supplement to this exercise, discuss the limits set by the thickness of the fundamental plane on the IMF *below* $M_{\rm TO}$.

Exercise 2.2: In this case the Hubble time is only $t_{\rm H} \simeq 6.6$ Gy. Correspondingly we have $t_{\rm H} - t_{\rm F} \lesssim 0.04/0.05 = 0.8$ Gy.

Exercise 3.1: Given the luminosity evolution of a SSP (cf. exercise 2.1), for given IMF (and, therefore, for given total number of SNII's) the luminosity of a 5 Gy old galaxy would be $\sim 3^{(4-x)/3} = 2.6$ times higher (for $x = 1.35$). The predicted IMLR would be lower by the same factor. Assuming galaxies are younger than 15 Gy would make it more difficult to produce the observed IMLR. Since most of the iron is made at early times, making galaxies younger just brightens them *now* for a given amount of iron produced.

Exercise 3.2: In paragraph 3˙2.4 we have seen that in the case of the solar SN proportion a value of $\vartheta_{\rm SN} f = 10$ is required. For $\vartheta_{\rm SN} = 1/3$ this requires $f = 30$, and extrapolating from table 3.1 we have $s \simeq 1.6$. The lower the present SNIa rate, the higher it should have been in the past.

Note added in proofs.

These lecture notes were completed in November 1992. Here are a few additions updating some of the topics.

The idea in exercise 2.1 is further expanded in a paper by RENZINI and CIOTTI (*Astrophys. J.*, **416**, L49 (1993)). Among other topics, the limits set by the narrow thickness of the fundamental plane on the age and IMF dispersions among elliptical galaxies are discussed.

An expanded version of the third section can be found in RENZINI et al. (*Astrophys. J.*, **419**, 52 (1993)). In the lecture as well as in the paper a Hubble constant $H_0 = 50$ was assumed. However, both the iron mass in the ICM and that within galaxies depend on the actual value of the Hubble constant, and do so in different ways. When properly taking this effect into account, the total cluster IMLR becomes

$$\frac{M_{\text{Fe}}}{L_{\text{B}}} \simeq 0.01 h^{1/2} + 0.03 h \quad \left(\frac{M_\odot}{L_\odot}\right),$$

where the first and second terms refer to the contribution of the ICM and galaxies, respectively, and $h = H_0/100$. Thus, if for $h = 1/2$ one has and IMLR of $0.015 + 0.015 = 0.03$, as adopted in paragraph 3˙1.2, then for $h = 1$ one would have an IMLR $= 0.01 + 0.03 = 0.04$. One can conclude that the total IMLR is not very sensitive to the adopted value of the Hubble constant, but the relative fraction of iron in the ICM and within galaxies is instead rather sensitive to it. In the above example, in one case the amount of iron in the two components is nearly the same, in the other case galaxies beat the ICM 3 to 1. I am grateful to R. TERLEVICH for having asked the question of the dependence of the IMLR on the Hubble constant.

At the time of writing this addendum (April 1994) the ASCA satellite has been collecting X-ray data for over one year. This includes a great deal of exciting data on clusters of galaxies that cannot be discussed here. Preliminary results have been reported at the meeting *New Horizon of the X-ray Astronomy - First Results from* ASCA held in Tokyo, March 1994, and at the Moriond meeting *Clusters of Galaxies*, March 1994.

The Role of Star Formation in Galaxy Formation.

J. SILK

Departments of Astronomy and Physics, and Center for Particle Astrophysics
University of California at Berkeley - Berkeley, Cal.

1. – Introduction.

Star formation is the key to understanding galaxy formation. Perhaps the easiest starting point is with disk galaxies, where one has the advantage of being able to apply an extensive body of knowledge of disk instability, both local and global, that has been used to study star formation in spiral arms. Moreover the present rate of star formation in gas-rich disks is not grossly different from that in the past, thereby giving one some confidence in the possibility of extrapolating to earlier epochs.

Gravitational instabilities in a self-gravitating gas-rich disk are long-lived provided that the gas heating is balanced by gas dissipation[1] or cold-gas infall[2], and they result in spiral density waves. These nonaxisymmetric instabilities are generated by amplification of stochastic noise via swing amplification[3], by tidal interaction with a companion, or via torquing by a central stellar bar or triaxial halo. This latter feature may have been present since formation of the galaxy or else be produced during a merger[4]. The density waves, greatly amplified by the inelastic response of the gas, are believed to be the principal trigger for galactic star formation.

Negative feedback must play a central role in star formation. Locally, molecular clouds form stars over many ($\sim$ 10–100) cloud free-fall times. Globally, the galactic star formation rate has somehow conspired not to exhaust the gas reservoir over the age of the disk, $\sim$ 50 dynamical times. Attempts to understand the rate of star formation must evidently incorporate feedback. Feedback models have previously been advocated to infer the rates both of massive-star formation (*e.g.*, [5,6]) and low-mass-star formation [7,8]. Disk self-regulation is undoubtedly an important and effective ingredient. A realistic model for the disk star formation rate must explicitly include feedback to the gas.

I describe below a formalism for the galactic star formation rate, in which gas dissipation and cooling maintains a cold unstable disk that continuously forms stars via gravitational instability acting on the gas component. A multicomponent model for the interstellar medium is incorporated into the gas distribution, and the feedback is accomplished by supernova heat input, which self-regulates the multiphase gas disk. Some of these ideas were presented in an earlier discussion [9], and a more detailed exposition of the star formation rate in disk galaxies is given elsewhere [10]. I describe here an extreme version of the feedback mechanism with application to starbursts. Applications are also described to the Tully-Fisher relation, galaxy evolution, galaxy clustering and starbursts. I conclude by reviewing the implications for galaxy formation.

2. – Gravitational instability of cold disks.

The condition that the Toomre parameter $Q_g \approx \kappa \sigma_g / \pi G \mu_g \lesssim 1$ demarcates the onset of instability to local axisymmetric perturbations of a thin, differentially rotating, gaseous [11] or stellar [12] disk: in the latter case, π is replaced by 3.36, μ_g by μ_*, and σ_g by the stellar-radial-velocity dispersion. Here κ is the disk epicyclic frequency, equal to $2^{1/2} \Omega (2 + d \ln \Omega / d \ln r)$ for rotation rate $\Omega(r)$, σ_g is the gas velocity dispersion, μ_g is the gas surface density, and μ_* is the stellar surface density. A comparable criterion is also an important (although not unique) indicator of disk instability to global nonaxisymmetric perturbations, resulting in bar formation or swing amplification of spiral density waves.

For the more realistic case of a galactic disk containing both gas and stars, usually with the gas scale height and velocity dispersion being less than that of the stars, the definition of Q may be generalized. Incorporation of the stellar component leads to a more complicated expression of the general form $1/Q \propto \mu_g / \sigma_g^2 + \mu_* / \sigma_*^2$: it should be evident that the cold gas will dominate the disk instabilities until the gas is depleted. One can use this expression to introduce a correction factor f that allows for the role of stars in weakening the instability by writing $\bar{\kappa} = f \kappa$; see [9, 10] for an explicit definition of f.

At the solar galactocentric radius, 8.5 kpc: $\mu_g \approx 13 M_\odot$ pc^{-2}, $\mu_* \approx 50 M_\odot$ pc^{-2}, $\sigma_g \approx 6$ km s^{-1}, $\sigma_r \approx 40$ km s^{-1}, and $\kappa = 1.37 \cdot 10^{-15}$ s^{-1}; hence $f \approx 0.45$. One can write $Q_g \equiv \mu_{cr} / \mu_g$, $\mu_{cr} = \bar{\kappa} / \pi G \approx 2^{1/2} f \Omega \sigma_g / \pi G = 8 M_\odot$ pc^{-2} at the solar circle, where $d \ln \Omega / d \ln r \approx -1$. The inferred local value of Q_g is 0.63: hence the disk is unstable at $r \lesssim 8$ kpc to axisymmetric gravitational instabilities. These are the plausible precursors to molecular-cloud formation, coagulation and fragmentation, resulting in star formation, as discussed in the following section. At large r, the gas scale height increases, μ_g falls, and the disk is stabilized.

3. – Star formation rate in disks.

The SFR is controlled by the rate at which gas clouds form and collapse. A cold stellar disk containing H I and H$_2$, and undergoing differential rotation, is unstable to axisymmetric density perturbations and nonaxisymmetric shear perturbations that concentrate the gas into dense clouds. The gas complexes grow by coagulation of smaller clouds, as discussed below, become gravitationally unstable and collapse once the magnetic flux-limited Jeans mass is exceeded [13], subsequently forming stars. Star formation can also proceed in smaller cloud cores that are below the magnetic Jeans mass, these clouds suffering flux loss by ambipolar diffusion and undergoing flux-limited contraction [14]. However, the flux diffusion rate is primarily controlled by the ionization level in the cold cloud, and, therefore, by processes mostly associated with molecule formation and photodissociation that control cloud opacity [15] as well as ionizing radiation and energetic particles generated by massive-star formation [16]. Feedback from massive stars further restricts the process to be inefficient by directly heating the gas, most of the gas being recirculated into the more diffuse HI/H_2 medium. Supernova-induced triggering of star formation can also occur in special circumstances [17], but appears not to be the predominant phenomenon regulating star formation in quiescent disks, although the situation may be quite different in starbursts (sect. 4). I, therefore, assume that supernova and wind heating is balanced by dissipation and cooling of the gas, maintaining the instability until the gas supply is depleted.

A simple model for the formation rate and radial distribution of molecular clouds [18] and for the global rate of star formation rate in disks [19] led to the expression $SFR = \varepsilon \mu_g \Omega$. This expression was found to satisfactorily account for several observational correlations. These included disk metallicity gradients, the relationship between metallicity and age for the solar vicinity, and the dependence of current star formation rate on gas content for different galaxy types. The Ω-dependence played a critical role in the model, as star formation proceeded more rapidly in the inner galaxy and in earlier morphological types, with more rapid rotation.

Precisely how nonaxisymmetric disk instabilities drive cloud growth has been studied by means of numerical simulations (*e.g.*, [20, 21]), but the relevant physics is fairly straightforward. Cloud collisions at $\lesssim 3$ times the cloud escape speed are nondestructive and result in cloud growth [22]. Gas dissipation, for example via cloud-cloud collisions, is strongly enhanced by noncircular orbital motions. The nonaxisymmetric modes, in particular via spiral density waves [23] and, more vigorously, via the formation of stellar bars [4], drive growth of large gas complexes that become Jeans unstable and then collapse to inefficiently form stars. The ensuing star formation is inefficient because, if stars form with a conventional Miller-Scalo IMF, ionizing photons from massive stars will dissociate much of the cold diffuse molecular gas before it has time to collapse.

A semi-phenomenological expression that improves on the preceding discussion by attempting to incorporate the relevant physics, at least locally, via the introduction of a star formation threshold to yield the global rate of star formation, is given by [9] SFR = $\varepsilon \mu_g / t_{inst}$. Here ε is an *empirically* determined efficiency for converting gas into stars, and the linear growth rate for disk instabilities can be very approximately expressed as $t_{inst}^{-1} \sim \pi (G \mu_g / \sigma_g)(1 - Q_g) \sim$ $\sim \sqrt{2} \Omega (1 - Q_g)/Q_g$ in linear theory with $Q_g \lesssim 1$ and a flat rotation curve. A study by KENNICUTT [24] of the radial dependence of the Hα flux in galactic disks has demonstrated that it is necessary to incorporate such a star formation threshold to account for the observations. Indeed, this is implicit in the Toomre criterion, which can be rewritten as $Q_g \lesssim 1 \Leftrightarrow \mu \gtrsim \mu_{cr} \equiv \kappa \sigma_g / \pi G = 8.3 \, \overline{\kappa}_\odot \sigma_6 M_\odot$ pc^{-2}, where $\overline{\kappa} \equiv 6 \cdot 10^{-16} \, \overline{\kappa}_\odot \, s^{-1}$ and $\sigma_g = 6 \sigma_6$ km s^{-1}. One can have disk instability even when the gas is subdominant, provided it is sufficiently cold. The surface density of cold gas (H I + H$_2$) remains near the critical value where star formation is occurring, and drops below μ_{cr} (or $Q_g > 1$) outside the radius where H II regions are observed. Near the Sun, $\overline{\kappa}_\odot = \sigma_6 = 1$, and the surface density of interstellar gas adds up to $\mu_g = 13 M_\odot$ pc^{-2}, including H I (8.3 $M_\odot$ pc^{-2}), H$_2$ (2.8 $M_\odot$ pc^{-2}) and H II (1.7 $M_\odot$ pc^{-2}). I conclude that the galactic gas disk is just unstable against nonaxisymmetric modes ($Q_g = 0.64$), but stabilizes outside the solar circle.

The theoretically calculated star formation efficiency (SFE) ε ranges from 5 percent [25] to (1–2) percent [26]. The efficiency may be inferred from the observational data via two approaches. Direct observation of star-forming molecular clouds suggests that $\varepsilon \approx 0.01$ in large cloud complexes, increasing to $\varepsilon \approx 0.03$–0.04 in typical cloud clumps and to $\varepsilon \approx 0.4$ in the most massive clumps [27]. Approximately 50 percent of the cloud mass may not be in clumps. A general estimate compares the observed star formation time scale, t_*, defined as the ratio of molecular gas mass, either locally or globally, to the corresponding observed star formation rate, to the time scale t_i for production of ionizing photons by OB associations. The latter time scale gives the destruction time for molecular clouds, the former the time scale for complete conversion into stars. With $t_i \approx 5 \cdot 10^6$ y and $t_* \approx 5 \cdot 10^8$ y (*e.g.*, [15]), one obtains $\varepsilon = t_i / t_* = 0.01$.

A further complication arises when the cold interstellar gas, treated as an ensemble of clouds, is compressed by passage through spiral arms. Within the inner spiral arms, the gas velocity dispersion is increased by a factor $\lesssim 2$ for an increase in surface density by a factor of 2 [26], thereby reducing Q_g by a factor of between 2 and 4 within (4–8) kpc. Hence the gas should be strongly unstable, especially in the inner spiral arms: moreover κ increases by 2 between 8 and 4 kpc, and μ_g rises by a similar factor.

Initially, when the disk is gas-rich, self-gravity of the gas should be capable of enforcing disk self-regulation, $Q_g \approx$ const. The preceding expression for the star formation rate is clearly deficient in that it takes no account of any feedback from stellar energy sources. If stars were to form at a sufficiently great rate, one would restrict the supply of gas. In terms of Q_g, the velocity dispersion of the gas would be enhanced by energy input from massive stars, and one would expect Q_g to consequently increase and thereby tend to stabilize the disk against forming stars. Conversely, if the star formation rate drops sufficiently, the gas velocity dispersion would diminish and thereby further destabilize the gas. The preferred mechanism may involve the competing effects of gas in destabilizing as well as in stabilizing spiral instabilities via infall of cold gas [2] or by gas dissipation [28]. The detailed coupling is most likely a combination of expanding H II regions, stellar winds and supernova remnants, which heat the gas, raise σ_g, increase the scale height, and reduce μ_g.

I shall take a more heuristic approach here that results in a star formation law with a weaker than quadratic dependence on the gas surface density: e.g., when any explicit dependence on σ_g is incorporated into Q_g, one obtains

$$\text{(1)} \qquad \text{SFR} = \varepsilon \mu_g \bar{\kappa} Q_g^{-1}(1 - Q_g).$$

I emphasize again that this relation is *empirical*: the microphysics of conversion of molecular clouds into stars is treated as a black box. The key input parameter is star formation efficiency ε, for which I now develop a simple model that explicitly incorporates supernova feedback, in order to enforce self-regulation by controlling the disk scale height. This may be achieved by writing

$$\text{(2)} \qquad \text{SFR} \cdot p_{\text{SN}} = \mu_g \sigma_g \Omega (1 - Q_g).$$

Here $p_{\text{SN}} = 2 E_{\text{SN}} (v_c m_{\text{SN}})^{-1}$ is the final specific momentum deposited into the interstellar medium by a supernova remnant of initial kinetic energy $10^{51} E_{51}$ erg, and $v_c = 208\, E_{51}^{1/14} Z^{-3/14} n^{1/7}$ km s^{-1} is the cooling velocity at which a supernova remnant expanding into a uniform medium of density n and metallicity Z (relative to solar) first enters the approximately momentum-conserving phase [29]. The mean mass in stars that forms for each supernova is m_{SN}, a typical numerical value being $m_{\text{SN}} \equiv 200\, m_{200} M_\odot$ with $m_{200} \approx 1$ for a present-day star formation rate of $\sim 6 M_\odot$ y^{-1} and disk supernova rate of ~ 0.02 y^{-1}. This assumes that about 30 percent of the mass involved in forming stars is returned to the interstellar medium. One obtains $p_{\text{SN}} = 900\, E_{51}^{13/14} n^{-1/7} m_{200}^{-1}$ km s^{-1} and $\varepsilon \approx (\sigma_g / p_{\text{SN}}) Q$. Thus the global SFE, equivalent to ε in the linear law, is also controlled by the gas velocity dispersion, σ_g. I infer that, in the feedback-dominated regime, the star formation rate is linear and decreases in proportion to μ_g: the e-folding time scale for gas depletion is $\sim (p_{\text{SN}}/\sigma_g) \Omega^{-1} \sim 150\, \sigma_6 \Omega^{-1}$. A satisfactory present-epoch gas density within the solar circle is now obtained, with $Q_g \approx$ const and $\varepsilon \simeq 0.01\, \sigma_6 m_{200} E_{51}^{-13/14} n^{1/7}$, an efficiency that is consistent with the feedback hypothesis as well as with direct observation of molecular-cloud complexes. At the solar circle, with a gas return fraction $R = 0.3$, one has $t_* = 5 \cdot 10^9$ y, and

$$\text{(3)} \qquad \text{SFR} = 3(\mu_g/13\, M_\odot\, \text{pc}^{-2}) \bar{\kappa}_\odot\, \varepsilon_{0.01} M_\odot\, \text{pc}^{-2} \text{Gy}^{-1},$$

where $\varepsilon_{0.01} \equiv \varepsilon/0.01$, in agreement with the observed value.

4. – Starbursts.

One can imagine that, if the star formation rate were greatly perturbed, disk self-regulation would break down. This is the situation, presumably, in a starburst, by definition taken to be a greatly elevated rate of star formation that is likely to be triggered by pressure enhancement of the interstellar medium. A high-pressure environment would include gas cloud interactions in galaxy mergers that utilize tidal interactions and torques to drive noncircular orbital velocities of gas clouds. The resulting inelastic encounters help concentrate gas within the central kiloparsec of the merged galaxy [30]. The gas pressure consequently rises by an order of magnitude or more relative to the fiducial local interstellar-medium value $p/k = 3600$ cm^{-3} K [31]. I shall argue that the high pressure restricts the volume swept out by supernova remnants, allowing more gas to accumulate in the cold, star-forming phase that controls the star formation rate. Supernovae thereby trigger further star formation, a process that may be actually required to guarantee efficient massive-star formation [32]. The process is self-limiting once the supernova rate has increased sufficiently that the porosity of the hot phase of the interstellar medium, associated with the swept-out remnant bubbles, becomes excessive.

To provide a simple model of this phenomenon, I utilize the 3-phase interstellar-medium model [33]. Consider a three-component model for the interstellar medium, consisting of cold ($T \lesssim 100$ K) clouds, diffuse warm ($T \sim (10^3$–$10^4)$ K) H I and hot ($T \sim 10^6$ K) intercloud gas. I identify μ_g with the H I/H$_2$ component, and μ_h with the hot intercloud medium. This latter phase has a large volume-filling factor f_v^h but contains a small mass fraction, f_m^h, where $f_m^h = 1 - f_m^c = \mu_h/(\mu_g + \mu_h)$. As the star formation and star death rates increase, supernova remnants begin to overlap and produce a hot coronal medium that pressurizes interstellar clouds. When the hot medium becomes sufficiently porous and dominates the volume, the cold-gas fraction, and hence the star formation rate, will diminish. The volume-filling factor of the hot component is prescribed in terms of the supernova rate via $f_v^h = 1 - e^{-P}$, where the porosity $P = S\nu_{SN}$, S is the supernova rate (in pc^{-3}y^{-1}), and ν_{SN} is the 4-volume (pc^3 y) occupied by an old supernova remnant that is halted by pressure of the ambient interstellar medium. From [15], $S = 4 \cdot 10^{-13}$ pc^{-3}y^{-1} for the Milky Way at the solar galactocentric distance, and the 4-volume

$$(4) \qquad \nu_{SN} = 7.82 \cdot 10^{12} \, p_4^{-1.36} n^{-0.11} E_{51}^{1.26} Z^{-0.204} \beta^{-2.72} \text{ pc}^3 \text{y}$$

is filled during the late stages of the expansion of a spherically symmetric supernova remnant of initial energy $10^{51} E_{51}$ erg into a uniform medium of density n (cm^{-3}), metallicity relative to solar abundance Z, sound velocity c_s, velocity dispersion $\sigma_g = \beta c_s$ and gas pressure $p_g/k \equiv 10^4 \, p_4$ cm^{-3}K.

I will assume that the cold-gas filling factor can be estimated by this idealized model that utilizes spherically symmetric expansion of supernova remnants into a uniform medium. It is likely in reality that the detailed mass spectrum and fate of the clouds that are overtaken by the supernova remnant bubbles will affect global estimates of porosity, but I suppress such considerations here. Pressure balance estimates that determine the final volume of the expanding remnants should be approximately valid in an inhomogeneous medium. However, the expression for disk star formation rates requires specification of the cold-gas mass fraction, f_m^c, whereas the porosity only yields the volume-filling factor f_v^h.

To cure this, I note that the cold-gas mass and volume fractions are related by $f_m^c \mu_g / H_g = f_v^c \mu_{cl} / R_{cl}$, where μ_{cl} is the surface density of a molecular cloud of radius R_{cl}, H_g is the cold disk gas scale height, and $f_v^c = 1 - f_v^h$. Empirically, it is found that galactic molecular clouds obey simple scaling laws $\sigma_{cl} \approx 0.7(R_{cl}/1\,\mathrm{pc})^{0.5}\,\mathrm{km\,s^{-1}}$, this relation being normalized to the giant molecular clouds [34] and somewhat dependent on assumptions about cloud density profiles [15, 35]. Such clouds are in near virial equilibrium, having a surface density $\mu_{cl} = 143\,M_\odot\,\mathrm{pc}^{-2}$, that has a dispersion measured to be as low as 10 percent. Typical extinctions amount to $A_v \approx 7.5$; for a spherical cloud, $\mu_{cl} = 22.3 A_v M_\odot \mathrm{pc}^{-2}$. This value of the extinction applies to warm star-forming cloud complexes: cold clouds, without any massive-star formation, characteristically have $A_v \approx 3.0$ [36]. Uncertainty in the effective value to be adopted for μ_{cl} means that f_m^c is correspondingly uncertain.

Let $\alpha_0 \equiv \langle \rho_{cl}/\rho_{gas} \rangle$ be the ratio of characteristic cloud density to average density of the cold disk gas, $f_v^c = e^{-P}$, and $R_{cl} \approx 50(M_{cl}/10^6\,M_\odot)^{1/2}$ according to the observed scaling. Utilizing a value of μ_g ($13\,M_\odot\,\mathrm{pc}^{-2}$) and scale height H_g (135 pc) appropriate to the cold (and warm)-gas components of the galactic disk at the solar circle, one obtains

$$(5) \qquad f_m^c = \begin{cases} 1 & (P < 3.4), \\ \alpha_0 e^{-P}, \ \alpha_0 \approx 30 & (P > 3.4). \end{cases}$$

The porosity may be large, driving a hot-gas filling fraction $f_v^h \approx 0.95$ for $f_m^c \approx 1$. There is a negligible mass fraction in the hot-gas component.

Cloud growth and, eventually, star formation occur if the disk is unstable to local axisymmetric perturbations, a condition that requires $Q \equiv \mu_{cr}/\mu_g f_m \lesssim 1$. This occurs if one has enough cold gas, which in turn requires that the porosity not be too large: $P \lesssim \mathcal{O}(1)$.

On the other hand, if $f_m \ll 1$, the triggering of star formation drives high porosity and ensuing gas outflows via a supernova-driven wind. With excessive porosity, $P \gtrsim \mathcal{O}(1)$, one can self-regulate star formation by disrupting the gas supply for star formation. Gas flows will develop initially via multiple supernova-driven chimneys [37] or supershells [38-40]. A typical value for $\langle \rho_{cl}/\rho_{gas} \rangle$ is 10, and $\langle \mu_0/\mu_{cr} \rangle$ may be of order a few. I conclude that self-regulation during a starburst restricts the porosity to $P \approx 2\text{-}3$.

The expression for star formation in the disk becomes a function of the porosity, which is proportional to the SFR via $P = \mathrm{SFR}\, \dot{\nu}_{\mathrm{SN}}/4Hm_{\mathrm{SN}}$. Here H is the scale height of the supernovae. The supernova rate per unit area of the disk is taken to be $S_{\mathrm{2D}} = \mathrm{SFR}/m_{\mathrm{SN}}$, and is related to the supernova rate per unit volume by $S = S_{\mathrm{2D}}/4H$. The latter expression incorporates a factor of 2 reduction because only about half of the supernovae are randomly located in the disk (cf. [41]). A typical numerical value is $H \approx 150 \,\mathrm{pc}$ (as appropriate for the scale height of the youngest pulsars).

The global star formation rate is obtained by using an effective disk area of 840 kpc^2 (appropriate for a disk exponential scale length of 4 kpc and a galactocentric distance of 8.5 kpc). This integrates to give

$$\dot{M}_* \approx 27 (\alpha_4^{-2} H_{150}\, m_{200}\, p_4^{1.36}\, n^{0.11})\, M_\odot\, \mathrm{y}^{-1} \tag{6}$$

for the global star formation rate with disk exponential scale length $\alpha^{-1} \equiv 4\alpha_4^{-1}$ kpc and scale height $H = 150\, H_{150}\, \mathrm{pc}$.

The negative-feedback aspect is now explicit: as the SFR increases, P increases, and the surface density of cold gas $f_m \mu_0 \propto e^{-P}$ drops below $\mu_{\mathrm{crit}} = \bar{\kappa} \sigma_g/\pi G$, thereby cutting off the star formation via disk instabilities and self-regulating the star formation rate. I infer that pressure enhancement can greatly stimulate the SFR.

5. – Implications for galaxies and galaxy evolution.

I now consider some rather speculative applications of the foregoing star formation rate derivations for galaxies, both local and distant.

5`1. *Tully-Fisher relation*. – The quiescent mode should be applicable to ongoing star formation in disk galaxies, with $\dot{M}_* = \int \mathrm{SFR}\, 2\pi r\, dr$. Consider a disk with total surface density (gas plus stars) satisfying

$$\mu_{\mathrm{d}} = \frac{\alpha^2}{2\pi} M_{\mathrm{d}} \exp[-\alpha r]. \tag{7}$$

The disk mass is M_{d}; the halo mass contribution within a disk scale length α^{-1} will be neglected. The maximum rotation velocity is attained somewhere in the range $0.1 \leq \alpha r \leq 0.8$ and satisfies [42] $v_{\mathrm{rot,\,max}} \approx 0.8(\alpha G M_{\mathrm{d}})^{1/2}$. Integration over the disk radial profile yields $\dot{M}_* \propto v_{\mathrm{rot,\,max}}^4 (\mu_g \mu_{\mathrm{d}})^2 (\kappa \sigma_g)^{-1}$. The central surface brightness is approximately constant for disks of spirals, being equal to 21.8 (± 0.8)$B\mu$ [43, 44]. Hence within a Hubble type, where $M/L \approx \mathrm{const}$, both μ_{d} as well as μ_g (e.g., $\mu_g(\mathrm{H\,I}/\mathrm{H}_2) = 21(\pm 3.2)\, M_\odot\, \mathrm{pc}^{-2}$ in Sc galaxies: [45]) are approximately constant. One might well expect the residuals to be correlated, and, with $\sigma_g \approx \mathrm{const}$, a tight correlation is predicted between the global star formation rate and $v_{\mathrm{rot,\,max}}$. For gas-rich systems, where feedback is important, a flatter relation is expected ($\dot{M}_* \propto v^2_{\text{rot, max}}$). For starbursts, the relation might be steeper, $\dot{M}_* \propto v^{5.6}_{\text{rot, max}} (p_\text{g}/p_\text{v})^{1/4}$, where p_v is the virial pressure (and includes the total self-gravitating mass density).

To infer the present luminosity, one, of course, has to integrate $\dot{M}_*$ over the past history of galactic star formation. To the extent that the near-infrared luminosity of a spiral galaxy is dominated by red supergiants, one would expect a correlation between infrared as well as blue luminosity and the maximum rotation velocity for a given Hubble type. More generally, it is apparent that the zero point of the Tully-Fisher relation, $L/v^4_{\text{rot, max}}$, must depend on environment via the proportionality to μ_g^2. In some circumstances, μ_g could be depleted, for example after ram pressure stripping by intracluster gas; in other circumstances, it could be enhanced, as might happen following a merger or a period of gas infall. Since calibration of the Tully-Fisher relation is based on a local sample, care should be taken in inferring any apparent peculiar velocities deduced from Tully-Fisher distances without due regard to both galaxy environment and history.

5˙2. *Galaxy clusters*. – As the hierarchy of structure develops via gravitational instability, clusters of galaxies grow by infall of gas-rich galaxies. In the case of the intracluster medium in a rich galaxy cluster, one typically has $T \approx 10^8$ K, $n \approx 10^{-3}$ cm^{-3}, so that $p/k = 10^5$ cm^{-3} K, and eq. (6) yields $\dot{M}_* \gtrsim 100 M_\odot$ y^{-1}. Ram pressure-enhanced star formation should occur over cluster and supercluster scales of ≤ 10 Mpc: for a galaxy peculiar velocity of 600 km s^{-1}, the resulting ram pressure is 5000 $(n/10^{-4}$ cm$^{-3})(v/600$ km s$^{-1})^2$ cm^{-3} K and enhances the star formation rate by a factor of $\gtrsim 2$ in galaxies within about twice the classical virialization radius, $GM/\sigma^2 = 4.3 M_{15} \sigma_{1000}^{-2}$ Mpc for a cluster of mass $10^{15} M_{15} M_\odot$ and velocity dispersion 1000 σ_{1000} km s^{-1}. The relevant radius exceeds the virialization radius by up to a factor of 2 or so, because of previrialization [46], a rich cluster forming by irregular, nonradial collapse of a clumpy distribution of galaxies and galaxy groups that begin to interact dynamically and shock heat the intracluster gas once turn-around commences, at $\sim 10\, \sigma_{1000} h^{-1}$ Mpc (here $h = H_0/100$ km s^{-1} Mpc^{-1}). The cluster tidal field also enhances internal cloud motions at a comparable radius [47]. The infalling galaxies usually are in small spiral-dominated groups. The pressure enhancement, both dynamic and static, experienced by a gas-rich galaxy undergoing first infall to a cluster directly induces enhanced star formation and eventually converts the spirals to S0's. Mergers presumably play an important role in the densest regions, forming ellipticals at an early epoch. Evidence for infall-induced star formation is provided by DRESSLER and GUNN [48], who found that the after-effects of starbursts are still apparent at $z \sim 0.5$-1, with a population of post-starburst galaxies at cluster peripheries. The spatial distribution of post-starburst galaxies can indeed be understood in the context of a first infall model if a starburst is induced by a rapid (over a galaxy dynamical time) pressure increase [49].

5'3. *Galaxy peculiar velocities.* – The postulated luminosity enhancement of galaxies near clusters makes galaxies appear to be closer, leading to an underestimate of distance and thereby to an apparent positive peculiar velocity. Samples of IRAS-selected galaxies are used, because of the almost complete sky coverage, to derive large-scale velocity fields via an iterative scheme that relies on the proportionality between light and acceleration [50]. However, enhanced star formation in high-pressure environments enhances galaxy detectability at, say, 60μ, and will lead to anomalous peculiar motions that are artefacts of this correlation over cluster scales. These are likely to be correlated with «peculiar motions» inferred from Tully-Fisher distances. While this would apply to spirals, systematically positive, and possibly spurious, peculiar velocities have been identified by GREGG [51] in a sample of ellipticals that displays morphological peculiarities, including deviations from elliptical isophotes and stellar population differences that are apparently due to recent (~ 1 Gy ago) star formation [52] and correspondingly enhanced luminosity. The zero-points of the distance indicators for both spirals and ellipticals may, therefore, contain a scatter that is due to an environment-linked variation in galaxy M/L as argued by SILK [53], with potentially serious consequences for the apparent large-scale streaming motions.

5'4. *Large-scale structure.* – Spatial variations in star formation rate between galaxies can affect the amplitude of the angular correlation function and the variance in galaxy cell counts that are determined from magnitude-limited samples of galaxies, and especially those samples selected in wavelength bands that are sensitive to star formation. Suppose, for example, that a fraction $\psi \lesssim 0.3$ of galaxies have enhanced star formation due to first infall occurring at $z \sim 1$. This is not an unreasonable guess, since ~ 20 percent of all galaxies are $S0$'s, most of which presumably formed via this mechanism, and many early-type spirals would also have similarly undergone enhanced star formation. The angular correlation function of these galaxies will be enhanced up to a level of as much as $\sim \psi^2$ or ~ 10 percent of the cluster-cluster correlation function on large scales. Also, by enhancing the luminosities of galaxies in distant superclusters, the variance in galaxy counts will be boosted, especially near the limit of a survey where the luminosity function of only the brightest galaxies is being sampled. Indeed, analysis of the QDOT survey reveals that some of the cells containing the largest numbers of galaxies coincide with known distant superclusters [54].

While the separation between clusters is imprinted onto the resulting galaxy correlations on large scales, galaxy cluster interaction-induced star formation should result in small-scale antibiasing. Low-redshift ($z \lesssim 0.5$) star-forming galaxies avoid the cores of rich clusters, whereas elliptical formation occurred at high redshift ($z \gtrsim 1$) when the cluster cores formed. This effect is seen for nearby galaxies, and even appears to extend well outside cluster cores as is

Nuova Tipografia Compositori, Bologna

Course CXIV
Industrial and Technological Applications of Neutrons
edited by M. FONTANA and F. RUSTICHELLI

Course CXV
The Use of EOS for Studies of Atmospheric Physics
edited by J. C. GILLE and G. VISCONTI

Course CXVI
Status and Perspectives of Nuclear Energy: Fission and Fusion
edited by R. A. RICCI, C. SALVETTI and E. SINDONI

Course CXVII
Semiconductor Superlattices and Interfaces
edited by A. STELLA

Course CXVIII
Laser Manipulation of Atoms and Ions
edited by E. ARIMONDO, W. D. PHILLIPS and F. STRUMIA

Course CXIX
Quantum Chaos
edited by G. CASATI, I. GUARNERI and U. SMILANSKY

Course CXX
Frontiers in Laser Spectroscopy
edited by T. W. HÄNSCH and M. INGUSCIO

Course CXXI
Perspectives in Many-Particle Physics
edited by R. A. BROGLIA, J. R. SCHRIEFFER and P. F. BORTIGNON

Course LXXXVIII
Turbulence and Predictability in Geophysical Fluid Dynamics and Climate Dynamics
edited by M. GHIL, R. BENZI and G. PARISI

Course LXXXIX
Highlights of Condensed-Matter Theory
edited by F. BASSANI, F. FUMI and M. P. TOSI

Course XC
Physics of Amphiphiles: Micelles, Vesicles and Microemulsions
edited by V. DEGIORGIO and M. CORTI

Course XCI
From Nuclei to Stars
edited by A. MOLINARI and R. A. RICCI

Course XCII
Elementary Particles
edited by N. CABIBBO

Course XCIII
Frontiers in Physical Acoustics
edited by D. SETTE

Course XCIV
Theory of Reliability
edited by A. SERRA and R. E. BARLOW

Course XCV
Solar-Terrestrial Relationships and the Earth Environment in the Last Millennia
edited by G. CINI CASTAGNOLI

Course XCVI
Excited-State Spectroscopy in Solids
edited by U. M. GRASSANO and N. TERZI

Course XCVII
Molecular-Dynamics Simulations of Statistical-Mechanical Systems
edited by G. CICCOTTI and W. G. HOOVER

Course XCVIII
The Evolution of Small Bodies in the Solar System
edited by M. FULCHIGNONI and L. KRESÁK

Course XCIX
Synergetics and Dynamic Instabilities
edited by G. CAGLIOTI and H. HAKEN

Course C
The Physics of NMR Spectroscopy in Biology and Medicine
edited by B. MARAVIGLIA

Course CI
Evolution of Interstellar Dust and Related Topics
edited by A. BONETTI and J. M. GREENBERG

Course CII
Accelerated Life Testing and Experts Opinions in Reliability
edited by C. A. CLAROTTI

Course CIII
Trends in Nuclear Physics
edited by P. KIENLE, R. A. RICCI and A. RUBBINO

Course CIV
Frontiers and Borderlines in Many-Particle Physics
edited by R. A. BROGLIA and J. R. SCHRIEFFER

Course CV
Confrontation between Theories and Observations in Cosmology: Present Status and Future Programmes
edited by J. AUDOUZE and F. MELCHIORRI

Course CVI
Current Trends in the Physics of Materials
edited by G. F. CHIAROTTI, F. FUMI and M. TOSI

Course CVII
The Chemical Physics of Atomic and Molecular Clusters
edited by G. SCOLES

Course CVIII
Photoemission and Absorption Spectroscopy of Solids and Interfaces with Synchrotron Radiation
edited by M. CAMPAGNA and R. ROSEI

Course CIX
Nonlinear Topics in Ocean Physics
edited by A. R. OSBORNE

Course CX
Metrology at the Frontiers of Physics and Technology
edited by L. CROVINI and T. J. QUINN

Course CXI
Solid-State Astrophysics
edited by E. BUSSOLETTI and G. STRAZZULLA

Course CXII
Nuclear Collisions from the Mean-Field into the Fragmentation Regime
edited by C. DETRAZ and P. KIENLE

Course CXIII
High-pressure Equation of State: Theory and Applications
edited by S. ELIEZER and R. A. RICCI

Course LVIII
Dynamics Aspects of Surface Physics
edited by F. O. GOODMAN

Course LIX
Local Properties at Phase Transitions
edited by K. A. MÜLLER and A. RIGAMON-TI

Course LX
C*-Algebras and their Applications to Statistical Mechanics and Quantum Field Theory
edited by D. KASTLER

Course LXI
Atomic Structure and Mechanical Properties of Metals
edited by G. CAGLIOTI

Course LXII
Nuclear Spectroscopy and Nuclear Reactions with Heavy Ions
edited by H. FARAGGI and R. A. RICCI

Course LXIII
New Directions in Physical Acoustics
edited by D. SETTE

Course LXIV
Nonlinear Spectroscopy
edited by N. BLOEMBERGEN

Course LXV
Physics and Astrophysics of Neutron Stars and Black Holes
edited by R. GIACCONI and R. RUFFINI

Course LXVI
Health and Medical Physics
edited by J. BAARLI

Course LXVII
Isolated Gravitating Systems in General Relativity
edited by J. EHLERS

Course LXVIII
Metrology and Fundamental Constants
edited by A. FERRO MILONE, P. GIACOMO and S. LESCHIUTTA

Course LXIX
Elementary Modes of Excitation in Nuclei
edited by A. BOHR and R. A. BROGLIA

Course LXX
Physics of Magnetic Garnets
edited by A. PAOLETTI

Course LXXI
Weak Interactions
edited by M. BALDO CEOLIN

Course LXXII
Problems in the Foundations of Physics
edited by G. TORALDO DI FRANCIA

Course LXXIII
Early Solar System Processes and the Present Solar System
edited by D. LAL

Course LXXIV
Development of High-Pozer Lasers and their Applications
edited by C. PELLEGRINI

Course LXXV
Intermolecular Spectroscopy and Dynamical Properties of Dense Systems
edited by J. VAN KRANENDONK

Course LXXVI
Medical Physics
edited by J. R. GREENING

Course LXXVII
Nuclear Structure and Heavy-Ion Collisions
edited by R. A. BROGLIA, R. A. RICCI and C. H. DASSO

Course LXXVIII
Physics of the Earth's Interior
edited by A. M. DZIEWONSKI and E. BOSCHI

Course LXXIX
From Nuclei to Particles
edited by A. MOLINARI

Course LXXX
Topics in Ocean Physics
edited by A. R. OSBORNE and P. MALANOTTE RIZZOLI

Course LXXXI
Theory of Fundamental Interactions
edited by G. COSTA and R. R. GATTO

Course LXXXII
Mechanical and Thermal Behaviour of Metallic Materials
edited by G. CAGLIOTI and A. FERRO MILONE

Course LXXXIII
Positrons in Solids
edited by W. BRANDT and A. DUPASQUIER

Course LXXXIV
Data Acquisition in High-Energy Physics
edited by G. BOLOGNA and M. VINCELLI

Course LXXXV
Earthquakes: Observation, Theory and Interpretation
edited by H. KANAMORI and E. BOSCHI

Course LXXXVI
Gamow Cosmology
edited by F. MELCHIORRI and R. RUFFINI

Course LXXXVII
Nuclear Structure and Heavy-Ion Dynamics
edited by L. MORETTO and R. A. RICCI

Course XXVI
Selected Topics on Elementary Particle Physics
edited by M. CONVERSI

Course XXVII
Dispersion and Absorption of Sound by Molecular Processes
edited by D. SETTE

Course XXVIII
Star Evolution
edited by L. GRATTON

Course XXIX
Dispersion Relations and their Connection with Causality
edited by E. P. WIGNER

Course XXX
Radiation Dosimetry
edited by F. W. SPIERS and G. W. REED

Course XXXI
Quantum Electronics and Coherent Light
edited by C. H. TOWNES and P. A. MILES

Course XXXII
Weak Interactions and High-Energy Neutrino Physics
edited by T. D. LEE

Course XXXIII
Strong Interactions
edited by L. W. ALVAREZ

Course XXXIV
The Optical Properties of Solids
edited by J. TAUC

Course XXXV
High-Energy Astrophysics
edited by L. GRATTON

Course XXXVI
Many-Body Description of Nuclear Structure and Reactions
edited by C. BLOCH

Course XXXVII
Theory of Magnetism in Transition Metals
edited by W. MARSHALL

Course XXXVIII
Interaction of High-Energy Particles with Nuclei
edited by T. E. O. ERICSON

Course XXXIX
Plasma Astrophysics
edited by P. A. STURROCK

Course XL
Nuclear Structure and Nuclear Reactions
edited by M. JEAN and R. A. RICCI

Course XLI
Selected Topics in Particle Physics
edited by J. STEINBERGER

Course XLII
Quantum Optics
edited by R. J. GLAUBER

Course XLIII
Processing of Optical Data by Organisms and by Machines
edited by W. REICHARDT

Course XLIV
Molecular Beams and Reaction Kinetics
edited by CH. SCHLIER

Course XLV
Local Quantum Theory
edited by R. JOST

Course XLVI
Physics with Intersecting Storage Rings
edited by B. TOUSCHEK

Course XLVII
General Relativity and Cosmology
edited by R. SACHS

Course XLVIII
Physics of High Energy Density
edited by P. CALDIROLA and H. KNOEPFEL

Course IL
Foundations of Quantum Mechanics
edited by B. D'ESPAGNAT

Course L
Mantle and Core in Planetary Physics
edited by J. COULOMB and M. CAPUTO

Course LI
Critical Phenomena
edited by M. S. GREEN

Course LII
Atomic Structure and Properties of Solids
edited by E. BURSTEIN

Course LIII
Developments and Borderlines of Nuclear Physics
edited by H. MORINAGA

Course LIV
Developments in High-Energy Physics
edited by R. R. GATTO

Course LV
Lattice Dynamics and Intermolecular Forces
edited by S. CALIFANO

Course LVI
Experimental Gravitation
edited by B. BERTOTTI

Course LVII
History of 20th Century Physics
edited by C. WEINER

PROCEEDINGS OF THE INTERNATIONAL SCHOOL OF PHYSICS «ENRICO FERMI»

Course I
Questioni relative alla rivelazione delle particelle elementari, con particolare riguardo alla radiazione cosmica
edited by G. PUPPI

Course II
Questioni relative alla rivelazione delle particelle elementari, e alle loro interazioni con particolare riguardo alle particelle artificialmente prodotte ed accelerate
edited by G. PUPPI

Course III
Questioni di struttura nucleare e dei processi nucleari alle basse energie
edited by C. SALVETTI

Course IV
Proprietà magnetiche della materia
edited by L. GIULOTTO

Course V
Fisica dello stato solido
edited by F. FUMI

Course VI
Fisica del plasma e relative applicazioni astrofisiche
edited by G. RIGHINI

Course VII
Teoria della informazione
edited by E. R. CAIANIELLO

Course VIII
Problemi matematici della teoria quantistica delle particelle e dei campi
edited by A. BORSELLINO

Course IX
Fisica dei pioni
edited by B. TOUSCHEK

Course X
Thermodynamics of Irreversible Processes
edited by S. R. DE GROOT

Course XI
Weak Interactions
edited by L. A. RADICATI

Course XII
Solar Radioastronomy
edited by G. RIGHINI

Course XIII
Physics of Plasma: Experiments and Techniques
edited by H. ALFVÉN

Course XIV
Ergodic Theories
edited by P. CALDIROLA

Course XV
Nuclear Spectroscopy
edited by G. RACAH

Course XVI
Physicomathematical Aspects of Biology
edited by N. RASHEVSKY

Course XVII
Topics of Radiofrequency Spectroscopy
edited by A. GOZZINI

Course XVIII
Physics of Solids (Radiation Damage in Solids)
edited by D. S. BILLINGTON

Course XIX
Cosmic Rays, Solar Particles and Space Research
edited by B. PETERS

Course XX
Evidence for Gravitational Theories
edited by C. MØLLER

Course XXI
Liquid Helium
edited by G. CARERI

Course XXII
Semiconductors
edited by R. A. SMITH

Course XXIII
Nuclear Physics
edited by V. F. WEISSKOPF

Course XXIV
Space Exploration and the Solar System
edited by B. ROSSI

Course XXV
Advanced Plasma Theory
edited by M. N. ROSENBLUTH

driven by multiple supernova outbursts and incorporates a multiphase interstellar medium. A dependence of star formation rate on ambient pressure is inferred, and leads to the prediction of strong low-redshift evolution for gas-rich galaxies in cluster and supercluster environments. A similar model is applied to elliptical-galaxy formation, a process that may be associated with galaxy merger-induced starbursts. Isolated gas-rich dwarf galaxies undergo inefficient star formation and are long-lived.

There naturally emerges a low-efficiency mode with a surface density threshold that applies to cold disks, and a mode of high efficiency that applies to starbursts when the porosity of the hot interstellar medium is large. Consequently the star formation *rate* is *bimodal*. This leads to a reasonably robust prediction: isolated galaxies should quiescently form disks, perhaps with continuing nondisruptive infall of smaller gas clouds, until external perturbations, here speculated to arise via mergers or strong tidal interactions, occur that drive up the star formation rate as a starburst develops.

Protoelliptical galaxies are envisaged as the high-redshift counterpart of current-epoch starbursts. The large-scale environment acts via the locally enhanced pressure near clusters of galaxies to modulate and enhance star formation, resulting in low-redshift evolution of star-forming galaxies. It is suggested that this effect could enhance the measured degree of large-scale power in the luminous-galaxy distribution by preferentially selecting the richest clusters and superclusters in the more distant radial shells of magnitude-limited samples of galaxies. Inferences about the large-scale power, and also large-scale peculiar velocities, may be systematically biased by the influence of environment on star formation rate.

* * *

This work was initiated during a sabbatical year at the Mount Stromlo and Siding Spring Observatories, where I gratefully acknowledge the hospitality provided by the Director, Prof. A. RODGERS. I thank A. EVRARD, C. LACEY, C. MCKEE, A. SZALAY, B. WANG, D. WEINBERG and R. WYSE for comments on this work. This research has been supported at Berkeley in part by grants from NASA and NSF.

REFERENCES

[1] R. H. MILLER, K. PRENDERGAST and W. J. QUIRK: Astrophys. J., **161**, 903 (1970).
[2] J. A. SELLWOOD and R. A. CARLBERG: Astrophys. J., **282**, 61 (1984).
[3] A. TOOMRE: in The Structure and Evolution of Normal Galaxies, edited by

unit volume. If self-regulation sets $P \sim 1$, I infer that $\dot{\rho}_* \sim p_g^{1.36} n^{0.1} m_{\rm SN}$. The supernova rate at any radius r determines the local porosity P in a shell at this distance. I set the star formation rate per unit volume equal to $(1-R)\dot{\rho}_* = \varepsilon e^{-P} \rho_{\rm gas} t_{\rm ff}^{-1}$, where R is the returned fraction of gas via stellar-mass ejection into the interstellar medium and $t_{\rm ff} = r/\sigma$ is the local free-fall time scale. This expression is motivated by the expectation that, for a spheroidal protogalaxy, star formation should occur over about a free-fall time scale in order to avoid gas dissipative formation of a disk.

Modelling the protoelliptical by an isothermal sphere with $\rho = \sigma^2/4\pi G r^2$, $T = \sigma^2 m_p/3k$, I infer that $\varepsilon_{\min} \equiv (\sigma/\sigma_{\rm cr})^{2.66}$, where $\sigma_{\rm cr} = 47 f_{0.1}^{-0.18} r_{10}^{-0.02} m_{200}^{-0.4}$ km s^{-1}. Star formation is inevitably efficient in deep potential wells. Only in *dwarf* galaxies could one have *inefficient* star formation. A shallow potential well is required for inefficient star formation. This is equivalent to the criterion given by DEKEL and SILK [74] who associated the suppression of star formation with the initiation of supernova wind-driven mass loss. In fact, gas-rich dwarf-galaxy winds would be radiatively unstable, and hence nonsteady (cf. [75]), forming an extensive gaseous envelope in the surrounding dark halo. The filling factor of the hot phase of the interstellar medium becomes dominant in dwarfs, but results in star formation with diminished efficiency, $\dot{M}_* \propto \sigma^{5.64}$. The low star formation rate and efficiency predicted for dwarfs suggest that a population of gas-rich dwarfs is produced, as also was inferred from (9).

7. – Conclusions.

The star formation rate is the key to understanding galaxy formation. Population synthesis models demonstrate that spiral galaxies formed inefficiently over at least 3 Gy, and had a relatively low rate of star formation, whereas ellipticals formed with much higher efficiency and had a high specific rate of early star formation. A successful theory for galaxy formation must at least commence with a description of global star formation in nearby galaxies.

I have shown that in quiescent disk galaxies which are unstable to gravitational instabilities, a semi-phenomenological expression for the global star formation rate depends on the cold-gas surface density and on the differential rotation rate. Energy input from supernova remnants is likely to self-regulate the rate of star formation in cold gas-rich disks, yielding a dependence of star formation rate on gas surface density that is between linear and quadratic, but with a threshold. The quiescent mode accounts for detailed aspects of nearby star-forming galaxies, including the dependence of star formation rate on gas surface density in nearby disk galaxies, and suggests that the global star formation rate is affected both by the *present* and *past* environments of the galaxy. Any resulting Tully-Fisher-like relation would depend on environment.

For starbursts, a porosity-limited star formation theory is developed that is gas-rich) dwarfs indeed appear to exhibit weaker angular correlations than do red (and presumably gas-poor) dwarfs [69].

6. – Galaxy formation.

According to schemes for hierarchical formation of structure from primordial Gaussian fluctuations [70], protogalaxies are gas-rich objects undergoing frequent mergers [71]. The inferred star formation rate per unit mass in protospheroids and protoellipticals resembles that in starbursts. I will accordingly adapt the starburst model developed above for the process of galaxy formation.

6`1. *Star formation rate in protogalaxies.* – I assume primordial Gaussian density fluctuations that undergo linear evolution in an Einstein-de Sitter universe described by evolving the power spectrum of the density field, with Fourier amplitude δ_k, as a function of wave number k and redshift z according to $|\delta_k|^2 \propto k^n (1+z)^{-2}$. For $\nu\sigma$ fluctuations, where $\sigma \propto M^{-1/2 - n/6}$ is the r.m.s. density fluctuation amplitude on comoving scale M and ν is the threshold Gaussian density peaks where galaxies are presumed to form [72], self-similar scaling yields

$$(8) \qquad \nu^2 = (v_{\text{circ}}/v_{\text{circ},0})^{n+3}(1+z)^{(1-n)/2} = (M/M_0)^{1+n/3}(1+z)^2 \, .$$

Here $v_{\text{circ},0}$ and M_0 are the galaxy circular velocity and mass chosen at $z = 0$ for normalization.

The cosmological implications for galaxy formation are worthy of note. The characteristic star formation time scale may be defined as $t_* \equiv M/\dot{M}_*$, and, making use of the previously derived scaling with v_{circ}, is found to vary as

$$(9) \qquad t_*/t \propto \nu^{-5.8/(3+n)}(1+z)^{(1.31-n)/(3+n)} \propto \nu^{-1.3} M^{-0.2(1-n)} \, .$$

With $n \approx -2$, as appropriate for protogalactic mass scales, $t_*/t \propto \nu^{-5.8}(1+z)^{3.9} \propto \nu^{-1.3} M^{-0.65}$. I deduce that, while rare fluctuations form stars more efficiently at a given epoch, star formation generally becomes less efficient at earlier epochs, especially for protodwarf galaxies. High-threshold cores do have smaller t_*/t, and result in the formation of dense stellar subsystems that would be efficient at transferring angular momentum, to form the nuclei of galaxies [73]. On the other hand, late-forming objects and low-threshold objects form stars inefficiently. These are protodisks, which are indeed inferred to have $t_* \approx$ const.

6`2. *Protoelliptical galaxies.* – Consider now a spherical protogalaxy. Suppose first that the porosity of supernova-driven winds self-regulates star formation in a multicomponent interstellar medium. I redefine porosity for a spherical system by $P = \nu_{\text{SN}} \dot{\rho}_*/m_{\text{SN}}$, where $\dot{\rho}_*$ is the star formation rate per apparent from the angular correlations of galaxies in the Pisces-Perseus supercluster [55] and the galaxy morphology-density relation [56, 57]. Since nearby samples show that spiral galaxies are more weakly correlated than ellipticals, by about a factor of 2 in correlation length, the enhanced correlations of star-forming spirals undergoing first infall, especially near distant ($z \gtrsim 0.5$) clusters, could boost the amplitude of their correlation function by a factor of up to 4 without generating excessive small-scale power. Both the blue-selected (and hence star formation rate sensitive) APM survey determination of the angular correlation function $w(\theta)$ to $\lesssim 10$ degrees [58] and the far-infrared selected (and hence star formation rate sensitive) QDOT survey variance in cell counts indicate that canonical cold dark matter (CDM) is deficient by an amount of this order in large-scale power on a scale $\sim 30h^{-1}$ Mpc. I predict that luminous IRAS galaxies should accordingly display stronger spatial correlations than intrinsically lower-luminosity galaxies. Indeed, an enhancement is observed in (blue) luminosity for the brightest spirals that results in enhanced galaxy correlations extending out to a scale of $\gtrsim 8$ Mpc [59].

5˙5. *Low-redshift galaxy evolution.* – Evidence for low-redshift evolution of field galaxies has emerged from studies of Mg II absorption line systems towards quasars. Star-forming galaxies are frequently found in the vicinity of, and at the same redshift as, the absorption line systems [60, 61]. Halos at $z \sim 0.5$ appear to be more gas-rich, fueling more intense star formation activity in the associated disks, than halos of nearby galaxies. The deep redshift surveys [62, 63] have found that a substantial fraction of field galaxies at a median redshift of up to 0.3 have enhanced rates of star formation, as inferred from the frequency of [O II] equivalent widths. These effects may be all attributed to the predicted strong evolution of the starburst galaxies near cluster peripheries. An IRAS-selected galaxy sample should show a substantially *steeper* slope in number counts than expected for an optically selected sample. There should be a high-luminosity tail in the deep IRAS sample, as indeed may be observed [64]. The third- and fourth-order angular correlations for a 1.2 Jy IRAS-selected galaxy sample [65] are consistent with predictions of secondary infall, implying that infrared-luminous galaxies may be undergoing infall into clusters. The deep pencil beam surveys also appear to preferentially sample superclusters [66]; this result may be in part due to a modest enhancement in galaxy luminosity at cluster and supercluster peripheries.

Low-surface-brightness gas-poor dwarf galaxies in nearby clusters exhibit a steepening (relative to the field) at the faint end of the galaxy luminosity function [67, 68]. If these low-surface-brightness galaxies are identified with the remnants of star-forming dwarfs, whose disruptive early bursts of star formation were triggered at first infall, one would expect these dwarfs to survive as gas-rich systems in the field. Star-forming, blue (and presumably